Molecular Orientation Analysis Using Polarized Vibrational Spectroscopy

Dennis K. Hore

Molecular Orientation Analysis Using Polarized Vibrational Spectroscopy

A Programmatic Approach

Springer

Dennis K. Hore
Department of Chemistry
University of Victoria
Victoria, BC, Canada

ISBN 978-3-032-04873-8 ISBN 978-3-032-04874-5 (eBook)
https://doi.org/10.1007/978-3-032-04874-5

This Springer imprint is published by the registered company Springer Nature Switzerland AG
The registered company address is: Gewerbestrasse 11, 6330 Cham, Switzerland

To my parents
Bärbel and Sukumar

and my family
Jill, Isobel and Isaac

Preface

As a chemistry student, I became fascinated in how polarized light could be used to provide a structurally-rich description of the manner in which molecules were oriented within materials. (A longer version of that story is provided in Chap. 1.) I've maintained that interest over the past three decades, and it has been fueled by

1. studying molecular vibrations (rather than electronic transitions) so that this topic becomes intimately connected to molecular conformation, as the orientation of sub-molecular units can be elucidated;
2. considering nonlinear spectroscopic experiments that provide a more refined picture of the distribution of molecular orientations;
3. learning about surface characterization, that offers the ability to study how molecules in bulk environments are arranged when they adsorb at surfaces, or are simply present as their bulk representatives at interfaces;
4. motorizing and automating experiments to maximize the information that can be recorded, thereby overcoming practical limitations of time and effort, and enabling more complex measurements to be performed;
5. modelling the results of such experiments, for the purpose of gaining a better understanding of what may be observed, to interpret the experimental results, and to design new experiments.

I cannot emphasize enough how much computer programming, or 'coding' as people prefer to say now, has been a part of this journey. Being able to write small simple scripts to carry out the above steps is a necessary and simultaneously gratifying experience. I've dabbled with many programming languages over the years, and settled on Python in the mid-1990s as a simple and flexible tool that accomplished many of the tasks that I was interested in, and with broad community support. Over the years, the number of scientific Python users has grown to the point where nearly any task can be performed in Python, providing a unified and highly flexible workflow. I should clarify that Python is not unique in its capabilities, and there are many suitable tools available. However, for newcomers without a strong preference, I recommend this language primarily as a result of the wealth of high quality add-on packages available for numerical methods, parallel computation, symbolic algebra, and fine control of plotting. For these reasons, I increasingly find that any discussion I'm having in my office about spectroscopy rapidly gets mixed with a discussion about implementing some ideas in Python. Hence, the first chapters are dedicated to the basics of the language, along with the foundations of plotting. Afterwards, those concepts are weaved throughout all of the subsequent chapters. Whenever something is introduced, I aim to illustrate the Python syntax alongside the new ideas. When I'm thinking about something I would like to understand better, nine times out of ten the answer comes in the form of a graph, and I appreciate being able to produce exactly the graphic that I want to see. I'm hoping that readers of this book will have a similar experience and opportunity.

This material is intended to fill the gap between basic introductions to spectroscopy, and more advanced material on molecular orientation distributions that is found only in research papers. It is suitable for a one-semester advanced undergraduate course. Following the introduction to coding in Python (Chap. 2), exploring some of its numerical (Chap. 3) and graphical (Chap. 4) capabilities, and introducing some basic principles of symbolic calculations (Chap. 5), the traditional spectroscopy content starts in earnest with a description of polarized light (Chap. 6) and how it interacts with isotropic materials (Chap. 7). After this, there are three major topics: understanding how and what type of molecular orientation information can be obtained from (a) polarized IR absorption measurements, (b) polarized spontaneous Raman scattering measurements, and (c) vibrational sum-frequency generation (SFG) spectroscopy. It makes sense for the discussion of IR spectroscopy to come next as Chap. 8, as it is an absorption spectroscopy, so is connected to the transitional dipole moment in the simplest way. The challenge is in deciding in what order to present the subsequent two chapters. The treatment of these subjects is tied to the order of the accompanying susceptibility tensor, which means that the SFG discussion could come before Raman. However, since IR and Raman are both capable of studying isotropic condensed phases, Raman (Chap. 9) is introduced before SFG. The SFG discussion in Chap. 10 then starts by introducing the general subject of characterizing molecules in non-centrosymmetric environments, with a focus on surfaces.

Finally, it probably comes as no surprise that this material was initially aimed at those interested in learning about the basics of orientation analysis using SFG spectroscopy. I mention this, as it motivated the way in which the topics in the IR and Raman chapters were selected and written. More comprehensive texts dedicated to describing the basics of IR and Raman spectroscopy would cover many topics that are neglected here, since I have a strong focus on molecular orientation. And there are some peculiar things: IR spectroscopists would be surprised that I bother to perform coordinate transformations on the polarizability, a rank-2 tensor, when the end-result of modelling IR spectra can be carried simply by transforming the squared elements of transition dipole moment vector. Likewise, Raman spectroscopists may find it egregious that I start by illustrating how a third-order susceptibility described by a rank-4 tensor is rotated in Cartesian coordinates, when it would be much more practical to work with the 3×3 'Raman tensor', the derived polarizability tensor whose nine elements can simply be squared to produce all elements of the sought rank-4 response. This is why I confess that this treatment is designed explicitly to provide those interested in SFG with a unified picture of the connection between closely related concepts in IR and Raman spectroscopy. As a result, my own study of such topics over the course of my SFG research has made me increasingly interested in IR absorption and spontaneous Raman scattering.

Victoria, Canada
July 2025

Dennis Hore

Acknowledgements

I would like to start by thanking my mentors, Peter Dawson (McMaster University), Paul Rochon (Royal Military College), Almeria Natansohn (Queen's University), and Geraldine Richmond (University of Oregon) who have given me the tremendous opportunity to be a student spectroscopist in their labs where I developed an interest in this material. I relish the opportunity to think about a problem to the point where I become thoroughly confused. Frequently, the hole is so deep that I need to be fished out by friends and colleagues. I am therefore grateful for enlightening and at times entangling conversations with Christopher Barret (McGill University), Mischa Bonn (Max Planck Institute for Polymer Research), Thierry Buffeteau (Université Bordeaux), Bertrand Busson (Université Paris-Saclay), Rogério de Sousa (University of Victoria), David Harrington (University of Victoria), Takeshi Hasegawa (Kyoto University), Sergei Kazarian (Imperial College London), François Lagugné Labarthet (Western University), Aaron Massari (University of Minnesota), Thomas Mayerhöfer (Leibniz Institute of Photonic Technology), Akihiro Morita (Tohoku University), Isao Noda (University of Delaware), Ulrike Stege (Univeristy of Victoria), Tahei Tahara (RIKEN), Eric Tyrode (KTH Royal Institute of Technology), and Hongfei Wang (Westlake University). I am deeply appreciative of the opportunity to discuss these topics with so many students on a near-daily basis over more than two decades. I thank Kenneth Howell and Sabrina Reba at Springer for their support and encouragement.

Victoria, Canada
July 2025

Dennis Hore

Acknowledgements

Contents

Acronyms

ATR	Attentuated total internal reflection
CP	Circularly-polarized
DAQ	Data acquisition
DC	Formally direct current, but more casually any quantity that is not time-dependent
DCM	Direction cosine matrix
DFG	Difference frequency generation
DOP	Degree of polarization
FTIR	Fourier transform infrared
FWHM	Full width at half maximum value
GNU	GNU's Not Unix! collection of free software
GUI	Graphical user interface
HD	Heterodyne-detected
IR	Infrared
IRE	Internal reflection (or reflecting) element
IR-MME	Infrared Müller matrix ellipsometer
KHD	Kramers-Heisenberg-Dirac
LCP	Left circularly-polarized
LP	Linearly-polarized
MD	Molecular dynamics
MEM	Maximum entropy method
MKS	Metre, kilogram, second system of units
MME	Müller matrix ellipsometer
MP	Most probable
NR	Non-resonant
OD	Orientation distribution
ODF	Orientation distribution function
PCSA	An ellipsometer with an optical path configured in the polarizer-compensator-sample-analyzer order
PDF	Portable document format
PEM	Photoelastic modulator

PNG	Portable network graphics (format)
RA	Reflection absorbance, or reflection absorption
RCP	Right circularly-polarized
RGB	Red green blue (colour model)
SF	Sum frequency
SFG	Sum frequency generation
SH	Second harmonic
SHG	Second harmonic generation
SI	International System of units
TDM	Transition dipole moment
TIR	Total internal reflection
USB	Universal serial bus
UV	Ultraviolet

Chapter 1
Introduction

1.1 Vibrational Spectroscopy as a Tool for Characterizing the Organization of Molecules in Materials

The properties of materials are intimately related to the way in which their constituent molecules are arranged. In some cases, the arrangement is random, or isotropic. In materials such as liquid crystals, molecules are ordered as a result of interactions between neighbouring molecules that propagate to macroscopic dimensions. Many devices, for example computer displays based on liquid crystals, operate on the basis of altering the molecular orientation. Here, applying a DC electric field through the material causes the permanent dipoles of each molecule to align in the direction of the field. The extent and speed to which molecules are able to do this ultimately affects the performance of the device. Being able to characterize the orientation distribution is critical to understanding how to optimize device performance, including guiding the chemical synthesis of new liquid crystals.

At surfaces and buried interfaces, the situation is even more fascinating, as the environment is always highly anisotropic, even when the two adjacent bulk phases are isotropic. That is because, at the interface, neighbouring molecules may be dissimilar, or the environment may be characterized by a lack of neighbours. At a polymer–water interface, chains at the surface will experience forces due to the bulk polymer phase on the one side, and water on the other—and will orient in response to both interactions. Furthermore, when surfaces are exposed to vapours or liquids like this, the arrangement can change when the environment changes. This same polymer may organize its surface chains differently when the ionic strength or pH of the aqueous phase changes, or when it is dry and therefore next to air, an ultra-hydrophobic environment. When molecules in solution are adsorbed to surfaces, this presents another angle of interest, as a dilute concentration in the bulk aqueous phase may form a monolayer at the surface. As an example, consider a 1 mM solution of a surfactant such as sodium dodecylsulfate (SDS). In the solution phase below the critical micelle concentration, practically all of the molecules adjacent

D. K. Hore, *Molecular Orientation Analysis Using Polarized Vibrational Spectroscopy*,
https://doi.org/10.1007/978-3-032-04874-5_1

to each SDS molecule are water. However, when a polymer surface is introduced and SDS adsorbs, not only does it experience interactions with the polymer, but a highly two-dimensional concentration of SDS molecules can result. Being able to characterize the organization of the molecules in such systems provides critical information on physical and chemical interactions that are at the heart of materials design, environmental science, and the self-assembly of biological structures.

There are many experimental methods that can be used to study the organization of molecules, both in bulk and at interfaces. Optical methods are interesting in that they can be nondestructive, can sometimes by used without extensive sample preparation, and can access buried interfaces as long as light can penetrate. We will see that when polarized light is used, the interaction between the controlled electric field direction of the electromagnetic waves preferentially interacts with one or more transition dipole moments. By recording spectra of such interactions, in a sense we are viewing a projection of the molecular orientation. If enough projections can be gathered, a 3D view of the molecular organization can be constructed or inferred. Many regions of the electromagnetic spectrum can be used for this, but there is a special place for mid-infrared wavelengths, from about 2–20 μm or 5000–500 cm^{-1}, as resonances in that region are often highly localized to specific chemical bonds. In the best case, performing an orientation analyses on separate parts of a flexible molecule (with one or more torsional angles that alter its shape) therefore leads to additional information on the conformation of that molecule. For rigid molecules or moieties, information from different functional groups can be coupled to provide additional information in cases where the orientation problem may otherwise be underdetermined. Finally, when vibrational resonances serve as intrinsic chromophores, no dyes, markers, or other tags need to be added, so there is a lower risk of perturbing the system we wish to study.

It is often straightforward to collect data, particularly polarized IR and Raman spectra, as a result of the capability and availability of commercial instruments. All what's needed then is an understanding of how the sought molecular orientation information is encoded in those spectra, a route to extracting it, and the tools to do so. In cases where custom intrumentation is to be developed, an understanding of the link between the material properties and the measured signals is used to optimize the design.

1.2 Coding as an Essential Component of a Spectroscopist's Toolkit

It's always exciting to learn new tools because they are enabling in the sense that they open up possibilities for new lines of investigation. Over the decades, my love of spectroscopy has been surpassed only by my love for computer programming, and Python has been an important part of that journey. One of the great things about optics and spectroscopy is that experimental results are well predicted by simple and

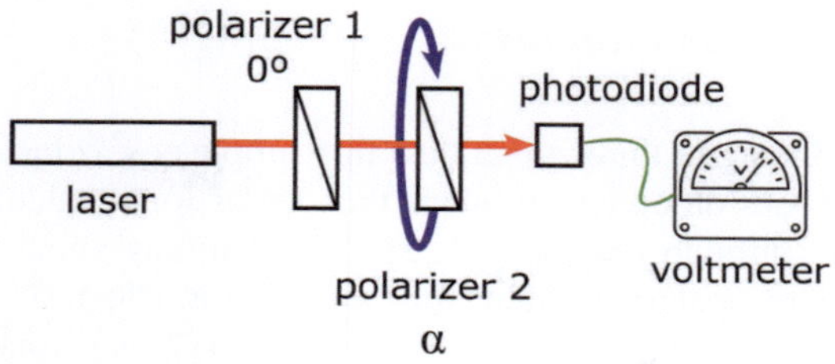

Fig. 1.1 An experimental scheme with a light source, two polarizers, detector, and voltmeter used to read the photocurrent. The second polarizer is installed in a mount that enables it to be rotated about the light propagation direction

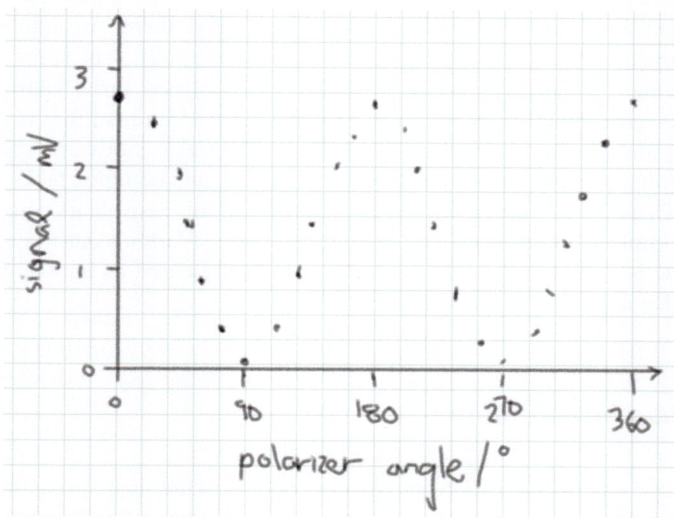

Fig. 1.2 Data collected from an experimental configuration such as that shown in Fig. 1.1, where the second polarizer is rotated through 360° in steps of 15°, and the voltmeter reading is recorded

accessible models. One of my favourite examples is Malus's Law stating that the intensity of light transmitted between two polarizers is proportional to the square of the cosine of the angle α between the transmission axes of the polarizers. In the simplest experiment, one could set up a laser pointer and two polarizers, the second in a rotating mount, and a photodiode connected to a voltmeter as shown in Fig. 1.1. One can record the photocurrent while manually rotating the second polarizer through 360° in steps of 15°, with pen on graph paper, and be delighted at the result that exactly matches $\cos^2 \alpha$ as seen in Fig. 1.2. I'm not sure if that would be everyone's idea of fun on a Saturday afternoon, but the first time I did exactly that in the summer of 1996 triggered something for me, in that the model matched the data *so* well. Moreover, the model was *simple*. However, like most things in life, things are better with Python.

1.2.1 Numerical Simulations

It is naturally intriguing to think about how this simple $\cos^2\alpha$ model comes about. Introductory textbooks on optics introduce the idea of Jones calculus for propagating polarized light through materials [1, 2]. After learning about the Jones vector description of linearly polarized light (that we'll see in Chap. 6) upon exiting the first polarizer, and the Jones matrix description (Chap. 7) of a rotated polarizer for the second element, things come together quickly, and the basis of $\cos^2\alpha$ can be understood. If we were to add phase-shifting elements known as retarders, even more sophisticated types of experiments can be performed. But how does one actually go about predicting the output of such optical systems? Spreadsheets get the job done, but they are clunky and one quickly outgrows them for such tasks. This is where practically any programming language works well, but some appear to be better suited to the task. In Chaps. 2 and 3 we will see that Python, together with the powerful NumPy [3] package, provides an environment that is fine-tuned for friendly and efficient numerical simulations of all sizes. Software projects often have the motto "Simple tasks should be simple to do, and complex tasks should be possible". Python and NumPy exceed expectations in this respect, as complex tasks are often simple with intuitive and streamlined code. Working out the solution to Malus' Law is a great beginner's problem, and is presented as an exercise in the problems that appear at the end of Chap. 3.

1.2.2 Data Visualization

Naturally, one can only go so far in writing code for numerical simulations before wanting to look at the result. In fact, throughout this book I will emphasize that data and model visualization and analysis go hand in hand. For simple projects, the code may culminate with making a graph. But for more complex tasks, having an efficient, flexible and powerful means for viewing the results earlier on, from the raw data stage actually guides one through decision making leading to the results. There are many fancy plotting packages available. Within the Python community several good options exist, each tailored to a slightly different application. One of my favourites is Matplotlib [4], again for its ease of use and flexibility. A common trait of powerful software is that there are often many ways to achieve the same thing. Although this may overwhelm newcomers, one quickly realizes that this offers streamlined options that can be well-matched to problem one is working on. Now one can create an array of α values from 0° to 360°, create a Jones calculus model of the intensity $I(\alpha)$ that gets stored in another array, and plot them as shown with points in Fig. 1.3. Then we can also plot $\cos^2\alpha$ with a solid line in Fig. 1.3 and see that this exactly matches the numerical prediction. We'll explore features of the Matplotlib package in Chap. 4.

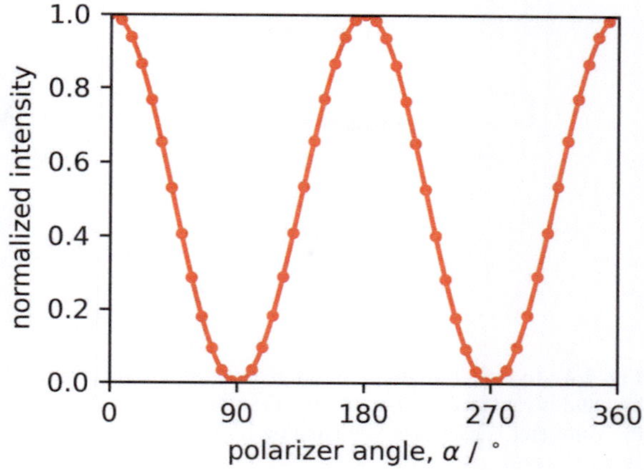

Fig. 1.3 The output of a numerical simulation based on Jones calculus may appear to be quite similar to a cosine squared curve, but the calculated intensity values is simply an array of numbers. When it is plotted (points), it can be seen that it exactly matches $\cos^2\alpha$ (solid line)

1.2.3 Symbolic Calculations

Although I'm generally a proponent of numerical approximations to everything, there are times where analytical solutions are important. For example, if one used only numbers in assembling the intensity values after multiplying the corresponding Jones vector and Jones matrix to determine $I(\alpha)$, it would not be apparent that the result was in fact proportional to $\cos^2\alpha$ unless it was plotted in a manner as demonstrated above. However, when working out the solution by hand, which is easy for this simple problem, one seems the analytical solution $I(\alpha) \propto \cos^2\alpha$ emerge. And I suppose it goes without saying that analytical solutions enable a deeper understanding of the underlying physics. In other words, there could be a rather complex functional form $f(\alpha)$ that has an appearance very similar to $\cos^2\alpha$. If the data is slightly noisy, it may be hard to tell. But the fact that we know that the function is exactly $\cos^2\alpha$ is important. The idea behind symbolic calculations that one can use something like α as a symbol, rather than being forced to assign it to a number of an array of numbers. In this way, mathematical operations can be performed on expressions containing such symbols, and return results in terms of those symbols. This is exactly what one does when working with pencil and paper (and, at least in my case, several high quality erasers). For decades, there have been two leading software packages that provide facilities for symbolic calculations, Mathematica™ [5] and Maple™ [6]. Those tools are truly amazing, and can do many things in addition to symbolic calculations. Before I became aware of the Python package SymPy [7], I would use those tools to work out the expressions of interest, and then convert the expressions into Python code, either by hand, or using the software's built-in code generation capabilities. So analytical expressions obtained there would be turned into NumPy expressions for evaluation, graphing, curve fitting, etc. Although SymPy offers only a small subset of what those commercial tools provide, the ability to work inside Python, and convert the symbolic expressions to NumPy

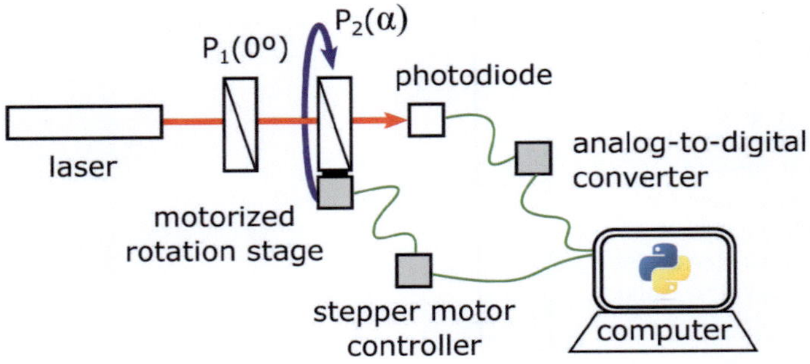

Fig. 1.4 In this configuration, the manual mount that allowed the second polarizer to be rotated by hand is replaced by a motorized rotation stage that can be controlled by a computer. Similarly, the voltmeter is replaced by an analog-to-digital converter that can read the photodiode output and store the result

expressions inside the code itself, has been a game changer. A brief introduction to some basic SymPy features is provided in Chap. 5.

1.2.4 Instrument Control

The next level of geekdom seeks data with higher angular resolution (when a 15° step size in α doesn't cut it), better signal-to-noise (more averaging at each α position), automated instrument control, and automated data logging. A schematic illustrating a slight modification to the setup is shown in Fig. 1.4. Needless to say, this is a slippery slope. And can you guess that I want to do this using Python?

Data acquisition code

```
import numpy as np
import voltmeter
import motor

# setup
stepsize = 5.
vm = voltmeter.initialize('COM1')
avg = 50
m = motor.initialize(1, speed=10, direction='cw')
m.set(0.)
alpha = np.arange(0., 360., stepsize)
intensity = []

# collect the data
for a in alpha:
    intensity.append(vm.read(avg))
```

```
# save the data
np.savetxt('data.txt', np.array([alpha, intensity]).T)
```

Some of the commands in the example above are not recognizable as Python, since they are custom functions, tucked away into user-defined libraries (that Python calls modules). Since I'm not showing all those details now, this is basically akin to pseudocode illustrating how straightforward it is to get the job done. To be honest, Python does not have any particular strengths or weaknesses for such tasks. But since I encourage you to master your Python skills, it makes a lot of sense to keep everything in this ecosystem.

1.2.5 Experimental Data Processing

Now that the data has been collected, how well does it fit the model? How can we visualize the fit compared to the standard deviation of the mean of triplicate experiments? We may wish to compare with

$$f(\alpha) = b\cos^2(\alpha - c) + d \tag{1.1}$$

where b is an amplitude scaling factor related to the intensity of the light source and the transmittance of each polarizer, considering reflection loses at the surfaces (Chap. 7). Here c is an angular offset that shows us how well $\alpha = 0°$ was initially set. It is typically hard to accurately set this at the beginning, either by maximizing the signal near $\alpha = 0°$ or minimizing the signal near $\alpha = 90°$, due to fluctuations in the signal level. It is therefore a better approach to introduce this constant c in the data fitting. If desired, the data can then be replotted by shifting all α values by c (and then a subsequent fit would have $c = 0$). The more interesting parameter here is d that controls the vertical shift of the data. This doesn't have any place in the model, but might be present if there is an offset in the detector response. An example of all of this is shown in Fig. 1.5. Of course, this is a very simple example of data processing, but already illustrates some basic operations that I would not want to perform without my beloved Python tools. Some spectral curve fitting examples will be discussed in detail in Chap. 4.

1.2.6 Scaling Up the Complexity

The above example illustrates all steps in a workflow from conceptualization, model development, data collection, and processing, but doesn't have anything to do with vibrational spectroscopy. That's easy to fix. Imagine a liquid crystal placed in

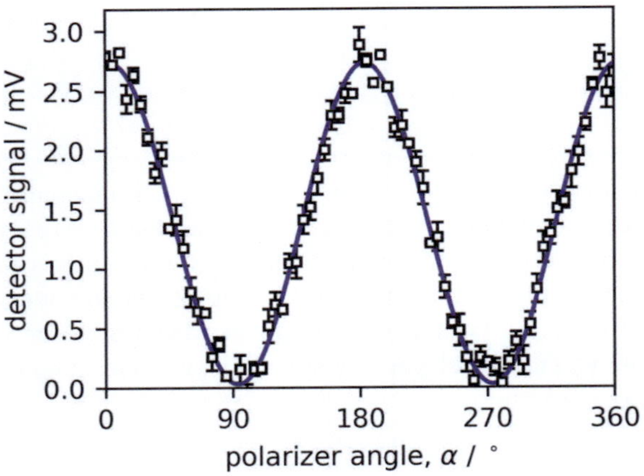

Fig. 1.5 Computer-acquired triplicate data, where the average is indicated with squares and the error bars are one standard deviation about the mean. A fit to Eq. 1.1 returns $b = 2.7$ as the unnormalized signal amplitude that is related to the absolute intensity through a calibration. It can be seen that the minima are slightly shifted from $\alpha = 90°$ and $\alpha = 270°$ due to an experimental offset in the angle of the polarizer. This is obtained in the fitting as $c = 3.5°$. Finally, one notices that the best-fit pattern is shifted up from zero, with $d = 0.03$

between the two IR polarizers. Now, instead of a photodiode, the transmission is monitored using an FTIR. The code to collect the data might look like:

- decide on a step size for α
- decide on the spectral data acquisition parameters, notably the desired spectral resolution and the number of averages
- move the second polarizer to position α_i inside a loop
- collect the FTIR spectra as a function of α.

Now, when the experiment finishes, we don't just have an array of N intensity values corresponding to N values of α. For each value of α we have an entire vibrational spectrum. On top of this, what if we had a way to raster the sample in the path of the IR beam, so we could perform spectral mapping over an area of say 20 mm × 20 mm. If we move the sample vertically and horizontally in increments of 2 mm, corresponding to the spot size of the IR beam, that would be 10 × 10 = 100 spectra per polarizer angle α. The hardware for such an experiment is illustrated schematically in Fig. 1.6. As you can see, the density of such data adds up quickly. One of the great advantages of incorporating Python tools in the workflow from the very beginning, at the data collection stage is that all the files and associated data structures can be immediately stored into an organized set, and subsequently available for batch processing. One can picture a situation where all the spectra are analyzed at once, and then spectral features are related back to spatial features, for example mapping the methyl symmetric stretch intensity (absorbance) at 2875 cm^{-1} as a function of the (x, y) spatial dimensions over the 20 mm × 20 mm area. This par-

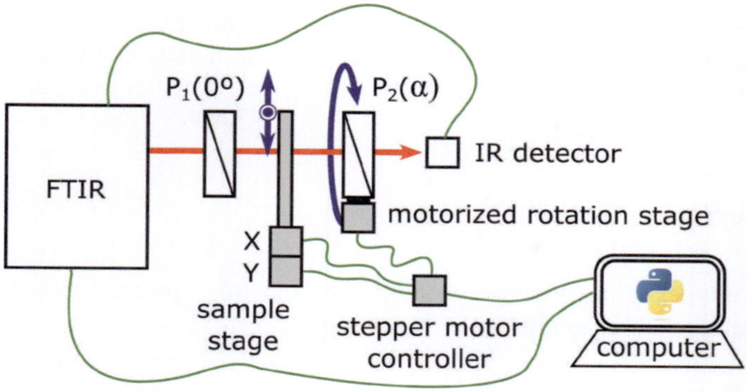

Fig. 1.6 A conceptually minor variation of the setup shown in Fig. 1.4 where the laser is replaced with an FTIR spectrometer, and the sample is rastered in the (x, y) direction perpendicular to the beam which propagates along z. The density of the data obtained in such an experiment is high, as it includes spectral and spatial dimensions

ticular example is edging towards what commercial software bundled with IR and Raman microscopes can do. However, there will always be a limit to the feature set in software that others write. If you can create the tools for spectral analysis yourself, the possibilities are endless.

1.3 Aims and Scope

The following chapters contain content appropriate for a single-semester (13-week) senior undergraduate course in molecular spectroscopy. There are many excellent books that cover IR absorption [8–12], Raman scattering [13–18], and vibrational SFG spectroscopy [18–20], and all of them provide material that is significantly more detailed and comprehensive that what is described here. My goal is to provide a unified treatment of these three phenomena and their associated experimental methods. At the heart of this motivation is illustrating that the IR absorbance is proportional to the square of the IR transition dipole moment precisely because two IR transition dipole moments couple to form the 3×3 polarizability tensor, and this therefore results in the sensitivity to molecular orientation that one expects for a rank-2 response function. Similarly, the treatment that we follow is expressly designed to highlight that the ability of spontaneous Raman spectroscopy to comment on both the mean orientation and width of the molecular tilt angle distribution arises solely from the fact that the measured response originates from a rank-4 quantity. Those two techniques are powerful in their range of applicability to the characterization of many types of materials. For newcomers to molecular orientation analysis based on vibrational SFG spectral data, the groundwork set up in the IR and Raman chapters will hopefully fill in some of details that is often missing from research papers or review articles.

1.4 Exercises

1. a. What types of instruments for vibrational spectroscopy exist around you, and what are their capabilities for data analysis?

    - IR spectrometer, Raman spectrometers, other
    - ability for polarized measurements
    - ability for imaging or mapping
    - spectral ranges
    - size of samples that can be accommodated
    - whether there is enough room in the sample area or sample compartment so that accessories can be added

   b. For each of the instruments you have found, describe the capabilities of the software provided by the instrument manufacturer. Consider whether it

    - is bare bones, providing data acquisition only
    - provides any built-in features for baseline correction
    - has any advanced data analysis capabilities such as peak picking, peak area determination, etc.
    - can easily export the data into formats that enable further analysis to be done with other tools

---

2. Many commercial including third-party accessories exist for FTIR instruments. However, if you wanted to simply retrofit your FTIR with a motorized polarizer, what parts would you need to procure, assemble, and program? Considerations include

    - stepper motor or DC servo motor?
    - what type of motor controller is required?
    - what types of polarizers exist for the mid-IR, what are their specifications including size and operating wavelength range?
    - where can such items be purchased, and how much do they cost?
    - What programming languages do the motor controllers support out of the box? (i.e. without any hacks)

- How can control of the motors be synchronized with the FTIR instrument control software? Think about a few options, starting from the ideal case to something that is clunky but would work.

---

3. If you have some experience writing code in a language other than Python, most of the material in Chap. 2 is likely too introductory, although it will provide a basic overview of Python syntax. You may instead be interested to find some tutorials on transitioning from the language you are familiar with to Python. If this applies to you, see what is available. Following that, you might be ready to dive straight into NumPy (Chap. 3) and Matplotlib (Chap. 4) and symbolic calculations using SymPy (Chap. 5).

References

1. E. Hecht, A. Zajac, *Optics* (Addision-Wesley, Reading, MA, 1979)
2. D.S. Kliger, J.W. Lewis, C.E. Randal, *Polarized Light in Optics and Spectroscopy* (Academic Press Inc., 1990)
3. C.R. Harris, K.J. Millman, S.J. van der Walt, R. Gommers, P. Virtanen, D. Cournapeau, E. Wieser, J. Taylor, S. Berg, N.J. Smith, R. Kern, M. Picus, S. Hoyer, M.H. van Kerkwijk, M. Brett, A. Haldane, J.F. del Río, M. Wiebe, P. Peterson, P. Gérard-Marchant, K. Sheppard, T. Reddy, W. Weckesser, H. Abbasi, C. Gohlke, T.E. Oliphant, Nature **585**, 357 (2020)
4. J.D. Hunter, Comp. Sci. Eng. **9**, 90 (2007)
5. Wolfram Research. Mathematica (2023). https://www.wolfram.com/mathematica/. Accessed: 2025-07-09
6. Maplesoft. Maple (2023). https://www.maplesoft.com/. Accessed: 2025-07-09
7. A. Meurer, C.P. Smith, M. Paprocki, O. Čertík, S.B. Kirpichev, M. Rocklin, A. Kumar, S. Ivanov, J.K. Moore, S. Singh, T.E. Rathnayake, B.E. Vig, B.E. Granger, R.P. Muller, F. Bonazzi, H. Gupta, S. Vats, F. Johansson, PeerJ Comp. Sci. **3**, e103 (2017)
8. N.J. Harrick, *Internal Reflection Spectroscopy* (Interscience Publisher, Wiley, New York, NY, 1967)
9. M. Milosevic, *Internal Reflection and ATR Spectroscopy* (Wiley, Hoboken, NJ, 2012)
10. D.C. Harris, M.D. Bertolucci, *Symmetry and Spectroscopy: An Introduction to Vibrational and Electronic Spectroscopy* (Dover Publications, New York, 1989)
11. T.G. Mayerhöfer, *Wave Optics in Infrared Spectroscopy: Theory, Simulation, and Modeling* (Elsevier, Amsterdam, 2024)
12. V.P. Tolstoy, I.V. Chernyshova, V.A. Skryshevsky, *Handbook of Infrared Spectroscopy of Ultrathin Films* (Wiley, Hoboken, New Jersey, 2003)
13. D.A. Long, *Raman Spectroscopy* (McGraw-Hill, New York, 1977)
14. D.A. Long, *The Raman Effect: A Unified Treatment of the Theory of Raman Scattering by Molecules* (Wiley, New Jersey, USA, 2002)
15. W.H. Weber, R. Merlin, *Raman Scattering in Materials Science* (Springer, New York, 2000)
16. R.L. McCreery, *Raman Spectroscopy for Chemical Analysis* (Wiley-Interscience, New York, 2000)
17. J.R. Ferraro, Z. Nakamoto, C.W. Brown, *Introductory Raman Spectroscopy* (Academic Press, San Diego, 2003)
18. G.J. Simpson, *Nonlinear Optical Polarization Analysis in Chemistry and Biology* (Cambridge University Press, Cambridge, UK, 2017)

19. Y.R. Shen, *Fundamentals of Sum-Frequency Spectroscopy* (Cambridge University Press, Cambridge, 2016)
20. A. Morita, *Theory of Sum Frequency Generation Spectroscopy* (Springer, Singapore, 2018)

Chapter 2
Python Fundamentals

2.1 Objectives and Motivation

This chapter will appear focused on parsing files to extract data, rather than performing much analysis. That part will come in Chap. 3. If you are the one that wrote the code that operated the instrument that produces the data (or you can modify it), then you are in control of the data format. You can set things up so that they are easy for you to parse in a programmatic way. But that is not always the case. Most instruments are bundled with software that export data in a human- (and therefore also computer-)readable format. In some cases, that format is slightly awkward. Learning how to parse such files is not only a practical, but a good way to start to learn any language. This chapter therefore has the following objectives:

1. Demonstrate that many common data manipulation tasks may easily be performed by writing a small script.
2. Demonstrate that repetitive tasks (such as selecting column 3 from 200 different files) may easily be performed by writing a small script.
3. Convince you that a programmatic approach is worth it in the long run for reprocessing data sets, in the case of new data or new analysis with old data.
4. Demonstrate that Python is an ideal language for these types of tasks, where these things are not only possible, but straightforward.
5. Illustrate that Python can hold its weight against commercial programming environments that have been in research environments for a long time. There is nothing wrong with using such a package, if your team or institution has a site license, or if you wish to purchase an individual license. However, the community of Python programmers is very large, and there is amazing community support available for nearly any kind of task.

Supplementary Information The online version contains supplementary material available at https://doi.org/10.1007/978-3-032-04874-5_2.

D. K. Hore, *Molecular Orientation Analysis Using Polarized Vibrational Spectroscopy*,
https://doi.org/10.1007/978-3-032-04874-5_2

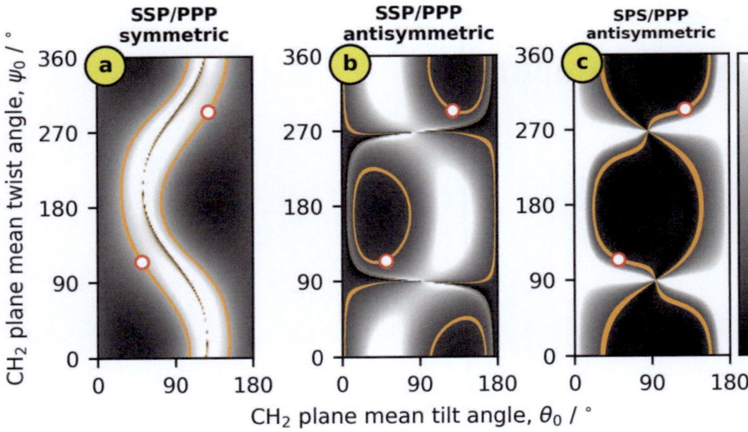

Fig. 2.1 Ratios of the second-order susceptibility values for different vibrational modes

These topics are presented assuming no previous programming experience. However, it does not provide any formal introduction to programming concepts, i.e. it does not introduce programming principles in a structured way. There are essentially two ways to learn how to code in a research environment: (1) learn the basics by reading good books [1, 2], picking up good habits, and understanding the underlying principles; (2) inheriting some code from a person who worked on a similar project before you, trying to figuring out how it works, and then trying to make minor modifications to suit your task without breaking it. I understand the second approach sounds sloppy and carries inherent risks, but I think it is realistic in a lab environment and it's the way many of us might have started. So I'm picturing a typical situation which is something in-between the two extremes. Perhaps you already know about enough general concepts to be dangerous, but want to fill in some gaps.

I have a deep interest in scriptable graphics, i.e. plots that are generated automatically with no interactive control from the user. As an example, imagine a calculation that produces thousands of plain text data files. There is obviously no ideal way of looking at the entire data set. A Python script that can retrieve a specific file for inspection could be useful. Consider that such a script can be run on the entire set of files, producing an image from each one. The resulting thousands of images can then be made into a movie for playback or scrubbing through. The challenge in this case is to have the code that generates the graphics make 'smart' decisions to maximize the clarity of the presentation. Figure 2.1 illustrates a panel of three dimensional projection plots with a few specific contours and points on the surfaces highlighted. Here we are trying to determine what tilt angle θ and twist angle ψ of phenylalanine's methyl group best describes our observed vibrational response in an sum frequency generation (SFG) experiment [3]. Large values of the ratio between two calculated observables are indicated with brighter intensity, the highest in white. Smaller values of the ratio have darker colour, zero being black. The specific elevations of these surfaces that are consistent with the experimental observables are highlighted in the orange contours. One can see that there are many possible solutions in this $\theta - \psi$ space. However, the only physically valid solutions are the ones that are coincident between all three panels. To aid in the visualization, any points where orange contours intersect in all three panels are indicated by the two dots. As a reminder, this graphic is one of approximately ten thousand in the dataset. We are interested in identifying any and all solutions that contain intersections. As a result, visualizing this one 'slice' is not part of any practical data analysis pipeline. Realistically, one would write a separate script to simply report the best-fit solution. However, to ensure that things are working correctly, it is extremely valuable to be able to point to one of these ten thousand slices and generate such an image. More details about this type of experiment and the associated data analysis will come in Chap. 10.

A second example comes from a molecular dynamics simulation of leucine in water adsorbing onto a hydrophobic surface [4]. Figure 2.2 illustrates some measured parameters obtained during a 10 ns window of a much longer trajectory. There can be millions of saved frames or snapshots from such trajectories. Visualization relies on being able to select a subset of this trajectory, as well as thinning out the data by viewing the results of every 100th frame or so. In the top panel, we plot the distance from leucine's centre of geometry (grey) and its isobutyl ispo carbon (black) to the solid surface. In cases where the centre of geometry is within 0.75 nm of the surface, the entire frame is shaded with a yellow background. At the same time, we were interested in monitoring the orientation of a few different parts of the molecule in the bottom panel, plotting the angles of the methyl C_3 axis (blue), carboxylate bisector (green), and NH_4^+ group (red) axes with respect to the surface normal. The manner in which such types of data can be used to augment the orienta-

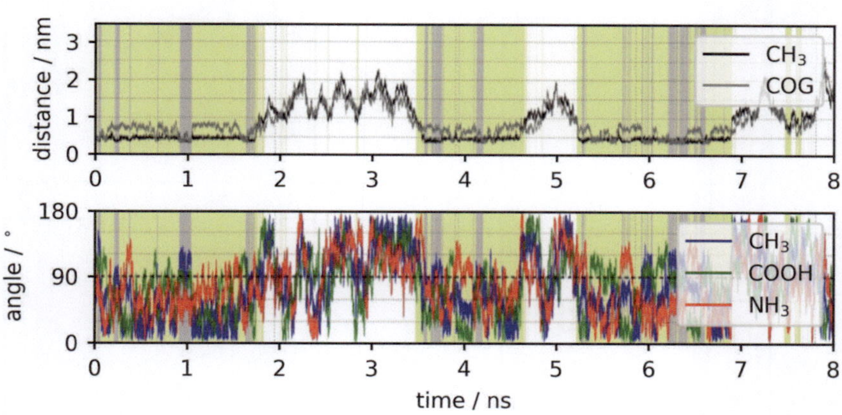

Fig. 2.2 Some useful metrics for amino acid adsorption onto a hydrophobic surface, as revealed by molecular dynamics simulations

tion analysis that can be performed using experimental data will be further discussed in Chap. 11.

A third example comes from a set of electronic structure calculations that ultimately provide the hyperpolarizability of specific vibrational modes. As a prerequisite to this calculation, the derivatives of the dipole moment vector $\partial \mu_k / \partial Q$ (Chap. 8) and linear polarizability tensor $\partial \alpha_{ij} / \partial Q$) (Chap. 9) with respect to the normal mode coordinate Q are evaluated at the equilibrium geometry [3]. The top three rows of Fig. 2.3 indicate values of the 3×3 polarizability tensor α_{ij} for 7 steps along

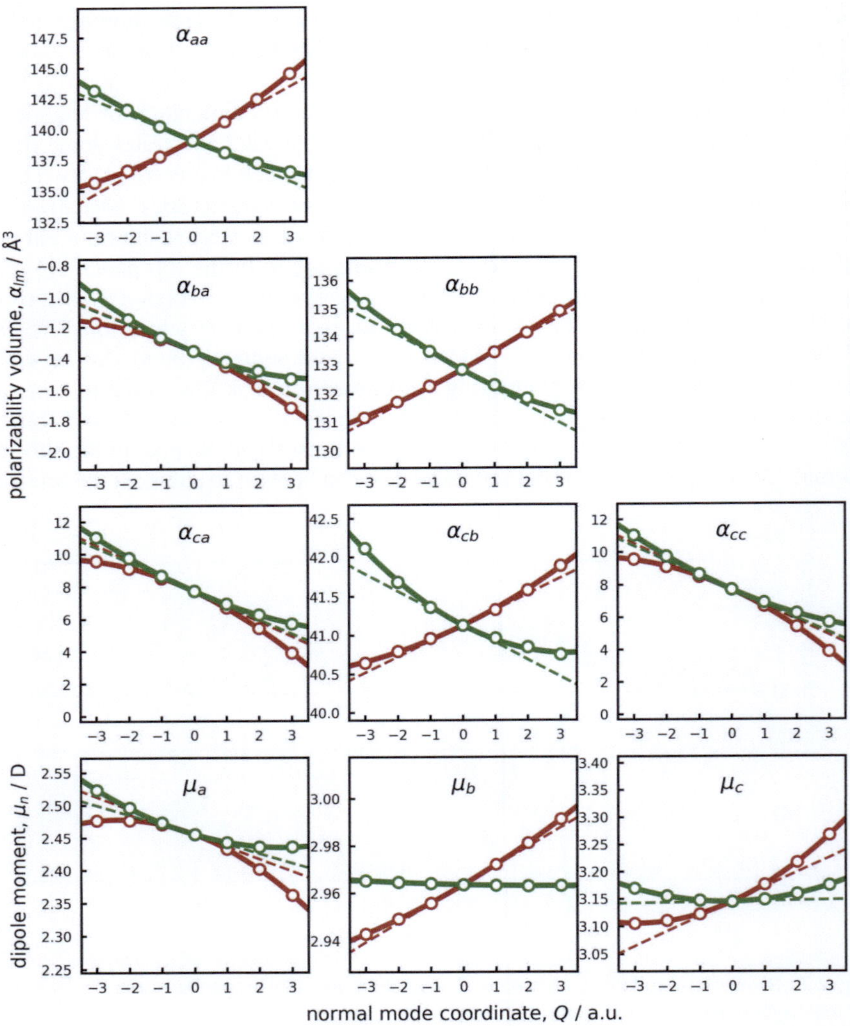

Fig. 2.3 Polarizability (rows 1–3) and dipole moment derivatives (row 4) for the methylene symmetric (red) and antisymmetric (green) stretching modes

Q, with $Q = 0$ representing the equilibrium geometry. Results for phenylalanine's methylene symmetric stretch are indicated with the red circles, and antisymmetric stretch with the green circles. The bottom row of Fig. 2.3 indicates values obtained for the three elements of the dipole moment vectors. In all plots, solid lines are a fit to a second-order polynomial, and the dashed lines indicate the tangent at $Q = 0$. Consider that there are $3N - 6$ vibrational modes, and we ultimately care about only these slopes. Once benchmarking of such a technique has been accomplished, such graphs are not required in most applications, as the values of the slopes are the only quantities to be carried forward in the application of interest. However, when in doubt, it is useful to pick an arbitrary mode, and have a Python script that reads the coordinate and Hessian data and makes such a graph. I note that this plot is not as attractive as it could be, with occasional crowded horizontal spacing of the plots owing to the variability in the width of the numerical field for the y-axis tick labels, and extents of the vertical axes that are not necessarily optimized for the presen-

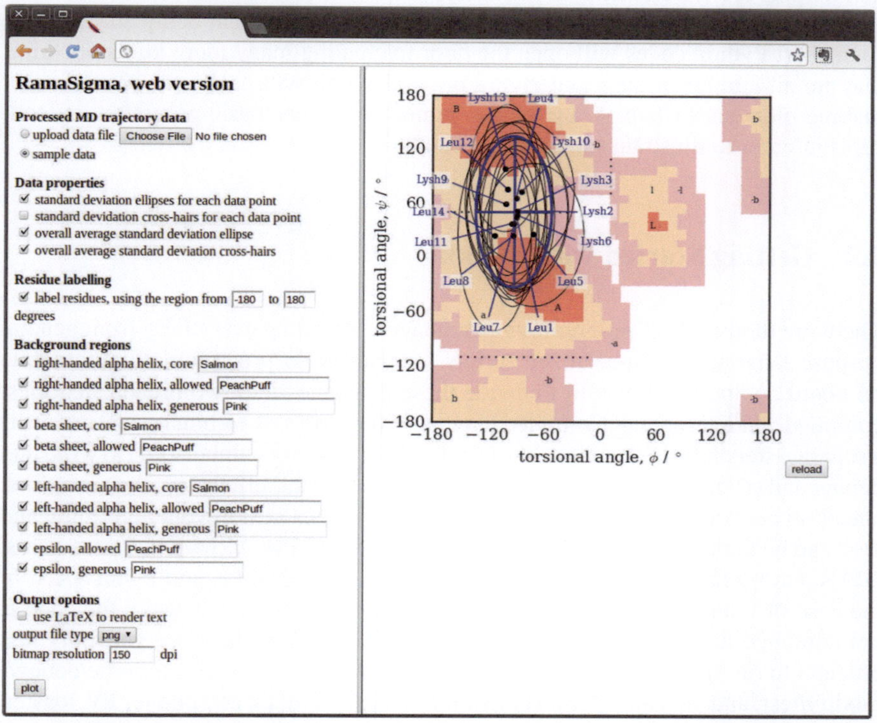

Fig. 2.4 Output of a Python script that generates Ramachandran plots for the visualization of dihedral angles in peptides and proteins. The Python code draws the interface that presents the options for selection of the plotting elements, does the calculation to produce standard deviation ellipses from molecular dynamics trajectories, and displays the results back in the same web browser window

tation of each plot. For a publication-quality plot, the script could be made more elaborate to take care of such issues using one-time customizations. But the main point here is that this is a prime example of the utility of a programmatic rather than interactive creation of the graphic, perhaps "for internal use only".

A final example is related to the structural dynamics of a peptide. Figure 2.4 illustrates a Ramachandran plot of the peptide torsional angles ψ and ϕ. Classically allowed regions are highlighted in shades of red. The average values of ϕ and ψ for each residue are indicated by black dots, each surrounded by a black ellipse who semi-major and semi-minor axes extend one standard deviation. The global average deviation is indicated by the blue ellipse, with blue cross hairs indicating the mean torsional angles. Although this graphic is scripted in Python, the goal here is not just autonomous plotting but also data exploration. Exploring the data is by nature an interactive process. For this reason, every aspect of this graph, including the colours and which elements are to be displayed are controlled by a graphical user interface (GUI). This GUI itself is simply generated by another Python script that outputs HTML elements to be rendered in a web browser.

I have selected these four examples in order to illustrate that being able to write such short Python codes will open the door to creating many more kinds of scripts that are all similar in their underlying principles. We will delay our discussion of making plots until Chap. 4, and now continue with some basic programming concepts in order to illustrate some fundamentals.

2.2 Getting Started with Python

There are thousands of programming languages, including everything from general purpose languages to specialized scripts for gaming to programming automobile on-board computers. It useful to divide these languages into two broad categories: compiled and interpreted. Compiled languages perform text-to-binary conversion as a separate step before execution, have a reputation of being cumbersome to write and debug, and efficient to run. There are two reasons for their efficiency: they can 'look ahead' to generate optimized code based on what comes next in your algorithm, and they can be highly optimized for specific hardware (x86 vs ARM-based processors, GPUs, etc.). Examples of compiled languages include C, C++, and FORTRAN. In the case of interpreted languages (also known as scripting languages), the code is not compiled first, but rather run 'one line at a time'. It is therefore generally less efficient to run (no looking ahead). But the tradeoff is that it is easy to write/debug/-modify/test, and cross-platform compatible (mac, PC, linux, cell phone, TV, toaster, blender ...). Examples of interpreted languages include Python, perl, BASIC, ruby, and TCL. Prior to the 1970s, when computers were slow and expensive, it was worth paying a team of people to work for months to write efficient code. These days, with fast and cheap computers, compiled languages are still chosen in several situations, for example when speed is the most important factor. However, for most common custom scientific applications that are researcher-led, interpreted languages such as Python are perfect, mainly due to their flexibility.

Here is a real-world example from my lab: I have a FORTRAN code that took two days to write (including debugging and testing) and 15 s to run. The same code, which I have now re-written in Python, took two hours to write and three minutes to run. One conclusion is that Python is 12 times slower. However, be careful to consider the future time required for modifying and maintaining the code. I'm using FORTRAN and Python as examples, but this is true for almost all compiled vs interpreted languages. The bottom line is that, for a project like looking for extra terrestrial life, it may be worth spending months writing code, since the program will run for 5 years (instead of 50 years). We are primarily interested in writing small programs for day-to-day tasks. Nowadays, many spectroscopists have encountered Python. It is intuitive, has a relatively clean syntax, cross-platform, and most importantly is widely-used. Dedicated modules are available for many tasks, and are easily located and installed.

2.2.1 *Base Python Installation*

Python is already installed on all unix- and BSD-based systems (including linux and MacOS). You may wish to install a more updated version of Python, and leave your system-installed Python untouched.

If you are using a mac, the homebrew environment allows you to install a large number of unix tools that are not bundled with MacOS. At the time of writing, you can install homebrew by opening a terminal window and typing:

```
/bin/bash -c "$(curl -fsSL https://raw.githubusercontent.
    com/Homebrew/install/HEAD/install.sh)"
```

If that doesn't work, consult a more up-to-date installation guide online. After homebrew is installed, you can install Python with the command

```
brew install python@3.12
```

where the number after the `@` symbol specifies the desired version number, here assuming you wish to install version 3.12. Somewhere along the way, you will be prompted to install XCode (it is huge, so will take a while if your internet connection is not fast), and agree to the terms of Apple's license agreement. Homebrew should guide you along those steps. If you are using a linux distribution (Fedora, Ubuntu, etc.), you can use the built-in package manager to install whatever version of Python you want using an analogous command. For those of you using Windows, head to the Microsoft store, and start typing `python` in the search bar. Select and install the desired Python version.

2.2.2 *Installing Python Modules*

In this chapter, we'll be primarily working with basic Python functionality that's included when any reasonably-recent version of Python is installed. In the next two chapters, we'll be exploring additional functionality related to math and scientific programming and plotting that will require some extensions in the form of third-party Python packages. We might as well set those up now since, after you learn a little Python, you'll be inspired to explore some of these and other packages.

It is highly recommended that you create a virtual environment in which you can install Python packages (to extend Python's functionality) in a manner that will leave the system-installed Python alone. In fact, in recent versions of MacOS and some linux distributions, this is a requirement. The overarching motivation behind virtual environments is that you can have a customized Python environment for individual projects. This is especially convenient when specific packages might clash with each other in a global sense, but the two offending packages are not needed for a single project. That way, each project can have only those packages installed that it needs. If this is new to you, or you don't have any immediate need to maintain multiple Python environments with different versions of packages, I can recommend a simpler approach. That is to have a single Python virtual environment that you use for everything. That will get us to the point where we can install the few (perhaps only three) packages we need for all of the tasks we're about to embark on. If this later doesn't continue to meet your requirements, you can always set up additional virtual environments. So for this single-environment case, we'll install what we need in our home directory. At a unix or MacOS terminal (or in Windows powershell), you can type

```
cd
python3 -m venv .venv
```

The first command `cd` tells the terminal that we wish to change directory, i.e. navigate to a different folder. Ordinarily one would specify the folder name (as in `cd folderName`) but, when none is specified, this takes us to our home directory, even if we are there already. We then create a virtual environment called `.venv` inside our home directory, although you can put this in any folder. In unix-based operating systems, file and folder names beginning with a period are hidden; it is merely a convention that settings are kept inside hidden folders. This makes things a bit tidier as they won't always be displayed when listing the contents of the parent folder. But if you wanted your virtual environment to be called `Bob`, you could type `python3 -m venv Bob`, and then a folder hosting some special files within `Bob` would be created in your current directory (perhaps the home directory).

Now the virtual environment needs to be activated. In unix-based operating systems (including Linux and MacOS), in a terminal window, one would type something like

```
source ~/.venv/bin/activate
```

if you followed my advice on the filenames and the location. Otherwise, you'll have to specify the location and a virtual environment name other than `.venv`. In Windows, the command is slightly different, and looks like

```
venv\Scripts\activate
```

You'll know that activation of the virtual environment is successful as the prompt will have changed, typically including `(venv)` to remind you.

Now we are ready to install some packages. Let's use NumPy (that we'll explore in Detail in Chap. 3) as an example. Still inside the terminal window (or Windows powershell) and inside our virtual environment, type

```
pip3 install numpy
```

If all goes well, the Python package manager (affectionately known as pip, with the Python 3 variant often called `pip3`) will find the requested package, download it, and install it. If you didn't receive any error messages, test it by issuing two commands

```
python3
import numpy
```

where the first command launches Python (at the command line) and then, once inside the Python interpreter, `import numpy` is the command to load the module. If everything was successful, you won't see anything. If the NumPy package wasn't loaded or wasn't installed correctly, Python will complain that it doesn't know about any packages called `numpy`. The error message typically looks like

```
Traceback (most recent call last):
  File "<stdin>", line 1, in <module>
ModuleNotFoundError: No module named 'numpy'
```

Now here is something to pay attention to: If everything worked, but you can't load NumPy the next time you are working with Python, or running a Python script that you previously wrote that uses NumPy features, it most likely means that you are not inside your virtual environment. We installed the NumPy package inside this environment, so you'll have to enter it in order for Python to be aware of its associated packages (think of them as plug-ins). While you're here and things are sailing smoothly, you might as well install three more packages

```
pip3 install scipy
pip3 install matplotlib
pip3 install sympy
```

You could install them all at once, but I generally prefer to install them one-by-one so I can see if any errors occurred.

We're all set. From here, we we can work on the Python command line, enter commands (such as `import numpy`) in the running Python interpreter directly. I'll illustrate a few such small commands shortly, but eventually you'll want to create a file that contains the Python commands (i.e. your program or Python script), and then pass this file on to Python. Not only does that enable you to save your work, but even for simple tests it is less annoying that way as you can easily correct any errors or make changes and then just run your program again.

2.2.3 Setting up a Python Development Environment

You can use any text editor to write your Python code. In fact, you can even use a word processor, so long as you save (export) your files into a plain text format. However, there are benefits to using an integrated development environment (IDE), such as syntax highlighting, built-in help, debugging, and many other features. There are many IDEs to choose from, including ones that have been popular with Python developers for decades. If you don't already have a strong preference, I highly recommend Microsoft Visual Studio (VS) Code, available on all platforms. If you are using Windows, install it from the Windows store or from the VS Code website. For mac users with homebrew installed, you can use the command

```
brew install --cask visual-studio-code
```

After launching VS Code, when you open a Python file (a file ending with `.py`) for the first time, or create a new Python file, the editor will detect that you are interested in Python, prompt you to install some additional packages (say 'yes'), and automatically detect the Python interpreter you have installed in the previous step. In some cases (like on the mac where you now have two different Python interpreters installed in different locations), VS Code will prompt you to select which one you want to use. Select the Python version that you have most recently installed. Most importantly, VS Code should detect that you are running inside of a virtual environment, with some indication of that in the bottom right corner of its window.

Before writing any programs in VS Code, we can play around with the Python interpreter at the command line using interactive sessions.

```
>>> a = 4
>>> b = 3
>>> a
4
>>> b
3
>>> a + b
7
>>> a
4
>>> b
3
>>> c = a + b
>>> c
7
```

In the above demonstration, the `>>>` symbol is meant to represent the prompt you are greeted with when you start the Python interpreter without specifying a file that contains your Python code. In a unix environment, that's what you'll get when you type `python` at the shell prompt, rather than `python3 codeFile.py`. Although working with the interpreter in an interactive manner like this is a bit of a pain for more complex tasks, it is convenient when you just want a calculator (with full scientific capabilities!) that stores variables. It occasionally also provides you with a way to investigate whether some syntax is correct, rather than running the whole code you are working on. In general, however, you will want to save your code in a file (for example, my unimaginative name: `codeFile.py`) and then pass that file to the Python interpreter. More on that a bit later. In the meanwhile, we'll go through a few more examples of basic Python syntax with some one-liners.

2.3 Operators, Variable Assignment, and Basic Arithmetic

In the previous section we have seen an example such as `c = a + b`, making use of the addition operator `+`. Commonly-used operators include

- `+` for addition,
- `-` for subtraction,
- `*` for multiplication,
- `/` for division,
- `**` for exponentiation (note that some languages use the ^ symbol instead),
- `=` for variable assignment.

The right-hand side is evaluated and then assigned to the left-hand side. So in case you find yourself doing things like:

```
>>> b = a + b
>>> b
```

it is important to keep in mind that `b` was still `b` during the evaluation, but then got overwritten during the last step, the assignment. The above example is more space-saving than:

```
c = a + b
b = c
```

We don't need `c`, so this is unnecessary as long as we don't later need to use the old value of `b`.

This is most commonly encountered when you have bookkeeping in the form of `count = count + 1`. This situation comes up fairly frequently as you keep track of how many instances of an event occurred. For example, you might be sifting through a file of molecular coordinates, counting how many oxygen atoms are in the molecule. By the way, all the cool kids (Pythonistas) write this as `count += 1`. One typically initiates things with `count = 0` (since `count += 1` adds 1 to the count, so `count` has to previously exist). Then one enters a loop, or processes the file line-by-line. At the end, the last value of the variable `count` is of interest. We will later see that there are better, built-in ways of counting things in Python.

These examples illustrate that Python makes a handy desk calculator for arithmetic. Let's suppose you are aligning a laser at the beginning of your experiment and want to know the average of 4 measurements you have recorded in your lab notebook. This can be done easily with

```
>>> (1.10 + 2.40 + 3.70 + 1.87)/4.
2.2675
```

I much prefer this to using an actual calculator as it is easy to correct any mistakes, and refer back to what I entered on the screen in case I have a doubt about what I typed. A bit later we'll talk about how to make the display more attractive, controlling the number of decimal places displayed in the output, and aligning numbers into columns. Notice that I used `4.` rather than `4` in the denominator. This has to do with *data types*. In many programming languages (including versions of Python prior to 3.0) the output of

```
>>> 3/4
>>> 3./4
>>> 3/4.
>>> 3./4.
```

is not the same. The short explanation is that `3` is an integer, while `3.0` (or simply `3.` for short) is a floating point number, or float. For clarity, I usually prefer `3.0/4.0` is it is more obvious to readers of the code that a division of two floats is the desired action. Starting with Python 3.0, the output of the above lines is the same, but it is good practice nonetheless to clarify if integer or floating point arithmetic is intended, by always using a decimal point for any non-integer values.

2.4 Data Types

Python has 4 simple and 3 compound data types. The simple ones are integer (such as `-4`, `7`, `0`, and `319`), float (such as `-4.`, `4.0`, and `4.17`), complex (such as `4.3 + 1.8j` or `0 - 2.0j`) and the logical types `True` and `False`. Note that in the case of floats, you can use scientific notation to write 4.13×10^3 as `4.13e+3` or simply `4.13e3`. Also note that `4e3` is considered to be a float, even though you may write this without an explicit decimal character. In the case of complex numbers, Python uses the engineering notation `j` to represent $\sqrt{-1}$ instead of i as is used in some other languages. Also note

```
>>> a = 0 - 2.0j
>>> type(a.real)
<type 'float'>
```

```
>>> a
-2j
>>> a.imag
-2.0
```

indicating that there are no integers defined as part of the complex type, even though you may not explicitly have entered any decimal characters. Here I am also illustrating the `real` and `imag` methods of complex data types. This may be different from your experience with other programming languages that might have used functions like `real(a)` to accomplish a similar thing. Python is revealing some of its objected-oriented nature by displaying that its variables inherit methods that are specific to the data type.

Finally, note the capitalization of the logical or 'boolean' `True` and `False`.

```
>>> type('True')
<type 'str'>
>>> type(True)
<type 'bool'>
>>> type(true)
Traceback (most recent call last):
  File "<stdin>", line 1, in <module>
NameError: name 'true' is not defined
```

In this case, `true` could have been defined as a variable of *any* type, but that would not be the best practice, as it could lead to confusion with `True` down the road.

For better or worse, Python is a *dynamically typed* language. In other words, we don't declare the type of a variable before assigning a value to it. As you have seen in the above examples, Python figures out the type based on the assignment; this can change in the program.

```
>>> peakArea = 4
>>> type(peakArea)
<type 'int'>
>>> peakArea = 4.2
>>> type(peakArea)
<type 'float'>
```

There are four basic compound data types: strings, lists, tuples, and dictionaries. In general, compound data types have elements or parts. We have already encountered strings.

```
group = 'methyl'
```

Lists are a set of data enclosed in square brackets. They can be of a single type, as in a list of integers

```
list1 = [4, 3, 1, 7]
```

or of mixed types

```
list2 = [4, 3.2, 'methyl', 17]
```

The same applies to tuples. For now, consider that tuples are like lists, but defined using parenthesis (round brackets)

```
a = (4, 3.2, 'methyl', 17)
```

There are some important differences between lists and tuples. For examples, lists can be rearranged, while tuples are immutable. Rather than going in to the technical difference further at this stage, I would just like to point out the data types so that you are aware they exist. Something you will notice in your first programs is that, at times you will have to supply Python with tuples while, in other instances, lists will work too. If you're just looking for a way to store your data, it may not matter what you use. If in doubt, go with lists as you'll see they are more flexible in terms of what you can do with them.

All compound data types can be indexed. In other words, you can refer to a particular element of a list by its integer position in the list. One must remember that, in Python, the list index starts at `0`. Compound objects therefore have their highest index one less than their length, easily obtained by the function `len`. In other words, if your list is called `a`, you can write `len(a)`. The actual indexing is achieved using square brackets immediately after the list or, more commonly, the name of the variable that holds the list. See if you can predict the result of

```
a[1]
a[1]*2.0
2.0*a[1]
type(a[1])
```

I mentioned that strings are also considered to be compound data types. That's because you can do things like

```
>>> b = 'methyl'
>>> len(b)
>>> b[4]
```

If you try to index a float or integer or other simple data type, Python will complain.

Any element of a list or a tuple may be another list or tuple, and so on.

```
molecularProperties = [4, 17, 21, 5, [3, 8], 9]
molecularProperties[3]
molecularProperties[4]
molecularProperties[4][0]

a = [4, 3.2, 'methyl', 17]
a[2][2]
```

For beginning programmers, most errors come from miscounting indices. Here is a more complicated example:

```
molecularProperties = [[1, 2], [[4, 5, 'ethyl', 8], 2, (3,
    2)]]
```

As a test, see if you can predict the output of

```
molecularProperties[0]
molecularProperties[1][0][2][2]
molecularProperties[1][1][0]
type(molecularProperties)
type(molecularProperties[1][2])
type(molecularProperties[1][2][0])
```

and check your answers in an interactive session of the Python interpreter.

There are functions to convert between variable types, and they are typically smart enough to work in all cases where it makes sense

```
>>> float(4)
4.0
>>> a = 4
>>> b = float(a)
>>> b
4.0
>>> float('4')
4.0
>>> float('   4   ')
4.0
>>> float('isopropyl')
Traceback (most recent call last):
  File "<stdin>", line 1, in <module>
ValueError: could not convert string to float: isopropyl
```

So you can see that the string 4, even with an arbitrary number of spaces around it, gets converted okay. This behaviour can be a real time-saver, especially when reading data files, as you don't have to first strip of any leading or trailing whitespace. However, as you can see, if Python, despite its best attempts, can't figure out how to do the conversion, it complains. Formally, it raises an exception. You can learn about what is known as exception handling if you are interested in modifying this behaviour.

Compound types can also be converted, most commonly between lists and tuples

```
d = [3, 4, 5.6]
e = tuple(d)
```

We now illustrate the dictionary data type. Whereas lists and tuples are indexed with integers that indicate the element's position, dictionaries are indexed with a keyword (key) string.

```
b = {'he':'James Clerk Maxwell', 'she':'Katherine Mary
    Dewar'}
>>> type(b)
<type 'dict'>
```

```
>>> b['she']
'Katherine Mary Dewar'
```

Now that we have seen the basic variable types, I wish to point out a couple of additional things that are useful. You can assign multiple variables at once

```
a, b = 3, 4
```

Behind the scenes, the right-hand-side is actually a tuple, even though we didn't explicitly write `(3,4)`. So tuples can be unpacked in this manner, as can lists.

```
>>> coolness = [3, 4]
>>> a, b = coolness
>>> a
3
>>> b
4
```

We will see that this especially handy when python functions return a list or a tuple. If those compound data types have only a few elements, you can then assign them directly without having to later index into them. We'll see some specific examples shortly.

2.5 Basic Introduction to Functions

We will be working extensively with functions: user-defined functions, ones built-in to Python (intrinsic functions), and ones available from modules. We have already seen and used a few intrinsic Python functions.

```
type(x)
len(x)
float(x)
```

In general, functions take anywhere from zero to many arguments, and the arguments can be of any type

```
a = functionName(b, c, d)
```

This function operates on `b`, `c`, and `d` and gives something back, i.e. it has a return value. The assignment operator (=) is used to store what is returned in the variable `a`. In Python, functions can return only a single thing. This is easy to get around since this object can be a compound data type, such as a list or tuple.

We have already encountered the functions `type` and `len`. Another commonly used one is `print` that displays the value of a variable

```
a = 4
print(a)
a = 'I love molecular spectroscopy.'
```

```
print(a)

print('please wait ....')
```

Another useful one, `input` gets user input from the screen.

```
>>> name = input()
Augustin-Jean
>>> name
'Augustin-Jean'
```

This function takes an optional argument, a string that is displayed ('echoed') to the screen as a prompt to help the user know what to input

```
>>> name = input('What is your name? ')
What is your name? Augustin-Jean
>>> name
'Augustin-Jean'
```

One must remember that, regardless of what gets entered at a `input` prompt, the data is always captured and stored as a string. So this may not be exactly what you want

```
>>> d = input('Enter the dipole moment in units of Debye:
    ')
Enter the dipole moment in units of Debye: 2.3
>>> d*3
'2.32.32.3'
```

This can be fixed in multiple ways. The two most common approaches are

```
d = float(input('Enter the dipole moment in units of Debye
    : '))
print(d)
```

or equivalently

```
print(float(d)*3)
```

If you know that you will always want to treat d as a float, then the first option is more convenient. This also illustrates a widely-used feature of Python and many other programming languages, nesting expressions. Inner nested expressions are evaluated first, and their return values are then the arguments of outer expressions.

```
type(float(input('atomic number:')))
```

2.6 Writing Python Scripts

I think that's enough playing with the interactive Python interpreter. I'm slowly getting tired of writing >>> all the time, and I imagine that you are probably wishing you had an easier way of correcting typos rather than having to re-enter commands. Although it remains convenient to quickly test ideas and syntax in an interactive session (and do your taxes or figure out your share of the restaurant bill), you will normally write your code ahead of time in a text editor (like VS Code), save it, and then feed it to Python. If you encounter any errors, you correct them, save and run again.

Using the text editor of your choice (VS Code!), write a small Python program. Something like

```
a = float(input('enter 1st number: '))
b = float(input('enter 2nd number: '))
c = a*b
print('The product is: ')
print(c)
```

and save it as `product.py`. Note that the output of the program may not look very nice. We'll later look at fancier ways to display it when we talk about formatted I/O. Now we want to run it. In VS Code, you can click the green arrow at the top of the screen. It is convenient if you assign a keyboard shortcut to that. If you are using a different IDE, there is most likely already a keyboard shortcut defined that runs the code. Note that, in VS Code, if any input is required, or if output is generated to the screen, a small terminal window opens up at the bottom of the screen. You can click in there to supply the arguments to the `input` functions.

2.7 Methods Associated with Lists

As we have mentioned, Python is an object-oriented language. This refers to its ability to define abstract data types and provide a way of encapsulation. We will talk more about OO programming later, but for now consider that several Python data types have associated *methods*. We have already seen that complex data types have two methods

```
>>> a = 3.2 + 4.7j
>>> a.real
3.2
>>> a.imag
4.7
```

Lists have many very useful methods. Mastering Python is all about understanding how to use and manipulate lists. Remembering that these methods exist (before setting out to write your code in a more complex way without time) can save you a lot of effort.

2.7.1 Appending Items to Lists

We will frequently encounter the situation where we want to increase the size of an existing list, most commonly adding items to the end. For example, let's say that we are processing an IR absorption spectrum and we want to determine the number of bands in a specific wavenumber range of interest, perhaps by inspecting the second derivative of the absorbance versus wavenumber. As our analysis progresses, the number of resonances we have located grows. At the end of the analysis, it would be handy to have all the located centre frequencies be contained in a list. The list method `append` adds a single item to the end of an existing list.

```
>>> a = [2, 4, 7]
>>> a.append(3)
>>> a
[2, 4, 7, 3]
```

If you want to grow a list from scratch, consider starting off with an empty list

```
>>> b = []
>>> b.append('hi')
>>> b.append('bye')
>>> b
['hi', 'bye']
```

In contrast, `extend` appends items from a list to the end of an existing list

```
>>> a = [2, 4, 7]
>>> b = [4, 8]
>>> a.extend(b)
>>> a
[2, 4, 7, 4, 8]
```

Just as a refresher, keep in mind that if we don't intend to use `b` later, it may not be that big of a convenience defining it in the first place.

```
>>> a = [2, 4, 7]
>>> a.extend([4, 8])
>>> a
[2, 4, 7, 4, 8]
```

Consider the difference between `append` and `extend`, just in case this is not what you wanted to do:

```
>>> a = [2, 4, 7]
>>> b = [4, 8]
>>> a.append(b)
>>> a
[2, 4, 7, [4, 8]]
>>> len(a)
4
```

2.7.2 *Inserting and Removing Items from Specific Locations in a List*

To be honest, I frequently create lists and append items to a growing list, but I seldom have need to remove items from a list. Nevertheless, this situation does arise, and we briefly investigate some basic functionality to do just that. I occasionally find myself in a situation where I want to add something to a position other than at the end, so `append` won't work. In such cases `insert(i, x)` can be used to insert an item `x` *before* position `i` in the original list.

```
>>> a = [2, 4, 7]
>>> a.insert(1, 1)
>>> a
[2, 1, 4, 7]
```

To insert an item at the front of the list, use index `0`.

```
>>> a.insert(0, 'wow')
>>> a
['wow', 2, 1, 4, 7]
```

An unusual thing about the `insert` list method is that the index can be beyond that of any index in the current list. The inserted item will then appear at the end.

```
>>> a.insert(105, 'last')
>>> a
['wow', 2, 1, 4, 7, 'last']
```

Of course, if this was really your intention, you could simply use the `append` method.

The `remove` method searches for the first occurrence of a specific item and then removes it

```
>>> a = [2, 4, 7, 3, 7, 3]
>>> a.remove(7)
>>> a
[2, 4, 3, 7, 3]
```

`pop` removes *and returns* an item of a list based on its index.

```
>>> a = [2, 4, 7, 3, 7, 3]
>>> a.pop(4)
7
>>> a
[2, 4, 7, 3, 3]
```

Since something is returned, this may be useful to you as you can assign it to a variable.

2.7.3 Finding Items in a List

Being able to quickly search through the items in a list is handy, but keep in mind that such operations typically make more sense for integers and character strings, rather than floating point numbers. In any case, `index` returns the index of the first match in the list

```
>>> a = [2, 4, 7, 3, 7, 3]
>>> a.index(7)
2
```

If the item you're asking for doesn't exist, Python will complain (raise an exception).

```
>>> a = [2, 4, 7, 3, 7, 3]
>>> a.index(3.1)
Traceback (most recent call last):
  File "<stdin>", line 1, in <module>
ValueError: 3.1 is not in list
```

The `count` method returns the number of times a specific item appears in a list

```
>>> a = [2, 4, 7, 3, 7, 3]
>>> a.count(7)
2
```

This is therefore a handy way of checking to see if something is there, without generating an error if it's not.

```
>>> a = [2, 4, 7, 3, 7, 3]
>>> a.count(13)
0
```

2.7.4 Changing the Order of List Items

We previously mentioned that, while Python tuples are immutable, the items of a list can be rearranged in a variety of ways. The most common rearrangement is to use the `reverse` method to flip the order of the elements of a list. One common situation is when parsing absorbance data in the mid-infrared, such as that created by a particular FTIR control software. Historically, the wavenumber axis is written from high to low values, for example 4000–400 cm^{-1}. That is slowly falling out of fashion, and I will plot all IR wavenumbers from low to high values. In case such a list is obtained directly from reading somebody else's file, you can reverse it (and the corresponding absorbance values.) It is important to note that this operation works *in place*. In other words, there is no return value.

```
>>> a = [2, 4, 7, 3, 7, 3]
>>> a.reverse()
>>> a
[3, 7, 3, 7, 4, 2]
```

The numbers in my example above are clearly not mid-IR wavenumbers in units of cm^{-1}, but I wanted to demonstrate that there is no sorting applied in this operation—it is a strict reversal of all indices in the list. Notice how, in an interactive session, I had to ask it to display the contents of the compound variable `a`? This is an indicator that the method worked in place, with no return value. Sometimes I forget, and end up making a mistake like this

```
>>> a = [2, 4, 7, 3, 7, 3]
>>> b = a.reverse()
>>> b
>>>
```

Since `a.reverse()` doesn't return anything, nothing is assigned to `b`. Incidentally, the list `a` is still reversed in the process, which may lead to more confusion if you were expecting it to stay the same, with only `b` containing the reversed copy!

This is similar behaviour to a bunch of other list methods, like `sort`

```
>>> a = [2, 4, 7, 3, 7, 3]
>>> a.sort()
>>> a
[2, 3, 3, 4, 7, 7]
```

If a method like `sort` wasn't available, think of what type of Python code you would have to write in order to achieve this. This is definitely a time-saver.

2.7.5 Indexing and Slicing Lists

We have already seen how a specific item of a list may be called by its index.

```
>>> a = [2, 4, 7, 3, 7, 3]
>>> a[1]
4
```

The same syntax may be used to modify this element of the list

```
>>> a = [2, 4, 7, 3, 7, 3]
>>> a[1] = 18
>>> a
[2, 18, 7, 3, 7, 3]
```

Every element from index 2 to (but not including) index 4 is obtained via

```
>>> a = [2, 4, 7, 3, 7, 3]
>>> a[2:4]
[7, 3]
```

From index 2 until the end

```
>>> a[2:]
[7, 3, 7, 3]
```

From the beginning until index 4

```
>>> a[:4]
[2, 4, 7, 3]
```

You can count from the back of the list by using a negative index

```
>>> a[:-1]
3
>>> a[-5]
4
```

The last 3 elements of the list are given by

```
>>> a[-3:]
[3, 7, 3]
```

I think you probably get the picture. One last thing, if you specify a third index, then this requests every nth value, with $n = 1$ being the default

```
>>> a[1:4:2]
[4, 3]
```

Seen enough? What would `a[::-1]` return?

2.8 Flow Control

Any program that you would write using only the concepts we have mentioned so far would have limited use, as the programs wouldn't be all that helpful beyond what a calculator could provide. One of our goals is to write programs to automate repetitive tasks, and make decisions for us based on set criteria. This is referred to as flow control. Even in the case of scripted graphics, many operations can be simplified using flow control structures.

There are two main types of flow control structures, looping for repetitive tasks, and conditional statements for decision making.

2.8.1 For Loops

`for` loops (also known as ‘do’ loops in some other languages) cause a bunch of Python code to repeat a specified number of times. In Python, `for` loops commonly work by iterating over each element of a list. (One can actually iterate over a variety of data structures in addition to lists, but we’ll stick with lists as a simple example for now.)

```
a = [2, 4, 7]
for b in a:
   print(b)
```

would produce the output

```
2
4
7
```

You’ll have noticed that the last line of the above code snippet was indented with respect to the first two lines. Some languages ignore indentation, leaving it as an option to the programmer to make the code easier to read by helping to identify flow control structures. In Python, indentation is mandatory as it is the *only* thing that signifies which block of code is part of the loop.

```
for b in [2, 4, 7]:
   print(b)
   print('hello')
print('done')
```

would produce

```
2
hello
4
hello
7
hello
done
```

The amount of indentation doesn’t matter, but Python will complain if you are not consistent. Be especially careful about mixing tabs with spaces, as these correspond to different ASCII characters, and will therefore raise an inconsistency, even if the code appears okay in your editor screen. Most programming editors allow you to customize the behaviour of the TAB key, offer to insert a set number of spaces as a convenience. Some people actually prefer to use tabs. My preference is for spaces (3 of them for each level of indentation), and so I configure my editor to insert 3 spaces upon pressing TAB.

Often you want to have a fixed number of iterations. Say you want to print something 8 times. A loop in Python always requires a list to iterate over. A quick way of making a list of 8 items is to use the `range` function.

```
>>> range(8)
[0, 1, 2, 3, 4, 5, 6, 7]
```

so the code could look like

```
for x in range(8):
   print('hello')
```

to produce

```
hello
hello
hello
hello
hello
hello
hello
hello
```

Since we are not actually using any of the list elements in this particular example, note that a list of *any* 8 elements such as

```
for x in [2, 4, 'dipole moment', 'polarizability', [0, 0],
     4.0-8.5j, -3, 'refractive index']:
   print('hello')
```

would produce exactly the same output, so `range` is merely a convenience.

If any of you are coming from a different programming language, beware of unnecessary indexing of loops. Here is an awkward example that is seldom necessary in Python:

```
a = [2, 4, 7]
for i in range(len(a)):
   print(a[i])
```

that produces the same output as our earlier example. Having said that, there are occasionally situations where you want to have a 'loop index', i.e. a variable that provides a count of the loop iterations. A simple example is

```
colours = ['red', 'green', 'blue']
a = [2, 4, 7]
for i in range(len(a)):
   print(a[i], colours[i])
```

There is nothing wrong with that, but Python provides a handy `enumerate` function that is useful in such situations. On its own, `enumerate` produces an object that contains an ordered tuple, where the first item is an index that starts at 0, and the second item is that of the supplied list.

```
colours = ['red', 'green', 'blue']
for x in enumerate(colours):
   print(x)
```

produces

```
(0, 'red')
(1, 'green')
(2, 'blue')
```

We can take advantage of this to write

```
colours = ['red', 'green', 'blue']
a = [2, 4, 7]
for i, x in enumerate(a):
   print(a, colours[i])
```

so we have a list index (counter) for situations where we need it.

A related function is `zip` that combines multiple lists into tuples where each element comes from the sequential combination of the lists. Although `zip` may be used to produce new data types of different shapes, it is most commonly seen when you need to temporarily create such a structure for iterating over the sets. This means that rather than

```
greetings = ['hello', 'hi', 'howdy']
luckyNumbers = [40, 35, 7]
for x in range(len(greetings)):
   print(greetings[x], luckyNumbers[x])
```

(nothing wrong with that, by the way) you could use

```
greetings = ['hello', 'hi', 'howdy']
luckyNumbers = [40, 35, 7]
for x in zip(greetings, luckyNumbers):
   print(x)
```

If you try the above example yourself, you'll notice that the output of the two versions is not quite identical, as `zip` returns a tuple. We could easily have printed a tuple `(greetings[x], luckyNumbers[x])` as well, but that's not the point. The goal is to illustrate that there are ways of working with ordered arrangements of objects in lists and tuples without having to resort to looping through each item individually, with the help of an explicit loop counter, to get what you want. This will all make more sense when we get into some spectroscopy examples.

2.8.2 While Loops

It is often desired to keep a particular loop running as long as a certain test condition remains true. The moment that condition becomes false, loop iterations stop and the program continues after the loop structure (designated by the change in indentation!).

As an aside, we consider the use of logical or Boolean expressions. These are commonly encountered in `while` loops where Python needs to evaluate whether an expression is true or false. Such an evaluation is typically based on a comparison using one of 6 operators

- less than, <,
- greater than, >,
- equal to, == (and note the distinction between the = operator used for variable assignment),
- less than or equal to, <=,
- greater than or equal to, >=, and
- not equal to, !=.

Consider the following example of a while loop

```
a = 0
while a < 4:
    print(a)
    a += 1
print('done')
```

that produces the output

```
0
1
2
3
done
```

Sometimes the value of the variable whose condition is checked is of interest, instead of or in addition to the actions that happen inside the loop. What would you expect to see if we asked to print the value of `a` at the end of the code?

2.8.3 Conditional Statements

Another essential structure in any programming language provides the ability to execute a certain block of code if a tested condition is true. There are three general schemes for such conditional decision making. The simplest one is a solitary 'if' construct

```
if (a < 4):
    print('okay, small enough')
```

Then there is the 'if–else' construct

```
if (a < 4):
    print('okay, small enough')
```

```
else:
    print('nope, too big')
```

and the nested 'if–(else if)–else' construct, where 'else if' has the syntax `elif` in Python

```
if (a < 4):
    print('okay, small enough')
elif (a < 10):
    print("okay, we'll make an exception")
    # note the snazy use of double quotes for a
    # string that is then allowed to contain single
    # quotes
else:
    print('nope, too big')
```

There can be as many `elif` causes as you need

```
if (a < 4):
    print('okay, small enough')
elif (a < 5):
    print('close enough')
elif (a < 10):
    print("okay, we'll make an exception")
else:
    print('nope, too big')
```

The order in which you place clauses within a multiple-clause `if` statement is important, as the moment one clause is found to be true, none of the following clauses are checked.

Keep in mind that if you want a subsequent condition to be evaluated, even if the previous one was already found to be true, you need to create separate (independent) checks.

```
if (a < 4):
    print('first is good')

if (b < 5):
    print('second is good')
```

As Python does not care about extra blank lines, I usually put one in between such statements to help draw attention to the fact that these are not part of a single flow control design. But keep in mind that's optional, as all that matters is the indentation.

Logical operations may be combined with `and` and/or `or`

```
if ((a < 4) and (b < 5)):
    print('both are good')
```

Note that once you start using `and/or` together with less/greater than, it's easy to get the logic confused. You may also have been noticing my liberal use of parenthesis. With some exceptions (like creating tuples), Python will allow you to use as many sets of nested parenthesis as you like for clarity in your code.

Ready for combining loops and conditional statements? As an exercise, carefully study the indentation to predict what the output of this program would be.

```
for x in [1, 12, 4, 18]:
   for y in [3, 18, 2]:
      if (x < y):
         print(x, 'is less than', y)
         print('fantastic')
   print('here I am')
print('done')
```

In summary, loops and conditions do all the tough work in most programs. Consider the following program that checks to see if the input word is a palindrome.

```
word = input('enter word: ')

answer = True
for x in range(int(len(word)/2)):
   if word[x] != word[-x-1]:
      answer = False

if answer:
   print('yup')
else:
   print('nope')
```

2.9 String Manipulation

Just like we saw for lists, strings (another compound data type) have their own properties and associated set of methods. There are many string methods; you should check out the online Python documentation to see all of them. It is a good idea to review such documentation periodically, in general, as some things evolve over time. I constantly discover built-in methods that are new (or new to me!) that would have eliminated lines of error-prone code in my old Python programs. I'll just point out a few of the commonly-used ones one now.

2.9.1 Concatenation

Multiple strings may be concatenated with the + operator.

```
>>> first = 'Neils'
>>> last = 'Bohr'
>>> greeting = 'Hello ' + first + ' ' + last + '!'
>>> greeting
Hello Neils Bohr!
```

Note that no extra spaces are inserted by default, only the contents of the individual strings appear, so the spaces between the words need to be manually inserted.

A somewhat odd but at times useful behaviour of Python is that strings can be multiplied by integers to create multiples of the original string.

```
>>> molecule = 5*'atom'
>>> molecule
'atomatomatomatomatom'
```

```
>>> atom*2.3
Traceback (most recent call last):
  File "<stdin>", line 1, in <module>
TypeError: can't multiply sequence by non-int of type '
    float'
```

2.9.2 Character Replacements

`capitalize` returns a copy of the string with the first character capitalized

```
>>> element = 'helium'
>>> element.capitalize()
'Helium'
```

It doesn't care if the first character is already capitalized.

```
>>> 'Silver'.capitalize()
'Silver'
```

When you look up Python documentation online, as with other programming languages, authors often use square brackets to indicate optional arguments. For example, the string method referred to as `center(width, [fillchar])` will centre a string in a field that is `width` characters wide

```
>>> 'dennis'.center(14)
'    dennis    '
```

where the default fill characters are spaces. We can change this by specifying `fillchar`

```
>>> 'dennis'.center(14, '*')
'****dennis****'
```

`expandtabs([tabsize])` replaces tabs in strings with a certain number of spaces. If this is not specified, the default is 8 spaces.

```
>>> '0.125\t0.343'.expandtabs()
'0.125   0.343'
>>> '0.125\t0.343'.expandtabs(2)
'0.125 0.343'
```

2.9.3 *Searching Inside Strings*

`count(sub, [start, [end]])` is especially useful as it counts the number of occurrences of a substring (optionally within a specified range) within a larger string. The optional arguments are best explained with some examples.

```
>>> 'spectroscopy'.count('s')
2
>>> 'spectroscopy'.count('sc')
1
>>> 'spectroscopy.count('s', 3)
1
>>> 'spectroscopy'.count('s', 3, 7)
0
```

The string method `endswidth(suffix, [start, [end]])` returns a `True` or `False` Boolean value.

```
>>> 'feb11a.txt'.endswidth('t')
True
>>> 'feb11a.txt'.endswidth('.txt')
True
>>> 'feb11a.txt'.endswidth('a')
False
```

`find(substring, [start, [end]])` returns the lowest index where a substring was found within the larger string.

```
>>> 'feb11a.txt'.find('11')
3
>>> 'feb11a.txt'.find('1')
3
```

If the substring was not found, `-1` is returned.

```
>>> 'feb11a.txt'.find('12')
-1
```

`isalnum` returns `True` if all characters in the string are either letters or numbers, and `False` otherwise.

```
>>> 'feb11a'.isalnum()
True
>>> 'feb11a.txt'.isalnum()
False
```

Similarly, `isalpha` returns `True` if all characters are letters, and `isdigit` reports on whether all characters are numbers. The `islower` and `isupper` methods can be used to detect whether all characters are lower or uppercase, respectively. The `upper` and `lower` methods will force the case of each character, regardless of what is was originally.

```
>>> 'Feb11a.TXT'.lower()
'feb11a.txt'
```

I find this especially handy for fixing the file names when some annoying programs like to save files with extensions like `.PNG` instead of `.png`. Another use is to simplify flow control based on user input to questions like yes/no. Rather than a construct like

```
if response == 'yes' or response == 'Yes' or response == '
    YES':
```

you could write

```
if response.lower() == 'yes':
```

to handle much of the variability. In some less frequent situations, you may find a use for the `swapcase` string method that toggles the case of each letter within the string.

Slightly off-topic, but it is at times useful to replace one substring with another with `replace(old, new, [count])`. If the `count` argument is not specified, the default is to replace all occurrences.

```
>>> 'equilibrium'.replace('i', 'ooo')
'equooolooobroooum'
>>> 'equilibrium'.replace('i', 'ooo', 2)
'equooolooobrium'
```

2.9.4 Tabs and Line Breaks

You can use the escape sequence `t` for each tab stop, and `n` for each line break, that you wish to insert into a string. This is especially useful when writing data to files (as we will discuss shortly) because files are often formatted in a regular manner, for example with data organized into columns. Creating such tabular output is typically less important when printing to the screen, at least for our spectroscopy applications, but there are occasions when you'll want things to look nice.

```
>>> print('first\nsecond\t\thello')
first
second          hello
```

Note that the `print` function automatically inserts a line break at the end. This may be suppressed with a comma *after* the function (yes, after closing the parenthesis). This is not used very often, as most people treat output to the screen as primarily for debugging, and only really care about how files are written. More on this below.

2.9.5 Splitting Strings

The need to divide strings up occurs so frequently that I would like to give the topic some attention. First consider the `partition(separator)` method. It separates a string based on a specified substring. It returns a 3-tuple containing the part before the first occurrence of the separator, the separator itself, and the part after the first occurrence of the separator.

```
>>> 'feb11a.txt'.partition('.')
('feb11a', '.', 'txt')
>>> 'feb11a.txt.zip'.partition('.')
('feb11a', '.', 'txt.zip')
```

Contrast this behaviour with that obtained by using `split([delimiter, [maxsplit]])`. Often you will observe the default behaviour

```
>>> 'sodium hydrogen carbonate'.split()
['sodium', 'hydrogen', 'carbonate']
```

where a list of items is returned that appears to be space-delimited. Actually, if the delimiter is not specified, then `split` operates on *all* whitespace (spaces, line breaks, carriage returns, tabs), *and* treats multiple delimiters as single delimiters. Quite often, this is exactly the behaviour that you want.

```
>>> '   sodium\t hydrogen\t\t    carbonate    '.split()
['sodium', 'hydrogen', 'carbonate']
```

Note that leading and trailing whitespace is also cleaned up nicely. Seems useful for reading data from a file, right? After the `split` operation, the only thing that you may have to deal with is that your numbers are stored as strings. Shortly we'll see how to clean this up; later we'll see that there are better ways of reading numbers from data files.

It may be convenient that you can specify the delimiter

```
>>> 'sodium hydrogen carbonate'.split('o')
['s', 'dium hydr', 'gen carb', 'nate']
```

Just be aware that the moment you specify the delimiter, multiple consecutive delimiters are no longer ignored.

```
>>> '3,4,5,,6'.split(',')
['3', '4', '5', '', '6']
```

so you end up with a 5-item list that contains a null string as the fourth item. If this is a nuisance for you, then there is likely a better way to read your data.

One of my favourite ways to make use of string splitting is when I'm making a list of labels for plots. Imagine that you have a 3×2 arrangement of 6 subplots in a figure, and you want to create inset labels a, b, c, etc. in the upper left corner of each plot, to be read and placed in a loop that iterates over the plot number. I find that typing

```
labels = ['a', 'b', 'c', 'd', 'e', 'f']
```

to be a pain, typing all of those quotes and commas; I much prefer writing

```
labels = 'a b c d e f'.split()
```

which creates the same list.

2.10 Formatted Output

So far, we've seen a few instances where Python output can be rather ugly

```
>>> a = 0.43
>>> b = 2
>>> c = a*b
>>> c
0.8599999999999996
```

Python is smart enough to know the answer, it just doesn't put in any effort in formatting the result unless it is asked to. What if you wanted the output to look like `0.86` or `0.8600` or perhaps even `8.60e-01`? Python offers a few options for doing this; we will take a look at the two most popular methods.

2.10.1 Traditional Syntax

In many languages, a special sequence of two or more characters is used to indicate the format of the designed result. Python accepts the widely accepted notation `%d` for an integer, `%s` for a string, `%f` for a float, and `%e` for a float in scientific notation.

This has now been largely replaced with an alternative method, that we will look at next. However, what I'm calling *traditional* syntax is still supported, and is very useful. Its general usage is

```
print(string % tuple)
```

Where the string contains one or more of the above-mentioned format specifiers, % is the string interpolation operator, and you also supply a tuple that contains the same number of elements as there are format specifiers in the string. It sounds complicated, but it's really not. A few short examples will best illustrate this.

```
>>> property = 'polarizability'
>>> n = 9
>>> print('The %s tensor has %d elements.' % (property, n)
    )
The polarizability tensor has 9 elements.
```

If there is only a single interpolation to be performed, the string to be filled in doesn't need to be presented as a 1-tuple.

```
>>> n = 9
>>> print('%n elements should be specified.' % n)
9 elements should be specified.
```

If you want to do any math or other manipulation (perhaps through a function) on the string before it is to be interpolated, then it *must* be presented as a tuple, even if there is only one string to interpolate. For example,

```
>>> n = 9
>>> print('%d real and imaginary components, in total' %
    (2*n))
18 real and imaginary components, in total
```

Leaving the parentheses off `2*n` would, in this case, produce a very different result (try it!), and in many other cases would generate errors. So plan to use tuples when you use the string interpolation operator.

In the case of integers and strings, often no width is specified. In that case, the interpolation increases the length of the string by the number of characters in the strings to be inserted, and the number of digits in the integers to be inserted. But the width may also be specified:

```
>>> n = 9
>>> '%5d % n'
'    9'
```

where Python has inserted 4 leading spaces since it saw that `n` would occupy 4 of the requested 5. This may be useful if you are creating a table and you want to right-justify to align numbers. The same behaviour occurs for strings, padding spaces on the left.

```
>>> '%10d' % 'dipole'
'    dipole'
```

Floats are a bit more interesting, as they may be formatted by specifying the number of decimal places desired in the output and the overall width of the field. To clean up an earlier example

```
>>> c
0.85999999999999996
>>> '%7.3f' % c
'  0.860'
```

Remember to count the decimal place as one of the characters. That's why we end up with 2 spaces added on the left to make up a total width of 7 characters. The same concept applies to formatting in scientific notation

```
>>> '%11.2e' % (-c)'
' -8.60e-001'
```

where I'm just showing off by making the number negative too. But it's also a reminder to account for all the space required by the 3-digit exponent, its sign, and the sign of the coefficient (even it's seldom negative!). At times the data collection in my experiment has been hampered as I've forgot to include space for a negative sign, and the value became negative in the middle of the night. Programming languages vary in terms of how they behave if you specify the width of a string that is not large enough to contain the output. In some cases, the width will automatically be expanded for you, pushing everything over. This may be merely an inconvenience, as your data won't line up with the other rows. If you intended to later parse this data with something like `string.split` then maybe you won't care. However, in some cases, data that's too wide will be chopped off to fit into the format specification. If you are really worried about this, you usually get 'everything' if you don't format the data, i.e. you use simply `%f` instead of something like `%3.2f` (when you should be using `%4.2f`). However, this isn't always the solution. You can't trust Python to switch into scientific notation when the numbers get 'small'. So collecting data that looks like `0.000000012` is not useful for you, when every data point looks the same, even though Python considers this number 'big enough' that it hasn't switched into scientific notation yet. In such cases, you'll really want to specify `%16.6e` if you want to be assured of getting 7 significant figures.

2.10.2 F-Strings

Another string formatting method that is gaining in popularity since the Python 3.6 is to use formatted string literals, affectionately known as f-strings. Here one prefixes the string with the `f` character to specify that the string contains one or more formatting placeholders to be filled in using the contents of previously-defined variables. This looks like

```
n = 9
f'{n} elements should be specified.'
```

Note that we didn't need to tell Python whether n was an integer, float, string, etc. But we could have done that using

```
f'{n:d} elements should be specified.'
```

or even include the width of the field

```
f'{n:5d} elements should be specified.'
```

And more complex examples can be built up with these principles

```
direction = 'x'
value = 2.8
print(f'The {direction}-component is {value:2.3f} D')
```

I recommend you try these for yourself, perhaps converting the examples we discussed using the traditional syntax to this method. As you can see, one advantage of this approach is that the variables are more directly appearing in the proper positions in the strings. This is helpful for long strings that require many substitutions, as there is a lower chance of errors due to incorrectly specifying the order of the items in the corresponding tuple.

2.10.3 Raw Strings

There are times when you want Python to ignore anything that it might otherwise interpret as a formatting instruction inside the string. This can be done by specifying a raw string, using the r character as a prefix. As a simple example, the n sequence inside a string would normally be interpreted as a line break.

```
print('First line.\nSecond line.')
```

But if you instead created the string as

```
print(r'First line.\nSecond line.')
```

Python would simply interpret every character literally, so n becomes nothing more than followed by n as two adjacent characters. The use of a raw string is necessary when your string contains characters that would cause the Python interpreter to complain that your syntax is in error. Raw strings are your way of saying "Trust me, I've got this." You'll see plenty of examples of that throughout this book, starting in Chap. 4. That's because I like to use LaTeX syntax when specifying symbols for use in axis labels, legends, and other plot annotations. It is nearly always the case that Python thinks that there is something wrong with those strings. By specifying them as raw, they get through to Matplotlib's (Chap. 4) font rendering which happily accepts them.

2.11 File Handling

Most tasks for the spectroscopist fall under one of two categories, modeling and function visualization, and data processing. Sometimes those are combined into a single code, as when a function is fit to a dataset to determine the values of some parameters that best fit the data. If you are fortunate enough to get to write Python code to control your spectrometer and save the data, then you have the ability to create output files in addition to reading them. As I mentioned, formatting output for the screen can make things look pretty, but formatting output to files is especially useful. We will now look at writing data to files, and reading data from files.

2.11.1 An Overlooked Alternative

Before we get into the real stuff, I would like to point out an often ignored but incredibly useful trick, and that is to simply redirect output from the screen to a file. In other words, if your output is not too complicated (multiple files, fancy buffering, etc.), let the operating system take care of all of this for you. Although this approach is technically OS-independent, it is generally geared towards those working on a command-line, such as in a unix terminal or in windows powershell. I will briefly describe this approach nonetheless: Imagine a Python program called `count.py` that contains the lines

```
for x in range(10):
   print(x)
```

you could then send the output to the file `output.txt` when you run the program from a command-line

```
python3 count.py > output.txt
```

I can’t emphasize enough how enthusiastic I am about this method. Not only do you avoid the overhead (that you’re about to see) for the Python file I/O, but this is also great for debugging. You can start off by running your code with the output sent to the screen

```
python3 count.py
```

and, after you fix all the errors or get the formatting just right, then run it one more time, redirecting to the file. Of course, this is most useful if you don’t require any interactive input from the user, i.e. you are not using any `input` statements. If you do need input from the user, I would still advocate the use of this method, but you’re better off getting the input in a different way—more on that later.

2.11.2 Writing Files Using Real File Handling

We can't always avoid file I/O within Python there are many cases in which real file handling much better. Furthermore, as I mentioned, your normal workflow may be inside an IDE, or by clicking on files, rather than working in a terminal window. Therefore, for a variety of reasons, we want to be able to write output to a file, and also read data from a file. Later, when we talk about some scientific packages that extend Python's base capabilities, you'll see really easy/slick ways of doing this. But for now, I think it is worthwhile if I demonstrate some methods that work in plain vanilla Python, as you'll always be able to fall back on them if the fancy ones aren't flexible enough for your requirements.

There are three steps to writing to a file. The first is opening the file, and creating a 'file handle'. This is an abstract Python object that can be assigned to any variable. For historical reasons, I like to use `hndl`, but call it anything you want. In the case of

```
hndl = open('output.txt', 'wt')
```

`hndl` is the variable that holds the file handle, the first argument supplied to `open` is the filename string, and the second argument is a two-character string `wt`. The `w` signifies that we wish to open the file for writing. Note that this would clobber the contents of file if it happens to already exist, without any warning. If you want to append onto an existing file, specify `a` (as in `'at'` instead). The append mode doesn't mind if the file happens to *not* exist, it will create a new file for you. Finally, the `t` indicates that we wish to store the data in plain text (ASCII) format. Specifying `b` would have resulted in a binary file. You will have to know what is best for your data, depending on the type (for example, numbers versus bitmapped images versus scalable vector graphics).

The second step is the actually writing itself. One simply uses the `write` method of the file handle object, and supplies a single string as the argument.

```
hndl.write('hello\n')
```

We have seen that `print` always tacks a line break onto the end of the string. So `print('hellon')` would cause two line breaks to appear. However, `write` (thank goodness) doesn't do this, so you'll have to be explicit about when you want to start a new line. This is really a good thing, as much of the pickiness about formatting strings is for writing to files. Just as a basic refresher, see that

```
name = 'Dennis'
hndl.write('hi %s\n' % name)
```

is equivalent to

```
formatted_string = 'Hi %s\n' % 'Dennis'
hndl.write(formatted_string)
```

The point is that the `write` function requires a single string to be supplied as its argument. It doesn't matter whether you do some interpolation on the fly or other things to assemble the string, or whether you assemble the string you want first.

The final step is to close the file using the `close` method of the file handle object.

```
hndl.close()
```

Forgetting to close the file has erratic consequences, depending on the operating system. (It may be just fine, it may be fine until the next time the program runs, the output may be truncated, the output may appear delayed until the OS as a courtesy closes the file for you. First time is free.)

I have shown only snippets so far. Here is a complete example that illustrates writing to a file.

```
hndl = open(input('output filename: '), 'wt')
for x in range(10):
   hndl.write('iteration %d of 10\n' % x)
hndl.close()
```

2.11.3 *Reading Data from Files*

After a file has been opened for reading, individual lines may be read, one at a time, using the `readline` method of the file object.

```
hndl.open('data.txt', 'rt')
for x in range(5):
   eachline = hndl.readline()
hndl.close()
```

As you can see, Python will keep track of where `readline` is in the file, automatically advancing to the next line with each use. An error will result if you try to read beyond the last line of the file. It would make sense if I illustrate some techniques to avoid this, such as looking for the end-of-file marker. But I will instead show another method that can be used in most instances, and that is to read all the lines in one fell swoop with `readlines`

```
hndl.open('data.txt', 'rt')
data = hndl.readlines()
hndl.close()
```

`readlines` returns a list (now stored in the list variable `data`), of length corresponding to the number of lines, with each item in the list a string corresponding to the individual lines. In case you only wanted the first 5 lines, as in the previous example, you could always access them with `data[:5]`. The performance of `readlines` is generally better than its single-line counterpart, as you can then manipulate the data list in memory, rather than going to the disk each time. It may also be a convenience in case you need to skip around in the lines of the file, or read through them many times. If you read all the lines at once, the file handle can be closed immediately, another possible benefit in some instances. The only drawback is for very large files. You need to make sure that you have enough memory to store everything. For example, if you are opening a molecular dynamics trajectory with millions of frames, it may be preferable to read one line at a time, and then throw it out as you replace the contents of that variable with the next line until you find the line you are looking for.

2.11.4 File Reading and Writing Example

I'll end this section with a complete example. Say you have a data file called `input.txt` that looks like

```
2.2     4.7
3.4     1.1
7.0     0.1
```

and you want to create a file, `output.txt` that contains the first column multiplied by two, and the second column multiplied by three. Here is one possibility that is not particularly elegant, but works just fine

```
# read the old data
hndl.open('input.txt', 'rt')
data = hndl.readlines()
hndl.close()

# write the new data
hndl = open('output.txt', 'wt')
for eachline in data:
   x = 2.0*float(eachline.split()[0])
   y = 2.0*float(eachline.split()[1])
   hndl.write('%4.1f\t%4.1f\n' % (x, y))
hndl.close()
```

Note that it was okay to use a variable called `hndl` for writing the file, as the previous file reading using the same variable was closed. If you need to open multiple files (reading or writing) at the same time, you will need to use variables with different names.

Here are a couple of random Python tidbits that we can use to improve on this example. First you can use an *implicit* loop to apply a function to each element of a list.

```
>>> a = ['dennis', 7.3, 4, 8]
```

```
>>> [type(i) for i in a]
[<type 'str'>, <type 'float'>, <type 'int'>, <type 'int'>]
```

Asking for the type may not seem that useful, but this is particularly handy when converting from one type to another as we'll see shortly. Second, recall that we can assign variables to items of a list or tuple with a single statement. This leads to a slightly improved version of the second half of our program.

```
hndl = open('output.txt', 'wt')
for eachline in data:
   x, y = map(float, eachline.split())
   hndl.write('%4.1f\t%4.1f\n' % (2.*x, 3.*y))
hndl.close()
```

Python Tidbit

The hash character (#) is handy for short comments. Longer comments (lengthy explanations, usage guides, documenting chunks of code) may use 3 sets of double quotes.

```
# we'll need these later
a = 7
b = 3

"""You can see that we have just assigned
   values of 7 & 3 to a & b, respectively.
   Now we'll continue with the program."""

c = 12
```

Remember that Python ignores blank lines, so feel free to use those liberally for extra clarity.

2.12 A Closer Look at Functions

We have already seen many situations where functions are used, and so their basic syntax may be clear to you already. However, there are a few additional things to discuss, both in terms of technicalities and style. Recall that in the function call

```
a = input('filename: ')
```

the function's return value is assigned to the variable `a`, `input` is the function's name and, in this case, there is a single string provided to the function. Built-in functions such as `input` are always available in any Python session, on any computer, when using any operating system.

It's easy to write your own functions. Sticking a block of code into a function has two advantages. The first is clarity, also known as code abstraction. The second is convenience, as multiple uses of the same code in the same program need to be defined only once. This is a bigger advantage than you might realize, as correction of errors or improvements need to be implemented only once. Both concepts are illustrated in the following pseudo code.

```
# define all the functions you need
def prepare_all_files():
   # many things

def generate_data():
   # lots of steps

def prepare_data():
   # fancy code

def process_data():
   # could be hundreds of lines of code here

def report(type):
   # note that one argument must be supplied

def erase_temp_files():
   # cleans up at the end

# main program starts here
prepare_all_files()
data = generate_data()
for x in data():
   prepare_data()
   process_data()
   if (quality < tolerance):
      report('files')
      report('graphs') write('finished step %d' % x)
erase_temp_files()
```

You'll have to use your imagination for the content of the functions such as `process_data` encountered above. By not showing you those details, I wish to emphasize that the main program (without the definitions of the functions) consists of only 9 lines of code. There could be hundreds of lines of code in the functions. Also note, that most functions are called only once. In such cases, someone else reading your code will have an easy time of understanding the overall flow, since the details of how the functions actually work are not interfering with the presentation of the outline. That's the code abstraction part. Also note that the `report` function is called twice. If you realize that you want to change the way something is handled in that function, you'll need to change it only once, in the function itself. It is worth remembering this as you write your own codes, and think of places that you could use functions to clean things up, and reuse code. Although this may seem like you are complicating some relatively simple tasks at first, you will later come to appreciate how streamlined your code is, easier to debug, modify, and maintain.

As another example, consider the following function

```
def report(count, filename):
    a = 6.2
    hndl = open(filename, 'at')
    hndl.write('event occurred %d times' % count)
    hndl.close()
    b = float(count)*a/26.5
    return b
```

Note that `a` and `b` are local variables. Assignment and changing their values does not affect variables with the same name outside of this function.

Also keep in mind that functions don't have to return anything. In other words, there is no obligation to have a `return` statement at the end.

```
def report(count, filename):
    a = 6.2
    hndl = open(filename, 'at')
    hndl.write('event occurred %d times' % count)
    hndl.close()
```

We have also seen examples where functions don't take any input arguments.

```
def report():
    hndl = open('log.txt', 'at')
    hndl.write('event occurred')
    hndl.close()
```

If the function does not do anything with variables passed as input arguments, then most likely it doesn't return anything either, although this doesn't need to be the case. Note that, when defining and calling this function, one still needs to type the set of parentheses, writing `report()` instead of `report`. You'll later see examples where the *names* of functions can be passed to other functions (so `report` would be the provided parameter). That occurs most often when using packaged optimization routines to fit data. Coming back to the point, here we want to *call* the function, i.e. have it execute, so we'll need to use `report()`. However, in most other places throughout this book we won't include the parenthesis in the text when we are referring to the name of a function. Any source code examples will of course show the proper syntax.

As we mentioned previously, in Python functions can return only a single item, but this can be any compound data type if you require it, for example a list or a tuple.

```
def circle_sphere(radius):
    pi = 3.14159265
    area = pi*r**2.0
    volume = 4.0/3.0*pi*r**3.0
    return (area, volume)
```

Keep in mind some of Python's flexibility in assigning multiple variables from a tuple or list

```
a, b, c = [3, 4, 3.2]
```

Although that has nothing to do with functions, it is very handy as you can write

```
area, volume = circle_sphere(7.1)
```

rather than the more cumbersome

```
area = circle_sphere(7.1)[0]
volume = circle_sphere(7.1)[1]
```

Something important to note is that, in the above code, the worst part is not having to write out the line twice to get the area and volume assigned. It is that the function `circle_sphere` is called twice unnecessarily. If that function took 3 min to run (picture a more extensive calculation!), it would produce the area and volume, but then you would have to wait a total of 6 min while it runs again, again producing the area and volume. Each time, you are throwing out one of the two quantities while assigning the other. This could be avoided with

```
data = circle_sphere(7.1)
area = data[0]
volume = data[1]
```

so the function ends up being called only once. But, unless you have a specific reason for this,

```
area, volume = circle_sphere(7.1)
```

is nicer. Here is a complete example so you can see something that is more than just a short snippet.

```
def report(count, filename):
    a = 6.2
    hndl = open(filename, 'at')
    hndl.write('event occurred %d times' % count)
    hndl.close()
```

```
    b = float(count)*a/26.5
    return b

for x in range(7):
    report(x, 'xfiles.txt')
    for y in range(200):
        report(y, 'yfiles.txt')
        if (x*y < 50):
            c = report(x*y, 'lessthan50.txt')
            print('great! %d', % c)
```

The value of variables inside functions are considered to be 'private' to those functions. This means that if we were to try to use the variables a or b outside of the report function, Python would complain, acting as if those variables didn't exist. That same point is less obvious (and a bit dangerous as it wouldn't generate the same error to alert you) if a and/or b also had values outside of the function. In that case, they still have their values in the main code, and maintain a separate (hence private) identity inside the function. Since that can be a recipe for disaster, it is generally good practice to give the variables different names. The code below demonstrates a related concept:

```
from math import sin

def signal(x):
    return 3.0*sin(x - 0.5) + 0.7

# read the time data from a file
hndl = open('time3.dat', 'rt')
data = hndl.readlines()
hndl.close()

# convert this into a list of time values
time = map(float, data)
# do some stuff
for each in time:
    print(each, signal(each))
```

The function `signal` expects a single argument (defined as `x`) and we are supplying it when we call the function `signal(each)`. Don't be bothered by the fact that `x` and `each` are separate variables—I've done that on purpose. As far as `signal` is concerned, it is expecting a variable as input, and we are providing one. Whatever the value of `each` was at the time of calling, that's the value of `x` inside of the function. If we replaced the name `each` with `x` everywhere, or vice versa, that would be okay too. Having them defined as separate variables just makes the point more clear, and draws attention to the point (in a good way) that `x` exists only inside the function.

The opposite to private is ‘global’. In Python any constants defined outside of a function are global. But be careful, functions can read global variables (without needing to have them explicitly passed as arguments), but if the function modifies the variable, the global variable is not modified (by default). There are ways to get around that, but there are almost always better approaches that avoid this.

2.13 Modules

Modules are a source of additional functions, constants, and data types. Modules are classified into two varieties. Standard modules are always available upon request in any Python session. External modules are available from third-party sources; one has to find, download, and install them.

2.13.1 Using Functions and Constants from Modules

There are several options for the manner in which functions are accessed from modules, and each has an associated syntax. Say you want to take the sine of a number

```
>>> sin(.1)
Traceback (most recent call last):
  File "<stdin>", line 1, in <module>
NameError: name 'sin' is not defined
```

Python complains that there is no function `sin` in the current namespace. But there is a sine function in the math module (among other places).

```
>>> from math import sin
>>> sin(.1)
0.09983341664682815
```

(Note that nearly all functions, especially trigonometric functions, work with and return angles in radians.)

Constants may be imported in the same way

```
>>> from math import pi
>>> pi
3.141592653589793l
```

Multiple items may be imported simultaneously:

```
>>> from math import sin, pi
>>> sin(27.2/180.*pi)
0.4570979270586942
```

and items may be renamed as they are imported

```
>>> from math import sin as hello
>>> sin(.1)
Traceback (most recent call last):
  File "<stdin>", line 1, in <module>
NameError: name 'sin' is not defined
>>> hello(.1)
0.09983341664682815
```

There are two ways to get everything the module has to offer. The first approach imports all functions and constants into the current namespace.

```
>>> from math import *
>>> sin(.2)
0.19866933079506122
>>> cos(.2)
0.9800665778412416
>>> tan(.2)
0.20271003550867248
```

The second approach keeps all items from the math module in their own separate namespace

```
>>> import math
>>> sin(.1)
Traceback (most recent call last):
  File "<stdin>", line 1, in <module>
NameError: name 'sin' is not defined
>>> math.sin(.1)
0.09983341664682815
```

Although the first approach seems to be more convenient, one must be careful with importing everything (using the `*` wildcard) from a module in your current namespace, as any functions or constants you previously defined (or that were built-in) will be clobbered if they have the same name. For example, if `function1` exists in `moduleA`, and then again as a different version in `moduleB`

```
>>> from moduleA import *
>>> from moduleB import *
function1(arg)
```

the definition of `function1` from `moduleA` was over-written; only the definition as appears in `moduleB` remains in your program. A simple fix, and the recommended solution, would be to keep the namespaces separate

```
>>> import moduleA
>>> import moduleB
>>> moduleA.function1(arg)
>>> moduleB.function1(arg)
```

Note that you could have written

```
import moduleA, moduleB
```

to have the same outcome.

2.13.2 Basic Math Capabilities

As we mentioned, there are some standard modules that are included with base Python installations. This means that you have to import them, just as you would for a third-party module, but you don't have to find and install them through the pip tool. There are many modules in this category, and our goal is not to summarize their capabilities; I encourage you to look at the Python documentation to see what exists. But I do want to give a brief introduction to three specific ones.

The `math` module provides many commonly-used functions including

```
math.sin(x), math.asin(x), math.sinh(x)
math.cos(x), math.acos(x), math.cosh(x)
math.tan(x), math.atan(x), math.atan2(y, x), math.tanh(x)
math.degrees(x), math.radians(x)
math.hypot(x, y)
math.exp(x), math.log(x, [base]), math.log10(x)
```

as well as the constants `pi` and `e`. Although these appear to be important for us, we will instead get our mathematical functions primarily from NumPy rather than math module, as we'll discuss extensively in the following chapter. However, it is good to know about `math` since it is an intrinsic module, so always available. If you have a small job to do in a location where NumPy is not installed, and you can't install it yourself, the functions provided in `math` may suffice for some simple operations. An example would be a simple file manipulation where you may need to take the square root of electric field intensity data in column 2 to turn it into something proportional to the magnitude of the electric field vector. However, as I mentioned, for more serious analysis of spectroscopic data, we'll want more capabilities. Over time, I find myself just using NumPy tools for everything, even simple tasks, as I have NumPy installed and I'm familiar with its math functions.

The module `random` provides tools for random number generation, including

```
random.seed(x)
random.getstate(x), random.setstate(x)
random.randrange(start, stop, step)
random.randint(min, max)
random.random()
random.uniform(min, max)
```

Once again, we'll later use the random number generating features provided by NumPy, but I also wanted to point this out as a base-Python option in case you want to use it. An example of such a situation typically occurs when your programming task is simply to read and write files in a specific format. Sometimes random numbers are required, sometimes basic math functions are required. You may not wish to install or import NumPy for such basic tasks.

2.13.3 System-Specific Information and Command-Line Argument Processing

The `sys` module provides tools for interacting with the operating system, and getting info about the Python interpreter currently running. These functions include

```
sys.exit('error message')
sys.platform
sys.version
```

I sometimes use `sys.exit(1)` in the middle of a code if I want to stop execution at that point, rather than commenting out all the remaining code. If you are using a sophisticated code editor (such as VS Code) breakpoints can be added for debugging purposes as an alternative. Another application is to process so-called command line arguments. I'm not going to go into this in as much detail as I'd like, as I recognize that it is not so common to work on the command line outside of a unix environment, even though it is certainly possible. But here is a short example of what I'm talking about. Imagine a code that needs to determine how many atoms are in a system, simply by multiplying the number of atoms in each molecule by the number of molecules. One option is to define these quantities within the code itself.

```
numAtoms = 12
numMolecules = 100
totalAtoms = numAtoms * numMolecules
```

That is straightforward, but "hard-coding" the values of 12 and 100 requires that they are changed inside the code when the values need to be updated. In some cases, that is fine and might be the preferred option. But there are situations when it may be desirable to not modify the code for this. An alternative that we have already seen is to ask the user to supply those two numbers at run time using the `input` function.

```
numAtoms = int(input('How many atoms in each molecule: '))
numMolecules = int(input('How many molecules: '))
totalAtoms = numAtoms * numMolecules
```

While the above method may seem like the obvious approach to solve the problem, it generally requires that the values of the two parameters be typed into the terminal. This is okay for occasional use, but not convenient if the code is to be automatically run to process many files. This is where command-line arguments come in handy.

```
import sys
numAtoms = int(sys.argv[1])
numMolecules = int(sys.argv[2])
totalAtoms = numAtoms * numMolecules
```

The idea is that the code is run like this:

```
python3 myCode.py 12 100
```

where `myCode.py` is the name of your Python file, and I'm assuming that `python3` is the name of your Python interpreter. `sys.argv` is simply a list of all the parameters (delimited by spaces) that were issued when the Python interpreter was called. The first item of the list, `sys.argv[0]` is the name of your program itself, so `myCode.py` in this case. The main advantage here is speed. Even if you are just using this program once, it is sometimes easier to type in those values of 12 and 100 as you execute the code. Imagine that you are creating a graphic whose appearance depends on those values. You can easily change them and see the appearance change. This may not seem like a huge time-savings compared to using the `input` function, but the real advantage comes from cases where you have another script (typically not written in Python) that automatically runs your Python code. That script can then easily supply the two parameters.

One potential disadvantage of using `sys.argv` is that there are no variable names (that you would see as `numAtoms` and `numMolecules` when hard-coding) or prompts (the string argument to `input`) displayed. So one has to know which arguments to supply and provide them in the correct order. One way to mitigate this is to have the Python program provide help on its syntax if the incorrect number of arguments is supplied. An easy way to do this is

```
import sys
if len(sys.argv) != 3:
   print('Usage:')
   print('myCode.py [numAtoms] [numMolecules]')
   sys.exit(1)
numAtoms = int(sys.argv[1])
numMolecules = int(sys.argv[2])
totalAtoms = numAtoms * numMolecules
```

The check at the beginning doesn't cover all scenarios that could cause the program to crash. For example, just because two command-line arguments were supplied (making the length of `sys.argv` equal to 3) doesn't mean that they can be successfully converted to integers on the next line, or that they were supplied in the correct order, where the number of atoms in each molecule needs to come first. However, one of the nice things about having the initial check like this is that one can intentionally trigger it by typing

```
python3 myCode.py
```

knowing full well that there are missing command-line arguments, and using this as an opportunity to view the usage statement. There is a lot of room for fine-tuning this interaction, using keyword rather than positional arguments, and combining multiple methods of supplying input. I mention this briefly just to demonstrate some of the possibilities in case this helps to provide some ideas for your large-scale data processing or simulation tasks.

2.13.4 Interacting with the Operating System

The built-in `os` module allows your Python code to interact with the operating system to do things like change directories, get the name of the current directory, etc. One of the things I do most with `os` is execute other programs with its `system` function. As an example, let's say that I've just created a graphic called `output.png` and now I want to convert it to a different image format `output.jpeg`. I could use my favourite command-line image processing tool ImageMagick to do this, with the command

```
convert output.png output.jpeg
```

Rather than having to do this each time my Python program exits, I would like this to be done automatically, so that the PNG and JPEG versions of the files exist. One option is to have a shell script that runs the Python code, and then executes the ImageMagick program. But since this is a simple task, I can do this within Python. The last few lines of my Python code could look like

```
import os
filePrefix = 'output'
pythonSaveFunction('%s.png' % filePrefix)
os.system('convert %s.png %s.jpeg' % (filePrefix,
    filePrefix))
```

where I'm hiding the Python detail of how the graphic was created (details in Chap. 4) inside a mystery function `pythonSaveFunction` for now. Then you can imagine an even more streamlined method

```
import sys, os
filePrefix = sys.argv[1]
pythonSaveFunction('%s.png' % filePrefix)
os.system('convert %s.png %s.jpeg' % (filePrefix,
    filePrefix))
```

that is ready for your automated processing!

2.13.5 Writing Your Own Modules

Now that you have mastered using and writing functions, you'll appreciate being able to package your functions into modules. The first advantage is clarity. Remember how one of my above examples mentioned that the function definitions may take hundreds or thousands of lines, and then the 'main' program consists of 9 lines of code? It would be even more clear if those functions were contained in a separate file. The second advantage is convenience. Just as it is convenient to makes changes to code in a function for multiple uses of that function, it is even more convenient to edit a function that is part of a module, for then many individual Python programs can benefit from your bug fixes or performance enhancements.

This is all easy to do, as a module is nothing more than the Python code you already know. Consider the contents of a file called `denmod.py`

```
def func1(x, y):
   # good stuff

def func2():
   # more good stuff

def func3():
   # amazing

# etc
```

and the Python scripted called `denprog.py`

```
from denmod import func1
apple = 16.0
a = func1(3.4, apple)
# etc
```

2.14 Summary and Next Steps

There are many interesting applications of Python programming for scientific applications, whether it is data processing, fitting data to a model, or simulations. If you are new to programming you will find that, in each case, a different subset of Python commands will be prevalent. For example, parsing data files involves extensive string manipulation which can be done using intrinsic Python functions (some of the basic ones were introduced here), and does not rely much on external modules. However, in the next chapter we will see that third-party Python packages such as NumPy and SciPy are ideally suited for a broad range of scientific computing, especially in our current focus of molecular spectroscopy. As we get into these topics, there is a need to visualize the results graphically, and we will use the capable Matplotlib (Chap. 4) package to do this.

2.15 Exercises

These exercises are designed to be written in pure Python, using only intrinsic modules. In the next Chapter, we will discover very convenient functions and data types provided by the NumPy module that make such tasks much easier. If you already know some NumPy/SciPy tricks, you'll find that avoiding them is rather painful. However, the purpose of these exercises is to develop basic Python programming skills. This is important, as you will then be able to extend the functions provided in modules, as well as modify them to suit your needs.

1. You have a mass spectrum of an unknown methylaluminoxane that has a molecular ion peak at $m/z = a \pm b$. Considering Al = 26.29815 amu, O = 15.9994 amu, $CH_3 = 15.0235$ amu, what are all the possible molecular formulas $Al_xO_y(CH_3)_z$ that would be consistent with your experimental data? Ask the user to input values for a and b and return the list of candidate molecules along with their actual masses.

---

2. Use a random number generator to 'experimentally' determine the value for the irrational number π. Using

   ```
   import random
   x = random.random()
   ```

   determine whether this random draw falls inside or outside of a unit circle. Repeat this experiment enough times to get some good statistics. Report your experimental value of π.

   Bonus: Just in case the above was too easy for you, make a plot that illustrates the accuracy as a function of the number of iterations. You may need a logarithmic scale to show this nicely.

---

3. From the supplemental information available online, download the archive `chapt2q3.zip` that contains 38 files named `data36.dat` thorough `data73.dat`. After you have unpacked these files, you'll notice that each contains two columns (let's call them x and y). *Assuming that the* x *column is the same in each file*, write a program that creates a new file called `data36-73avg.dat` that contains 2 columns: x and the average of the y data from each file. To make your program more generally useful, design it so that it can average an arbitrary number of files. For example, it could prompt the user for the file name pattern:

```
        file name prefix: data
first file in the series: 36
 last file in the series: 73
        file name suffix: .dat
```

Based on the user input, it could also generate the output file name `data36-73avg.dat`, so you don't have to prompt the user to enter that.

---

4. Think about how you might modify your program if you cannot (or don't want to) rely on the assumption that the x columns are the same. For example, imagine that the x series of the different files are slightly shifted from each other, or if the data collection interval is slightly different (or non-uniform) between the data points. *Don't actually write this program*, just describe how your program would best achieve this goal. Such a program would be very useful, but would be (unnecessarily) painful to write using only the Python constructs we have described so far. We will see easier ways to achieve this, but the algorithm that you have in mind will still apply.

---

5. From the online archive `chapt2q5.zip`, unpack the five Gaussian™ [5] electronic structure calculation files

```
6-311Gdp0.out
6-311Gdp1.out
6-311Gdp2.out
6-311Gdp3.out
6-311Gdp4.out
```

Write a program that reads these files (automatically and sequentially, in a loop) and prints out two columns: the total dipole moment *vs* average S–O bond length. Think about style, clarity, readability here: Inside the loop, use one or more functions to do all the messy work. In order to come up with a strategy, open one of the files in a text editor/viewer first, and have a look to see how the information is organized, and where the information you want exists in the file. The best strategy is to assume that the information isn't always on the same line number, but that the headers that get printed around the text you wish to parse are consistent. Figuring out how to do this efficiently is the goal here.

---

6. You are using IR absorption spectroscopy to monitor a reaction and have recorded the following data.

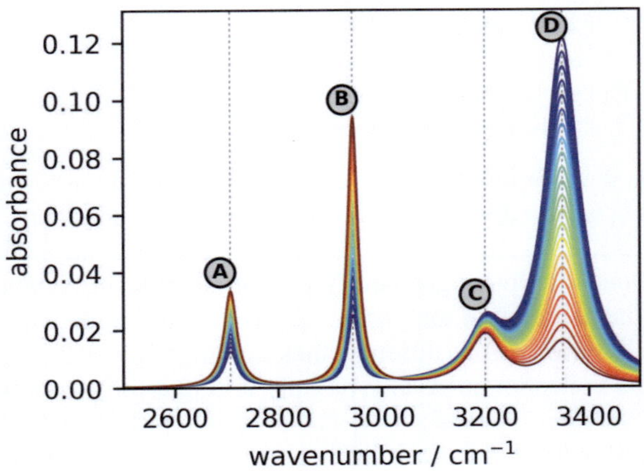

The instrument saved the data in files that are named:

```
spa00007.txt
spa00008.txt
spa00009.txt
.
.
spa00028.txt
```

The anatomy of each file is

```
0.09100
    2.50000e+03       4.00872e-04
    2.50100e+03       4.02330e-04
    2.50200e+03       4.03800e-04
    2.50300e+03       4.05282e-04
    2.50400e+03       4.06777e-04
    2.50500e+03       4.08285e-04
    2.50600e+03       4.09805e-04
    2.50700e+03       4.11338e-04
    2.50800e+03       4.12884e-04
```

etc., where the first line indicates the time (in units of hours). Starting on line 2, there are two columns. The first is the IR wavenumber; the second is the absorbance. Your task is to produce a table with 2 columns, the first being time (in hours), and the second the ratio of the area of peak A to the area of peak B. (In order to process these files, see if you can think of a nifty way of dealing with the zeros in the filename, rather than using lots of `if` statements.) You'll find the data in the file `chapt2q6.zip` available online.

References

1. M. Lutz, *Learning Python*, 3rd edn. (O'Reilly Media, Newton, MA, 2007)
2. C. Hill, *Learning Scientific Programming with Python* (Cambridge University Press, Cambridge, 2015)
3. S.A. Hall, A.D. Hickey, D.K. Hore, J. Phys. Chem. C **114**, 9748 (2010)
4. T.G. Trudeau, D.K. Hore, Langmuir **26**, 11095 (2010)
5. M.J. Frisch, G.W. Trucks, H.B. Schlegel, G.E. Scuseria, M.A. Robb, J.R. Cheeseman, J.A. Montgomery, Jr., T. Vreven, K.N. Kudin, J.C. Burant, J.M. Millam, S.S. Iyengar, J. Tomasi, V. Barone, B. Mennucci, M. Cossi, G. Scalmani, N. Rega, G.A. Petersson, H. Nakatsuji, M. Hada, M. Ehara, K. Toyota, R. Fukuda, J. Hasegawa, M. Ishida, T. Nakajima, Y. Honda, O. Kitao, H. Nakai, M. Klene, X. Li, J.E. Knox, H.P. Hratchian, J.B. Cross, V. Bakken, C. Adamo, J. Jaramillo, R. Gomperts, R.E. Stratmann, O. Yazyev, A.J. Austin, R. Cammi, C. Pomelli, J.W. Ochterski, P.Y. Ayala, K. Morokuma, G.A. Voth, P. Salvador, J.J. Dannenberg, V.G. Zakrzewski, S. Dapprich, A.D. Daniels, M.C. Strain, O. Farkas, D.K. Malick, A.D. Rabuck, K. Raghavachari, J.B. Foresman, J.V. Ortiz, Q. Cui, A.G. Baboul, S. Clifford, J. Cioslowski, B.B. Stefanov, G. Liu, A. Liashenko, P. Piskorz, I. Komaromi, R.L. Martin, D.J. Fox, T. Keith, M.A. Al-Laham, C.Y. Peng, A. Nanayakkara, M. Challacombe, P.M.W. Gill, B. Johnson, W. Chen, M.W. Wong, C. Gonzalez, J.A. Pople, Gaussian 03, Revision C.02 (2004)

Chapter 3
Tools for Numerical Calculations

3.1 Introduction

Our basic introduction to Python in the previous chapter demonstrated some essential features of the language, most notably how to construct lists and then use a loop to iterate through the items in a list. Armed with only that knowledge, one can write many programs to parse data files in order to search for and extract items of interest, or to rearrange the contents. Being able to write such codes will be useful in the handling of experimental data, but those features will not be the central focus for what we are about to study. We need a way to perform mathematical operations with Python, both numerically and symbolically.

We have seen that some basic math functions are available through the intrinsic `math` module, functions such as `math.sin(x)`. Although we'll be able to get by with mostly basic functions, there is a need to have some more advanced features available to us, such as

- root-finding,
- interpolation,
- optimization and curve fitting,
- numerical integration and differentiation,
- Fourier analysis,
- linear algebra, and
- digital signal processing tools.

The primary packages that provide such features are NumPy [1, 2] and SciPy [3, 4]. Historically, there were a few other packages in existence, and they had overclapping features. Recently, there is a move to avoid such duplication, especially for packages that people are likely to use together. For example early releases of SciPy included some of the basic features of NumPy. But now Python programmers are encouraged to use NumPy for fundamental numerical techniques, and hence the underlying NumPy functions are being removed from other packages, so that they

D. K. Hore, *Molecular Orientation Analysis Using Polarized Vibrational Spectroscopy*,
https://doi.org/10.1007/978-3-032-04874-5_3

are available only in NumPy. NumPy and SciPy contain a huge number of features, covering many of the numerical recipes that have been developed and optimized over the years, and typically available for compiled languages such as C and FORTRAN [5]. We're only going to cover the tip of the iceberg, exposing some of the most critical features for spectroscopists.

The first thing to notice is that practically everything from the `math` module is available in NumPy, so there's no need to import `math`; we can get things done using essentially the same or closely analogous NumPy functions. And some of the basic constants such as π are also available when we need them.

```
>>> import numpy as np
>>> np.pi
3.141592653589793
>>> 3
3
>>> np.sin(.1)
0.09983341664682815
```

From this point forward, I'll want to show you some code snippets, without having to repeatedly type the `import numpy as np` command. Some functions may have slightly different names than what you are used to, but chances are they are still there somewhere.

```
>>> np.asin(.1)
Traceback (most recent call last):
  File "<stdin>", line 1, in <module>
NameError: name 'asin' is not defined
>>> np.arcsin(.1)
0.1001674211615598
```

So what was `math.asin` is now `np.arcsin`. For the most part, the function names are obvious. When in doubt, consult the online documentation for an up-to-date and comprehensive list of all functions, and their required and optional keyword arguments. In some cases, passing optional parameters can control the type of output the function returns.

3.2 The NumPy Array Data Type

The most useful and life-altering thing that NumPy provides is a new data type called `array`. (Keep in mind this is `numpy.array` or `np.array` if you've imported NumPy as `np`.) Simply stated, an array is an ordered set of objects, usually numbers, much like we have already seen in a list or tuple. So while we can always write `a = [1, 2, 3]` or `b = (1, 2, 3)` since those are basic Python data types available everywhere at all times, we now introduce `c = np.array([1, 2, 3])` which, at the outset, looks much the same. What's special is that most functions that take scalar arguments, when given an array, return an array of the same size, with the function being applied to each element. Here is a small example:

```
>>> import numpy as np
>>> a = np.array([0., pi, pi/2.0])
>>> np.sin(a)
array([  0.00000000e+00,   1.22464680e-16,   1.00000000e
    +00])
```

Note that many NumPy functions may do this for you automatically when a list is supplied

```
>>> a = [ 0., pi, pi/2.0]
>>> type(a)
<type 'list'>
>>> b = np.sin(a)
>>> type b
<type 'NumPy.ndarray'>
```

although it may be a bit dangerous/sloppy to rely on such behaviour, so it's best to create arrays ahead of time if possible. Arrays are amazing for many reasons. A top one is that, in many cases, they eliminate the need for looping constructs such as

```
# imagine you have a list called 'data'
result = []
for x in data:
   result.append(somefunction(x))
```

since you can now go directly to

```
result = somefunction(data)
```

which is actually the same as our `np.sin` example, but now in a more general context as it applies to user-defined functions as well. In other words, it is not the NumPy version of the `sin` function that is special, rather the NumPy array data type has enabled this behaviour. The beauty of this is that you can design your function to operate on a single number, then it will behave the same way regardless if a number, or a multi-dimensional array is passed to the function. In other words, there is no need for the function to detect the input data type and then act differently just to produce the expected result. This is a huge time-saver, and enables very short and clean codes to be written.

In the previous chapter, we have seen that a quick way of creating an ordered list was by using the intrinsic `range` function. In NumPy we have the same thing available that automatically creates an array data type, the function `arange`.

```
>>> b = np.arange(4)
>>> b
array([0, 1, 2, 3])
>>> type(b[0])
```

```
<type 'NumPy.int64'>
>>> c = np.arange(4.)
>>> c
array([ 0.,  1.,  2.,  3.])
>>> type(c[0])
<type 'NumPy.float64'>
```

`np.arange` is actually more flexible than `range`, since `range` was limited to working with integers. In `arange`, the data type is automatically determined by the type of the arguments. Integers, floats, and complex numbers are supported.

The Matrix Data Type

NumPy technically offers another unique multi-dimensional data type called `matrix` that at first looks to be convenient for rank-2 tensors. In practice, it is less flexible than the array type, and was designed mainly for those making the transition from other programming languages to Python. One limitation is that, unlike `array` elements that can have an arbitrary number of dimensions, the `matrix` data type is restricted to be exactly two-dimensional.

```
>>> a = np.matrix([[1,2],[3,4]])
```

When using the matrix data type, element-wise multiplication for matrices is achieved with the `multiply` function.

```
>>> b = np.matrix([[4,5],[6,7]])
>>> np.multiply(a, b)
array([[ 4, 10],
       [18, 28]])
```

However, use of the `matrix` data type is discouraged, and it is in the process of being deprecated.

Here is something about lists we didn't really discuss, although you may have discovered it accidentally:

```
>>> a = [1,2,3,4]
>>> print(a*2)
>>> [1, 2, 3, 4, 1, 2, 3, 4]
```

So it follows that multiplying a list by anything other than an integer doesn't work

```
>>> print(a*2.2)
Traceback (most recent call last):
  File "<stdin>", line 1, in <module>
TypeError: can't multiply sequence by non-int of type '
   float'
```

However, the expected behaviour is achieved when using the array data type:

```
>>> a = np.array(a)
>>> print(a*2.2)
[2.2, 4.4, 6.6, 8.8]
```

A similar thing happens with addition

```
>>> a = [1,2,3,4]
>>> print(a + 3)
Traceback (most recent call last):
  File "<stdin>", line 1, in <module>
TypeError: can only concatenate list (not "int") to list

>>> a = np.array(a)
>>> print(a + 3)
[4, 5, 6, 7]
```

This provides a convenient way of initializing an array variable to a certain value, such as zero. One doesn't have to start by creating an array of zeros, as once the array is added to the single floating point zero, element-wise addition will take place. Note that we're actually describing the above example again, but this time in the reverse order

```
a = 0.
a += np.array([1., 2., 3.])
```

This is an especially convenient way of making sure that we have a complex-valued data type, even though the starting set of array values may be real

```
a = 0j
a += np.array([1., 2., 3.])
```

You might be curious as to why this is practical or necessary. Let's suppose that, in some situations, your Python code ends up with a complex number, or an array of complex numbers. Some functions, for example inverse trigonometric functions, behave differently when given real or complex-valued input. Knowing that all of your data is complex-valued can create a more streamlined workflow.

The array data type therefore provides a way of defining vectors, but can also be used for matrices as a multi-dimensional array may be built using a 'list of lists' notation where each list defines a row.

```
>>> a = np.array([[1,2], [3,4]])
>>> a
array([[1, 2],
       [3, 4]])
```

When it comes to indexing, NumPy offers two ways of addressing specific elements in a multidimensional array. The first is the traditional Python syntax

```
>>> a[0][1]
2
```

and the second is a Matlab-like [6] syntax `a[0,1]`. It's a bit messy having both of these options available, and you'll have to remember that the second syntax won't work for a 'traditional' list-of-lists, only for array data types when you are using NumPy.

3.3 Linear Algebra

There is a very large amount of linear algebra support available within NumPy, and if you have a project that requires this or can benefit from this, you'll have to dig into those details on your own. In order to fulfill our objective of analyzing molecular orientation distributions from spectroscopic data, we'll just need a few basic operations in this category. The first is matrix multiplication. NumPy defines the `dot` function for matrix multiplication using arrays.

```
>>> a = np.array([[1,2],[3,4]])
>>> b = np.array([[4,5],[6,7]])
>>> np.dot(a,b)
array([[16, 19],
       [36, 43]])
```

This can also be achieved with the convenient `@` operator,

```
>>> a @ b
array([[16, 19],
       [36, 43]])
```

Note that the conventional `*` operator results in element-wise multiplication

```
>>> a*b
array([[ 4, 10],
       [18, 28]])
```

Generally there is no need to make a distinction between column and row vectors, as the appropriate operations can be specified, regardless of the shape of the arrays. For example, the (row) vector

$$\mathbf{a} = [1 \;\; 2]$$

can be specified in the conventional way using `a = np.array([1, 2])`. Then it is possible to create the column vector

$$\mathbf{b} = \mathbf{a}^{\mathrm{T}} = \begin{bmatrix} 1 \\ 2 \end{bmatrix}$$

by writing `b = np.array([[1], [2]])` but this typically serves no purpose. If we wish to evaluate

$$\mathbf{ab} = [1 \;\; 2]\begin{bmatrix} 1 \\ 2 \end{bmatrix}$$

this can just as easily be achieved using the dot product `np.dot(a, b)`, that has the additional advantage of returning 5 as a scalar quantity. If we had instead used the ‘double index’ notation (writing things like `[[1]]` for the elements in one of the vectors) the result would be numerically correct, but have the shape of a multi-dimensional array even though it contains only a single value. That is not an issue, but may require an extra step to clean up.

The transpose is readily obtained using the transpose method of the object.

```
>>> a = np.array([[1,2],[3,4]])
>>> a.T
array([[1, 3],
       [2, 4]])
>>> a
array([[1, 2],
       [3, 4]])
```

but note that the transpose method doesn’t operate in place. That is, the original matrix method is not altered in the operation. (If that is what you actually wanted, you can always do something like `a = a.T`.)

The complex conjugate of any number or array can be obtained in two ways, either by using the NumPy function `conjugate`

```
>>> b = np.array([[1+3j,2-1j],[3,0-4j]])
>>> np.conjugate(b)
array([[-3.j,  2.+1.j],
       [ 3.-0.j,  0.+4.j]])
```

or by accessing a built-in method associated with a NumPy data type, `b.conj`. However, note that `.conj` is available only for true NumPy objects, for example not floats that might look like a NumPy float. That’s why `.conj` will work on any array, since arrays are available only as NumPy objects. When in doubt, the `np.conjugate` function will always work. If it is the conjugate transpose you are after, that can be achieved by combining two operations as in

```
>>> b.conj().T
array([[ 1.-3.j,  3.-0.j],
       [ 2.+1.j,  0.+4.j]])
```

Many other linear algebra features are available in NumPy. However, other than these basic ones along the lines of matrix-style multiplication of arrays, most are tucked away into the `linalg` subpackage. For example the inverse may be obtained using

```
>>> np.linalg.inv(a)
array([[-2. ,  1. ],
       [ 1.5, -0.5]])
```

and determinants are accessed via

```
>>> a = np.array([[1,2],[3,4]])
>>> np.linalg.det(a)
-2.0
```

Eigenvalues and eigenvectors can also be obtained from that sub-module.

```
>>> a = np.array([[1,2],[3,4]])
>>> results = np.linalg.eig(a)
```

Where we are capturing the ouput of `eig`, a 2-element tuple. The eigenvalues are in the first element

```
>>> results[0]
array([-0.37228132+0.j, 5.37228132+0.j])
```

And the eigenvectors are in the second element

```
>>> results[1]
array([[-0.82456484, -0.41597356],
       [ 0.56576746, -0.90937671]])
```

3.3.1 Prefabricated Arrays

There are many applications where it convenient to start with an array of zeros. For example, when you are about to sift through a lot of data and build up some statistics that you want to display in a histogram. The `zeros` function can do that.

```
>>> np.zeros(4)
array([ 0., 0., 0., 0.])
```

For a multidimensional array of zeros, the argument can be a list or tuple.

```
>>> np.zeros([4,3])
array([[ 0., 0., 0.],
       [ 0., 0., 0.],
       [ 0., 0., 0.],
       [ 0., 0., 0.]])
```

Speaking of histograms, say you are making a 2D histogram to keep track of the occurrences of some events. You could do something like

```
count = np.zeros([rows,cols])
for x in range(rows):
```

```
    for y in range(cols):
        if (data[x][y] > threshold):
            count[x,y] += 1
```

The idea is that some values of x and y will not receive any counts, and so those values will remain zero. So if it weren't for zeros, you would have to first build a multidimensional array for which every element is set to zero. As you become more familiar with NumPy data types, and methods for referencing them, you'll soon figure out that this example can also be handled in a single line, and without creating the empty zero-valued array `count` in advance with

```
count = (data > threshold).astype(int)
```

The magic is that `data > threshold` filters the values of data that meet this criteria, returning an array of the same size and shape as `data` with each value set to either `True` or `False`. Appending the `.astype(int)` simply converts the logical true/false values to 1 and 0. Nevertheless, you will find uses for `zeros`.

Another handy prefab function is `ones` that can create an array of any shape like this

```
>>> np.ones(5)
array([ 1.,  1.,  1.,  1.,  1.])
```

If you wanted to create a 5×5 identity matrix, i.e. a matrix with ones along the diagonal and all off-diagonal elements zero, this can be done using

```
>>> np.identity(5)
array([[ 1.,  0.,  0.,  0.,  0.],
       [ 0.,  1.,  0.,  0.,  0.],
       [ 0.,  0.,  1.,  0.,  0.],
       [ 0.,  0.,  0.,  1.,  0.],
       [ 0.,  0.,  0.,  0.,  1.]])
```

noting that we specify the size with only a single element (`5` rather than `[5, 5]`) since the identify matrix by definition is a square matrix.

3.4 Reading and Writing Regularly Formatted Data

In the previous chapter, we encountered a very general way of reading any kind of data from a file. In the case where the file contains say 2 columns and 1000 rows, there is a NumPy function `loadtxt` that is extremely powerful due to its ease of use and flexibility. Rather than using the plain Python method of

```
# read the data from the file
hndl = open('data.txt', 'rt')
data = hndl.readlines()
hndl.close()
```

```
# parse the data to assemble lists
x = []
y = []
for eachline in data:
    x.append(float(eachline.split()[0]))
    y.append(float(eachline.split()[1]))
```

We can do this in one (yes, one) line

```
x, y = np.loadtxt('data.txt', unpack=True)
```

There are three things to pay attention to here. First, you can specify either a filename or file handle. When a filename is given, the file is opened, read, closed, and the (hidden) file handle is discarded. Second, string representations of numbers are automatically converted into floats, as will be discussed further below. Lastly, specifying the keyword argument `unpack=True` requests that the data array be transposed to allow direct assignment to variables. If the data inside that file looked like a 1000×2 arrangement, `loadtxt` by default would return a 1000-element array, each with 2 items. By transposing the data, there are two elements in the 'list of lists' picture, each with 1000 items. Just to be clear, if you left out the `unpack =True`, the above line of code would generate an error, as you are trying to assign what is returned by `loadtxt` to only two variables (`x` and `y`), and Python would complain about an assignment mismatch, i.e. it is expecting another 998 variable names from you!

Pretty cool, eh? But wait, it gets better. Additional keywords greatly increase the flexibility of this convenience. Say the 1000 rows of 2 columns started on row 6, and the first 5 rows were some header where the instrument writes the settings for that particular run. Think you are out of luck, and have to resort to `readlines`? Think again!

```
x, y = np.loadtxt(filename, skiprows=5, unpack=True)
```

What if there are 1000 rows, but 200 columns (plus the 5-line header). If you want *all* of the data, you aren't going to bother specifying variable names.

```
data = np.loadtxt(filename, skiprows=5, unpack=True)
```

will give you everything returned as a single two-dimensional array called `data`, and you'll have to index into the array using `data[138, 880]` to get the specific items you want. You'll probably want to use the list slicing tricks we learned earlier to grab entire rows or columns. But if you knew from the beginning that you were interested in only two of these 500 columns (say the first one and column 4), you could use

```
x, y = np.loadtxt(filename, skiprows=5, usecols=(0,5),
    unpack=True)
```

Other useful keywords include `dtype`, where you can change the data type of the resulting array from its default `dtype=float`; `comments` where you can specify the starting characters of any line in the file that you wish to skip. For example, `comments = '#'` will skip any rows beginning with the hash character. This is handy, as it offers more flexibility than `skiprows` for avoiding non-numerical data. You'll have noticed that the parsing of the file, extracting numbers and converting them to floats is handled automatically. Multiple spaces are treated as single space, and tabs are treated as spaces. The behaviour can be modified by setting the `delimiter` keyword, which is seldom necessary.

Finally, the `converters` keyword allows for some fancy on-the-fly processing of rows (or columns when `unpack=True`) using built-in or user-defined functions. The idea is that we supply `converters` with a dictionary that maps row (column) numbers onto the name of a function that will convert the raw strings (as read) into anything you want. Consider the text file `data.txt` with contents

```
1 2
2 4
3 6
4 8
```

The Python code

```
import numpy as np

def twice(w):
   return 2*w

def thrice(p):
   return 3.0*float(p)

x, y = np.loadtxt('data.txt', converters={0:twice, 1:
    thrice}, unpack=True)
print(x)
print(y)
```

would produce the output

```
[ 11.  22.  33.  44.]
[  6.  12.  18.  24.]
```

There is an equivalent NumPy function `savetxt` that writes one- and two-dimensional arrays to files. One can use either the syntax

```
np.savetxt(filename, data)
```

where `data` is a 2D array, or

```
np.savetxt(filename, x, y, z, ...)
```

where a series of 1D arrays (of equal length) are specified. Some useful keywords include `fmt` where the format may be specified, for example `fmt='%8.3e'`, and `delimiter` which defaults to a single space.

3.5 Interpolation

There are instances in which you will need to interpolate some data, that is you'll want to know the value of a certain quantity in between two values. In our field of study, this commonly occurs for the refractive index. When modeling a quantity (such as the transmittance) as function of wavelength or frequency, the refractive indices of all phases also need to be known for the same set of frequencies. Some refractive index information is available in the form of a dispersion relation, a function whose parameters have been fit to experimental data to produce approximately correct values over a specific frequency range. If such a function is available in our frequency range of interest, then any frequency can be input to get the refractive index we need. However, sometimes experimental refractive index values are available for only a discrete set of frequencies. For example, $n = 1.414$ at $3100\,\text{cm}^{-1}$, $n = 1.416$ at $3200\,\text{cm}^{-1}$, and $n = 1.419$ at $3300\,\text{cm}^{-1}$. Let's say we want to know the value of the refractive index at $3145\,\text{cm}^{-1}$. We could fit these three points to a polynomial and come up with our dispersion relationship, but we can also just interpolate between the closest two data points.

A variety of interpolation functions are available in the SciPy module, within the `interpolate` sub-module. The simplest one is a 1D-interpolation function called `interp1d`. In order to use it, we create a special interpolation object by passing the function the x- and y-arrays of the data we wish to interpolate. For example,

```
import scipy
nint = scipy.interpolate.interp1d([3100, 3200, 3300],
    [1.414, 1.416, 1.419])
print(nint(3145))
```

where we would see a value of 1.4149 returned. One of the cool things about simulataneously working in the NumPy and SciPy environments is that, rather than passing a single x value to `nint` and getting back the interpolated y-value, we can pass it an entire array of x-values at once:

```
import numpy as np
import scipy
nint = scipy.interpolate.interp1d([3100, 3200, 3300],
    [1.414, 1.416, 1.419])
newx = np.linspace(3140, 3280, 20)
newy = nint(newx)
```

Note that since `newx` has 20 elements (it is an array of 20 evenly-spaced values between 3140–3280 cm^{-1}) 20 interpolated values have been determined and assigned as `newy`.

One final point, is that this example was a bit unrealistic since the experimental data had only 3 points. Most likely, you'll find a source of data that contains hundreds or thousands of points. That's good, because it will likely contain refractive index values close to the frequencies that you are interested in. However, it probably still won't have values at your exact frequencies of interest. So, rather than specifying the input data as `[3100, 3200, 3300]` and `[1.414, 1.416, 1.419]` let's read those thousands of points from a file, and assume that we want to know the refractive index (`newy`) at the same 20 points we called `newx`. This is easily accomplished with something like this

```
import numpy as np
import scipy

# read the literature data
wn, n = np.loadtxt('datafile.txt',
 unpack=True)
nint = scipy.interpolate.interp1d(nw, n)

# wavenumbers for which we want to know the refractive
   index
newx = np.linspace(3140, 3280, 20)

# get them!
newy = nint(newx)
```

where this is based on knowing that the file `datafile.txt` contained data in 2 columns. In one of the end-of-chapter exercises, you are presented with the situation that both real and imaginary components of the refractive index data is available as a function of wavelength, either as separate files, or as a single file with three columns. You can perform two interpolations and return an array of complex-valued refractive indices.

3.6 Numerical Integration

The subject we are gearing up to study involves determining the orientation distribution of molecules from polarized vibrational spectra of different types. As part of this analysis, we will frequently require integrals to be solved. In some cases, this will be done analytically, as will be demonstrated in Chap. 5. However, in other situations the functions may not have analytical solutions for their definite integrals, and so a numerical approach will be required.

There are, in general, many techniques available for this. The corresponding functions are all inside of the `scipy.integrate` sub-package, and I encourage you to refer to the documentation on the SciPy website for a complete list of available integration functions, their underlying principles, and instructions on their use. For general-purpose evaluation of a definite integral by Gaussian quadrature, consider

```
from scipy import integrate
```

```
integrate.quad(function, a, b, args=(), ...)
```

where a and b are the limits of integration. A complete example would look like

```
from scipy import integrate

def function(x):
   return 2.0*x + 7

print(integrate.quad(function, 3, 8)[0])
```

The output of quad is a list that includes details of the performance. The first element of the list contains the result, hence the list index 0 in the above example. You can verify that this produces the output -90.0. There are additional keyword arguments that can be supplied to improve performance, handle special cases, and generally control how the routine works.

While there are many efficient routines in the SciPy integrate package, I admit that I have not widely explored them. Instead I typically evaluate integrals through a Riemann sum

```
data = np.array([...])
dx = data[1] - data[0]
integral = np.sum(data) * dx
```

That method is fast, and easily extensible to multi-dimensional arrays

```
data = np.array([[1, 2, 3], [3, 4, 5], [5, 6, 7]])
dx = data[0][1] - data[0][0]
dy = data[1][0] - data[0][0]
integral = np.sum(data) * dx * dy
```

The result is accurate so long as the interval (dx, dy etc.) is sufficiently small. My examples had the data in arrays with a fixed interval. Typically, we wish to integrate an array of numbers that represents the evaluation of a function over a fixed domain, so we can get away with passing a sufficiently small value of dx in order to generate that array. When in doubt, you can always try varying the size of dx until you are satisfied that your result converges. This way you can achieve a balance between accurate results (small enough integration step size) memory usage, and speed.

3.7 Speeding Up Your Python Codes

At this point we've covered enough ground that you might have already fully implemented some code to solve your own spectroscopy problem using Python. Or perhaps you had code written in a different language, and you've now ported it over to Python. However, you might be in situation where your code is just not running fast enough. A faster computer might speed it up by a factor of two, or even ten, but sometimes that is still not satisfactory.

The best approach here, in general, requires some practice and experience. There are many options, including parallel computation involving different cores or different computers, multithreading, GPU computation, etc., and there is Python support for all of those options. A good approach to start with, however, is to ensure that you have used as many NumPy array operations as possible, and thereby eliminate traditional Python structures such as loops. We have previously discussed that being able to operate on an array automatically in an element-wise manner is a major convenience, as functions don't need to be written in a way that handles float, 1D-, 2D- or *n*D arrays differently. The major advantage here, in addition to the convenience, is speed. For relatively small programs with simple calculations, it may not be apparent just how much faster your code executes when it is array-based. But for larger more complex codes, the difference can be dramatic. I also encourage you to use NumPy operations whenever possible, for example converting a float in degrees to radians using `np.radians(x)` instead of `x/180.*np.pi`. By embracing NumPy methods and NumPy data structures your code will become smaller, simpler and therefore easier to understand and debug, and noticeably faster.

There is another approach, which I won't advocate for that strongly, but is interesting enough to mention. That is to use a package such as Numba that includes facilities for compiling Python into machine code. It includes features that enable you to take existing functions you have written and pass them to Numba for just-in-time compilation. In some cases these work without a single modification to your code and provide an impressive speed up of approximately two orders of magnitude. If you write code for curve-fitting or other types of optimization, such a dramatic speed-up is especially appreciated, as an optimization routine may call a function thousands of times. That said, I encourage you to to push the limits of Python/NumPy before diving too far outside the standard Python boundaries. A drawback of tools such as those provided by Numba, especially for less experienced users, in that not all Python/NumPy data types and operations are currently supported.

3.8 Summary and Next Steps

We have seen how NumPy greatly expands the basic capabilities of Python, most importantly by adding the `array` data type. This allows for 'vectorized' processing where functions are automatically applied element-wise to each item of an array. This simultaneously simplifies and dramatically shortens the code by eliminating the need for loops, and speeds up its execution. Other NumPy functions satisfy the core needs for scientific data processing. More specialized features (curve fitting, integration techniques, etc.) are provided by the SciPy package. The examples presented in this chapter have covered only a very small fraction of the vast NumPy and SciPy space. Some additional features, and plenty of additional examples of the discussed functions, will be presented as part of our investigation of the spectroscopic techniques. First we'll talk about generating plots, and ways to perform analyses symbolically in the next two chapters.

3.9 Exercises

1. When light approaches glass at an angle θ_1 with respect to the normal, a beam is reflected at angle θ_1, and a beam is transmitted into the glass at an angle θ_2 that can be obtained from

$$n_1 \sin\theta_1 = n_2 \sin\theta_2 \tag{3.1}$$

where $n_1 = 1$ in this case (incident from air). If the glass is infinitely thick, that's the end of the story. But if the glass has a finite thickness d_1, then there will also be reflection and transmission from the back glass-air interface.

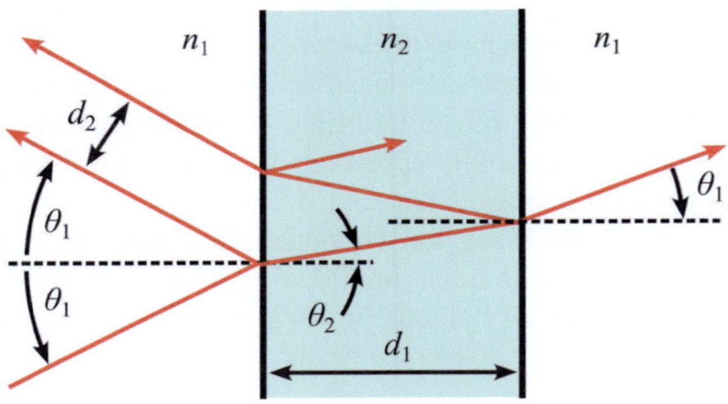

The beam reflected from the back is particularly interesting, as it has the possibility to be transmitted through the top, so it travel parallel to the first reflected beam in air, according to the figure below.

Work out an expression for the distance d_2 in terms of n_2, θ_1 and θ_2. It would be extra nice if you express θ_2 in terms of θ_1 and n_2 using Eq. 3.1 with $n_1 = 1$.

---

2. According to Jones calculus, light traveling along the z-direction may be described in terms of the electric field along x and y with the two-element Jones vector

$$\mathbf{E} = \begin{bmatrix} E_x \\ E_y \end{bmatrix},$$

so horizontally-polarized light looks like

$$\mathbf{E} = \begin{bmatrix} 1 \\ 0 \end{bmatrix}$$

and a polarizer set to transmit horizontally polarized light (an azimuth of 0°) has the form

$$\mathbf{P}(0^\circ) = \begin{bmatrix} 1 & 0 \\ 0 & 0 \end{bmatrix}.$$

If horizontally-polarized light encounters a polarizer at an azimuthal angle θ, the electric field of the transmitted light is described by

$$\begin{aligned} \mathbf{E}_{\text{out}} &\equiv \begin{bmatrix} E'_x \\ E'_y \end{bmatrix} \\ &= \mathbf{P}(\theta) \cdot \mathbf{E}_{\text{in}} \\ &= \mathbf{R}(-\theta) \cdot \mathbf{P}(0^\circ) \cdot \mathbf{R}(\theta) \cdot \mathbf{E}_{\text{in}} \end{aligned}$$

where the rotation operator is given by

$$\mathbf{R}(\theta) = \begin{bmatrix} \cos\theta & \sin\theta \\ -\sin\theta & \cos\theta \end{bmatrix}.$$

Once you know the output electric field, the intensity of the transmitted light may be obtained from the product of the field with its transposed conjugate

$$I = \begin{bmatrix} E'^*_x & E'^*_y \end{bmatrix} \cdot \begin{bmatrix} E'_x \\ E'_y \end{bmatrix}.$$

Write a Python program that calculates the intensity as a function of polarizer azimuth. Let the azimuth vary from 0–360° in steps of 1°. Make a plot of the result.

Hint: Although this seems contradictory to some of the core NumPy programming principles we discussed, this task is easiest if you use a `for` loop to iterate through all of the desired azimuthal angles. Inside the loop, use the NumPy `array` data type to do the matrix calculations, figure out the intensity for that particular azimuthal angle, and append this to a growing list.

Optional: If you want more of a challenge, see if you can make a 100% vectorized version of the code that doesn't use any explicit loops. In the end, which version of your code do you prefer? Think about the advantages and disadvantages of both methods.

---

3. As an introductory Python exercise in Chap. 2, I proposed that you write some code to numerically determine the value of π.

   a. Are there any NumPy or SciPy features that could help you to speed up the code, or implement things in a more compact way?
   b. See if you can speed up your execution time using a just-in-time (JIT) compiler for Python, such as that offered by the Numba module.

---

4. Take a look at some of the data available online at `refractiveindex.info` studying how it is organized based on the types of material. For example, you will find entries for water under the category of oxides. Within each entry, you will see refractive index data available from different literature sources. Some have only the real part (n) listed, and others contain both the real and imaginary (κ) parts of the refractive index. Also note that different sources have data available for specific wavelength regions.

   a. Find an entry that contains both real and imaginary components of the refractive index for water in the mid-infrared region.
   b. Download the data and inspect the type of file you have retrieved, how it is organized into columns, the delimiter, etc. Write a small Python code to read the file(s) using `np.loadtxt`. You might have to supply some keyword arguments to do the parsing properly, ignore columns you don't want, skip headers, etc.
   c. Write a Python function that takes a single IR wavenumber (in units of cm^{-1}) as input and returns the complex refractive index (as a Python complex data type). Typically refractive index data is tabulated in terms of wavelength in microns, so you'll have to decide whether you want to convert all of the data into cm^{-1} first, prior to the interpolation, or whether your interpolation function just does that on the fly. Test this by supplying the value of $3100\,\mathrm{cm}^{-1}$, and confirm that you are getting the right result back, either by manually looking through the file(s) you have downloaded, or by inspecting the plots available on the `refractiveindex.info` website.
   d. Now, instead of a single IR wavenumber, try passing the array `wn = np.array([2900, 3000, 3100, 3200])` to your function. If things are working properly, you should get back an array of 5 complex refractive index values. Being able to do this quickly and efficiently will be extremely useful for your future work.

References

1. C.R. Harris, K.J. Millman, S.J. van der Walt, R. Gommers, P. Virtanen, D. Cournapeau, E. Wieser, J. Taylor, S. Berg, N.J. Smith, R. Kern, M. Picus, S. Hoyer, M.H. van Kerkwijk, M. Brett, A. Haldane, J.F. del Río, M. Wiebe, P. Peterson, P. Gérard-Marchant, K. Sheppard, T. Reddy, W. Weckesser, H. Abbasi, C. Gohlke, T.E. Oliphant, Nature **585**, 357 (2020)
2. T.E. Oliphant, Comput. Sci. Eng. **9**, 10 (2007)
3. P. Virtanen, R. Gommers, T.E. Oliphant, M. Haberland, T. Reddy, D. Cournapeau, E. Burovski, P. Peterson, W. Weckesser, J. Bright, S.J. van der Walt, M. Brett, J. Wilson, K.J. Millman, N. Mayorov, A.R.J. Nelson, E. Jones, R. Kern, E. Larson, C.J. Carey, İ Polat, Y. Feng, E.W. Moore, J. VanderPlas, D. Laxalde, J. Perktold, R. Cimrman, I. Henriksen, E.A. Quintero, C.R. Harris, A.M. Archibald, A.H. Ribeiro, F. Pedregosa, P. van Mulbregt, Nat. Methods **17**, 261 (2020)
4. E. Jones, T. Oliphant, P. Peterson, et al. SciPy: Open source scientific tools for Python (2001–). URL http://www.scipy.org/
5. W.H. Press, B.P. Flannery, S.A. Teukolsky, W.T. Vetterling, *Numerical Recipes in FORTRAN 77* (Cambridge University Press, New York, NY, 1992)
6. The MathWorks Inc. MATLAB version r2024a (2024)

Chapter 4
Data Visualization

4.1 Introduction

In this chapter we'll talk about graphics. Plotting goes hand-in-hand with numerical calculations, partially because being able to see what you're doing as an intermediate step is an extremely useful debugging and data exploration technique. Take for example a code that plots the reflectance versus wavelength, given the dispersion of the refractive index. As an initial step, I would open the refractive index data in Python to see if I can read its first entry, comparing it with what I see when I manually open the file. As a subsequent step I would read all of the index data and interpolate it so that I can estimate the refractive index at any arbitrary wavelength. At this point, it's great to be able to quickly plot the index *vs* wavelength, even though that is not our ultimate goal. That allows us to spot something that may be off in a way that is more obvious than seeing numbers printed onto the terminal screen. Finally, the reflectance is calculated, and the plot is updated to display that. From there, the plots are customized with axis labels and other annotations to fulfil our objectives.

As another example, consider a data processing pipeline where data from solution-based experiments is compared to surface adhesion markers from nonlinear vibrational spectroscopy (Chap. 10) and with microscope images that are analysed to generate trends as shown in Fig. 4.1. Many steps are involved, including loading the raw image data, thresholding the images, and ultimately generating scatter plots with control of the line styles, markers, labels, and axes. These are then related to specific time points in the time-resolved spectra [1]. In the ideal scenario, what you see isn't something that was created just for display in a paper or presentation, but is the culmination of an iterative process where the data is treated in a stepwise manner, with each intermediate step visualized to help in the decision-making that leads to the final analysis. While it may be true that the best exploratory tools are not necessarily those that make the most attractive "static" images, at least there shouldn't be a large difference in their codebase. As an example, when determining the appropriate normalization method for the data, the extent of the vertical axis

D. K. Hore, *Molecular Orientation Analysis Using Polarized Vibrational Spectroscopy*,
https://doi.org/10.1007/978-3-032-04874-5_4

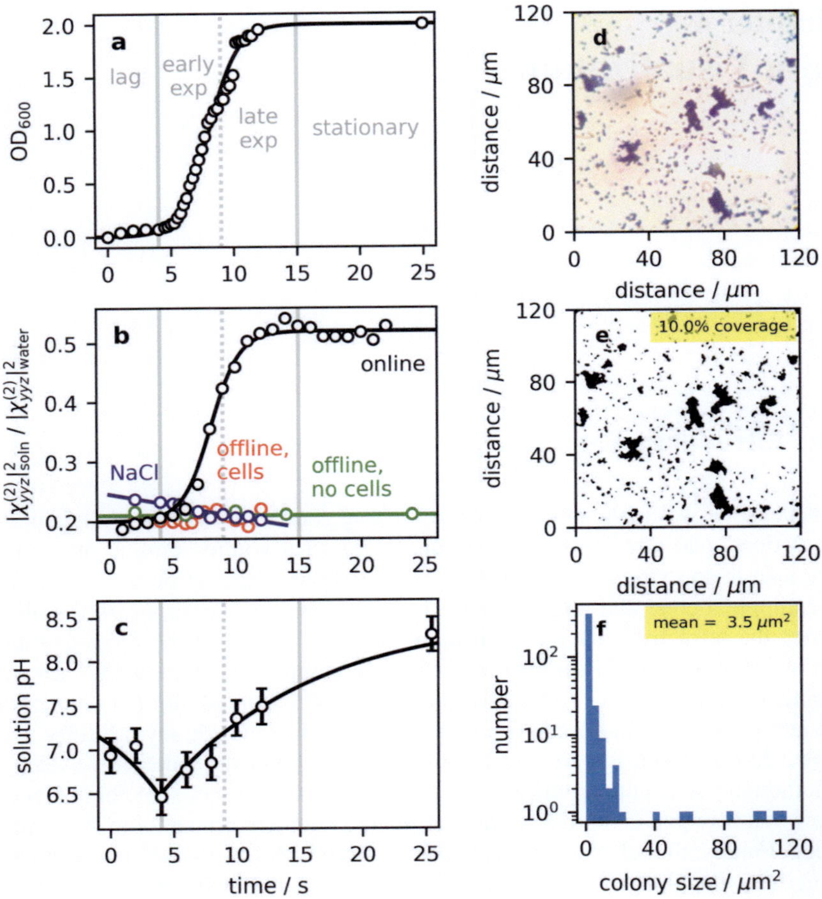

Fig. 4.1 An example of an imagine analysis task, where **a** light scattering from bacteria in solution is compared with **b** spectroscopic monitoring of cell adhesion characteristics and **c** solution pH changes as a function of time. This is related to the cell surface density obtained from **d** microscope images that are subsequently **e** thresholded in order to obtain **f** statistics about the distribution of surface colony sizes. Data from Ref. [1]

changes. When the graphics tool is left to its own, it therefore chooses reasonable values for the tick marks. However, when the image is finalized, it may be desirable to manually set the tick marks for the best effect. Perhaps these are too many details to consider when we haven't started yet, but I wanted to provide a sense of my requirements for scriptable scientific graphics and where we're headed.

A lot of effort has been put into the development of several graphics modules for Python, some of them Python ports of existing packages for other languages. After about a decade, many of the contenders that offered core plotting functionality have disappeared, and the Matplotlib package [2] emerged as the defacto choice, at least for 2D plots that are not interactive. Although Matplotlib has 3D capabilities, and there is the ability to capture user input using a keyboard and mouse, and create basic dashboard-style interfaces, those are not its strong suit. Modules that provide such features are available, but are outside of the scope of our current discussion. When Matplotlib was developed, it had a motto shared by many software tools: that simple tasks should be simple to do, and that difficult tasks should be possible. Achieving this requires access to lots of options for controlling the finest details of plots, but also a good set of reasonable defaults so one doesn't have to get buried in these options just to make a simple scatter plot of y vs x.

4.2 Basic Plotting

4.2.1 The Pyplot Interface to Matplotlib

There are two different ways to use the Matplotlib package. One is through its `pyplot` interface. In order to understand the motivation behind the syntax of the pyplot-style commands, one needs to realize that the "Mat" in Matplotlib is a tribute to Matlab™ [3]. Python and Matlab™ already have many syntactic similarities, and people appreciated the powerful and flexible plotting capabilities of Matlab™. I don't want to get into the historical details too much, mainly because it's largely irrelevant to Matplotlib now, but occasionally there are some design choices that persist today that come from this.

To get started, let's say you want to visualize a polynomial

$$y = ax^3 + bx^2 + cx + d$$

for the choice of coefficients $a = -0.0004$, $b = 0.001$, $c = 0.003$ and $d = 0$ over the range $x = -2$ to 5. The actual polynomial *evaluation* has nothing to do with Matplotlib and is a basic NumPy feature that is accessed using the `np.polyval` function

```
x = np.linspace(-2, 5, 1000)
y = np.polyval([-0.0004, 0.001, 0.003, 0], x)
```

The plot can then be created and decorated with some labels using simple and intuitive commands.

```
import numpy as np
from matploblit.pyplot import *
x = np.linspace(-2, 5, 1000)
y = np.polyval([-0.0004, 0.001, 0.003, 0], x)
plot(x, y, 'k-')
xlim(-2, 4.5)
ylim(-0.005, 0.01)
```

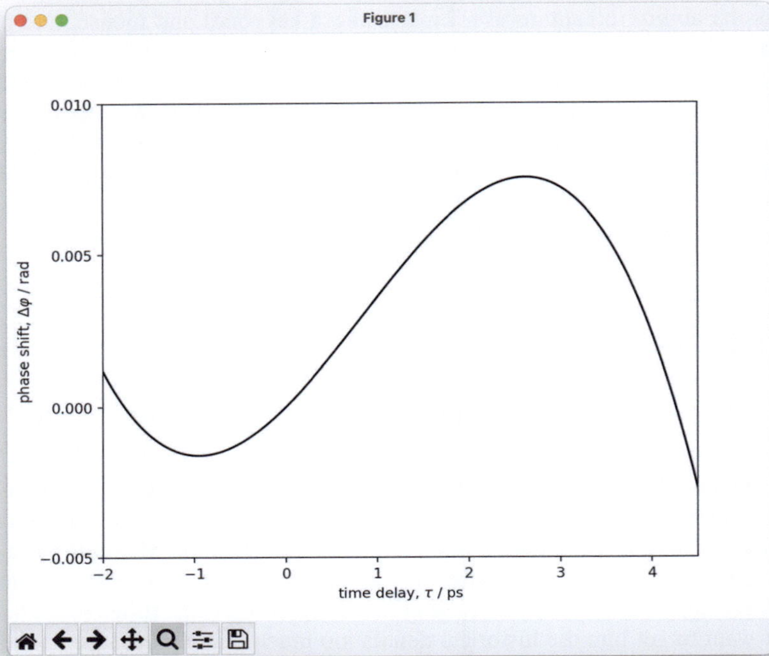

Fig. 4.2 A very basic plot created using the `pyplot` interface to Matplotlib

```
yticks(np.arange(-0.005, 0.011, 0.005))
xlabel(r'time delay, $\tau$ / ps')
ylabel(r'phase shift, $\Delta \varphi$ / rad')
show()
```

Here I've also illustrated setting the horizontal and vertical axis labels, setting the extent of the axes, and overriding the default vertical tick mark spacing. Note that any LaTeX syntax can be used in the strings for the labels by enclosing the LaTeX syntax with a set of `$` characters, and this sometimes requires specifying that those are raw strings (with the `r` prefix). Otherwise Python may be confused by things like `\tau` since the escape sequence `\t` represents a tab in regular Python string formatting. This creates the interactive plot window shown in Fig. 4.2. The first thing you'll notice in that code example is that all pyplot commands were loaded into the current namespace, a departure from what we have been practicing until this point. This is never a good idea with large packages, but I wanted to show this example because you'll encounter things like this when you search online for examples to help you with complex plotting tasks. That part can be remedied with a minor change in syntax:

```
import numpy as np
from matploblit import pyplot
x = np.linspace(-2, 5, 1000)
y = np.polyval([-0.0004, 0.001, 0.003, 0], x)
pyplot.plot(x, y, 'k-')
pyplot.xlim(-2, 4.5)
pyplot.ylim(-0.005, 0.01)
pyplot.yticks(np.arange(-0.005, 0.011, 0.005))
pyplot.xlabel(r'time delay, $\tau$ / ps')
pyplot.ylabel(r'phase shift, $\Delta \varphi$ / rad')
pyplot.show()
```

and all functions then remain in their respective namespaces. Inspecting the screenshot in Fig. 4.2, one notices that panning, zooming, and the Cartesian coordinates of the cursor (in the scale of the data!) are available with mouse controls. The figure may also be saved to a (relatively low resolution) bitmap file format using the buttons at the bottom of the plot window. Although this is convenient, I discourage you from using the `pyplot` interface *in this way*. In order to explain why, I like to tell the story of a presentation given by Matplotlib creator John Hunter at the 2007 SciPy conference. He started his presentation by asking for a show of hands for how many people start their day by typing `from matplotlib.pyplot import *` . (It was called something different back then, but I'm translating.) He then said "I'm sorry, that's my fault". He said that he developed that interface as a way to win over Matlab™ [3] users, but regretted it soon after. That's because (my words now) Python has so much to offer that the plotting capabilities of Matplotlib would be better integrated using a more "pythonic" interface. From the beginning, both were available, but the native Python mode was always slightly more awkward to use. However, the payoff was great, as it offered a way to dig deep into the fine details of plot objects like tick mark lengths, default colour sequences, etc. The formally-prescribed alternative, that you can read about on the Matplotlib website, is to `import matplotlib`, and do away with the limitations of `pyplot`. However, there are many convenient `pyplot` functions, so I have found a middle ground between the two approaches that enables all of the Python goodness of Matplotlib without letting go of those few parts of `pyplot`, when desired. Let's take a third pass at writing the above code.

```
import numpy as np
from matplotlib import pyplot

fig = pyplot.figure()
ax = fig.add_subplot()

x = np.linspace(-2, 5, 1000)
```

```
y = np.polyval([-0.0004, 0.001, 0.003, 0], x)
ax.plot(x, y, 'k-')

ax.set_xlim(-2, 4.5)
ax.set_ylim(-0.01, 0.01)
ax.set_xlabel(r'time delay, $\tau$ / ps')
ax.set_ylabel(r'phase shift, $\Delta \varphi$ / rad')
pyplot.show()
```

While the above code certainly appears slightly more cumbersome, it has several advantages moving forward. Simply stated, the Python-styled interface to Matplotlib offers more possibilities for customization. Looking at the code, you can see that I created a figure instance and then an axis within that figure. We then see that `plot` is no longer a function on its own, but is now a method of the axis object. I call this a hybrid approach, as we're still free to call any pyplot function we want, as in the `show` command. Again, I struggle to decide how much of these historical things to mention, as this could be an opportunity to introduce a single approach with a clean slate. However, I do think it will be important for you to understand the differences between the codes that I show throughout all the remaining chapters and code snippets you might find as examples on the internet.

4.2.2 Choosing a Bitmap File Rather Than Screen Output

Although `show` is convenient as it offers some interactive controls, I rarely use it. I much prefer to save the figure to a file for two reasons. Matplotlib has several "backends" that it can use to render the graphics. Some of them output to the screen, and some create files. At the same time, I set the figure size and changed the default font size. So the final iteration looks like

```
import numpy as np
from matplotlib import pyplot

# setup
pyplot.rc('font', size=8)
fig = pyplot.figure(figsize=(3.25, 2.5))
ax = fig.add_subplot()

# prepare the data
x = np.linspace(-2, 5, 1000)
y = np.polyval([-0.0004, 0.001, 0.003, 0], x)

# plotting
ax.plot(x, y, 'k-')
```

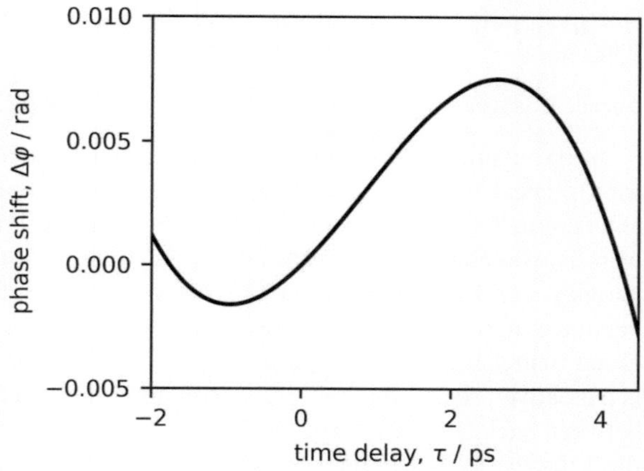

Fig. 4.3 Creating a `.png` file by using Matplotlib's PNG backend, rather than displaying the output to the screen using `show`. This is also an opportunity to control the aspect ratio, font size, and other fine details of the appearance of the plot

```
# pretty
ax.set_xlim(-2, 4.5)
ax.set_ylim(-0.01, 0.01)
ax.set_xlabel(r'time delay, $\tau$ / ps')
ax.set_ylabel(r'phase shift, $\Delta \varphi$ / rad')
fig.set_tight_layout(True)
pyplot.savefig('plotSimpleExample.png', dpi=600)
```

and produces the plot shown in Fig. 4.3. In comparison to what is displayed on screen (Fig. 4.2), this looks completely different, and much better in my opinion. We'll get into the details of the fonts and line widths below, but the main point I want to make here is that, even though one can customize all aspects of the plot, it doesn't make much sense to do so when using `show`, and the way the results are displayed on screen may not be consistent across different operating systems. When png output is selected, I believe it is more worthwhile to customize and fine tune the appearance of graphical elements since we can be guaranteed that those features will then be displayed in the same way on screen when viewing or embedding the `.png` file in a presentation, or when printed in a paper or poster.

After a figure has been created (whether or not it has also been displayed on the screen), the `savefig` command can be used to create a vector-based or bitmap version of the file. For example

```
pyplot.savefig('plotSimpleExample.ps')
```

will create a postscript file, where the postscript backend is implied by the extension provided in the filename. Additional keywords can always be used to override this behaviour. Commonly available backends (for most Matplotlib installations) include eps, jpeg, pdf, png, ps, and svg. Over the past 25+ years, my default is a 600 dpi PNG file.

```
pyplot.savefig('dataset14_15.png', dpi=600)
```

In our example, you can see that when the backend is not specified, `savefig` tries to guess based on the extension of the provided output file name. Not all backends are installed by default on all systems, so you'll have to ensure that what you want is available to you. Backends differ in the way they render text, fonts, transparency, etc. I therefore prefer to see the graph exactly the way I expect the final version to appear, as I'm working on it. The second reason that I always save images rather than using screen output is that the layout, choice of colours, line styles, etc., is a creative process that requires many iterations, running the Python interpreter in between each step to refresh the image. Since `show` is a blocking call (i.e. it interrupts the Python script until the window is closed), I don't want to have to close the interactive plot window each time I make a change and a new one pops up. I instead prefer to use an image viewer that can watch the file and automatically refresh it when the file changes, preferably even in the background when the image window doesn't have the mouse focus. VS Code is good at this.

Finally, you might think that a postscript backend would be superior to creating a fixed-resolution bitmap graphic, since a vector-based format such as postscript (or pdf or svg) is resolution-independent and can therefore be scaled to an arbitrary size. In theory that's true, but I find working with such files a bit too risky. For reasons unknown to me, there are sometimes errors in rendering, either for screen display or during printing. Nothing is worse than previewing a poster on screen, but then having it come back from the printer with lines or random letters from text missing. For this reason, I stick with bitmap files of sufficiently high resolution (600 dpi), with PNG as my preferred format.

4.2.3 Figure Size, Font Size, and Line Widths

I have set the graphics size to 3.25" wide by 2.5" tall, and decreased the default font size to match. This corresponds to the single-column width of most two-column journals, and also scales nicely (lines and fonts large and thick enough to look great) when filling a slide for a presentation. If you change nothing else except for the figure and font sizes, you can create some very professional looking plots with most of the Matplotlib defaults for all other parameters. In my opinion, nothing is worse than plots in papers or presentations with fonts that are too small or lines that are too thin. It's also distracting when the font size in figures differs by a factor of ten within the same article. The general guideline is: if writing a paper, design your figure for its intended final size, i.e. around 3.25" wide for a single column, or 6.75" wide to span two columns (considering a 0.5" gutter between the columns), and make all of the text as close in size as possible to the running text of the article, after it is typeset. If the figure's first use is for an on-screen presentation, I find that the above size recommendations still work surprisingly well, as 600 dpi contains enough pixels to stretch the image without it appearing pixelated when scaling it up to be full screen. I prefer this to setting the figure size larger, since that would also require setting the fonts to be larger. Note that, even if you doubled both the figure size and the font size, this still wouldn't look great in my opinion, as the default axis lines and tick lines would then be too thin. Of course, you could change those too, but why bother? My magic dimensions of 3.25" wide and however tall, together with an 8-point font and the default Matplotlib line widths, create a pleasing combination that can then be scaled up when larger images are needed. Note that units of inches are required, so if you would like to work in metric units, you'll have to do the conversion.

I mentioned that there are many options for changing Matplotlib defaults. The above example shows a combination of these methods in use. Black dashed lines are specified with the plot function positional argument `k--`. The spacing of the dashes is specified with the plot function keyword argument `dashes=[2,2]`. The default font size could be changed in the system-wide global `matplotlibrc` settings file, the user's personal `matplotlibrc` file, a session specific `matplotlibrc` file, of through use of pyplot's `rc` method with a very similar syntax to what would be specified in the files.

Finally, I've waited until now to reveal that I'm also a big fan of LaTeX (surprise). It's possible to tell Matplotlib to use your installed LaTeX distribution to render all text simply by specifying `pyplot.rc('font', usetex=True)`, and then there are other options for specifying the LaTeX preamble, fonts, loading special packages, etc. Those who are interested should consult the online documentation. I prefer to use Matplotlib's native text rendering instead, so I don't do this. However, I love how superscripts, subscripts, and all math rendering can be specified using the intuitive LaTeX syntax. If you're already a TeXpert, you've likely stopped reading and are busy making plots. If this syntax is unfamiliar to you, don't worry about it; just understand that's what I'm indicating if you see all the `$` symbols in my (raw) strings.

4.3 Further Customization

If our end goal was to make fancy plots using Matplotlib and this was our last chapter together, there would be a lot more to say. However, I'll keep it brief here as there will be plenty of opportunities to demonstrate different plotting options in our discussion of vibrational spectroscopy. That is one of my main motivations for providing complete code examples (rather than code snippets) for each of the plots in the remaining chapters. I also chose to provide those directly in the relevant sections, right after the plots are discussed, rather than relegating them to an Appendix, since I want you to be able to look at the plots and the code next to each other. Having said that, there are a few things we can discuss here before we get distracted by the transition dipole moments.

4.3.1 Basic Plotting Options

After the figure elements have been established with `plot` and associated commands, but before it has been generated with `show` or `savefig`, many annotations are possible such as specifying titles, labels for the horizontal and vertical axes, and a legend. Remember all of the Python string manipulation tools that you have at your disposal and use these to create programmatic labels as required. For example, nothing prevents you from writing

```
ax.set_ylabel(r'%s at %d cm$^{-1}$' % (quantity, int(wn)))
```

We have seen that there is an optional third argument to `plot`, a string that specifies quick combinations of line and marker styles. I have used `ax.plot(x, y, 'k-')` to indicate that I want to use lines (`-`) drawn in black (`k`). Similarly, red dashes could have been specified using `r-` and green circles with `'go'`. If just a few basic colours will satisfy you, then you can use the single letter codes

```
b  : blue
g  : green
r  : red
c  : cyan
m  : magenta
y  : yellow
k  : black
w  : white
```

There are fancier ways to specify the colour, including the use of RGB values and official HTML colour names. Some quick line styles available in this manner are

```
-      : solid line
--     : dashed line
-.     : dash-dot line
:      : dotted line
.      : points
,      : pixels
```

In the case of markers, basic ones include

```
o      : circle
^      : triangle up
v      : triangle down
<      : triangle left
>      : triangle right
s      : square
+      : plus
x      : cross
```

```
D     : diamond
d     : thin diamond
1     : tripod down
2     : tripod up
3     : tripod left
4     : tripod right
h     : hexagon
H     : rotated hexagon
p     : pentagon
|     : vertical line (symbols)
_     : horizontal line (symbols)
```

If you're picky like me, these options are okay for a quick display of results, but are not good enough for publication-quality plots. Finer control is available by using keyword arguments rather than the above quick combos. For example

```
ax.plot(x, y, marker='o', markersize=15, markerfacecolor='
    g', markeredgecolor='k', markeredgewidth=2, color='m',
     linewidth=3, linestyle='--', dashes=[2,2])
```

specifies round markers of a given size that are green inside with a thick black border. In addition to the markers, the points specified by (x, y) are connected by a magenta dashed line. When any type of dashed line style is used, both the length of the dash and the space between dashes can be specified with the `dashes` keyword. Since I frequently make use of such options, I appreciate that shorter versions of the keywords are also available

```
ax.plot(x, y, marker='o', ms=15, mfc='g', mec='k', mew=2,
    color='m', lw=3, ls='--', dashes=[2,2])
```

A much higher degree of customization is possible utilizing further keywords; I encourage you to familiarize yourself with the various plot options in the Matplotlib online documentation, as `plot` is one of the most basic commands as part of this plotting module.

4.3.2 *Plotting Multiple Series*

An arbitrary number of series may be displayed with a single plot command such as

```
ax.plot(x, y1, x, y2, color='r')
```

When using keyword arguments, those options apply to all series issued by that plot command

```
ax.plot(x, y1, x, y2, color='r')
```

A way to get around this is to take advantage of matplotlib's capability to overlay graphs with `hold()`.

```
ax.plot(x1, y1, color='r')
ax.hold(True)
ax.plot(x1, y2, color='k')
ax.hold(False)
```

Hold is enabled by default, so you'll have to set it to `False` if you don't want this. Without any arguments passed to the axis method, `hold()` toggles the hold state. As a result, I never combine the plotting of different series in a single command and opt for

```
ax.plot(x1, y1, color='r')
ax.plot(x1, y2, color='k')
```

where the default hold state does the job.

4.3.3 *Creating Legends*

There are many ways in which a legend can be created, and many options for controlling the style of the legend. The one that I use in almost all cases is connected with the `label` keyword of the plot command.

```
ax.plot(x1, y1, color='r', label='4 mV')
ax.plot(x2, y2, color='k', label='15 mV')
ax.legend()
```

On its own, `label` won't do anything, but if `ax.legend()` is later called, it finds all the plots that have legend entries specified by their labels, and creates a simple basic legend. If any label is defined as `_nolabel_`, then that series is excluded from the legend only (i.e. it is still plotted). Additionally, any `plot` command that does not include the `label` keyword will be excluded from the legend. An optional second argument is the location of the legend that may be either a string or its code equivalent

```
'best' : 0,
'upper right' : 1,
'upper left' : 2,
'lower left' : 3,
'lower right' : 4,
'right' : 5,
'center left' : 6,
'center right' : 7,
'lower center' : 8,
```

```
'upper center' : 9,
'center' : 10
```

Alternatively, the location may be specified using a 2-element tuple of fractional axis coordinates where `(0, 1)` is the top left, `(0.5, 0.5)` is the centre, etc.

4.3.4 Setting Axis Limits

Many plot attributes are queried/modified using a get/set syntax. For example

```
a, b = ax.get_xlim()
```

and

```
ax.plot(x, y1, 'k-')
ax.plot(x, y2, 'r-')
ax.set_xlim(2.5, 3.5)
```

If you want to change only one of these limits, and keep the other one as is (using the default value Matplotlib has determined, based on the extent of the data, major tick intervals, etc.), you can supply a keyword argument instead

```
ax.set_ylim(top=3.5)
```

Alternatively, you can use the `axis()` method that returns a tuple arranged as x_{min}, x_{max}, y_{min}, y_{max} Similarly, you can change these values by supplying a 4-element list or tuple

```
ax.axis([-5, 5, -5, 5])
```

or alter one or more values with keyword arguments

```
ax.axis(xmax=3.5, ymax=10.)
```

You'll notice that, by default, matplotlib automatically extends both axis limits beyond the data. While this typically looks good when plotting points (since the full markers then appear, rather than half of the marker being clipped by the edge of the axis), I find it doesn't have the right look when I'm plotting a function that I want to display as a line. There the "illusion" of a continuous function is better represented by a line that extends all the way to both axis limits in the horizontal directions.

```
x = np.linspace(5, 25, 1000)
y = f(x) # some function defined elsewhere
ax.plot(x, y, 'k-')
ax.set_xlim(np.min(x), np.max(x))
```

I could have simply written `ax.set_xlim(5, 25)` but by using `np.min` and `np.max` things will always look good, even when I later decide to change the x-axis range that is specified with the `linspace`. This is, of course, a matter of preference.

4.4 Creating Figures with Multiple Axes

Although we are now armed with the ability to overlay multiple data sets on a single axis, there are many situations where multiple axes are to be created within the same figure. An obvious case is when two or more different data sets don't share the same horizontal axis, or if data sets have widely different vertical extents. More commonly, different types of plots can be stacked on top of each other or beside each other. In general, individual axes within a multi-axis figure are referred to as subplots. As you might imagine, there are many mechanisms by which one can specify the layout of the subplots. Alternatively, the location and shape of each can be arbitrarily and precisely defined by specifying the coordinates within the figure in which the axes should be drawn. Here we will see the most basic implementation, that will suffice for the majority of our needs, that is simply by providing arguments to the `add_subplot` method of the figure object. Consider the following example:

```
import numpy as np
from matplotlib import pyplot

# setup
pyplot.rc('font', size=8)
fig = pyplot.figure(figsize=(3.25, 5))
x = np.linspace(-2, 5, 1000)

# first set of data
ax1 = fig.add_subplot(211)
y = np.polyval([-0.0004, 0.001, 0.003, 0], x)
ax1.plot(x, y, 'k-')
ax1.plot(0, 0, marker='o', ms=12, mew=1.5, mec='r', mfc='
    none')
ax1.plot(0, 0, marker='o', color='k')
ax1.set_ylim(-0.01, 0.01)
ax1.set_ylabel(r'phase shift, $\Delta \varphi$ / rad')
ax1.label = 'a'

# second set of data
ax2 = fig.add_subplot(212)
y = np.polyval([-0.0003, 0.002, 0.003, 0.002], x)
ax2.plot(x, y, 'k-')
ax2.set_ylim(-0.005, 0.03)
ax2.set_ylabel('amplitude variation / a.u.')
ax2.label = 'b'

# pretty
```

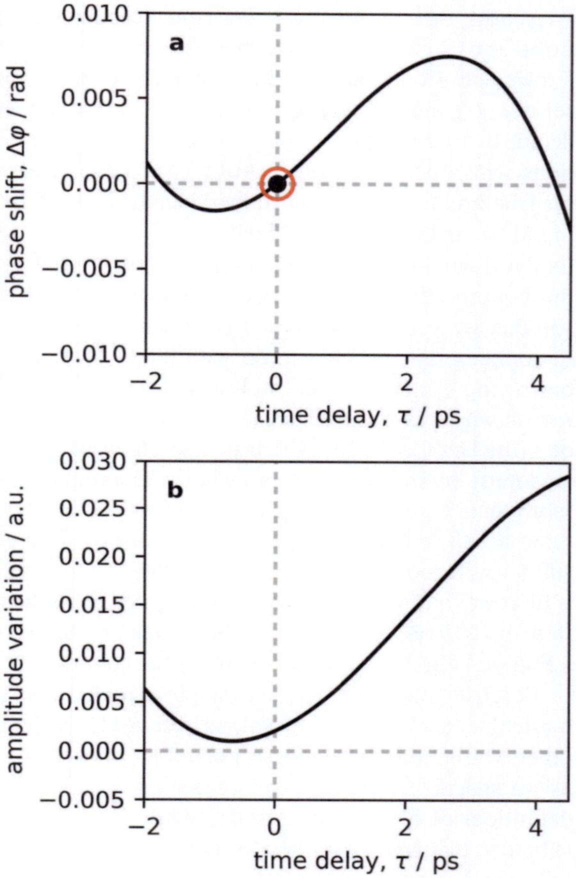

Fig. 4.4 A simple example that illustrates the creation and use of two subplots arranged in a single column, with some basic annotations

```
for ax in [ax1, ax2]:
    ax.set_xlim(-2, 4.5)
    ax.axhline(0, color='0.7', ls='--', dashes=[2,2],
        zorder=1)
    ax.axvline(0, color='0.7', ls='--', dashes=[2,2],
        zorder=1)
    ax.set_xlabel(r'time delay, $\tau$ / ps')
    ax.text(0.05, 0.89, ax.label, fontweight='bold',
        transform=ax.transAxes)
fig.set_tight_layout(True)
pyplot.savefig('subplotsExample.png', dpi=600)
```

The output of the above code is shown in Fig. 4.4. There are a few things to note here. First, we have already seen the subplot notation early on, when creating only a single axis using `fig.add_subplot()` which is short for `fig.add_subplot(111)`. Perhaps the notation makes more sense now in that `subplot(i,j,k)` specifies that we want a regular grid arrangement of plots with i rows and j columns, and that we are explicitly creating the kth plot within that set of $i \times j$ plots. There is no need to use all of the locations. For example, we can decide that we want a 3×3 layout of 9 plots, and can specify only the diagonal positions with `subplot(331)`, `subplot(335)`, and `subplot(339)`. Notice that the labelling for k starts at 1 (rather than the usual Python convention of indices starting at 0), and the order is specified as left-to-right, top-to-bottom, in the sequence that you would read words on a page. Just to make sure you are following along, the bottom-left plot, if desired, would be in location `subplot(337)`. The reason that we start the index at 1 (and hence the source of a great deal of confusion in complex layouts!) has to do with the history of the development of matplotlib, borrowing syntax from the well-established Matlab™ community. This is also the reason why it is possible to use the rather unusual `subplot(111)` syntax, instead of `subplot(1,1,1)`, which is also allowed. As you might have guessed, commas must be used in some cases to avoid ambiguity. For example, any layout with more than 9 positions must use commas, for example `subplot(3,4,12)` since there are then 2-digit integers. (If you are really creating such a large number of subplots, I encourage you to look up other, more flexible, layout options.) Similarly, if there are any variable names for the integers in the mix, commas must be used as in `subplot(3,3,x)`. So, if this is your first time around, you may want to adopt a Python-styled notation of including the commas.

Referring back to my code example in relation to Fig. 4.4, you can see that once a particular axis has been created, we can use all of the regular plotting commands. In other words, Matplotlib doesn't care that more than one set of axes exists, so long as we specify which axis we are dealing with. You will also note that I used my definitions of `ax1` and `ax2` to explicitly define things that were specific to the two subplots, like control of their vertical tick placements, vertical extents, and vertical axes labels. However, since I wanted to use the same horizontal axis settings for both plots, I decided to set those properties in a loop at the bottom of my code, rather than repeat those commands for each axis instance. I took advantage of that opportunity to do other things to both axes, like define a set of crosshairs using the `axhline` (axis horizontal line) and `axvline` (axis vertical line) commands.

I further illustrated that shades of grey can be specified by passing a number between 0 (black) and 1 (white) as a *string* in place of any colour. I also created a fancy bullseye-type marker simply by overlaying two circle markers. In order to create the red circle, I needed to turn off the marker fill colour (or set it to white, which would have produced a slightly different effect that hides the line segment inside the red circle). I did this by specifying no fill using `mfc='none'`. Don't confuse this with the reserved Python word `None`. Setting `mfc=None` (capital N and no quotation marks) indicates that we don't want to set the `mfc` keyword, i.e. it is the same as not including any `mfc` specification, and so we would end up with the default marker fill colour. Finally, I have used a convenient method for adding the "a" and "b" inset labels that is described below.

4.5 Text Annotations

Text can be added at any point within a specified axis using

```
ax.text(x, y, string)
```

where x and y are the bottom-left coordinates of the text string in the units of the axes themselves, i.e. the values bounded by `xlim` and `ylim`. If you want to place some text somewhere in the figure, not limited to what is displayed in the axis portion of the figure, you can use relative ($0 \rightarrow 1$) figure coordinates instead, with the syntax

```
fig.text(x, y, string)
```

where `x` and `y` must now be specified in fractional figure coordinates. In the above two examples, I'm of course assuming that your figure has been called `fig` and that you have named at least one of your axes `ax`. I'm merely illustrating that `text` is a method of the Matplotlib axis or figure object. While axis text annotations are more common than figure text annotations, there are situations in which you want to use relative ($0 \rightarrow 1$) coordinates within the axis, rather than the absolute axis units. A typical example is the placement of the labels a and b in the top left corner of two subplots. If both subplots had a horizontal axis that extended from 2800 cm^{-1} to 3800 cm^{-1}, and a vertical axis that extended from 0 to 2, then the labels might be placed with commands such as

```
ax1.text(2820, 1.8, 'a')
ax2.text(2820, 1.8, 'b')
```

However, if either the horizontal and/or vertical extents differ, you still would like the labels to look as it they were the same distance from the axis frames. Of course you could determine the appropriate values of x and y to pass along to `text()` by trial and error (that's when you're hoping your figure updates automatically!), or by calculating the position

```
def label(ax, label):
    xfrac, yfrac = 0.09, 0.85
    xmin, xmax = ax.get_xlim()
    ymin, ymax = ax.get_ylim()
    ax.text(xfrac*(xmax - xmin) + xmin, yfrac*(ymax - ymin)
           + ymin)

label(ax1, 'a')
label(ax2, 'b')
```

where I have determined the extent of each subplot, and then figured out the coordinates that correspond to 9% along the horizontal axis, and 85% along the vertical axis. A better way to achieve this is to specify that you want the axis text to use fractional coordinates within the extents of the axis. This is easily achieved by

```
ax1.text(0.09, 0.85, 'a', transform=ax1.transAxes)
ax2.text(0.09, 0.85, 'b', transform=ax2.transAxes)
```

A more complete discussion of transformations is beyond the scope of what we are interested in, but you will see me use the code above in many of my subsequent examples, so I just wanted to point it out here.

If I'm creating a graphic with many subplots, one of my favourite ways to add inset labels, if they are all to be positioned using the same fractional coordinates, is to create a new property inside the axis object. For example, if there are four subplots labelled `ax1` through `ax4` then I'll create a property called `label` (a logical but arbitrarily chosen name) and use it as in

```
ax1.plot(...)
ax1.label = 'a'
ax2.plot(...)
ax2.label = 'b'
# etc
for ax in [ax1, ax2, ax3, ax4]:
   ax.set_xlim(0, 200)
   ax.set_xlabel('time / s')
   ax.text(0.09, 0.85, ax.label, transform=ax.transAxes)
```

where my example illustrates that I loop through all four axis objections in order to set some common properties (limits, major tick intervals, etc.) including the inset labels. This has been illustrated in the example I provided in Fig. 4.4.

4.6 Customizing Your Matplotlib Environment

Earlier I mentioned that there are three methods for setting parameters. We have been working with the first method, passing values to functions, either as positional or keyword arguments, provided by the pyplot interface to Matplotlib. Although this is straightforward, routinely setting many properties to the same personal preferences with lots of keywords quickly becomes tedious. Another option is to set/change default values in a file. In unix-like operating systems, this is typically

```
$HOME/.matplotlib/matplotlibrc
```

In Windows, your personal settings are stored in

```
C:\Documents and Settings\username\.matplotlib
```

Let's now take a look at what this settings file looks like, and how we might use it. You may have noticed that the default line colour for the first series is blue if unspecified. Now let's see from where matplotlib is capable of retrieving this default setting. You can either modify the default settings for all users of that computer by editing `matplotlibrc`. The first time around, there is no personal copy, so open the system-wide file and save a copy in your own directory. This is a pretty big file with lots of settings. Scroll down and look for the section that handles drawing lines.

```
### LINES
#lines.linewidth : 1.0 # line width in points
#lines.linestyle : - # solid line
#lines.color : blue
#lines.marker : None # the default marker
#lines.markeredgewidth : 0.5 # the line width around the
    marker symbol
#lines.markersize : 6 # markersize, in points
#lines.dash_joinstyle : miter # miter|round|bevel
#lines.dash_capstyle : butt # butt|round|projecting
#lines.solid_joinstyle : miter # miter|round|bevel
#lines.solid_capstyle : projecting # butt|round|projecting
#lines.antialiased : True # render lines in antialised
```

Notice that *all* of the options here are commented out. (There is yet another level of global defaults, but we are about to override some of these.) If we were to change the line

```
#lines.color : blue
```

to

```
lines.color : red
```

And then simply issuing `ax.plot(x, y)` will draw the line in red. To reiterate, the relevant line in `matplotlibrc` was commented out, so the global (rather than user-level) `matplotlibrc` was used, as the line colour was considered to be unspecified. If we were to merely remove the hash (#) symbol from the line, we would now be specifying the line colour for all subsequent plots explicitly, but keeping it blue. To change the default, we therefore need to uncomment the line and change its value.

Another option is to change one of the matplotlibrc settings for the current session only.

```
import numpy as np
from matplotlib import pyplot
```

```
pyplot.rc('lines', color='red')
x = np.arange(0.,360.)/180.*pi
y = np.sin(x)
ax.plot(x,y)
pyplot.show()
```

Multiple parameters can be set via

```
pyplot.rc('lines', linewidth=2, color='r')
```

There is also an alternative dictionary-type syntax

```
from matplotlib import rcParams
rcParams['lines.linewidth'] = 2
rcParams['lines.color'] = 'r'
```

In summary, there is no shortage of options for customizing every imaginable attribute of your plots. You want to change every fifth tick mark to red? You want to change the axis label font colour so it becomes darker as your data gets smaller? There are ways to do all of these things, some more straightforward than others. For simple things, the Matplotlib online documentation is the best place to start looking. There is even a gallery of user-contributed plots. If you see something that matches what you want to create, or has some attribute that you want to implement, you can look at the corresponding code to investigate how it was done. For more complex or rarer tasks, the online community has archived a wealth of information. Just make sure that the advice you are following applies to the current version of Matplotlib you are using. New features are added all the time, and there are kludgy solutions to implement often-requested features in old versions, that now have built-in simple implementations. Once you figure out what you like, you now know many ways to change the current and default settings to suit your needs and tastes. Have fun exploring.

4.7 Polar Plots

Although plotting in polar coordinates creates a substantially different-looking result compared to plotting in a Cartesian coordinate system, the basic syntax is much the same. First set the type to polar using a keyword argument at the time the subplot is created via

```
ax = fig.add_supblot(projection='polar')
```

and then use the `plot` command as usual, but specifying (θ, r) instead of (x, y) as the first two arguments to `plot`. The example

```
theta = np.linspace(0., 2*np.pi, 500)
r = np.abs(np.sin(5*theta) - 2*np.cos(theta))
ax.plot(theta, r)
```

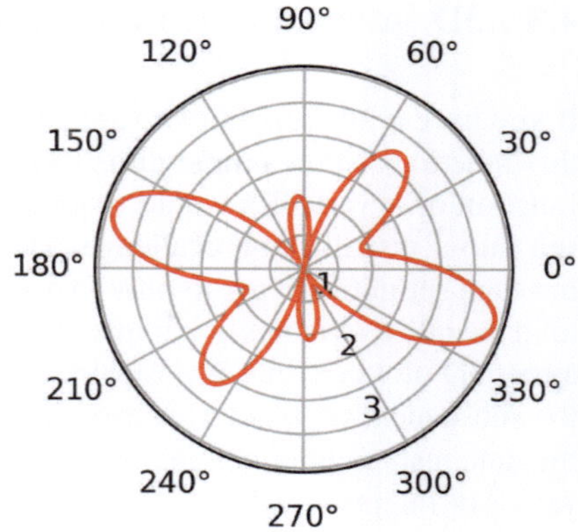

Fig. 4.5 An example of plotting using polar coordinates. Some additional features have been demonstrated to change the default tick marks and labels. See the code example in this section for details

produces a plot as shown in Fig. 4.5. Control of the grid lines is possible using the `rgrids()` and `thetagrids()` functions. If you require fine control of the plot appearance (adjusting tick intervals, r or θ gridlines, etc.) consult the online documentation. There you will see that practically all of the options available for standard Cartesian plots are available for polar plots, albeit with slightly different names to reflect (r, θ) rather than (x, y). In some cases, property names still refer to x and y. Some of these features that resulted in the appearance of Fig. 4.5 are shown in the code example below.

Code for Fig. 4.5

```
import numpy as np
from matplotlib import pyplot

# setup
pyplot.rc('font', size=9)
fig = pyplot.figure(figsize=(3.25, 2.5))
ax = fig.add_subplot(projection='polar')

theta = np.linspace(0., 2*np.pi, 500)
r = np.abs(np.sin(5*theta) - 2*np.cos(theta))
ax.plot(theta, r, 'r-')

# pretty
ax.set_rticks(np.arange(0, 3.1, 0.5), ('', '1', '', '2', '', '3', ''))
ax.set_xticks(np.radians(np.arange(0, 360, 30)))
ax.set_rlabel_position(-70)
fig.set_tight_layout(True)
pyplot.savefig('polar.png', dpi=600)
```

4.8 3D Image and Contour Plots

If you have a function of two variables $f(x, y)$ then a plot illustrating the values of this function requires three dimensions, for example $z = f(x, y)$. In general, any function of n variables would require $n + 1$ dimensions to display, so we quickly run into a visualization challenge. The most common approach is to keep slicing our data set until we get to only two variables that we wish to explore. That is, we hold the other variables or parameters fixed at a particular set of values. At this point, a true 3D plot certainly is possible, and we will see a small number of examples in the subsequent chapters. However, it is usually preferred to plot a 2D projection of the data, using either colours, or contour lines, or both to indicate the $z = f(x, y)$ values of interest.

Matplotlib offers several plotting functions for such data sets, but in each case, care must be taken with respect to setting the origin, not confusing x with y (i.e. which one appears on which axis), and the relationship between the axis scaling and the aspect ratio. We'll take a look at all of these options with a single example that we will build up. First note that if the 2D array of values comes from *experimental* data, then we will likely know which direction in the array corresponds to what property. Let's say that you have collected spectra as a function of time, recording at 100 different wavenumbers and 500 different time points to yield a 100×500 array. Regardless of whether this is stored as a text file with 100 columns and 500 rows, or 500 columns and 100 rows, there is a minimal chance of confusion if you collected the data, as you know how it was stored, and perhaps even what the data should look like when plotted.

However, when plotting a *function* of two variables, when you can decide how many x and y values there are, things can be much less obvious in terms of which variable is plotted along which axis. In some instances, it is clear based on the *values* stored in the 2D array you are plotting, and you will recognize if things are transposed as soon as you see the plot. But you would be surprised how many cases are not so obvious, and I'm always nervous about making the wrong conclusions based on an unintentional transposition. How could this happen? I mentioned that two-dimensional arrays in Python are handled as lists of lists. Depending on how you build these, or what functions return these, the "inner" lists might need to be interpreted as rows or columns. In other words, is the value that you expect to find at `data[x, y]` really at `data[y, x]`? You will need to verify this for your own data, and a suggestion for how to go about this follows.

Let's suppose you have a 2D data set that contains IR wavenumbers from 2850–3000 cm^{-1} along one axis, and time from 0–4 ps along another axis. Momentarily putting aside any preference for which axis gets displayed along the horizontal, at very least we want to make sure that we are not mislabelling them. What we have not considered is the crucial piece of information that describes the data set: how many points there are. If there are 500 points in the IR wavenumber range, evenly spaced 0.3 cm^{-1} apart to cover the 150 cm^{-1} span, and 500 points in the time range even spaced 0.008 ps apart, then we have a square data matrix. The crux of the matter is that the array values contain the intensity at every wavenumber and time point, but the *extent* of the axes is not contained anywhere in the 2D array itself; the $500 \times 500 = 250,000$ z values are stored in an *indexed* coordinate system. You can think of them in terms of pixel numbers in your image. At the very least, we know what they are, based on what we recorded in our laboratory notebook (2850–3000 cm^{-1}, and 0–4 ps). But if the pattern of the signals is not obvious to us (and in many cases it is really not obvious), we don't have any way of knowing whether we are looking at wavenumber along the horizontal, or wavenumber along the vertical direction of our image, as there are many opportunities for transposing the data upon creating the original array, saving it to file, or reading it. I have a suggestion that consists of a series of steps to minimize the chance of making a false assumption.

4.8.1 *Step 1: Having a Quick Preliminary Look at the 2D Array Data*

As an example, let's say we want to explore the relation

$$I(\omega, t) = \sum_j \frac{A_j \Gamma_{1,j}}{(\omega_j - \omega)^2 + \Gamma_{1,j}^2} \frac{\Gamma_{2,j}}{(t_j - t)^2 + \Gamma_{2,j}^2} \tag{4.1}$$

in order to understand its behaviour as a function of frequency ω and time t for a given value of the 5 parameters associated with each vibrational mode. Let's consider two vibrational modes and the 10 parameters

```
w0 = [2910, 2976]
t0 = [1.3, 2.4]
A = [1.1, 0.8]
Gamma_w = [15.2, 11.8]
Gamma_t = [1.4, 2.8]
```

Now let's say that you want to evaluate IR wavenumbers in the range 2850–3000 cm^{-1}, and times in the range 0–4 ps. We can set these up using `np.linspace`, and the number of points will determine how smooth the functions will look. That can be adjusted later, but to get things started, I urge you not to use the same number of points in both dimensions; make them obviously different, like by a factor of two, but any easily recognizable difference will do.

```
_wn = np.linspace(2850, 3000, 500)
_t = np.linspace(0, 4, 1000)
wn, t = np.meshgrid(_wn, _t)
```

Here I could have just used the variable names `wn` and `t` directly to create the 2D grid using `np.meshgrid` such as

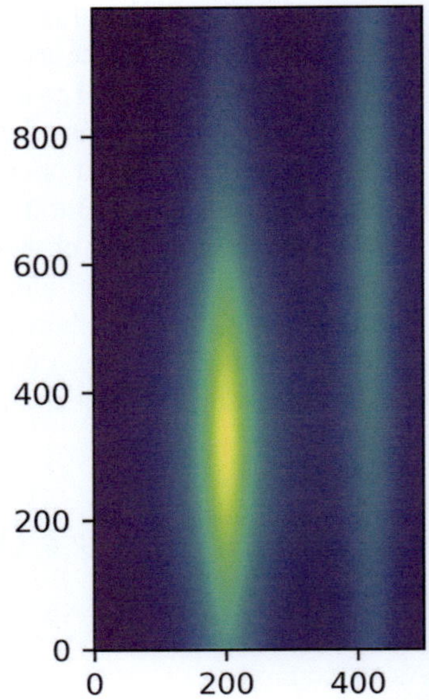

Fig. 4.6 An initial plot of the data array where we haven't specified any options to `imshow` other than telling it we want the origin to be in the bottom-left corner. However, the most important thing to observe at this stage is that the vertical direction contains twice as many data points as the horizontal direction. We should also be looking to verify that 1000×500 is in fact the size and shape of the data array that we have created

```
wn, t = np.meshgrid(np.linspace(2850, 3000, 500), np.
    linspace(0, 4, 1000))
```

rather than first defining the temporary names `_wn` and `_t`, but I did this in case I want to have access to the 1D arrays at a later point for convenience. The important point here is that I set up the wavenumber array with 500 points and the time array with 1000 points.

Now all we have left to do is define the function corresponding to Eq. 4.1

```
def f(wn, t):
    signal = 0.
    for i in range(len(w0)):
        signal += A[i]*Gamma_w[i]/((w0[i] - wn)**2 + Gamma_w
            [i]**2) * Gamma_t[i]/((t0[i] - t)**2 + Gamma_t[i
            ]**2)
     return signal
```

and set up the subplots (I'll show the complete code example later—just trust me for now) and plot it with

```
g = f(wn, t)
ax.imshow(g, origin='lower')
```

that results in the plot shown in Fig. 4.6. At this point, consider that the 2D array called g contains only the values of $f(\omega, t)$ at each point; it does not in itself have any indication of what the ranges of ω and t are. For this reason, `imshow` has used the indices of the points in the array to scale the two axes. From this, it is immediately obvious that the wavenumber is plotted along the horizontal axis (since 500 points appear along the horizontal) and the time axis is vertical (with 1000 points).

By the way, if you try this with your own *data* rather than my *function* example, you will notice that the result looks suspiciously smooth. That is because the default behaviour of `imshow` is to interpolate the data; this can be disabled with keywords. Such smoothing typically doesn't matter when you are plotting functions that are continuous, but it may or may not be desirable when visualizing experimental data.

Finally, In case you are curious, the reason that the origin defaults to the top left rather than bottom left is because `imshow` was intended for image data such as that read from a `.jpeg` or `.png` file. images have the origin defined in the upper left corner. If you find yourself using `imshow` a lot, you can change the default behaviour. As a review of setting Matplotlib plot options, I reiterate that this can be achieved

1. with a keyword argument to `imshow`, by issuing `ax.imshow(data, origin='lower')`, as we have done here,
2. with a session change of an "rc" parameter `pyplot.rc('image', origin='lower')` or `pyplot.rcParms['image.origin'] = 'lower'`
3. by changing *your* matplotlibrc file,
4. by changing the system-wide matplotlibrc file, therefore changing the defaults for all users on that computer.

4.8.2 Step 2: Commit to Your Choice of Horizontal and Vertical Axes

If you didn't like the way the plot looked, i.e. if you prefer to have the wavenumber on the vertical axis, you could swap that now by transposing the data with `g.T`. But let's suppose that things are good as they are. At this stage, I immediately add the axes labels before I lose track of them.

```
ax.set_xlabel(r'IR wavenumber / cm$^{-1}$')
ax.set_ylabel('time / ps')
```

After that one can feel free to change the number of points, perhaps making the grid contain 500 points in each dimension, and we don't have to worry about distinguishing the axes, as we've figured it out and committed our labels to remind us.

If you were to update your plot now, things would look worrisome, and for good reason, since the axis labels are appearing, along with the units we've put in the text strings, but those don't have proper numbers associated with them (see Fig. 4.7).

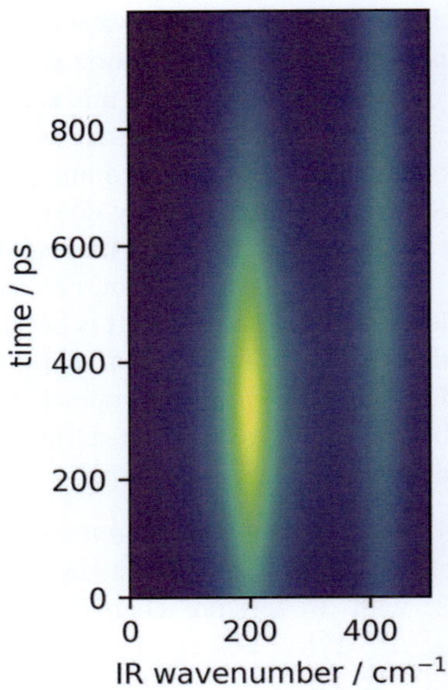

Fig. 4.7 After the vertical and horizontal directions of the 2D array data have been identified, we can commit to this assignment (and not be confused by the transposed array) by setting the axis labels along the appropriate directions. However, although our labels have anticipated the units we know about, it is important to remember that the actual values of the tick marks correspond to the number of data points alone, since we have not presented the plotting function with any scaling information

4.8.3 Step 3: Specify the Extents of the Data Array

Now we get into the topic of the extent, i.e. the ranges of the data along both dimensions. We have seen that, if left to its own devices, `imshow` will scale one of the axes from 0–500, and the other from 0–1000. That's not what we want. We want the axis ticks to show 2850–3800 (no units in the tick *marks*) and the other 0–4. But since this information is nowhere in our rectangular array of numbers (i.e. x and y are lost/implied, and our matrix contains only z values), we have to provide this information to `imshow`. This can be fixed in a few different ways. One is to tell `imshow` what the extent of the axes is with the `extent` keyword

```
ax.imshow(g, origin='lower', extent=[2850, 3000, 0, 4])
```

where the list of 4 numbers has the order $x_{\min}$, $x_{\max}$, $y_{\min}$ and $y_{\max}$. If you are following along, and update your plot, don't be alarmed if you see something like in Fig. 4.8. Things are actually better than they appear; it's just that the aspect ratio is off.

Specifying the extent via this keyword argument is great, as we immediately achieve what we want; it is also the most dangerous and risky operation if you rush into this command. That's because, if our data was incorrectly rotated (inadvertently transposed), the 1000 points along the time direction would now be scaled and labelled from 2850–3000 (there should be 500 points along that axis), and the 500 points along the wavenumber axis would be subjected to the 0–4 scaling. So, if we could not recognize any distinguishing features of our plots that would confirm the orientation, or alert us to a transposed data set, we would be out of luck. Because

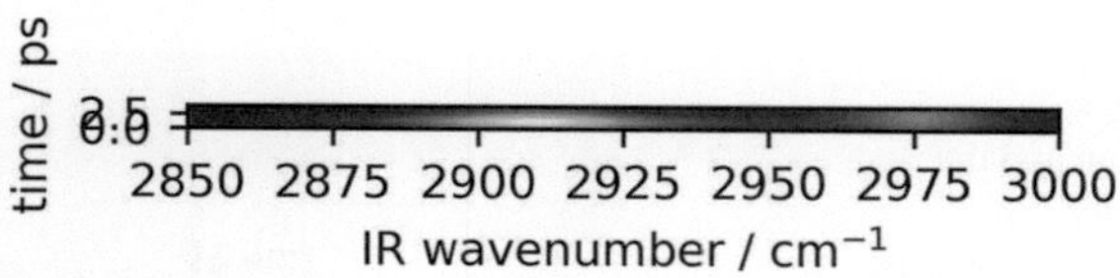

Fig. 4.8 What you would see with the extents of both axes specified, but no aspect ratio override. I've cropped out a significant amount of white space above and below the frame. That existed because we had specified an image that is 3.25" wide by 3.25" tall at 600 dpi, so a png file of 1950 pixels × 1950 pixels is created, regardless of whether we are drawing only in the centre of it

of this fear, whenever I make a plot, I hold off on specifying the extent, and always follow the sequence of these first three steps.

4.8.4 Step 4: Set the Desired Aspect Ratio

Previously, the aspect ratio was determined by the combination of the image (axis) dimensions and the number of points along each dimension. Now we have specified that the vertical axis has an extent of 4, and the horizontal axis extends $3000 - 2850 = 150$ in their respective (irrelevant) units, so the aspect ratio is 4/150. If we wanted this to appear as a square, we would have to adjust it to 150/4.

```
ax.imshow(g, origin='lower', extent=[2850, 3000, 0, 4],
    aspect=150./4.)
```

Keep in mind that you could write `aspect=(np.max(_wn)-npl.min(_wn))/(np.max(_t)-np.min(_t))` rather than hardcoding these values if you wanted things to change dynamically if you later change the extent of the ranges. Now we are all set, and have a plot that appears as in Fig. 4.9. The default colour scheme is not that attractive, and you'll likely want a colour bar that shows how the values of g are mapped onto colours. There are many such customizations available. I don't want to devote too much additional space to these points here. Instead, I'll provide complete examples of code in the subsequent chapters, and you'll see how I have set things up and styled them to create the accompanying figures.

4.8.5 Step 5: Continue to Dress Things Up

Now that we have taken care of all the preliminaries, we can proceed to making the plot look exactly how we want. Perhaps this consists of annotating certain points or regions, adding text labels, etc. All of those things are a pain (or impossible) to do at an earlier stage. Note that `imshow` isn't the only game in town, and contour lines are available as an alternative to the colour map using

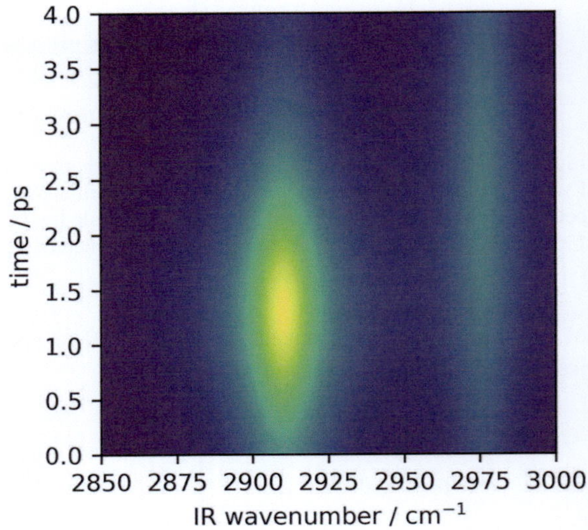

Fig. 4.9 A semi-final version of the plot, with the axes extents enabled and the aspect ratio corrected

```
ax.contour(g, origin='lower', extent=[2850, 3000, 0, 4],
    aspect=150./4.)
```

There are a lot of options to explore, such as what to do with negative values (use dashed lines, for example), setting the contour intervals or the number of contours (that don't have to be evenly spaced), whether to add labels to the lines, how the colours of the lines are to be determined, the formatting of the numbers, whether the numbers are draw at angles following the contour lines (geographic map or road atlas style), etc. Contouring elevations is as much an art as it is a science, and making attractive contour plots (where the labels don't interfere with the data) is a challenging and rewarding exercise. A simple example is shown in Fig. 4.10 where I have used

```
c = ax.contour(g, levels=10, origin='lower', extent=[2850,
     3000, 0, 4], aspect=150./4.)
ax.clabel(c, c.levels[:3])
```

in order to label only the first few elevations. Note that filled contours are available with `contourf`. You will have to explore the online documentation for the current version of Matplotlib to get a sense of how flexible the contour plotting capabilities are.

I must admit that one of my favourite weeknight combos is to plot an imshow using a nice colour scheme, and then superimpose a monotone set of contours on top.

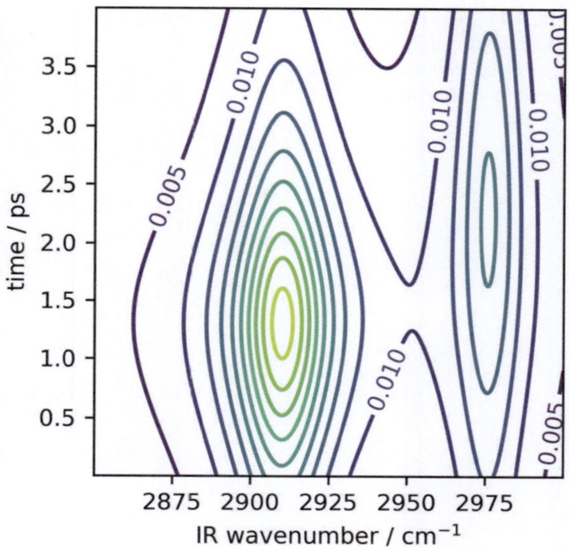

Fig. 4.10 The same data set, but plotted using contour lines in the default colour scheme. A few of the contours have been labelled

```
g = f(wn, t)
ax.imshow(g, origin='lower', extent=[2850, 3000, 0, 4],
    aspect=150/4, cmap=pyplot.cm.gist_stern)
c = ax.contour(g, levels=10, colors='k', origin='lower',
    extent=[2850, 3000, 0, 4])
ax.clabel(c, c.levels[:3])
```

This has a bit of a softer appearance than filled contours, as gradations of colour occur in between the contours. The final result is shown in Fig. 4.11 where I've also thinned out the marker spacing a bit along both axes (I thought it was a bit too crowded for my desired font size) and added some crosshairs.

Code for Fig. 4.11

```
import numpy as np
from matplotlib import pyplot

def f(wn, t):
    signal = 0.
    for i in range(len(w0)):
        signal += A[i]*Gamma_w[i]/((w0[i] - wn)**2 + Gamma_w[i]**2) *
            Gamma_t[i]/((t0[i] - t)**2 + Gamma_t[i]**2)
    return signal

# setup
pyplot.rc('font', size=8)
fig = pyplot.figure(figsize=(3.25, 3.25))
```

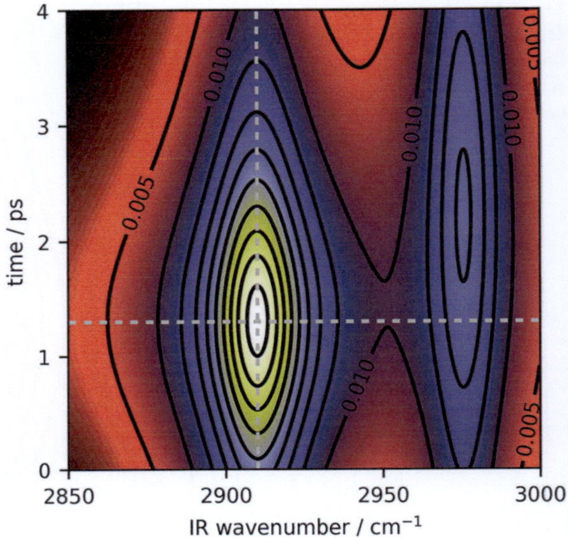

Fig. 4.11 The final image created with an `imshow` base layer with a set of black contours drawn on top

```
ax = fig.add_subplot()
_wn = np.linspace(2850, 3000, 500)
_t = np.linspace(0, 4, 1000)
wn, t = np.meshgrid(_wn, _t)
w0 = [2910, 2976]
t0 = [1.3, 2.4]
A = [1.1, 0.8]
Gamma_w = [15.2, 11.8]
Gamma_t = [1.4, 2.8]

g = f(wn, t)
ax.imshow(g, origin='lower', extent=[2850, 3000, 0, 4], aspect=150/4, cmap=
    pyplot.cm.gist_stern)
c = ax.contour(g, levels=10, colors='k', origin='lower', extent=[2850,
    3000, 0, 4])
ax.clabel(c, c.levels[:3])

# crosshairs
ax.axhline(t0[0], color='0.7', ls='--', dashes=[2,2])
ax.axvline(w0[0], color='0.7', ls='--', dashes=[2,2])

# pretty
ax.set_xlabel(r'IR wavenumber / cm$^{-1}$')
ax.set_ylabel('time / ps')
ax.set_xticks(range(2850, 3001, 50))
ax.set_yticks(range(0, 5))
fig.set_tight_layout(True)
pyplot.savefig('imshowDemo.png', dpi=600)
```

Since this is what I had in mind all along, now you can understand why I chose to evaluate `g = f(wn, t)` once in advance, rather than inside the `imshow()` function. In this case, the function is trivial. But just in case the function is more expensive to compute, I didn't want to have to re-evaluate it to plot the contours.

In summary, some best practices for preparing to plot 3D data (where the values are stored in a 2D array) are

1. At least initially, avoid square data arrays. Then use `imshow` without passing the `extent` argument. You can always have a square *data grid* (not necessarily square aspect ratio) later, once the rest has been figured out.
2. Observe the orientation of the rectangle when plotted with square (equal aspect ratio) pixels. If it is correct based on which axis you know contains more points, then leave it; otherwise, transpose the data. This can be as simple as `imshow(data.T, origin='lower')`.
3. Label the axes right away, as soon as you have identified them (after you have decided whether to transpose or not), according to which one you want to be horizontal and vertical. Don't delay this step!
4. Apply the `extent` keyword to get the tick marks right.
5. Adjust the aspect ratio, since this is determined by the extent, when specified.
6. Proceed to add any embellishments.

Following these steps, you can ensure that you are not misinterpreting your 2D data sets.

4.9 Curve Fitting

The topic of what is generally known as "curve" fitting is vast, with entire works devoted to its intricacies [4–6]. Here "curve" is a general term for what we should more broadly define as any activity where we wish to use experimental data to solve for the parameters in a function. In other words, there is no need for the function to be 1D; we could just as easily fit parameters to the 2D function that we plotted in the previous section. There are even more general scenarios where one might wish to explore the best functional form. But in most situations, the form of the function is known, for example

$$I(\omega) = \frac{A\Gamma}{(\omega_0 - \omega)^2 + \Gamma^2} + b \tag{4.2}$$

and we wish to determine the value of the four parameters A, ω_0, Γ, and b that best describe a particular set of experimental data. Although selecting a good strategy in terms of which algorithm to use, and perhaps a good initial guess of the parameters, can be a delicate issue, the good news is that the actual work of the fitting is easily performed in a single line of code, calling one of the many methods available in the `scipy.optimize` sub-module. This is therefore a topic that could have been introduced in the previous chapter, but I waited until this point because I have some tips for successful curve fitting that rely on being able to visualize the result at intermediate stages of the process, so I wanted to talk about basic plotting using Matplotlib first. We will now take a look at a simple 1D example using a stepwise approach.

4.9.1 Step 1: Plot the Experimental Data on Its Own

It is easy to get confused when too many things need to be debugged at the same time. Often, there are issues with the data that need to be resolved before any fitting is considered. Or there are some ambiguities with respect to how the data files should be parsed. For this reason, I recommend to simply read and plot the experimental data as a first step. I take advantage of this opportunity to set up the plotting parameters, tick marks, axis labels, etc. I'll show a complete example of all of those dress-up features at the end. But for now, the important bits are

```
wn_expt, intensity_expt = np.loadtxt('fitting.dat', unpack
    =True)
ax.plot(wn_expt, intensity_expt, marker='o', lw=0, ms=4,
    mew=1, mfc='w', mec='k')
```

You'll note that I'm not connecting the markers with lines, as I want to reserve the use of lines for plotting the function evaluated for a given (including the optimal) set of parameters. An example of the plot at this stage is shown in Fig. 4.12. If you're lucky, your own data will be less noisy than this, but this is an example of the typical signal-to-noise for experiments performed in my lab where the signals are often weak.

4.9.2 Step 2: Write a Function that Contains the Parameters of Interest, and Manually Explore their Values

We now need to create a function that describes this data in terms of the desired number of adjustable parameters. In this case, Eq. 4.2 can be written as

```
def f(x, amplitude, resonance, width, baseline):
   return amplitude*width/((resonance - x)**2 + width**2)
      + baseline
```

There are many types of optimizers available for the final step. Some require an initial guess of the function parameters, while others require only ranges of the

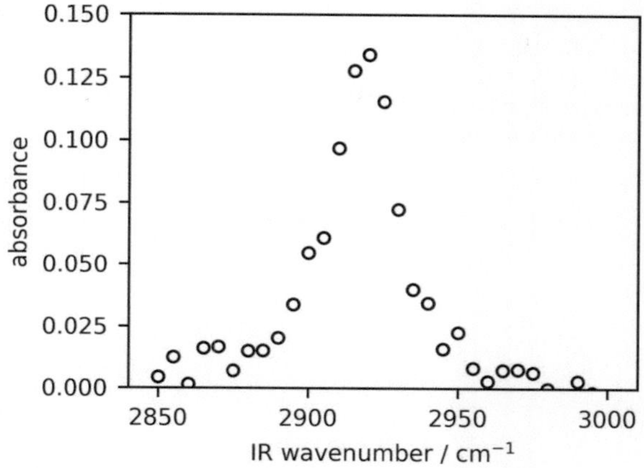

Fig. 4.12 A recommended first step in fitting data is to inspect the data itself

parameters that will be searched. Let's assume that we will use an optimizer that needs both. Now is a good time to manually play around with the parameters to determine reasonable values we can use as starting guesses and bounds. For example, just looking at the data in Fig. 4.12 we can guess that the resonance frequency is in the neighbourhood of $\omega_0 = 2920$ cm^{-1} and that the full width at half the maximum intensity (FWHM, 2Γ) is approximately 30 cm^{-1} so $\Gamma \approx 15$ cm^{-1} is a decent guess. The baseline offset term $b \approx 0$, but it is included to improve the quality of the fit in case the experimental baseline correction needs some fine adjustment. Finally, we need to have a guess at the value of the amplitude A. Looking at Eq. 4.2, we see that on resonance (when $\omega = \omega_0$) we expect a value of $A\Gamma/\Gamma^2 = A/\Gamma$. So any estimate (and the best-fit value) of A is linked to the value of Γ. Looking at the data, we see that the most intense point has a value of roughly 0.14, so if we have an initial guess of $\Gamma = 15$ (we'll leave the units out here, since those are technically divided out as indicated by the horizontal axis label in Fig. 4.12), then that corresponds to an initial guess of $A = 0.14\Gamma \approx 2.1$. Let's create a list of those values, pass them to our function, and do a quick visual inspection in comparison with our experimental data using

```
guess = [2.1, 2920, 15, 0.]
wn_fine = np.linspace(2840, 3050, 1000)
ax.plot(wn_fine, f(wn_fine, *guess), 'g-')
```

with the result displayed in Fig. 4.13. If I were to be 100% honest, I would admit that I wasn't observant enough initially to pick something as good as $\Gamma = 15$. But that's the whole point of this step. Before we invoke the "automatic" parameter estimation, we want to have a sense of how our function behaves. This is essential, not only for understanding reasonable bounds of our parameters, but in ensuring that we haven't made any mistakes in defining our function. If we had an error in our function and passed it directly to an optimizer, it would then be difficult to know what happens after that. In this example, the result displayed in Fig. 4.13 has me convinced that

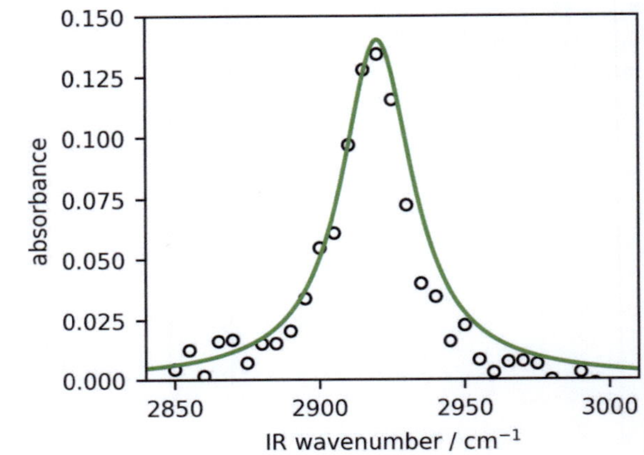

Fig. 4.13 Testing out our function and evaluating good initial guess values of the associated parameters that get us close to what we want

I'm on the right track, and now I just want an optimization routine to take over and tweak my initial guess to provide the best-fit values of the four parameters.

4.9.3 Step 3: Pass the Function to an Optimization Routine

We have already mentioned that the world of optimization is vast, with lots of different algorithms available at different levels of performance and complexity. I highly encourage you to take a look at the `scipy.optimize` documentation to get a sense of what is available. As a minimal introduction, I will say that the ideal optimizer is not overly sensitive to the initial values (so you don't have to worry too much about making a bad guess), is not too easily fooled by local minima (so can reliably reach the global minimum), is fast, and easy to implement. For very challenging problems with a large number of parameters, you basically have to pick any two of those four wishes!

There are a few different categories of optimizers available in `scipy.optimize` that are grouped according to their requirements. Even though our simple example could be solved more directly using a method other than what I'm about to illustrate, I wish to show you how to access a broad range of optimizers that accept a function and an initial set of parameters to adjust, and then return the set of parameters that produce the *smallest* value of a floating point number returned by that function. The amazing thing about this is that there are no restrictions about the kind of things we can do *inside* that function, as long as the function returns a single number.

Let's say that the returned number will be the sum of the squares of the residuals, i.e. the sum of the squares of the differences between the experimental absorbance value and the value of the function at those particular IR wavenumbers that correspond to the experimental data points. One option is to define the function to be handled by the optimizer as

```
def error(params, datax, datay):
    model = params[0]*params[2]/((params[1] - x)**2 +
        params[2]**2) + params[3]
    residuals = datay - model
    return np.sum(residuals**2)
```

There is nothing wrong with this definition of the `error` function, and it will work perfectly well. However, I prefer not to do this, and instead define my *model* function separately as I've illustrated for `f`, and then write the function that will be passed to the optimizer (that returns a single value) to look like

```
def error(params, datax, datay):
    residuals = datay - f(datax, *params)
    return np.sum(residuals**2)
```

for two reasons. The first reason why I wish to separate the model evaluation and "error" function is so that I can call the model function on my own in Step 2, to get a sense of what the parameters will be. I therefore don't want to have that code live solely in the `error` function, and I don't want to have the code appear in both places, since that's asking for trouble (remembering to make changes to the model in both places). I'll talk about the second reason below, but let's get back to our task at hand and invoke the optimizer.

In this example, we'll use Newton's method [7, 8]. This famous algorithm is based on a root-finding principle that proceeds by evaluating the tangent of the function at an initial guess value (that is, hopefully in the vicinity of an actual root), and then finding the root (zero crossing) of the tangent. The conjecture is that the root of the tangent is a better estimate of the true root than your initial guess. The algorithm then proceeds in an iterative way, approaching the root with successive tangents based on the previous iteration until one of the convergence criteria has been satisfied. Typically, this corresponds to either (1) the current value of the function differs by the previous value by less than the defined tolerance, or (2) the maximum number of iterations (controlled by a keyword) has been reached (in which case the optimizer gives up). There are other stopping criteria, and I encourage you to consult the documentation for whichever optimizer you decide to use. All of this is invoked with essentially one line of code

```
from scipy.optimize import fmin_tnc
opt = fmin_tnc(error, guess, approx_grad=True, args=(
    wn_expt, intensity_expt))
```

One of the convenient features of the truncated Newton's method (`fmin_tnc`) is that it also accepts bounds for each of the parameters. This is useful in case you don't want the optimizer to explore and potentially return unphysical values. You can see how to implement that in the full code listing at the end of this example. Another note about this particular optimizer is that it expects you to pass it not only the function it will try to minimize (`error` in this case) but also functions that calculate the derivates of the said function with respect to each of the parameters. Since I'm lazy to write those functions, I instead pass the keyword argument `approx_grad=True` which tells `fmin_tnc` that it should satisfy its requirement for derivatives numerically, by using extra function evaluations to evaluate the derivatives via a finite difference approach. This takes some more time in each iteration, but means that we can get away with a single (`error`) function.

Now, some explanation about how `fmin_tnc` (and many other optimizers in the `scipy.optimize` family) work. A commonality in the syntax of the optimizer function call is that the first parameter (`error` in this case) is the name of a *function* that the optimizer will call. Not all optimizers require an initial guess, but `fmin_tnc` does, and expects this as the next argument that must be supplied. All other arguments are optional and are in the form of keyword arguments. We have already mentioned that `approx_grad=True` is important for us, or else `fmin_tnc` will expect additional functions that can be used to evaluate the derivatives. A note about the `args` keyword: Remember that we have written a function that we've named `error`. Based on how we chose to write it, `error` has three required input arguments

```
error(params, datax, datay)
```

the *list* (or tuple) of parameters, the x-values of the experimental data, and the y-values of the experimental data. All three of these parameters must be supplied to `error` or your program won't run. This essentially means that `fmin_tnc` needs to be told how to call the `error` function. By default, it will pass the *second* argument supplied to `fmin_tnc` (i.e. `guess`) on to `error` as the first argument of `error`, as in `error(guess)`. That's a good start, as such a list of parameters (in the correct order) needs to be supplied as the first argument to `error`, but that's not enough, and that's what the `args` keyword is for. `args` supplies the *additional* arguments the function needs.

When this code runs, `fmin_tnc` produces some output to the screen by default

```
  NIT   NF   F                       GTG
    0    1  3.485821597484470E-03   1.67663541E-01
    1    4  1.368944216957995E-03   2.21088353E-04
tnc: fscale = 36.6003
    2    7  1.188099625183249E-03   7.01834287E-07
    3   16  1.119414878635898E-03   4.65010646E-03
    4   20  8.842600956813141E-04   2.43457097E-04
    5   23  8.547555134010329E-04   1.55116844E-06
    6   26  8.505500880606723E-04   1.55995638E-04
    7   29  8.476757614730076E-04   2.98045213E-07
    8   32  8.464606594643434E-04   4.78614896E-06
    9   36  8.461375933589483E-04   5.42728196E-07
tnc: fscale = 1277.02
```

```
  10    41  8.461011061111200E-04    4.03571249E-08
  11    50  8.459202828661251E-04    3.71689449E-07
  12    52  8.457854403096489E-04    5.61669279E-06
  13    59  8.456344187906187E-04    5.26642878E-07
tnc: |fn-fn-1] = 1.93562e-09 -> convergence
  14    62  8.456324831725080E-04    8.77075833E-08
tnc: Converged (|f_n-f_(n-1)| ~= 0)
```

We can change this behaviour, asking for more or less output (or none at all) with additional keywords. The first column (NIT) lists the iteration number. Iteration 0 corresponds to evaluating the function using the supplied initial guess for the parameters. The second column (NF) is a running total of how many times the function has been evaluated. So the initial guess required just one function evaluation, since none of the values of the parameters had been adjusted at this point. The third column (F) lists the return value of the (`error`) function. So that means that the sum of the squares of the residuals between experimental data points and the values of the `model` function is $\approx 3.49 \times 10^{-3}$. The optimizer will now try to vary the four parameters to reduce this number.

I'm using the term *function* a lot in this discussion, but we should not confuse the two functions we have written. The one that is more scientifically interesting to us is what we have called `f` as it represents our model that is intimately connected with the parameters of interest. But it is important to keep in mind that the optimizer doesn't care about that function—it is solely occupied with minimizing the output of the function passed to it, namely `error`. I have already emphasized that we can do whatever we want inside `error` as long as a single number is returned by that function. So the fact that `error` itself calls `f` is irrelevant to `fmin_tnc`. In its first attempt to deviate from the initial guess of the parameter values, we can see that the output of `error` has already been reduced to 1.37×10^{-3}. By the time we have reached the 14th iteration, the altered set parameters have resulted in a sum of the squares of the residuals of 8.46×10^{-4}. Since this was within the (default) tolerance range of the output from iteration 13, the search stopped.

If we look at what the optimizer actually returns to the code using `print(opt)` we see

```
(array([ 1.76703313e+00,  2.91815936e+03,  1.26264394e+01,
     -1.44399284e-03]), 310, 1)
```

The first item of this tuple is a list of the optimal values of the parameters. The second item is the total number of function evaluations, and the last item is a flag that indicates the reason why the optimizer stopped its search. In this case, `1` indicates that we have satisfied one of the convergence criteria. Refer to the `fmin_tnc` documentation online to see a full explanation of the return values and performance metrics. A quick inspection shows that our initial guess of 2.1 for the amplitude has been modified to 1.77; the resonance frequency initially assumed to be 2920 is now 2918, and the width Γ has been changed from 15 to 12.6. Recall that the amplitude and width must scale together to match the extent of the data. Visually, we could also see that not much baseline offset was required (so we passed an initial value of $b = 0$); the optimizer returned a suitably small value of $b = -1.4 \times 10^{-3}$. This kind of reality check is useful, as we have already noticed that our initial guess of the parameters was in reasonable agreement with the data, so we don't expect the optimizer to return values that differ substantially from that guess.

4.9.4 *Step 4: Visualize the Function Plotted with the Optimal Parameters*

The last step is to evaluate the model function `f` one last time, this time giving it the set of optimized parameters

```
ax.plot(wn_fine, f(wn_fine, *opt[0]), 'k-', lw=2)
```

and Fig. 4.14 shows this result. In a normal situation, you would comment out (or simply delete) the line of code that previously plotted the initial guess in green. But I have left it here to show you the improvement of the agreement between the experimental and the model when using the optimal set of parameters (as shown in the solid black line).

At this point, I can reveal the second reason why I prefer to separate the model `f` and the function handled by the optimizer `error` into two separate functions. From a technical perspective, only the values of the function evaluated at points (along the wavenumber axis) for which there is corresponding experimental data matter. That's because those are the only points at which the residual can be evaluated. However, it bugs me when I see the resulting best-fit plotted with a jagged line that isn't smooth because it matches the data interval. I much prefer to have all models looking like smooth lines, and only the experimental points being (necessarily) discrete. This is entirely cosmetic, but I also prefer if the model is displayed slightly outside of the boundary of the data. You can see in Fig. 4.14 that the data points are in the region 2850–2995 cm^{-1} but I'm plotting the model from 2840–3010 cm^{-1}. In my opinion that gives everything a more *normal* appearance, and better communicates the concepts of data *vs* model to your audience. So in separating the code for the model `f`, I am not only able to evaluate that function myself (otherwise only `fmin_tnc` would be able to call that function) to inspect some initial parameters, but I can also

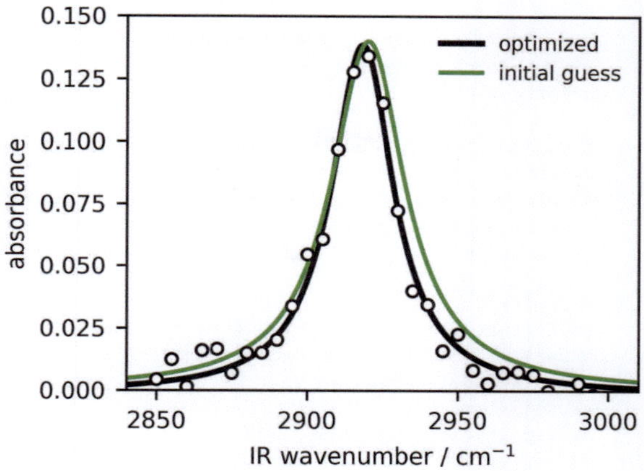

Fig. 4.14 The final plot of the data together with the model evaluated with the optimal set of parameters (black). For illustration purposes, the initial guess is still displayed (green), but that would be removed from the plot in an actual application since it was merely a debugging and data exploration step

pass a separate `np.linspace` that I've named `wn_fine` that contains more finely-spaced points, that extend slightly beyond the data coordinates. So now `fmin_tnc` can call `f` using the experimental (coarse) points along the wavenumber axis, and then I can display the output of the optimization using the same function, but passing it a different array of values. This was actually demonstrated in Fig. 4.12 to plot the green curve as well, but I didn't draw attention to this point until now. Happy curve fitting!

Code for Fig. 4.14

```
import numpy as np
from matplotlib import pyplot
from scipy.optimize import fmin_tnc

def f(x, amplitude, resonance, width, baseline):
    return amplitude*width/((resonance - x)**2 + width**2) + baseline

def error(params, datax, datay):
    residuals = datay - f(datax, *params)
    return np.sum(residuals**2)

# setup
pyplot.rc('font', size=8)
fig = pyplot.figure(figsize=(3.25, 2.5))
ax = fig.add_subplot()
wn_fine = np.linspace(2840, 3050, 1000)

# read the experimental data
wn_expt, intensity_expt = np.loadtxt('fitting.dat', unpack=True)

# fitting
param_ranges = [(0, 5), (2890, 2950), (10, 20), (-0.5, 0.5)]
guess = [2.1, 2920, 15, 0.]
```

```
opt = fmin_tnc(error, guess, bounds=param_ranges, approx_grad=True, args=(
    wn_expt, intensity_expt))
print(opt)

# compare the data and some function evaluations
ax.plot(wn_fine, f(wn_fine, *opt[0]), 'k-', lw=2, label='optimized')
ax.plot(wn_fine, f(wn_fine, *guess), 'g-', label='initial guess')
ax.plot(wn_expt, intensity_expt, marker='o', lw=0, ms=4, mew=1, mfc='w',
    mec='k')

# pretty
ax.legend(loc='upper right', frameon=False, fontsize=7)
ax.set_xlabel(r'IR wavenumber / cm$^{-1}$')
ax.set_ylabel('absorbance')
ax.set_xlim(2840, 3010)
ax.set_xticks(range(2850, 3001, 50))
ax.set_ylim(0, 0.15)
ax.set_yticks(np.arange(0, 0.151, 0.025))
fig.set_tight_layout(True)
pyplot.savefig('fitting3.png', dpi=600)
```

I generally prefer the look of markers on top of lines, rather than the other way around. In Matplotlib there are fancy ways of controlling the order in which objects are displayed, but an easy way to do this is simply by plotting objects in the order in which you want them to appear. That's the reason why, although I load the data in the beginning, I delay the actual plotting of the data point until I have first plotted the model results.

4.10 Histograms

When histograms are required, there are many ways to calculate the bins and draw them, depending on the style of the plot desired. For quick work, one can consider

```
data = np.random.random(1000)
ax.hist(data, 20)
```

where we have specified that we want the data divided into 20 bins. Although I'm not showing the result, a simple bar graph is created. I find the default output of `hist` is seldom attractive enough on its own but, like any matplotlib object, every aspect of its appearance can be customized.

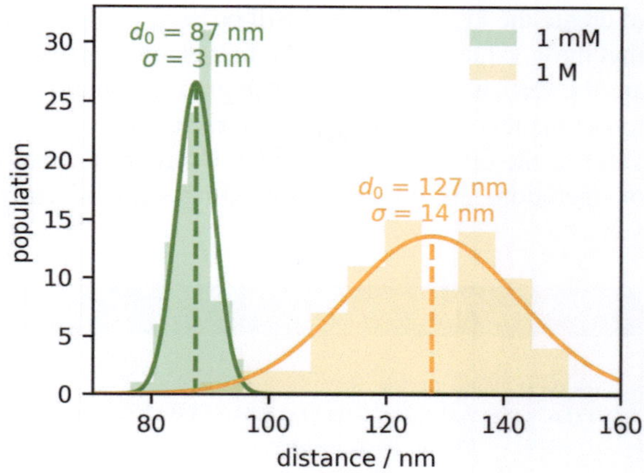

Fig. 4.15 Two histograms where the population values of the bins and the mean distance (centre of each bin) are captured for use in subsequent curve fitting steps

In most scientific applications, histograms are rarely plotted alone, and are usually part of a data analysis pipeline. I will present a slightly more complex example, with data taken from a study that uses total internal reflection fluorescence (TIRF) to determine the distance between algae and solid surfaces as a function of ionic strength [9]. Many cells were characterized and their fluorescence readings were converted into distances, both at 1 mM and 1 M NaCl. Histograms of the two data sets are seen in Fig. 4.15. In the full code example that follows, you can see that I used the command

```
histdata = ax.hist(data, nbins, fc=colour, alpha=0.2,
    label=concentration)
```

where `nbins=10` was specified in a global variable, the face colour (indicated by the `fc` keyword) was passed to a function that handles both concentrations, and the histograms were drawn with an 80% transparency (specifying `alpha=0.2` as the opacity).

After that, I wanted to fit the histogram data (not the raw data) to a Gaussian function. I won't go into the details of the curve fitting here since I want to keep the emphasis on the plotting. What I do want to illustrate is how to capture the vertical extents and horizontal positions of the bins. Remember that we didn't specify those as we would do in a typical plot. Instead, we just passed a single 1D array `data` to `hist` and told it to create a histogram with 10 bins. This is okay, as `hist` also returns some output that informs us of the choices that it made. You'll see that, in the code statement above, I've captured that output into a variable `histdata` that gets populated with a tuple. The first member contains the value (extent) of the histogram bins. I parse the output accordingly using

```
x = histdata[1][:-1] + 0.5*(histdata[1][1] - histdata
    [1][0])
y = histdata[0]
```

to create the (x, y) data that will be used in the subsequent curve fitting step. Note that the y values came from the first element of the returned tuple directly, as those are the vertical extents of the histogram rectangle elements. However, I didn't want to use the left or right edges of the rectangles, instead choosing the average (centre) value as the distance of the x data. Using these kinds of manipulations, many types of operations are possible, converting standard array data to histograms and vice versa.

Code for Fig. 4.15

```
import numpy as np
from scipy.optimize import curve_fit
from matplotlib import pyplot

def f(x, amp, pos, width):
    return amp*np.exp(-(x-pos)**2/(2*width**2))

def fit(x, y):
    opt = curve_fit(f, x, y, p0=[20, 80., 10.])
    return opt[0]

def distance(p, F):
    return p[0]*F**3 + p[1]*F**2 + p[2]*F + p[3]

# setup and load data
pyplot.rc('font', size=8)
fig = pyplot.figure(figsize=(3.25, 2.5))
ax = fig.add_subplot()
Flow = np.loadtxt('histo_low.dat')
Fhigh = np.loadtxt('histo_high.dat')
dlow = distance([1.92123723e+01, -6.06336123e+01, 1.40616181e+02,
     -9.94925282e-02], Flow)
dhigh = distance([105.10007788, -63.32338086, 141.0494608, -0.76081972],
     Fhigh)
nbins = 10
xfine = np.linspace(50., 200., 500)

def analysis(data, concentration, colour):
    # histogram
    histdata = ax.hist(data, nbins, fc=colour, alpha=0.2, label=
         concentration)

    # locate and mark the bin values
    x = histdata[1][:-1] + 0.5*(histdata[1][1] - histdata[1][0])
    y = histdata[0]

    # fitting & annotations
    opt = fit(x, y)
    ax.plot(xfine, f(xfine, *opt), color=colour)
    ax.vlines(opt[1], ymin=0, ymax=opt[0], color=colour, linestyle='dashed
         ')
    ax.text(opt[1], opt[0]+3, '$d_0$ = %d nm\n$\sigma$ = %d nm' % (int(opt
         [1]), int(opt[2])), ha='center', va='center', color=colour)
```

```
analysis(dlow, '1 mM', 'g')
analysis(dhigh, '1 M', 'orange')

# pretty
ax.legend(loc='upper right', frameon=False)
ax.set_xlabel('distance / nm')
ax.set_ylabel('population')
ax.set_xlim(70, 160)
ax.set_ylim(0, 33)
fig.set_tight_layout(True)
pyplot.savefig('histogram.png', dpi=600)
```

4.11 Loading and Manipulating Image Data

Historically, Matplotlib could do some basic image processing, but similar to the trend where NumPy functions are steadily being removed from SciPy and Matplotlib, so too are the image processing capabilities that formally are part of the Python Imaging Library (PIL) [10], now maintained as the fork Pillow [11]. This is a huge module that has a lot of fancy capabilities, but a basic feature is to be able to open a bitmap image and convert it into a NumPy array.

```
from PIL import Image
a = np.array(Image.open('onion.png'))
```

One can then take over in Matplotlib and use `imshow(a)` to produce something very similar to what is displayed in Fig. 4.16a. Technically, `Image.open` on its own produces a special image object type that Matplotlib can display directly with `imshow`, so the conversion to a NumPy array isn't strictly necessary. But doing so enables us to index that array later to play around with the RGB values. I have dressed up the plot a bit with fancy inset labels and removed the tick marks and labels, but we can talk about that later.

Remember that when working with images, the origin is in the upper left corner. So if you do a lot of image analysis, you may not want to change your global default value. In the above example `imread()` returns an $m \times n \times 4$ array, that contains RGB data in the first 3 channels, and transparency data (the alpha value) in the last channel. If we want to extract a single one of the RGB channels this can be accomplished with

```
flat = a[:,:,2]
ax.imshow(flat)
```

with the result displayed in Fig. 4.16b. Here note that it appears that the single channel data still has some colour, but keep in mind that this is now a false colour image. The notation `a[:,:,2]` slices the `a` array so as to maintain all elements in the first two dimensions (corresponding to the spatial coordinates specified in terms of

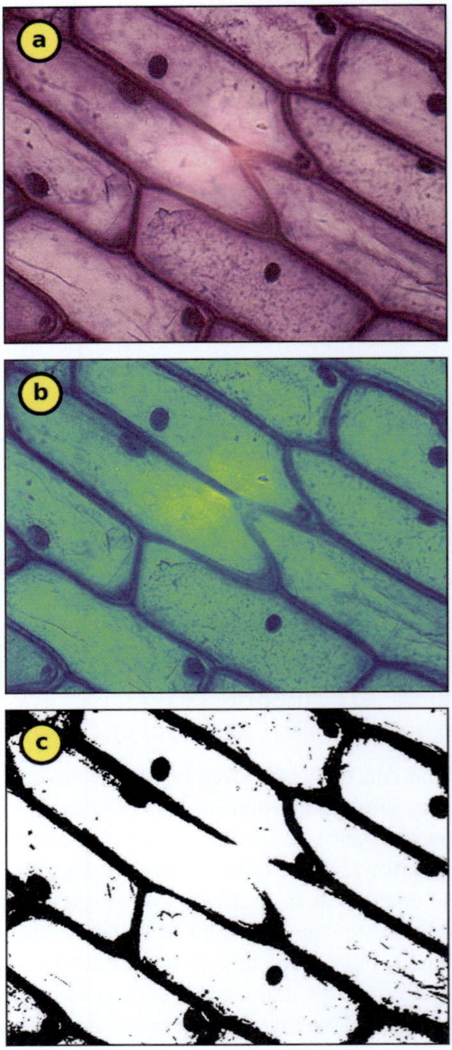

Fig. 4.16 An image of stained onion cells taken through a microscope plotted **a** straight from `imshow`, **b** with only the blue channel from the RGB data, **c** after some thresholding has been applied to turn the data into a binary set

pixels), but element two of RGB that is blue (since element 0 is red). The shades of yellow, green and blue visible in Fig. 4.16b come from the default colour mapping that is used to represent the intensity. This could be changed to a binary colour map, if desired.

Now that you have the data, originally in the form of a png file, imported into a NumPy array, the possibilities are endless. If you want to threshold the image to produce a black and white map with a specific intensity cutoff value, a simple function to filter and reassign the RGB values could be used, as shown in Fig. 4.16c.

The complete code listing is provided below. Since this is our last opportunity to get to focus on plotting before getting distracted with spectroscopy, I provided a few embellishments. First you'll notice that the axis tick marks and tick labels are removed. This is done conveniently using `ax.xticks([])` and a similar command for the `yticks`. The idea is that an empty list is passed as the location to put both the marks and labels, so both are disabled. Note that I also plot large yellow circles using simple markers. This could also be done using a Matplotlib circle patch object, but no need just for a small circle. When the text labels are added, I specified that the text should be centred at the coordinates supplied, rather than the default where the coordinates are for the bottom left of the text box. That way I could use the same coordinates for the yellow circle and the text. You'll notice that I specified those objects using fractional axis coordinates rather than the (hidden but still accessible) axes coordinates. That was accomplished via the `transform=ax.transAxes` keyword. Finally, you'll see that I've created a new axis property that I called `label`. That enabled me to easily assign the string using `ax1.label = 'a'` etc., and then loop through all axis objects and retrieve that string through the `label` property.

Code for Fig. 4.16

```
import numpy as np
from matplotlib import pyplot
from PIL import Image

# setup
pyplot.rc('font', size=8)
fig = pyplot.figure(figsize=(3.25, 7.5))
ax1 = fig.add_subplot(311)
ax2 = fig.add_subplot(312)
ax3 = fig.add_subplot(313)
a = Image.open('onion.png')

# original
ax1.imshow(a)
ax1.label = 'a'

# grayscale
flat = np.array(a)[:,:,2]
ax2.imshow(flat)
ax2.label = 'b'

# threshold
cutoff = 0.5 * np.max(flat.flatten())
flat[flat > cutoff] = 128
flat[flat <= cutoff] = 0
ax3.imshow(flat, cmap=pyplot.cm.gray)
ax3.label = 'c'

# pretty things
for ax in [ax1, ax2, ax3]:
```

```
    ax.plot(0.08, 0.9, marker='o', ms=20, mew=2, mec='k', mfc='yellow',
        transform=ax.transAxes, clip_on=False)
    ax.text(0.08, 0.9, ax.label, fontweight='bold', fontsize=11, ha='center
        ', va='center', transform=ax.transAxes)
    ax.set_xticks([])
    ax.set_yticks([])

fig.set_tight_layout(True)
pyplot.savefig('imageData.png', dpi=600)
```

If you are going to be doing a lot of image analysis beyond just thresholding and basic manipulations like cropping, resizing, rotating, colour conversion, you should look into the pillow (PIL) package more deeply. The functions included with PIL can make complex selections based on colours and other attributes, use these selections to define masks, and then apply NumPy and SciPy functions to the masks, along the lines of what you may be accustomed to using professional image editing programs such as GIMP [12].

4.12 Summary and Next Steps

The power of NumPy and SciPy tools go hand-in-hand with the availability to visualize the results using a capable and easy-to-use package like Matplotlib. This is important, not just in producing the final graphic that you might display in a paper, poster, or presentation, but for visualizing the results at intermediate stages of your analysis for data exploration and convenient debugging of complex tasks where visual feedback is invaluable. We will now use these tools in our exploration of some vibrational spectroscopy topics.

Up to this point, the code examples have been presented primarily in snippets. When we get into the spectroscopy, the demonstrations will be complete standalone codes. Just before that, there is one more topic to cover, related to the use of Python for symbolic calculations. This is a departure from the traditional use of Python together with NumPy and Matplotlib, although the tools are all interoperable.

4.13 Exercises

1. Consider an electron ($m = 9.11 \times 10^{-31}$ kg) in a well of length L and depth V_0. How many allowed energy levels n are there, and what are their values, E_n? In case you are unfamiliar with this problem, we're looking for solutions to either

$$\sqrt{\frac{2m(V_0 - E)}{\hbar^2}} = \sqrt{\frac{2mE}{\hbar^2}} \tan\left(\frac{L}{2}\sqrt{\frac{2mE}{\hbar^2}}\right)$$

the so-called even solutions, or the odd solutions

$$\sqrt{\frac{2m(V_0 - E)}{\hbar^2}} = -\sqrt{\frac{2mE}{\hbar^2}} \cot\left(\frac{L}{2}\sqrt{\frac{2mE}{\hbar^2}}\right)$$

where $\hbar$ is the reduced Planck's constant $h/2\pi$. For example, if $L = 9.5$ Å (0.95×10^{-9} m) and $V_0 = 1.5 \times 10^{-18}$ J, my graphical solution indicates that there are 5 allowed energy levels

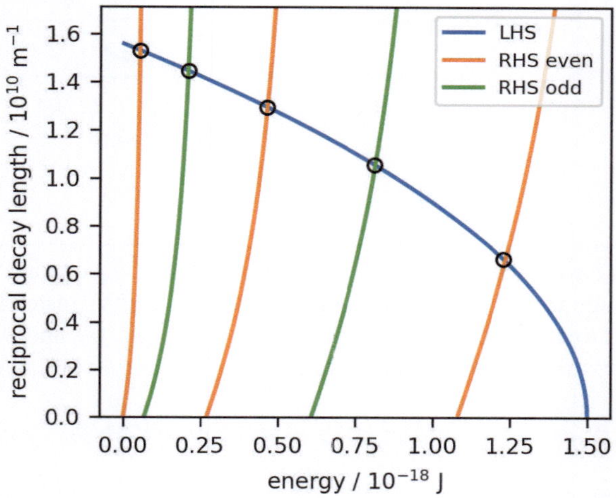

and these are at

$E_1 = 5.255 \times 10^{-20}$ J $\qquad$ $E_4 = 8.108 \times 10^{-19}$ J
$E_2 = 2.087 \times 10^{-19}$ J $\qquad$ $E_5 = 1.230 \times 10^{-18}$ J
$E_3 = 4.640 \times 10^{-19}$ J

This question is actually (a bit) trickier than it looks. Try to create a list of solutions (doesn't need to be perfect—you'll see what I mean once you start), or a graphical solution. If you want more of a challenge, try to produce the solutions *and* graphical output, and make it look "correct".

---

2. Using data files available online (the same `chapt2q6.zip` you might have downloaded already) write a spectral highlighting program that asks the user for an input file and a cutoff value. Your program should display a plot of the data, with a horizontal line indicating the cutoff value. Use some creative, artistic means of highlighting regions with data above the cutoff value. An example is shown below. For an additional challenge, figure out how to identify how many contiguous regions there are that exceed the threshold value.

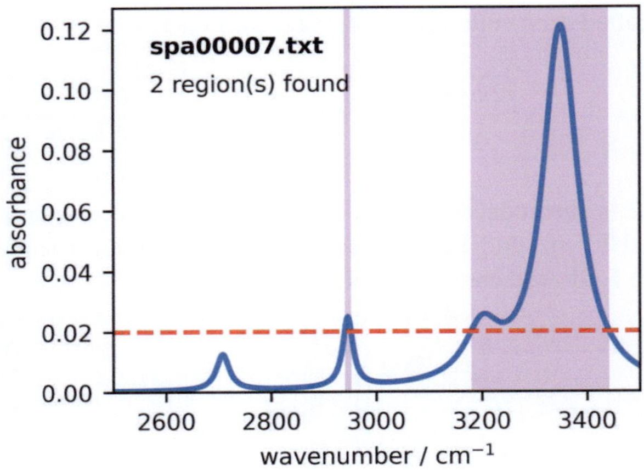

---

3. Say you have 9 data sets of mass transducer current I_m *vs* cone voltage V_c, where the voltage runs from −4 to 8 mV in steps of 2 mV. Instead of me providing data sets to download, just set the parameters p_0, p_1, p_2, and p_3 to random numbers, each uniformly distributed in the range −0.1 to +0.1, and use the relationship

$$I_m = p_3 V_c^3 + p_2 V_c^2 + p_1 V_c + p_0$$

Have your code

- display each data set in a subplot (create a loop for the 9 axes);
- attempt fitting with a second-order polynomial;
- If the sum of the squares of the residuals is less than 20, plot the fit result in blue. Otherwise plot the fit in red, and then plot a third-order polynomial fit in blue.
- Use some creative/artistic means to draw attention to the data sets that required a second round of fitting.

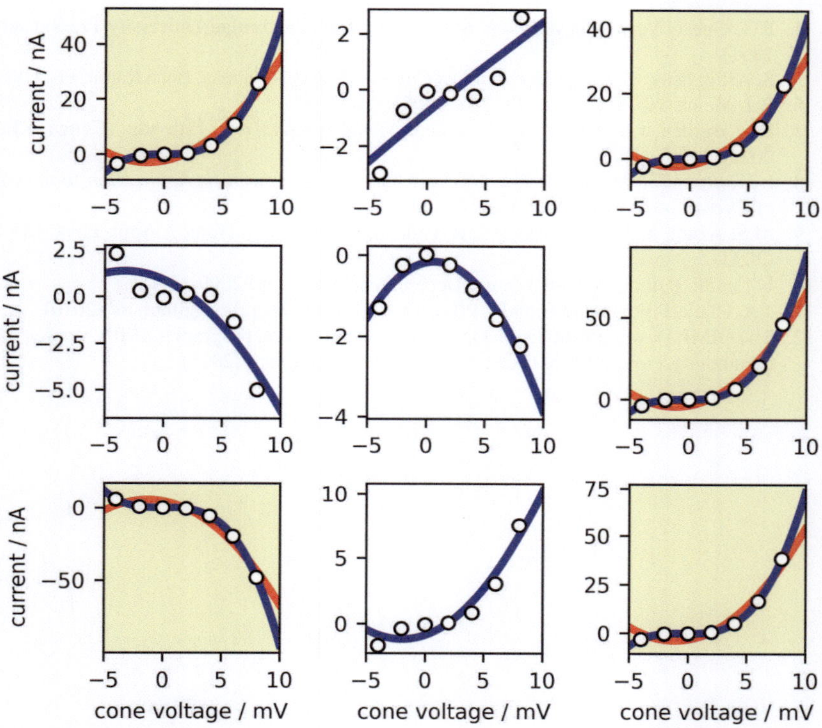

I've provided an example above. Your results will look different each time you run your code, since the data is based on random numbers.

Optional: If the subplots are created in a loop, how can you label the horizontal axes of only the bottom row, and the vertical axes of only the leftmost column?

Additional challenge: Notice how the y-axis label of my first row is tucked in to the right further than that of the other three rows? If you are really picky, you may like all the transducer current (vertical) axis labels aligned in the horizontal direction, irrespective of the width of the vertical axis tick number fields. How could you achieve that while still maintaining auto scaling along the vertical axes?

References

1. T.A. Jarisz, S. Lane, L. Gozdzialski, D.K. Hore, J. Chem. Phys. **148**, 222825 (2018)
2. J.D. Hunter, Comp. Sci. Eng. **9**, 90 (2007)
3. The MathWorks Inc. MATLAB version r2024a (2024)

4. P.G. Guest, *Numerical Methods of Curve Fitting* (Cambridge University Press, Cambridge, 1961)
5. S. Arlinghaus, *Practical Handbook of Curve Fitting* (CRC Press, Boca Raton, FL, USA, 2023)
6. R.J. Meier, Vib. Spectrosc. **39**, 266 (2005)
7. R.L. Burden, J.D. Faires, A.M. Burden, *Numerical Analysis*, 11th edn. (Cengage Learning, 2022)
8. I. Newton, *The Method of Fluxions and Infinite Series; with its Application to the Geometry of Curve-Lines* (Henry Woodfall, London, 1711)
9. M.S. Azam, X. Lu, B.S. Rathore, R. Padhan, Z. Wan, D.K. Hore, J. Appl. Phys. **133**, 215304 (2023)
10. F. Lundh. Python imaging library. Deprecated; last release 2009-11-15
11. J.A. Clark. Pillow—the friendly PIL fork. https://python-pillow.github.io/ (2010)
12. The GIMP Development Team. GNU image manipulation program (GIMP), version 3.0.0-rc3. community, free software (license gplv3). https://gimp.org (2025)

Chapter 5
Symbolic Calculations

5.1 Performing Calculations with Symbols Instead of Numbers

So far, we have been living in a numerical world, where we create a range of numbers x (conveniently achieved with `np.linspace`) and then pass that array to a function `f` to evaluate. Perhaps we make a plot of $f(x)$ versus x afterwards. Almost every situation that we will encounter in our study of vibrational spectroscopy can be dealt with in this manner, effectively using numerical and thereby discrete approximations to continuous variables. In fact, often the point of having many closely spaced values in a `linspace` (for example, `x = np.linspace(0, np.pi, 1000)`) is to plot something in a manner that creates the illusion that we have a smooth line. In reality, that line is made up of many (in this case 1000) individual line segments, but the transitions between segments aren't visible. There are some situations, however, where we want to work with the actual continuous version of the function itself, primarily to have additional insight into the relationships we are plotting.

The idea behind symbolic calculations is easy to explain, as it is what somebody would naturally do if they were using pencil and paper. Consider the case of Snell's Law of Refraction

$$n_1 \sin\theta_1 = n_2 \sin\theta_2 \tag{5.1}$$

where n_1 and θ_1 are the refractive index and angle of incidence (with respect to the surface normal) in the incident medium from which the beam is approaching, and n_2 corresponds to the index of the refracted medium in which the beam is transmitted, at an angle θ_2. A typical application is one where n_1, n_2, and θ_1 are known, and one wishes to solve for the refracted angle θ_2. The approach is to first solve for the angle of interest

$$\theta_2 = \arcsin\frac{n_1 \sin\theta_1}{n_2} \tag{5.2}$$

D. K. Hore, *Molecular Orientation Analysis Using Polarized Vibrational Spectroscopy*,
https://doi.org/10.1007/978-3-032-04874-5_5

and then substitute numerical values of the quantities of interest. When using NumPy, one usually starts with Eq. 5.2 and uses a code such as

```
import NumPy as np
n1 = 1.33
n2 = 1.52
theta1 = np.radians(70.)
theta2 = np.arcsin(n1*np.sin(theta1)/n2)
print(np.degrees(theta2))
```

where we note that the calculation is performed in radians, and we display the result in degrees. However, there are many situations where it is advantageous to be able to carry out the rearrangement that produced Eq. 5.2 with the help of a computer. In complex cases, where the expressions are very long, even if working the result out by hand is possible, the likelihood of making an error is high. Of course, there are many other reasons for symbolic calculations, most importantly to reveal the underlying physics of a problem, perhaps to understand the relationship between two quantities. As we said at the outset, there is no reason strictly speaking why such relationships cannot be explored numerically. But there are simply times when we want to be able to come up with the analytical expression that we are plotting, i.e. the basis of our function, rather than using it as a starting point. In some cases that function is not known, so we need to establish the relationship between the variables of interest, before we can do anything else. We'll encounter plenty of examples of such situations in our study of vibrational spectroscopy. We'll begin with some straightforward examples to introduce the tools available for this in the SymPy package [1].

5.2 Defining Variables

5.2.1 Simple Variables

First we need to distinguish what symbols will remain as traditional Python variables, and which ones we wish to manipulate symbolically. We can define the symbols in a straightforward way

```
import sympy as sp
n1, n2, q1, q2 = sp.symbols('n_1 n_2 theta_1 theta_2')
```

where we demonstrate a means of defining multiple symbols on a single line, where SymPy parses the string to figure out all the different variable names. There are a couple things to note. The first is that the items on the left can be named anything that we want. If our variable and the associated symbol were `x` and x, we would simply write `x = sp.symbols('x')`. However, the fact that we can decouple the Python variable we wish to use and its symbolic name is often handy. In our example, it's convenient to write `q1` for θ_1. But then what's the use of defining symbols like `theta_1` when we could be writing `q1 = sp.symbols('q1')`. Of course that would work too, but the beauty of our method is that it enables SymPy to use a better or more descriptive name in its output. We could make a request such as

```
print(q1)
```

and receive the result `theta_1`. There's nothing wrong with that, and it's frequently what I use just to get the job done. But it's also just as easy to request some nicer formatting in the output. SymPy has several "pretty printing" options, depending on the environment you're in. The easiest one, that is available in almost all instances, is

```
sp.pprint(q1)
```

in which case you will get something on the screen that very closely resembles θ_1. The degree to which θ is rendered to look like the proper Greek letter, and to which the superscript appears smaller depends on the capabilities of the particular code editor or terminal program that you are using to display the output. Inside VScode with the standard Python plugins installed, it looks good.

It is sometimes necessary to limit the range of values for particular variables. For simple algebraic manipulation, that is seldom necessary. For example, the function $f(x) = 2x$ may be valid over all x values, even though we intend to use it over only positive x values. Or there could be a situation where negative values of x are unphysical. Still, there is often no need to impose such restrictions as we can later analyse or plot the result keeping the valid variable or parameter range in mind. However there are situations, most commonly encountered in the evaluation of integrals, where it is worth specifying that parameters have a specific sign, or cannot be complex, in order to get a simpler solution. Otherwise, SymPy will simplify the result only as far as describes the most general case. Examples of this syntax is

```
a, b = sympy.symbols('a b', real=True, positive=True)
```

If different constraints apply to different symbols, then there is no need to group them together. One can instead write

```
a = sympy.symbols('a', real=True)
b = sympy.symbols('b', positive=True)
```

for added flexibility.

5.2.2 *Matrices*

Matrices or tensors of rank-2 can be defined using the `sp.Matrix` function, which has a syntax that is very similar to `np.array`, except that it is limited to rank-1 (vectors) or rank-2 ($n \times m$) objects. As an example, one can write

```
aa, ab, ac, bb, bc, cc = sp.symbols(('alpha_11 alpha_12
   alpha_13 alpha_22 alpha_23 alpha_33'))
alpha = sp.Matrix([[aa, ab, ac], [ab, bb, bc],
       [ac, bc, cc]])
```

and then `sp.pprint(M)` will return

$$\boldsymbol{\alpha} = \begin{bmatrix} \alpha_{11} & \alpha_{12} & \alpha_{13} \\ \alpha_{12} & \alpha_{22} & \alpha_{23} \\ \alpha_{13} & \alpha_{23} & \alpha_{33} \end{bmatrix}.$$

Vectors can readily be defined using a similar syntax, for example,

```
v = sp.Matrix([sp.cos(q), 0, -sp.sin(q)])
```

to produce

$$\mathbf{v} = \begin{bmatrix} \cos\theta \\ 0 \\ -\sin\theta \end{bmatrix}.$$

`Matrix` objects are mutable, which means that we can change the values of elements after they are defined. As an example, we can write `alpha[2,2] = -1` to create

$$\begin{bmatrix} \alpha_{11} & \alpha_{12} & \alpha_{13} \\ \alpha_{12} & \alpha_{22} & \alpha_{23} \\ \alpha_{13} & \alpha_{23} & -1 \end{bmatrix}.$$

For small matrices such as the 3×3 polarizability $\boldsymbol{\alpha}$ defined above, there is nothing embarrassing about manually setting each of the 9 α_{ij} elements manually as we have done above. However, there are situations where you would be pleased to have SymPy create the names for you. There is a feature where you can provide the name of the variable you wish to use, the number of indices, and the shape of the matrix (since it doesn't have to be square). This could be accomplished via the slightly awkward but highly effective operation

```
alpha = sp.MutableDenseNDimArray(sp.symbols(r'a(:3\,:3)'),
    shape=(3,3))
```

to produce a matrix called `alpha` (automatically receiving a SymPy data type) that contains nine elements that can be indexed using `a[0,0]` through `a[2,2]`. This could alternatively be accomplished using a feature from one of SymPy's tensor tools

```
from sympy.tensor.array.expressions import ArraySymbol
alpha = ArraySymbol('a', (3, 3))
```

Although the SymPy object called `alpha` that is created using either of these two methods has some similarities, and its elements can be indexed in the same way, there are some technical differences that affect the manner in which some operations are performed, and how the output is displayed. I suggested looking up the details as one of the end-of-chapter exercises.

5.2.3 Multidimensional Arrays

Starting in Chap. 8 we'll more formally introduce the topic of tensors that provide a universal way of describing and working with mathematical representations of physical quantities corresponding to numbers (rank-0 tensors), vectors (rank-1 tensors), matrices (rank-2), and arrays with higher dimensions (rank-3 and above). It will be more straightforward to deal with those topics when we see them in turn. However, while we're here, I'd like to point out that the syntax described above for manually building and auto-generating array data types in SymPy can readily be extended to tensors of higher dimensions.

5.3 Simplifying Output

A typical situation that one encounters in nearly every step of a more realistic calculation is that SymPy has come up with the correct result, but it is either not obvious (if you know the answer, and it doesn't match what you expect) or is surprisingly long (when something more compact was anticipated). In such cases, simplification can be performed. As an example consider the expression

$$f(\theta) = \frac{\theta^3 + \theta^2 - \theta - 1}{\theta^2 + 2\theta + 1} \tag{5.3}$$

that can be cleaned up using

```
sp.pprint(sp.simplify((q**3 + q**2 - q - 1)/(q**2 + 2*q +
    1)))
```

to show that $f(\theta) = \theta - 1$. Simplifying expressions is an art and SymPy, like all symbolic math software, has several tricks up its sleeve for what it can try in the simplification process. If you look at the online documentation, you'll see that you can also provide advice in the form of optional arguments to the `simplify` function that specify techniques to try. However, at the end of the day, there are some expressions that simply cannot be simplified, or ones for which proper simplification requires techniques for which I have no experience. In such cases, keep in mind that one can result to comparing expressions to determine if they are equal. I'm often trying to reproduce expressions written in books and papers to see if I understand their derivations. Then, I arrive at something that looks different and I wonder if it's actually the same. If only numerical tools were available, I could evaluate Eq. 5.3 and $g(\theta) = \theta - 1$ over some range of θ, and then plot the result to see if they are the same. However, it might be more obvious to plot the difference $f(\theta) - g(\theta)$ to see if the result is zero, or the ratio $f(\theta)/g(\theta)$ to check if the result is fixed at unity. Those approaches work in SymPy too, and there's no shame in trying `print(f-g)`or even `print(sp.simplify(f-g))`. Here I find the difference to be better than the ratio, as the ratio would result in a rather long calculation if $f(\theta)$ and $g(\theta)$ are in fact different. Note I'm using `print` rather than `sp.pprint` here, as I'm just interested to see if the result is zero or not, so no fancy formatting of the output is needed.

5.4 SymPy Intrinsic Functions

Just like NumPy provided basic math operations such as `np.sin` and `np.cos`, and some built-in constants such as `np.pi`, these are also available in SymPy, for example, `sp.sin`, `sp.cos`, and `sp.pi`. If we wish to use them, we need to make sure that we are using the SymPy versions of these functions. Therefore, it is a good idea to keep NumPy and SymPy functions in separate namespaces as we are doing here with the short forms `np` and `sp`. And if we define something new such as

```
a = sp.cos(q1)
```

then `a` is automatically a SymPy object, as all items on the right are SymPy objects; there is no need to define `a` as a SymPy object in advance. Incidentally, if we try something like

```
sp.pprint(a)
```

we will get $\cos\theta_1$ returned. SymPy doesn't know how to evaluate this any further, since we haven't assigned any values to the variables (symbols) yet. As you can tell, I'm starting to be a bit sloppy in what I'm referring to as a variable and a symbol, but I'm hoping that you get the picture. To make a long story short, we can use standard Python expressions such as `*` for multiplication, `+` for addition, and `**` for an exponent to combine SymPy objects, just as we would have done to combine NumPy arrays. We just need to keep in mind that whenever we have the idea that we want to have a symbolic result at the end, we need to use exclusively SymPy objects in our expressions.

5.5 Rearranging and Solving Equations

Let's return to our example where we start with Eq. 5.1 and we want to solve for θ_2. We re-write this as

$$n_1 \sin\theta_1 - n_2 \sin\theta_2 = 0 \tag{5.4}$$

and then ask SymPy to do the hard work for us

```
ans = sp.solve(n1*sp.sin(q1) - n2*sp.sin(q2), q2)
sp.pprint(ans)
```

and we get two solutions returned as a list

$$\left[\arcsin\frac{n_1 \sin\theta_1}{n_2},\ \pi - \arcsin\frac{n_1 \sin\theta_1}{n_2}\right]. \tag{5.5}$$

Considering the coordinate system that we use, where angles are expressed with respect to the *closest* normal, we can select the first solution using typical Python syntax for lists, `ans[0]`.

This same feature can be used to solve a system of linear equations. As an example, given

$$x + y = 4 \tag{5.6a}$$
$$2x - y = -2 \tag{5.6b}$$

we rearrange them into the standard form

$$x + y - 4 = 0 \tag{5.7a}$$
$$2x - y + 2 = 0. \tag{5.7b}$$

If the equations are more complex, it is sometimes convenient to define them using a separate variable first. We'll use `eq1` and `eq2` for those

```
import sympy as sp
x, y = sp.symbols('x y')

eq1 = x + y - 4
eq2 = 2*x - y + 2

solns = sp.solve([eq1, eq2])
sp.pprint(solns)
```

where we see the output

```
{x: 2/3, y: 10/3}
```

is in the form of a Python dictionary. The values of the individual variables can be accessed via `solns[x]` and `solns[y]`. Of course, the standard clean-up procedures may be needed, such as simplification.

5.6 Matrix and Vector Multiplication

Although many capabilities for symbolic linear algebra exist, for our purposes, it will usually suffice to be able to multiply matrices and form the product between a matrix and a vector. In SymPy, this is all accomplished with the `*` operator.

```
import SymPy as sp
aa, ab, ac, bb, bc, cc = sp.symbols('alpha_11 alpha_12
    alpha_13 alpha_22 alpha_23 alpha_33')
q = sp.symbols('theta')
M = sp.Matrix([[aa, ab, ac], [ab, bb, bc], [ac, bc, cc]])
v = sp.Matrix([sp.cos(q), 0, -sp.sin(q)])
```

Issuing the command `print(M*v)` results in the output

```
Matrix([[alpha_11*cos(theta) - alpha_13*sin(theta)], [
    alpha_12*cos(theta) - alpha_23*sin(theta)], [alpha_13*
    cos(theta) - alpha_33*sin(theta)]])
```

while the nicer `sp.pprint(M*v)` returns

$$\begin{bmatrix} \alpha_{11}\cos\theta - \alpha_{13}\sin\theta \\ \alpha_{12}\cos\theta - \alpha_{23}\sin\theta \\ \alpha_{13}\cos\theta - \alpha_{33}\sin\theta \end{bmatrix}.$$

5.7 Differentiation

Symbolic evaluation of derivatives can be accomplished using the `diff` function. We can illustrate this with a simple polynomial example

```
x = sp.symbols('x')
y = 5*x**3 - 3*x + 7
sp.pprint(sp.diff(y, x))
```

where the first argument to `diff` is the expression that we wish to differentiate, and the second argument is the variable with respect to which we would like to take that derivative. Being able to specify the variable is especially important when working with expressions that contain many variables or parameters. The above code produces the expected result $15x^2 - 3$. A third optional argument to the `diff` function allows higher derivatives to be calculated in a single step. For example, `sp.pprint(sp.diff(y, x, 3))` would provide 30 as the third derivative.

5.8 Integration

Evaluating integrals is what I find to be the most useful feature of SymPy for the type of work that we are interested in. Expressing the indefinite integral

$$\frac{1}{2}\int \sin\theta \mathrm{d}\theta$$

is easily accomplished with

```
sp.pprint( sp.integrate(sp.sin(q), q)/2 )
```

which, needless to say, returns $-\frac{1}{2}\cos\theta$, assuming we have already defined `q` as one of our symbols. The second argument to `sp.integrate` specifies that we want to evaluate the integral with respect to θ (named `q` here). When the value of a definite integral is desired, such as

$$\frac{1}{2}\int_0^{\pi} \sin\theta \mathrm{d}\theta$$

the syntax is slightly different

```
sp.pprint( sp.integrate(sp.sin(q)/2, (q, 0, sp.pi)) )
```

Here we have passed the variable with respect to which we want to integrate and the bounds as a 3-tuple in the second argument to `sp.integrate`. In the case of multi-variable integration, one simply passes a 3-tuple for each variable that one wishes to integrate over. An example of this will be provided later below.

Note the use of `sp.pi` rather than `np.pi` since we are not interested in numbers at this stage, and wish to keep the evaluation symbolic. When `sp.pi` is used, a value of `1` was returned. If `np.pi` had been used, a value of `1.00000` would have been returned, highlighting that the result was converted into a float. There is an important note here, concerning writing fractions such as $\frac{1}{2}$ as `1/2`. Using a decimal point in the numerator like `1./2`, denominator `1/2.` or both `1./2.` forces the calculation to use floats rather than integers, even if the result can be expressed as an integer. As an example `6/2 = 3` with the result stored as an integer. But if we had evaluated `6/2.` the result is a float. Whenever something turns into a float, SymPy breaks out of its "exact" result that it typically aims for, and gives a numerical approximation. When things like `1/2` are used, SymPy does its best to interpret what you mean, assuming that you aren't ready for floats quite yet. However, if you prefer not to leave things to chance, you can write `sp.Rational(1, 2)` so the expression becomes

```
sp.pprint( sp.Rational(1,2) * sp.integrate(sp.sin(q), (q,
    0, sp.pi)) )
```

and you will be guaranteed the exact result of `1`. To tell you the truth, I often don't bother with this, as I find it unnecessarily cumbersome, and I typically can pick out what I'm looking for at the end anyways. However, the use of `sp.Rational` is sometimes handy for larger, multi-step operations, where you want to ensure that things aren't converted to floats somewhere in the middle. The moral of the story is that once things become floats, there is no going back, and the output will always be a floating point approximation.

5.9 Converting Functions Between SymPy and NumPy

One of our primary motivations of using a symbolic calculation tool such as SymPy is to arrive at expressions that will subsequently be used in a numerical simulation, to fit experimental data, or to visualize the results for trial values of the parameters. First, it is important to mention that some of those tasks don't strictly require converting SymPy objects to NumPy functions and NumPy arrays. However, it is my choice to do so in most situations. For example, SymPy has some limited but useful plotting features. If you end up doing a lot of SymPy work, you may wish to explore those on your own, as they can be a fast way to see things as you are working. I won't return to that point, as we are more interested in ultimately having a numerical representation that we can plot using all the fine control that Matplotlib offers for 2D graphics. So I'm going to assume that a NumPy object is our end goal. Consider a function of the variable x whose value also depends on two parameters, a and b

$$f(x; a, b) = a \sin \frac{x}{2} + b. \tag{5.8}$$

The complete SymPy code is merely

```
import SymPy as sp
x, a, b = sp.symbols('x a b')
f = a * sp.sin(x/2) + b
```

Now say that we wish to convert from a symbolic function f to a numerical function `g`. This is done using

```
g = sp.lambdify([x, a, b], f)
```

where we have specified that the SymPy variables `a` and `b` are to be parameters of the function. Now we can use this function directly, either with numbers or NumPy arrays as arguments. For example,

```
x = np.linspace(-1, 2, 500)
g(x, 3.4, -1.7)
```

would create a 500-element NumPy array with $-1 \leq x \leq 2$, $a = 3.4$, and $b = -1.7$.

5.10 Example

I now present a more serious example that illustrates many of the concepts related to most of the basic SymPy commands we have outlined in this chapter. We consider a distribution of molecules where the population of tilt angles (with θ as the angle between the molecular long axis and a reference direction in the sample) is described by

$$f(\theta) = N \exp\left[\frac{\lambda_2}{2}\left(3\cos^2\theta - 1\right) + \frac{\lambda_4}{8}\left(35\cos^4\theta - 30\cos^2\theta + 3\right)\right] \tag{5.9}$$

where N is a normalization constant defined such that

$$\int_0^{\pi} f(\theta)\ \sin\theta\ \mathrm{d}\theta = 1 \tag{5.10}$$

so we ensure that the probability of finding a molecule with some tilt angle $0° \leq \theta \leq 180°$ is 100%. This function has two parameters, λ_2 and λ_4. These are simply constants, not to be confused with the wavelength. They are given the symbol λ for historical reasons, as we will discuss in more detail when we study Raman scattering (Chap. 9). We'll later see that Eq. 5.9 is in fact the orientation distribution that one would construct based on an analysis of a Raman vibrational mode in different beam polarizations. For now, our goal is simply to visualize the shape of $f(\theta)$ for different combinations of the parameters λ_2 and λ_4. We'll ultimately want to plot those functions in Matplotlib with values stored in NumPy arrays, but there are several interesting features of the functions that can be understand from Eq. 5.9 before we resort to numerical approximations.

5.10.1 SymPy Implementation of the Distribution Function

To get started, we define the function in terms of the symbols we need

```
import sympy as sp
q = sp.symbols('theta', real=True, nonnegative=True)
l2, l4 = sp.symbols('lambda_2 lambda_4', real=True)
```

```
P2 = sp.Rational(1,2)*(3*sp.cos(q)**2 - 1)
P4 = sp.Rational(1,8)*(35*sp.cos(q)**4 - 30*sp.cos(q)**2 +
    3)
f = sp.exp(l2*P2 + l4*P4)
```

where we have left out the normalization factor N for the time being.

We can later use the normalization condition in Eq. 5.10 to evaluate

$$\frac{1}{N} = \int_0^{\pi} f(\theta)\ \sin\theta\ \mathrm{d}\theta.$$

Based on what we have shown earlier in our study of definite integrals, you might think that this could be accomplished in SymPy using

```
sp.integrate(f * sp.sin(q), (q, 0, sp.pi))
```

While that syntax is certainly correct, in this case, we run into a function whose integral is not straightforward to evaluate, and so SymPy struggles to find a symbolic solution. For our goal, however, we have another option available to us: we can ask SymPy to consider the integral without expressly evaluating it via

```
sp.Integral(f * sp.sin(q), (q, 0, sp.pi))
```

noting the subtle `Integral` rather than `integrate`. You'll see this in the complete code listing provided below, but we can continue to leave out the normalization factor for a moment longer while we inspect the local minima and maxima of $f(\theta)$.

5.10.2 *Characterization of the Extrema*

We can locate the extrema by setting $\mathrm{d}f/\mathrm{d}\theta = 0$. We'll do this with a combination of the SymPy `diff` and `solve` functions.

```
extrema = sp.solve(sp.diff(f, q), dict=True)
```

After printing this list (a complete example of the code is provided at the end) we see that there are three extrema, located at $\theta_1 = 0$, $\theta_2 = \pi/2$, and

$$\theta_3 = \arccos\sqrt{\frac{3}{7} - \frac{6}{35}\frac{\lambda_2}{\lambda_4}}. \tag{5.11}$$

Since the two λ parameters are real and θ is real, we cannot have a negative result under the square root in Eq. 5.11. This alerts us that the third extremum is observed only when $\lambda_2/\lambda_4 < \frac{5}{2}$.

Whether these are local minima or maxima can be determined by evaluating the second derivative at each of the three points.

```
d2 = sp.simplify(sp.diff(f, q, 2))
```

We can then loop through each of the solutions contained in our variable `extrema`, and substitute them into the expression `d2` for the second derivative.

```
for i in extrema:
    sp.pprint(sp.simplify(i))
    symbol, value = list(i.items())[0]
    cleanSolution = sp.solve(symbol - value, q)[0]
    sp.pprint(sp.simplify(d2.subs(q, cleanSolution)))
```

The above code appears slightly more complicated that it probably could be, but the challenge is that the results contained in `extrema` are in the form of a list of dictionaries. In this case, each dictionary contains only a single entry. We therefore use the `items` method of a dictionary object to return the key (that we call `symbol`) and value (what we hope to be the θ locations of the extrema). However, one additional snag is that not all of the solutions have the form $\theta = \text{value}$. In some cases, we need to rearrange the output to isolate θ. Once this is done, the value of the individual extremum is then contained in `cleanSolution`. Finally, this is what is substituted into the second derivative. This would all make more sense and be more logical if you were actually doing it, rather than reading this paragraph!

Finally, from this analysis we see that

$$\frac{\mathrm{d}^2 f(\theta_1)}{\mathrm{d}\theta^2} = N\left(-3\lambda_2 - 10\lambda_4\right) e^{\lambda_2+\lambda_4} \tag{5.12}$$

which tells us that, in order for this to be a maximum where $\mathrm{d}^2 f(\theta_1)/\mathrm{d}\theta^2 < 0$ we would require that the ratio of our two parameters $\lambda_2/\lambda_4 > -\frac{10}{3}$. We next find that

$$\frac{\mathrm{d}^2 f(\theta_2)}{\mathrm{d}\theta^2} = N\left(3\lambda_2 - \frac{15\lambda_4}{2}\right)\exp\left[-\frac{\lambda_2}{2} + \frac{3\lambda_4}{8}\right] \tag{5.13}$$

which reveals that a local maximum at θ_2 exists whenever $\lambda_2/\lambda_4 < 5$. Finally, we consider the third possibility that is governed by

$$\frac{\mathrm{d}^2 f(\theta_3)}{\mathrm{d}\theta^2} = \frac{6N}{35\lambda_4}\left(-6\lambda_2^2 - 5\lambda_2\lambda_4 + 50\lambda_4^2\right)\exp\left[\frac{1}{7}\left(-\frac{9\lambda_2^2}{10\lambda_4} + \lambda_2 - 3\lambda_4\right)\right] \tag{5.14}$$

from which we can learn that negative values of $\mathrm{d}^2 f(\theta_3)/\mathrm{d}\theta^2$ exist when

$$-6\left(\frac{\lambda_2}{\lambda_4}\right)^2 - 5\frac{\lambda_2}{\lambda_4} + 50 < 0, \tag{5.15}$$

namely, when $\lambda_2/\lambda_4 < -\frac{10}{3}$ and also when $\lambda_2/\lambda_4 > \frac{5}{2}$.

5.10.3 Switching Over to NumPy and Plotting

There were several points along this journey where we could have converted our SymPy object into a NumPy function. However, the most interesting and useful place to do this is not for a solution involving a particular set of (λ_2, λ_4) values, but, in general, so we can later supply any combination (λ_2, λ_4) to a NumPy function and not have to return to SymPy for anything, including normalizing the distribution function. All the magic happens in a single line

```
func = sp.lambdify([q, l2, l4], f/norm)
```

where we now use the previously defined but un-evaluated SymPy expression `norm`. At this point, SymPy has more leverage when solving the integral, as it is preparing for a numerical solution. Now we can simply call `func` passing three arguments to that function. The first is either a value of θ in radians, or an entire NumPy array of tilt angle values. The next two parameters are the desired values of λ_2 and λ_4 that NumPy will use in the evaluation. I'll skip setting up the Matplotlib plot, defining attractive axis limits, specifying the legend position, etc. (see complete code listing below) and first highlight the key statements

```
a = 1.5
for b in [1., 1.5, 2.]:
    p = ax.semilogy(thetaDeg, func(np.radians(thetaDeg), a
        , b), label=r'$\lambda_2 = %3.1f, \lambda_4 = %3.1
        f$' % (a, b))
    theta0 = np.arccos(np.sqrt(3./7. - 6*a/(35.*b)))
    ax.axvline(np.degrees(theta0), color=p[0].get_color(),
         ls='--', lw=0.75)
```

for plotting three functions that all have $\lambda_2 = 1.5$, but with different λ_4 values. The resulting graphic is shown in Fig. 5.1. If this had been plotted on a linear scale, the maximum values a $\theta_1 = 0°$ would have been evident, with only a small hint of the other features. However, since the probability density $f(\theta)$ is plotted on a logarithmic scale, the features at $\theta_2 = 90°$ and θ_3 defined in Eq. 5.11 (indicated by the vertical dashed lines in Fig. 5.1) are clearly visible as well. We point out that it is no surprise that there are three extrema, as we knew this already from our symbolic calculations using SymPy. If we had skipped that step, and implemented this entire solution in NumPy from the start, we may have noticed this (on a log scale), but some of the small features may not have been detected unless we had a reason to look for them. Lastly, for these three particular (λ_2, λ_4) values, we already know based on our symbolic calculations that the extrema at $\theta_1 = 0°$ and $\theta_2 = 90°$ will be local maxima, and that the extremum at θ_3 will be a local minimum. Inspecting the plot, we see that the distribution function is symmetric about $\theta = 90°$, so there is another peak at $\theta = 180°$ which is also a maximum. That feature originates from the definition of $f(\theta)$, and will be explored in more detail in Chap. 9 when we talk about the types of distributions that are supported from Raman data.

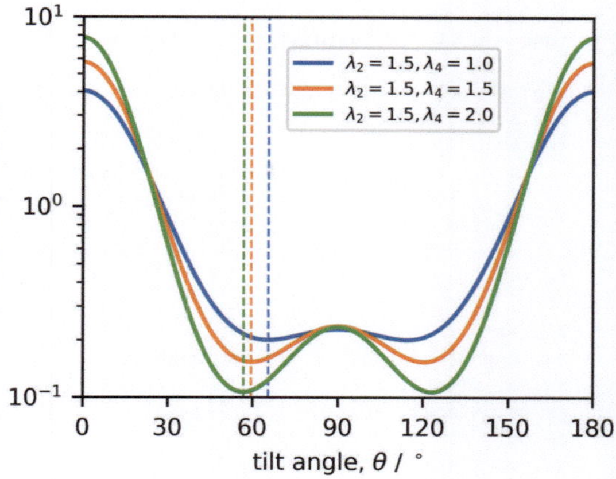

Fig. 5.1 Three tilt angle distributions created by a NumPy array that is a `linspace` of θ values along with a set of λ_2 and λ_4 parameters. For these three particular sets of (λ_2, λ_4) the features at $\theta_1 = 0/180°$ and $\theta_2 = 90°$ are always local maximum, while the feature at θ_3 is always a local minimum

Code for complete analysis including Fig. 5.1

```
import numpy as np
import sympy as sp
from matplotlib import pyplot
import sys

# setup
q = sp.symbols('theta', real=True, nonnegative=True)
l2, l4 = sp.symbols('lambda_2 lambda_4', real=True)

# setup
P2 = sp.Rational(1,2)*(3*sp.cos(q)**2 - 1)
P4 = sp.Rational(1,8)*(35*sp.cos(q)**4 - 30*sp.cos(q)**2 + 3)
f = sp.exp(l2*P2 + l4*P4)
pyplot.rc('font', size=8)
fig = pyplot.figure(figsize=(3.25, 2.5))
ax = fig.add_subplot()
thetaDeg = np.linspace(0, 180, 500)

# second derivative
d2 = sp.simplify(sp.diff(f, q, 2))

# locate the extrema
extrema = sp.solve(sp.diff(f, q), dict=True)
for i in extrema:
    # sp.pprint(sp.simplify(i))
    symbol, value = list(i.items())[0]
    cleanSolution = sp.solve(symbol - value, q)[0]
    sp.pprint(cleanSolution)
    # sp.pprint(sp.simplify(d2.subs(q, cleanSolution)))

# determine the normalization constant
norm = sp.Integral(f * sp.sin(q), (q, 0, sp.pi))

# convert to a NumPy function
```

```
func = sp.lambdify([q, l2, l4], f/norm)

# pick some (l2, l4) combinations and plot them
a = 1.5
for b in [1., 1.5, 2.]:
    p = ax.semilogy(thetaDeg, func(np.radians(thetaDeg), a, b), label=r'$\
         lambda_2 = %3.1f, \lambda_4 = %3.1f$' % (a, b))
    theta0 = np.arccos(np.sqrt(3./7. - 6*a/(35.*b)))
    ax.axvline(np.degrees(theta0), color=p[0].get_color(), ls='--', lw
         =0.75)

# pretty
ax.legend(loc=(0.4, 0.7), fontsize=6)
ax.set_xlim(0, 180)
ax.set_xticks(range(0, 181, 30))
ax.set_ylim(0.1, 10)
ax.set_xlabel(r'tilt angle, $\theta$ / $^\circ$')
ax.set_ylabel('probability density')
fig.set_tight_layout(True)
pyplot.savefig('sympyExample.png', dpi=600)
```

5.11 Summary and Next Steps

Using a combination of NumPy and Matplotlib in the previous chapter, we have seen how complex calculations can be performed in a very simple way, how data can be fit to determine the values of parameters in functions, and how all of the these steps can be visualized. Now we have added symbolic calculations to our repertoire, so we can derive the expressions of interest. Even though this short chapter has just scratched the surface of what SymPy can achieve, we will put these tools to good use in subsequent chapters. For example, imagine a 27-element $3 \times 3 \times 3$ tensor (picture a Rubik's cube) that we wish to rotate. Each element in the new coordinate system consists of a linear combination of all the elements in the old coordinate system, so we have 27 expressions each with 27 terms. Working this out by hand is certainly tedious and prone to mistakes. Using SymPy, we can perform simple elementary transformations to make up the result we want, and then simplify the expressions in the end. Plenty of examples are coming your way!

5.12 Exercises

1. Evaluate the following derivatives. Remember to use at least `sp.pprint()` to view your answers so that the output is formatted more closely to what you would write by hand, rather than in the style of code. If the answer is not what you expect, consider `sp.simplify` to clean up the result. You can combine the two commands with `sp.pprint(sp.simplify(x))`.

   a. $\frac{\mathrm{d}}{\mathrm{d}x}\ln(x)\tan(x)$
   b. $\frac{\mathrm{d}^2}{\mathrm{d}x^2}\left(x^2e^x + \frac{3}{x}\right)$
   c. $\frac{\mathrm{d}}{\mathrm{d}y}\left(e^y + \sin y\right)$
   d. $\frac{\mathrm{d}}{\mathrm{d}x}\frac{x(x^2+2)}{\sqrt{x+3}}$

---

2. Use the definitions of the spherical harmonics

$$Y_1^1(\theta,\varphi) = -\sqrt{\frac{3}{8\pi}}\sin\theta \mathrm{e}^{i\phi}$$

$$Y_1^{-1}(\theta,\varphi) = \sqrt{\frac{3}{8\pi}}\sin\theta \mathrm{e}^{-i\phi}$$

   a. to show that $p_x = -\frac{1}{\sqrt{2}}\left(Y_1^1(\theta,\varphi) - Y_1^{-1}(\theta,\varphi)\right) = \sqrt{\frac{3}{4\pi}}\sin\theta\cos\varphi$
   b. to show that each of $p_x(\theta,\varphi)$, $Y_1^1(\theta,\varphi)$ and $Y_1^{-1}(\theta,\varphi)$ are normalized such that

$$\int_0^{2\pi}\int_0^{\pi}|f(\theta,\varphi)|^2\sin\theta \mathrm{d}\theta \mathrm{d}\phi = 1.$$

   Note that $|f|^2 = f^*f$ so `sympy.conjugate(f)*f` is the same as `sp.Abs (f)**2`. Also note that the names of many SymPy functions are capitalized, differing from their NumPy counterparts.

---

3. a. The hydrogen 1s wavefunction can be described by $\Phi(r) = \mathrm{e}^{-\zeta r/a_0}$, but this form is not normalized. Re-write the wavefunction $\Phi(r)$ so that

$$\int_0^{2\pi}\int_0^{\pi}\Phi(r)^*\Phi(r)\sin\theta \mathrm{d}\theta \mathrm{d}\phi = 1.$$

   Note that, formally, `sp.conjugate()` provides the complex conjugate Φ^*, although in this case our function is real so it is not required.

b. The H_2 molecular orbital can be constructed from a linear combination of hydrogen atom atomic orbitals. Provide normalization constants for $\Psi_+ = \Phi_1(r_1) + \Phi_2(r_2)$ and $\Phi_- = \Phi_1(r_1) - \Phi_2(r_2)$.

---

4. Using the function

$$p_x(\theta, \varphi) = \sqrt{\frac{3}{4\pi}} \sin\theta \cos\varphi$$

evaluate

$$-\frac{\hbar^2}{2\mu r_0^2}\left[\frac{1}{\sin\theta}\frac{\partial}{\partial\theta}\left(\sin\theta \frac{\partial}{\partial\theta} p_x(\theta, \varphi)\right) + \frac{1}{\sin^2\theta}\frac{\partial^2}{\partial\varphi^2} p_x(\theta, \varphi)\right].$$

---

5. SymPy offers some basic plotting features that are worth exploring. Being able to plot in SymPy directly can be a great time-saver as you won't have to convert your SymPy functions into NumPy arrays just to use Matplotlib. For example, you could have a function

$$f(x) = ax^2 + bx + c$$

and quickly assign values to the three parameters a, b, and c in order to make your plot. But there are other times when you want to leave the world of SymPy, and move to the numerical realm of NumPy/SciPy and Matplotlib. Write a small piece of code that experiments with defining a SymPy function such as the simple polynomial above, and then convert this into a numerical function that can be called using all the standard (numerical) Python/NumPy routines.

---

6. If you just need to define a small matrix, you can always define symbols in SciPy such as a_{11} and a_{12}, and then "manually" create something like

$$\begin{bmatrix} a_{11} & a_{12} & a_{13} \\ a_{21} & a_{22} & a_{23} \\ a_{31} & a_{32} & a_{33} \end{bmatrix}.$$

However this is a pain for anything larger. For example, a rank-4 tensor with Cartesian coordinates would have $3 \times 3 \times 3 \times 3 = 81$ elements. SymPy has a way to automatically create indexed variable names when defining such a matrix or rank-n tensor. Learn how to use this feature.

---

7. Functions such as Legendre polynomials and spherical harmonics are included in SciPy and SymPy. Explore the documentation for these functions in each of the two modules, and compare the differences in the syntax and any potential differences in the definitions of the functions. For example, even though the functions should have the same set of variables and parameters, are they passed in the same order? Although defining your own functions is straightforward, being able to make use of such built-in functions is a huge time-saver, and has the added benefit of ensuring that there are no errors in your expressions.

Reference

1. A. Meurer, C.P. Smith, M. Paprocki, O. Čertík, S.B. Kirpichev, M. Rocklin, A. Kumar, S. Ivanov, J.K. Moore, S. Singh, T.E. Rathnayake, B.E. Vig, B.E. Granger, R.P. Muller, F. Bonazzi, H. Gupta, S. Vats, F. Johansson, PeerJ Comp. Sci. **3**, e103 (2017)

Chapter 6
Polarized Light

6.1 Electromagnetic Radiation

Throughout the ages, people have been fascinated by light, and rightfully so. The propagation of light, and the way it interacts with materials in everyday ways (reflection from water, bending of light rays through glass, etc.) was surprising, and made people question what light was actually made of. The fact that light could propagate in a vacuum was especially puzzling, as the wave-like behaviour of light was then in sharp contrast to the propagation of sound waves. In the mid-nineteenth century, James Clerk Maxwell and Katherine Mary Dewar combined some of the prevailing theories to formulate what has become known as Maxwell's equations [1, 2]. When there are no free charges, as most commonly encountered in non-metallic materials, the relations have the form

$$\nabla \cdot \mathbf{E} = 0 \tag{6.1a}$$

$$\nabla \cdot \mathbf{B} = 0 \tag{6.1b}$$

$$\nabla \times \mathbf{E} = -\frac{\partial \mathbf{B}}{\partial t} \tag{6.1c}$$

$$\nabla \times \mathbf{B} = \mu_0 \varepsilon_0 \frac{\partial \mathbf{E}}{\partial t} \tag{6.1d}$$

where $\mathbf{E}$ is the electric field vector and $\mathbf{B}$ is the magnetic field vector. Here we also introduce two fundamental physical constants, μ_0 the magnetic permeability of vacuum, and ε_0 the electric permittivity of vacuum. Recall from the previous chapter that the dot product between two NumPy arrays `a` and `b` is `np.dot(a, b)`, and that their cross product is given by `np.cross(a, b)`. This is assuming you are using the `np.array` data type and not the `np.matrix` data type, which is a good idea in general as it is more flexible. Here the dot and cross products in Eq. 6.1 are between the electric/magnetic field vectors and the vector quantity ∇, the gradient operator. When applied to a vector function f,

D. K. Hore, *Molecular Orientation Analysis Using Polarized Vibrational Spectroscopy*,
https://doi.org/10.1007/978-3-032-04874-5_6

$$\nabla f = \frac{\partial f}{\partial x}\hat{\mathbf{x}} + \frac{\partial f}{\partial y}\hat{\mathbf{y}} + \frac{\partial f}{\partial z}\hat{\mathbf{z}}.$$

The gradient operator is often written in traditional vector form

$$\nabla = \begin{bmatrix} \partial/\partial x \\ \partial/\partial y \\ \partial/\partial z \end{bmatrix}$$

keeping in mind that, since ∇ needs to operate on something, it doesn't make sense on its own. The dot product $\nabla \cdot \mathbf{E}$ is called the divergence of the vector field **E**, also written as div **E** since it describes how much the electric field lines are spreading out (positive value of the divergence) or accumulating (negative divergence) at any particular point. Likewise, the cross product $\nabla \times \mathbf{E}$ is called the curl of **E** as it describes the extent and direction in which the electric field rotates at a given point in space. Although the diverge operation results in a scalar field (a number that represents the degree to which the field diverges), the curl operates on the vector field **E** and produces another vector. The direction of the curl determines the rotation axis, and the magnitude of the curl is the extent. Note that $\nabla \times \mathbf{E}$ is frequently written as curl **E**. The same applies to the magnetic field in Eq. 6.1. These are both topics in vector calculus that we will not need to be so intimately familiar with for the purpose of our primary interest. Having said that, the most important conclusion is that by expanding the relations in Eq. 6.1a–d and combining them results in

$$\frac{\partial^2 \mathbf{E}}{\partial z^2} = \mu_0 \varepsilon_0 \frac{\partial^2 \mathbf{E}}{\partial t^2} \tag{6.2a}$$

$$\frac{\partial^2 \mathbf{B}}{\partial z^2} = \mu_0 \varepsilon_0 \frac{\partial^2 \mathbf{B}}{\partial t^2} \tag{6.2b}$$

where it is apparent that the form of these equations is that of a wave, as the wave equation can be written

$$\frac{\partial^2 \psi}{\partial z^2} = \frac{1}{v^2}\frac{\partial^2 \psi}{\partial t^2} \tag{6.3}$$

where ψ is a general wavefunction, and v is the propagation speed. Comparing with Eq. 6.2, we can see that the speed of light in a vacuum is

$$c = \frac{1}{\sqrt{\mu_0 \varepsilon_0}}. \tag{6.4}$$

This highlights one of the most important consequences of Maxwell's equations, that **B** supports **E** and vice versa, enabling light to propagate, even in the absence of a medium.

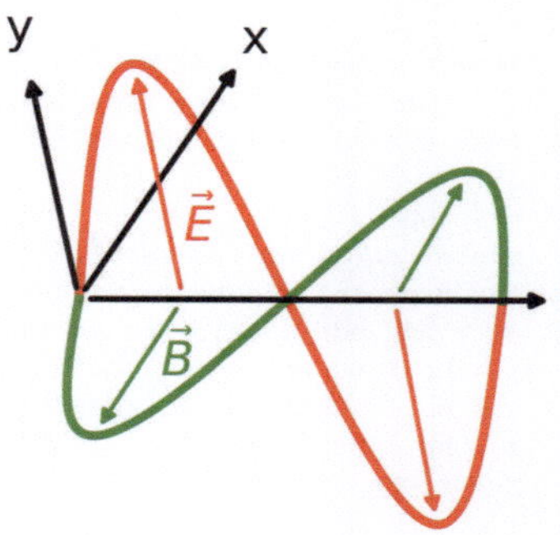

Fig. 6.1 An electromagnetic wave propagating along the z axis. The electric field vectors can be anywhere in the xy-plane, but here are drawn oriented along y. The magnetic field vectors, in this case, are oriented along x

We are not so interested in **B** since magnetic field interactions with most materials are very weak in comparison to the effects of the electric field. Our main interest is therefore Eq. 6.2a, a second-order differential equation that has a solution

$$E(z, t) = E_{\text{max}} \cos(\mathbf{k} \cdot \mathbf{r} - \omega t) \tag{6.5}$$

where E_{max} is the amplitude of the wave, $k = 2\pi/\lambda$ is the propagation vector or wavevector (the *wavenumber* multiplied by 2π), λ is the wavelength, and $\omega = 2\pi\nu$ is the angular frequency. The relationship between these quantities is illustrated in Fig. 6.1 where the propagation vector that was generalized as **r** in Eq. 6.5 is now **z**, and so $\mathbf{k} \cdot \mathbf{r}$ becomes $\mathbf{k} \cdot \mathbf{z} = k_z z$. In situations where the wavevector is entirely along **z** we can simply write k instead of k_z for this component so Eq. 6.1 becomes

$$E(z, t) = E_{\text{max}} \cos(kz - \omega t). \tag{6.6}$$

In Fig. 6.1, the light wave is *plane* polarized since all of the **E** vectors are restricted to the xy-plane. This polarization state is more commonly referred to as *linearly* polarized since, when looking directly along the direction of **k**, the **E** vectors are all pointing up/down along a single line (in the direction of $\hat{y}$). Note that, historically, both $kz - \omega t$ and $\omega t - kz$ conventions exist. If the wave is propagating along the general direction r, the wave equation requires that the spatial part kz (formally $\mathbf{k} \cdot \mathbf{r}$) and the temporal part ωt have opposite signs. We will use the convention as defined in Eq. 6.6 throughout.

Code used to generate Fig. 6.1

Here we make use of some of the limited but sufficient 3D capabilities of Matplotlib in order to generate the 3D perspective sketches of the electric field propagation. It is sometimes useful to have such a code handy in order to quickly visualize the relative magnitude and phase between electric field components.

```
import numpy as np
from matplotlib import pyplot
from dennis import arrow3 as arrow

# setup
pyplot.rc('font', size=8)
fig = pyplot.figure(figsize=(3.25, 2.5))
ax = fig.add_subplot(projection='3d')
x = np.linspace(0, 1, 100)
y = np.linspace(0, 1, 100)
z = np.linspace(0, 1.5, 100)
zeros = np.zeros(len(x))

# draw unit vectors
arrow(ax, [0, 2], [0, 0], [0, 0], lw=1, color='k')
arrow(ax, [0, 0], [0, 1.], [0, 0], lw=1, color='k')
arrow(ax, [0, 0], [0, 0], [0, 1.7], lw=1, color='k')
ax.text(2.15, 0, 0, 'x', ha='center', va='center')
ax.text(0, 1.15, 0, 'y', ha='center', va='center')
ax.text(0, 0, 1.75, 'z', ha='center', va='center')

# electric field
Ey = np.sin(y*2*np.pi)
ax.plot(zeros, Ey, z, 'r-')
arrow(ax, [0, 0], [0, 1], [1.5/4., 1.5/4.], color='r', lw=0.75)
arrow(ax, [0, 0], [0, -1], [3*1.5/4., 3*1.5/4.], color='r', lw=0.75)
ax.text(0, 0.25, 0.42, r'$\vec{E}$', color='r')

# magetic field
Ex = np.sin(x*2*np.pi + np.pi)
ax.plot(Ex, zeros, z, 'g-')
arrow(ax, [0, -1.], [0, 0], [1.5/4., 1.5/4.], color='g', lw=0.75)
arrow(ax, [0, 1.], [0, 0], [3*1.5/4., 3*1.5/4.], color='g', lw=0.75)
ax.text(-0.4, -0.1, 0.4, r'$\vec{B}$', color='g')

ax.axis('off')
ax.set_xlabel('$x$')
ax.set_ylabel('$y$')
ax.set_zlabel('$z$')
ax.view_init(20, 150, -90)
fig.set_tight_layout(True)
pyplot.savefig('wave1.png', dpi=600)
```

This code doesn't do much other than plot simple sine waves (only two lines in the above code): most of the code is dedicated to the decorations, letting the axis labels, turning off the tick marks, etc. The magic happens when creating the axis using the usual `add_subplot()` method of the figure object. The only difference is that here we add the keyword argument `projection='3d'`. After that, many aspects of the conventional plotting syntax that take x and y arguments now take an additional z argument. For example, `text(x, y, label)` is now `text(x, y, z, label)` with most of the usual keywords still available for customization.

Support for drawing arrows in 3D

Be aware that not all of the 2D features of Matplotlib are extended into 3D. For example, there is currently no built-in way to make arrows that originate from one point (x_1, y_1, z_1) to another point (x_2, y_2, z_2) in 3D space. However, it is straightforward to extend the 2D arrow drawing features using `from dennis import arrow`, a separate module that contains the code:

```
import numpy as np
from mpl_toolkits.mplot3d import proj3d
from matplotlib.patches import FancyArrowPatch

class Arrow3D(FancyArrowPatch):
    def __init__(self, xs, ys, zs, *args, **kwargs):
        super().__init__((0, 0), (0, 0), *args, **kwargs)
        self._verts3d = xs, ys, zs

    def do_3d_projection(self, renderer=None):
        xs3d, ys3d, zs3d = self._verts3d
        xs, ys, zs = proj3d.proj_transform(xs3d, ys3d, zs3d, self.axes.M)
        self.set_positions((xs[0], ys[0]), (xs[1], ys[1]))
        return np.min(zs)

def arrow3(ax, xs, ys, zs, color=None, lw=1, alpha=1):
    a3 = Arrow3D([xs[0], xs[1]], [ys[0], ys[1]], [zs[0], zs[1]], lw=lw,
         color=color, mutation_scale=5, arrowstyle='-|>')
    ax.add_artist(a3)

def arrow2(ax, xs, ys, color=None, lw=1, alpha=1):
    a2 = FancyArrowPatch([xs[0], ys[0]], [xs[1], ys[1]], lw=lw, color=color
        , mutation_scale=10, arrowstyle='-|>')
    ax.add_artist(a2)
```

Putting this code into a separate file, thereby creating a Python module, lets us reuse it in other examples without having to type or copy those lines each time.

For our purposes it will be more convenient to write the wave equation using a complex exponential notation. As a quick review, we consider an Argand diagram in Fig. 6.2 that depicts complex numbers as vectors in a complex plane, where the horizontal axis is the real component, and the vertical axis is the imaginary component. We can use this notation to illustrate that the complex number c may be written as $c = a + ib$ where the imaginary unit vector is $i = \sqrt{-1}$. This same figure shows that the length or magnitude is readily obtained from $|c| = \sqrt{a^2 + b^2}$, and that the phase angle $d = \tan^{-1}(b/a)$ is defined in a counterclockwise sense from the positive real axis. Using Euler's identity $e^{i\theta} = \cos\theta + i\sin\theta$, we can write our complex number in polar notation as $c = |c|e^{id}$.

We will frequently compute the phase angle, or simply the *phase*, of a complex-valued quantity by taking the arctangent of the ratio of its imaginary to its real part. Be aware that a challenge associated with the inverse tangent operation is the proper

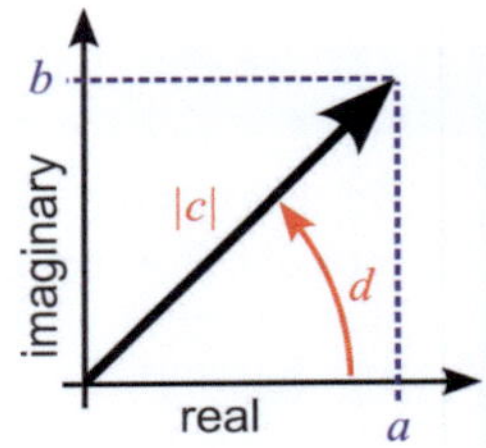

Fig. 6.2 Argand diagram showing how the Cartesian and polar forms of a complex number c are related. This description is useful to keep in mind, as it also helps in understanding how the two forms can be interconverted

assignment of the quadrant of the phase angle, i.e. 0–$\pi/2$ rad, $\pi/2$–π rad, etc. This is simply because the ratio $\mathbb{I}\mathrm{m}\{c\}/\mathbb{R}\mathrm{e}\{c\}$ is positive if $\mathbb{I}\mathrm{m}\{c\}$ and $\mathbb{R}\mathrm{e}\{c\}$ are both positive *or* both negative. Likewise, the ratio is negative if either one of the quantities is positive *and* the other is negative. The solution to this problem is simple: We know the signs of $\mathbb{I}\mathrm{m}\{c\}$ and $\mathbb{R}\mathrm{e}\{c\}$ individually before forming the ratio, so we know in advance what the quadrant of the angle d should be. Most programming languages offer a *quadrant-preserving* arctangent function, typically—as in the case of Python/NumPy—called `arctan2`. So while `d = np.arctan(b/a)` has a 50% chance of getting the quadrant of d correct, one can instead use `d = np.arctan2(b, a)`. Since the `arctan2` function takes $\mathbb{I}\mathrm{m}\{c\}$ and $\mathbb{R}\mathrm{e}\{c\}$ as separate input arguments, it can inspect their signs before dividing them, and then return d with the proper phase.

Getting back to the wave equation, this can now be written as

$$\begin{aligned} E(z,t) &= \frac{1}{2}E_{\text{max}}e^{i(kz-\omega t)} + \frac{1}{2}E_{\text{max}}e^{-i(kz-\omega t)} \\ &= \frac{1}{2}E_{\text{max}}e^{i(kz-\omega t)} + \text{c.c.} \\ &= \begin{bmatrix} E_x \\ E_y \end{bmatrix} e^{i(kz-\omega t)} \end{aligned} \tag{6.7}$$

where c.c. indicates the complex conjugate of the previous term(s). As one rarely considers the absolute magnitude of E (since intensities in the lab are typically measured in arbitrary units as intensity ratios are commonly used in quantitative applications) the above expression is frequently written without the factor of one half, as indicated in the last line above, and the complex conjugate is also dropped in this notation. Here we can also see that E_{max} in general has x- and y-components of the field, so we can express the full vector form of these amplitudes.

The intensity (a.ka. brightness or, more formally, irradiance) in a vacuum (or, practically speaking, air) is given by

$$
\begin{aligned}
I = \frac{c\varepsilon_0}{2}|E|^2 &= \frac{c\varepsilon_0}{2} E^H E \\
&= \frac{c\varepsilon_0}{2} \begin{bmatrix} E_x^* & E_y^* \end{bmatrix} \begin{bmatrix} E_x \\ E_y \end{bmatrix} \\
&= \frac{c\varepsilon_0}{2} \left(E_x^* E_x + E_y^* Ey \right) \\
&= \frac{c\varepsilon_0}{2} \left(|E_x|^2 + |E_y|^2 \right).
\end{aligned} \tag{6.8}
$$

Here $|E|^2$ refers to the magnitude squared of the complex-valued electric field vector, and the superscript H denotes the conjugate transpose, also known as the Hermitian transpose. It is obtained by first forming the transpose of the matrix (or vector in our example) and then taking the complex conjugate of each element, as denoted by the asterisk. A note about NumPy syntax here: in the previous chapter, we saw that the `matrix` data type has a method `.H` for this. Unfortunately there is no such method for the `array` data type, but we can form the conjugate transpose easily in two operations with `.conj().T`. So if we define the electric field vector as `E = np.array([[Ex], [Ey]]` (assuming that `Ex` and `Ey` are defined), then $E^H E$ is obtained from `np.dot(E.conj().T, E)[0,0]`. Here the final `[0,0]` is used to convert this from a 2-dimensional array data type to a float. Ultimately the choice of the data type used needs to match the manner in the multiplication is performed to compute the intensity.

6.2 Polarization of the Electric Field

Polarization of light refers to the orientation of the electric field vector as it propagates through space. The ability to create polarized light in the lab and to detect small changes in the polarization state is a key element in the toolbox of techniques that enable us to determine the structure of molecules. We will later see that this is because the interaction of molecules with light depends on the relative orientation of the electric field vector and the transition dipole moment. We first study a few examples of polarization states that will also introduce some notation.

In Fig. 6.3, we can see two waves, both propagating along $\hat{z}$. In comparing $\mathbf{E}_1$ and $\mathbf{E}_2$, the relative phase or phase shift $\Delta\phi$ is more relevant than the phase due to $kz - \omega t$, since the contribution to the propagation phase is common to both waves, and the absolute phase (linked to the definition of $z = 0$ and $t = 0$) is arbitrary. We can describe the field with a 2-element vector, the Jones vector

$$
\begin{aligned}
E_x &= A_x e^{i(\phi_x - \omega t)} \hat{x} \\
E_y &= A_y e^{i(\phi_y - \omega t)} \hat{y}.
\end{aligned} \tag{6.9}
$$

By the way, if you define Python variables as `Ex`, `Ax`, `phix`, `omega` and `t`, then the expression for the x-polarized component of the field is simply `Ex = Ax * np.`

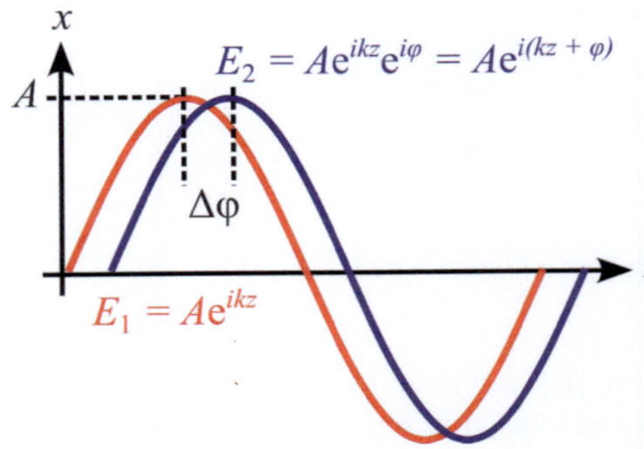

Fig. 6.3 Two waves with the same wavelength, amplitude, and polarization state, but shifted in phase

`exp(1j*(phix - omega*t))`. This is also a reminder that Python uses `j` for the imaginary unit $\sqrt{-1}$. The resultant field is given by the vector

$$\mathbf{E} = E_x + E_y = e^{-i\omega t}\begin{bmatrix} A_x e^{i\phi_x} \\ A_y e^{i\phi_y} \end{bmatrix}. \tag{6.10}$$

If you wanted to create this vector in Python, you could first define `Ex` and `Ey` as illustrated above and then use `E = np.exp(-1j*omega*t)* np.array([[Ex], [Ey]])` but keep in mind that if `omega` or `t` are arrays themselves, this will not work, since you cannot use this specific syntax to set the element of an array to be another array. Other schemes are necessary for this, and we will describe them when they come up. But for now, one option is to loop through your array of ω values, and then use this approach to calculate the corresponding intensity values one by one.

6.2.1 *Linearly Polarized Light*

The Jones vector defined in Eq. 6.10 can be used to describe any polarization state. For example, horizontal polarized light may be described as

$$\mathbf{E} = \begin{bmatrix} A_x e^{i(\phi_x - \omega t)} \\ 0 \end{bmatrix}. \tag{6.11}$$

Typically, Jones vectors are used in their standard form where the time dependence is not written explicitly, and they are normalized to have a magnitude of 1. The same linearly-polarized state is then written

$$\mathbf{E} = \begin{bmatrix} 1 \\ 0 \end{bmatrix}. \tag{6.12}$$

We can also use a 4-element Stokes vector to encode polarization information. This vector **S** is defined as

$$\mathbf{S} = \begin{bmatrix} S_0 \\ S_1 \\ S_2 \\ S_3 \end{bmatrix} = \begin{bmatrix} I_{\text{total}} \\ I_0 - I_{90} \\ I_{45} - I_{-45} \\ I_{\text{rcp}} - I_{\text{lcp}} \end{bmatrix} = \begin{bmatrix} E_x E_x^* + E_y E_y^* \\ E_x E_x^* - E_y E_y^* \\ E_x E_y^* + E_y E_x^* \\ i(E_x E_y^* - E_y E_x^*) \end{bmatrix}. \tag{6.13}$$

Although it is more cumbersome and computationally expensive to keep track of 4 elements for these vectors, the Stokes vector is often a convenient choice since all of the components represent intensities and so are readily determined in the lab. S_0 is the total intensity; S_1 is the intensity difference between horizontally- and vertically-polarized components; S_2 is the difference in intensity between components polarized at $+45°$ and $-45°$; S_3 is the difference between right- and left circularly-polarized light intensities. Although the components of the Stokes vector are customarily defined in terms of intensities, it is the relative intensities that are used in practice where S_j are rendered dimensionless after dividing by S_0.

In our example of the plane-polarized (linearly polarized) wave shown in Fig. 6.1, the normalized Stokes vector is

$$\frac{\mathbf{S}}{S_0} = \begin{bmatrix} 1 \\ S_1/S_0 \\ S_2/S_0 \\ S_3/S_0 \end{bmatrix} = \begin{bmatrix} 1 \\ 1 \\ 0 \\ 0 \end{bmatrix}. \tag{6.14}$$

Consider another example as shown in Fig. 6.4. At first glance, although this is reminiscent of the way we showed electric and magnetic waves travelling together, here we are drawing an x-polarized electric field and another electric field, polarized along the y direction. There are two equivalent ways to think about this. One is that we are interested in the resultant field that describes the superposition of these two waves. The other perspective is that the resultant field is in fact the only field propagating here. We can arbitrarily take any form of polarized light and express it as a linear combination of two orthogonal electric field basis vectors. In this case, the two basis vectors happen to be linearly polarized, but there are other possibilities. Since the two component fields are phase shifted by exactly $\Delta\phi = \pi$ rad, the resultant field remains linearly polarized. Note that this is true whenever $\Delta\phi$ is any multiple of π rad. In this case, since $\Delta\phi$ is an odd multiple of π, the waves combine in a manner that produces a resultant field that is in the negative x and positive y quadrant. If the amplitudes $A_x = A_y$, then this field is polarized at $-45°$ or $+135°$. We refer to this angle as the azimuth or inclination of the linearly-polarized light. You might imagine that if $A_x \neq A_y$ but $\Delta\phi$ is still a multiple of π, the resultant remains linearly polarized, and the exact value of the azimuth depends on the relative magnitudes of A_x and A_y. We will provide a general quantitative expression for the azimuth below, after we describe one more special case.

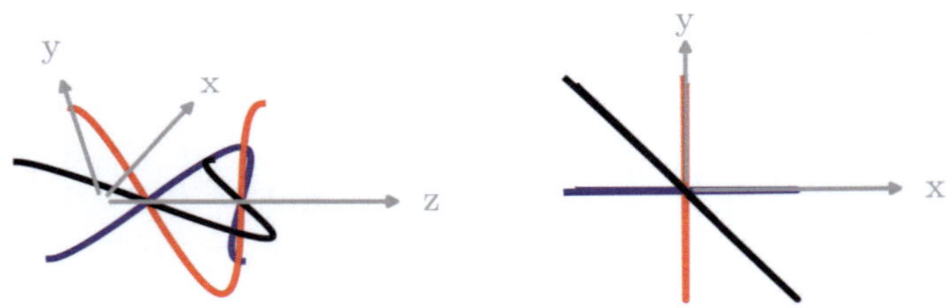

$$\vec{E} = \begin{bmatrix} -0.7071 + 0.0000i \\ +0.7071 + 0.0000i \end{bmatrix} \qquad \vec{S} = \begin{bmatrix} 1 \\ 0.000 \\ -1.000 \\ 0.000 \end{bmatrix}$$

Fig. 6.4 Three-dimensional and plane projection of linearly-polarized light with an azimuth of 135°

6.2.2 *Circularly Polarized Light*

A unique situation occurs when the superposition of two orthogonal linearly-polarized fields, for example E_x and E_y as shown in Fig. 6.5 are phase shifted by exactly 90° (equivalent to $\pi/2$ rad or $\lambda/4$) and have equal amplitudes $A_x = A_y$. The resultant vector is a bit tricky to draw accurately in 3D, but we can take some shortcuts and draw the resultant at positions along z where either $E_x = 0$ or $E_y = 0$. Figure 6.5 shows that these four locations end up with net **E** vectors pointing along $+y$, $+x$, $-y$, and $-x$. If you were to figure out the resultant at locations in between these positions, you could convince yourself that the net field rotates as it propagates along z, so the tip of the **E** vector traces out a helix. Since $A_x = A_y$, the projection along z is a perfect circle, hence the name circularly-polarized light. Note that the sign of relative phase $\Delta\phi$ between the two constituent waves, i.e. $\Delta\phi = \pm 90°$, affects whether the helix progresses in a right-hand or left-hand manner.

There are conventions for assigning the handedness of the polarization state. Consider that our wave is travelling along the positive z-axis. The y component is leading the x component by $\pi/2$ rad. Since the convention $\mathbf{E} = E_0(x, y)e^{i(kz-\omega t)}$ is being used, this phase shift is written as $\Delta\phi \equiv \phi_y - \phi_x = -\pi/2$ rad. Historically there are two methods of assigning the handedness of this circularly-polarized state. One method is to describe the tip of the field vector in time as it passes by an observer fixed in space. Another method is to freeze the wave in time and describe the helix in space. Since ωt and kz always have opposite signs, these two conventions will give opposite results! The convention adopted here is the one recommended by Kliger [3] based on the original proposal by Jones [4], and that is to use the spatial description. The normalized Jones vector associated with this RCP light is

$$\mathbf{E} = \frac{1}{\sqrt{2}} \begin{bmatrix} 1 \\ -i \end{bmatrix}. \tag{6.15}$$

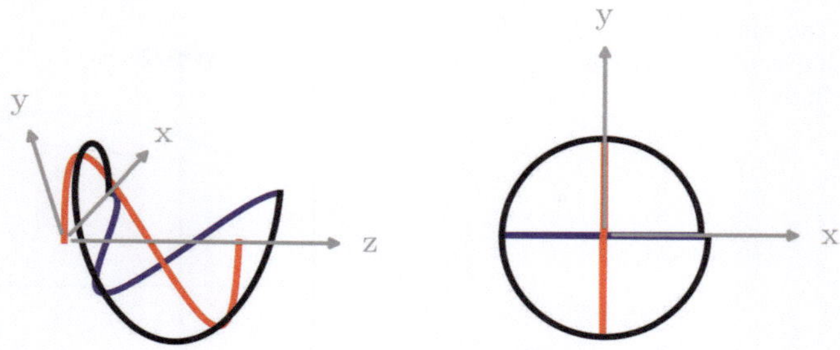

$$\vec{E} = \begin{bmatrix} +0.7071 + 0.0000i \\ -0.0000 - 0.7071i \end{bmatrix} \qquad \vec{S} = \begin{bmatrix} 1 \\ 0.000 \\ 0.000 \\ 1.000 \end{bmatrix}$$

Fig. 6.5 A snapshot of right circularly-polarized light frozen in time in a 3D representation, the field for many time points projected into the xy-plane, and the associated normalized Jones and Stokes vectors

The derivation of this expression is presented as one of the end-of-chapter exercises. The corresponding normalized Stokes vector follows in a straightforward manner from the definition of its elements

$$\mathbf{S} = \begin{bmatrix} 1 \\ 0 \\ 0 \\ 1 \end{bmatrix}. \tag{6.16}$$

6.2.3 *Elliptically Polarized Light*

The most general case (where $A_x \neq A_y$ and $\Delta\phi$ is not a multiple of π rad) is that of a helical progression of the electric field vector, similar to that of circularly-polarized light, but with a projection that forms an ellipse rather than a circle. This projection is shown in Fig. 6.6, and illustrates that this general polarization state is characterized by the specification of three properties: its intensity (magnitude), ellipticity, and azimuth. The ellipticity is defined as the ratio between the semi-minor (labelled b in Fig. 6.6) and semi-major (a) axes by the definition $\tan\omega = b/a$. The ellipticity $\tan\omega$ therefore has an associated *ellipticity angle* ω that is related to the electric field amplitudes as [3]

$$\omega = \frac{1}{2}\arcsin\left(\frac{2A_x A_y \sin\Delta\phi}{A_x^2 + A_y^2}\right). \tag{6.17}$$

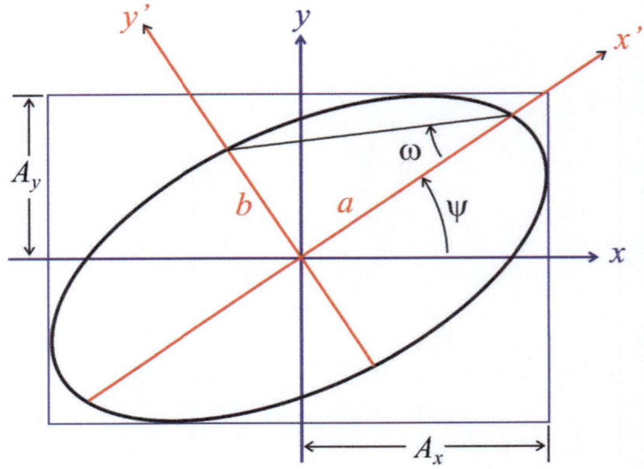

Fig. 6.6 The polarization ellipse in the laboratory reference frame

The azimuth, the orientation of the long axis of the ellipse measured counterclockwise from the positive x-axis of the laboratory frame, is given by

$$\psi = \frac{1}{2} \arctan \left(\frac{2A_x A_y \cos \Delta\phi}{A_x^2 - A_y^2} \right). \tag{6.18}$$

In the case of a right-handed ellipse, $\phi_y < \phi_x$ and so $\Delta\phi < 0$. For a left-handed ellipse, $\Delta\phi > 0$. These definitions can be used to characterize any polarization state. For the special case where $\Delta\phi = 0$ or if $\Delta\phi$ is any multiple of π radians, the orientation of $\mathbf{E}$ is fixed in space and time, and the wave is described as linearly polarized with azimuth ψ. When $\Delta\phi \neq n\pi$ $\mathbf{E}$ rotates as the wave propagates. If $\Delta\phi = \pm\pi/2$ and $A_x = A_y$, then looking along k_z, $\mathbf{E}$ traces out a circle and then the wave is circularly polarized.

The general form for elliptically polarized light is

$$\mathbf{E} = \begin{bmatrix} \cos\psi \\ e^{i\Delta\phi} \sin\psi \end{bmatrix} \tag{6.19}$$

recalling that $\Delta\phi \equiv \phi_y - \phi_x$ is the phase difference between x- and y-components with respect to the laboratory frame and ψ is the azimuth of the ellipse measured counterclockwise from the positive x-axis. Instead of arbitrarily including all of the phase in the y-component, this state may be expressed as

$$\mathbf{E} = \begin{bmatrix} e^{-i\Delta\phi/2} \cos\psi \\ e^{i\Delta\phi/2} \sin\psi \end{bmatrix}. \tag{6.20}$$

From this definition, note the commonly encountered special cases of linearly-polarized light

$$\mathbf{E} = \begin{bmatrix} \cos\psi \\ \sin\psi \end{bmatrix} \tag{6.21}$$

(since $\Delta\phi = 0$) and also circularly-polarized light: right circularly polarized

$$\mathbf{E} = \frac{\sqrt{2}}{2}\begin{bmatrix} 1 \\ -i \end{bmatrix} \tag{6.22}$$

with $\Delta\phi = -\pi/2$ rad; and left circularly-polarized light

$$\mathbf{E} = \frac{\sqrt{2}}{2}\begin{bmatrix} 1 \\ i \end{bmatrix} \tag{6.23}$$

with $\Delta\phi = \pi/2$ rad.

Drawing arbitrary polarization states

We now illustrate some Python code that, given any (unnormalized) Jones vector, will return the normalized Jones and Stokes vectors, a 3D drawing of the wave propagating along z, and the 2D projection in the xy-plane.

```
import numpy as np
from matplotlib import pyplot
from matplotlib.transforms import Bbox
from dennis import arrow3

# setup
x = np.linspace(0., 1., 100)
y = np.linspace(0., 1., 100)
z = np.linspace(0., 1., 100)
zeros = np.zeros(len(x))
pyplot.rcParams['text.usetex'] = True
pyplot.rcParams['text.latex.preamble'] = r'\usepackage{{amsmath}}'
pyplot.rcParams['font.family'] = 'serif'

# wavevector for unit wavelength
k = 2*np.pi

# input electric field components
E0x = 4.7 - 2.3j
E0y = 2.1 + 0.2j

# normalize them
norm = np.abs(E0x)**2 + np.abs(E0y)**2
E0x /= np.sqrt(norm)
E0y /= np.sqrt(norm)

Ex = E0x * np.exp(1j*k*z)
Ey = E0y * np.exp(1j*k*z)
pyplot.rc('font', size=8)
fig = pyplot.figure(figsize=(3.25, 2.5))
```

```
# 3D view
ax1 = fig.add_subplot(121, projection='3d')
arrow3(ax1, [0, 1.5], [0, 0], [0, 0], lw=1, color='0.6')
arrow3(ax1, [0, 0], [0, 1.], [0, 0], lw=1, color='0.6')
arrow3(ax1, [0, 0], [0, 0], [0, 1.6], lw=1, color='0.6')
ax1.text(1.65, 0, 0, 'x', ha='center', va='center', color='0.6')
ax1.text(0, 1.15, 0, 'y', ha='center', va='center', color='0.6')
ax1.text(0, 0, 1.7, 'z', ha='center', va='center', color='0.6')

ax1.plot(Ex.real, zeros, z, 'b-')
ax1.plot(zeros, Ey.real, z, 'r-')
ax1.plot(Ex.real, Ey.real, z, 'k-')
ax1.axis('off')
ax1.view_init(30, 150, -90)

# 2D view
ax2 = fig.add_subplot(122, projection='3d')
arrow3(ax2, [0, 1.5], [0, 0], [0, 0], lw=1, color='0.6')
arrow3(ax2, [0, 0], [0, 1.1], [0, 0], lw=1, color='0.6')
ax2.text(1.65, 0, 0, 'x', ha='center', va='center', color='0.6')
ax2.text(0, 1.15, 0, 'y', ha='center', va='center', color='0.6')

ax2.plot(Ex.real, zeros, z, 'b-')
ax2.plot(zeros, Ey.real, z, 'r-')
ax2.plot(Ex.real, Ey.real, z, 'k-')
ax2.axis('off')
ax2.view_init(90, 0, 90)

# display Jones vector
E0x = complex(E0x)
E0y = complex(E0y)

ax1.text(0, -0.2, 0.2, r'${\vec{E}} = \left[ \begin{array}{c} %s \\ %s \end
      {array} \right]$' % ("{num.real:+0.04f} {num.imag:=+8.04f}i".format(
      num=E0x), "{num.real:+0.04f} {num.imag:=+8.04f}i".format(num=E0y)),
      transform=ax1.transAxes)

# display Stokes vector
S0 = np.conjugate(E0x)*E0x + np.conjugate(E0y)*E0y
S1 = np.conjugate(E0x)*E0x - np.conjugate(E0y)*E0y
S2 = np.conjugate(E0x)*E0y + np.conjugate(E0y)*E0x
S3 = 1j*(np.conjugate(E0x)*E0y - np.conjugate(E0y)*E0x)
S1 = (S1/S0).real
S2 = (S2/S0).real
S3 = (S3/S0).real
ax2.text(2, -0.2, 0, r'${\vec S} = \left[ \begin{array}{c} 1 \\ %6.3f \\
      %6.3f \\ %6.3f \end{array} \right]$' % (S1, S2, S3), transform=ax2.
      transAxes)

fig.set_tight_layout(True)
pyplot.savefig('polarizationEllipse.png', dpi=600, bbox_inches=Bbox
      ([[0,0.15], [3.25,2]]))
```

The code starts by importing the usual modules we have been using, `numpy`, `pyplot`, and `dennis` for the 3D arrow hack as described earlier. Here we also import the bounding box (`Bbox`) specification from Matplotlib since we want to crop the image when we export it. Unlike the case of simple 2D plots, there is no equivalent "tight layout" specification that adjusts the whitespace around the plot when 3D elements are present.
The axes are drawn using all the same commands as for the previous 3D examples. The only difference is that we now have a second set of axes with a different viewing angle. The 3D projection is rendered using `ax1.view_init(30, 150, -90)` while the 2D projection uses `ax2.view_init(90, 0, 90)`. This choice of angles for `ax2` enables us to look straight into the z axis, with the positive x axis pointing to the right, and the positive y axis pointing up in a familiar orientation. The arbitrary input state defined by `E0x = 4.7 - 2.3j` and `E0y = 2.1 + 0.2j` defines the (unnormalized) Jones vector

$$\mathbf{E} = \begin{bmatrix} 4.7 - 2.3i \\ 2.1 + 0.2i \end{bmatrix}.$$

The resulting output is shown in Fig. 6.7.

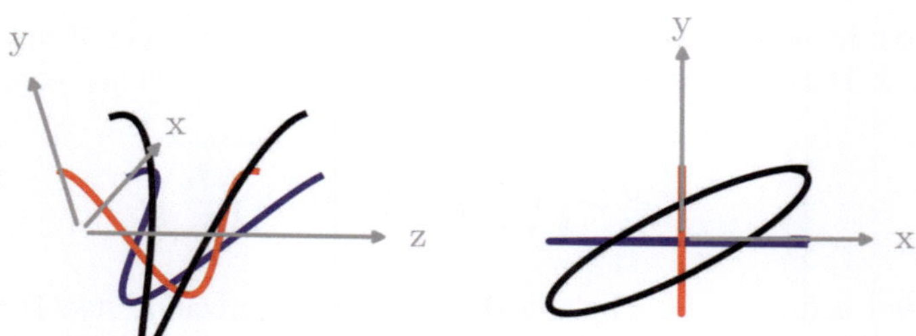

$$\vec{E} = \begin{bmatrix} +0.8331 - 0.4077i \\ +0.3722 + 0.0354i \end{bmatrix} \qquad \vec{S} = \begin{bmatrix} 1 \\ 0.720 \\ 0.591 \\ -0.363 \end{bmatrix}$$

Fig. 6.7 3D view and 2D projection of x (blue) and y (red) components of the electric field, along with the net field plotted in black. The normalized Jones and Stokes vectors are also indicated

6.3 Degree of Polarization

Although Jones vectors offer some convenience in that they are more compact than 4-element Stokes vectors, they can be used only in situations when all of the light is 100% polarized in some state. It is worth pausing to think about our definition of "polarized", as it differs from some colloquial usage. We say that light is polarized when there is a fixed phase relationship between orthogonal electric field components, and therefore, the shape of the polarization ellipse (including the edge cases of linear and circular polarization states) is stable with respect to time. People casually refer to light as polarized when they mean linearly polarized. (Hence, linear polarizers are simply called polarizers.) So our technical definition of what constitutes polarized light is more general. Stokes vectors are more versatile as they can be used even when only a fraction of the light is polarized. A beam with random polarization, also known as an unpolarized source, can be described as

$$\mathbf{S} = \begin{bmatrix} 1 \\ 0 \\ 0 \\ 0 \end{bmatrix}. \tag{6.24}$$

The Stokes vector elements obey the relation

$$S_1^2 + S_2^2 + S_3^2 \leq S_0^2 \tag{6.25}$$

where the inequality accounts for the fact that only a fraction of the light may be polarized. The degree of polarization (DOP) may then be defined according to

$$\text{DOP} = \sqrt{\left(\frac{S_1}{S_0}\right)^2 + \left(\frac{S_2}{S_0}\right)^2 + \left(\frac{S_3}{S_0}\right)^2}. \tag{6.26}$$

Using this definition, fully polarized light in any polarization state (linear with any azimuth, circular of either handedness, or elliptical) has DOP $= 1$, and unpolarized light has DOP $= 0$; naturally any value of the DOP between these extremes is also possible. An important conclusion is that one can always know the DOP precisely if all elements of the Stokes vector (whether it is normalized or not) are measured.

All polarization states can be mapped using a representation known as the Poincaré sphere[5–8] as shown in Fig. 6.8. Unpolarized light is represented by the point in the centre, inside the sphere defining the origin at $(S_1, S_2, S_3) = (0, 0, 0)$. Fully polarized light is represented by a vector that starts at the origin, and ends somewhere on the surface of the sphere. The north pole represents right-hand circularly-polarized light; the south pole left-hand circularly-polarized light. Points along the equator are all linearly polarized, with the meridian angle encoding the azimuth of the electric field. All other points on the surface of the northern hemisphere are right-hand polarized ellipses. Corresponding left-handed ellipses (with

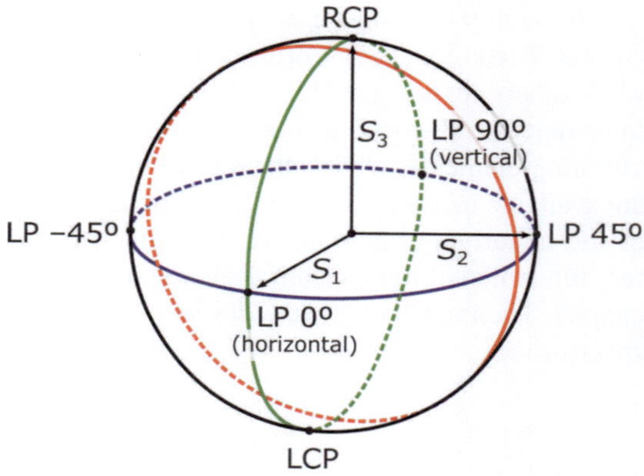

Fig. 6.8 All possible polarization states and Stokes vector combinations mapped using a Poincaré sphere. States representing fully polarized light (DOP of unity) are found on the surface of the sphere

the same azimuth and ellipticity) are located on the surface of the southern hemisphere. Partially polarized light, for example DOP $= 0.42$ would have the point at the tip of the vector somewhere inside the sphere. Based on the orientation of this vector, the exact polarization state of the portion that is polarized may be described.

6.4 Polarimetry

In its most general context, polarimetry is the discipline associated with measurement of the polarization state of light, and so is aptly named. Within chemistry, however, there is some confusion as the term polarimetry has a more specific meaning, as it is commonly used for the measurement of optical rotation, a direct result of circular birefringence. In that case, the detection of the azimuth of a particular linearly-polarized state is rotated upon passing through a chiral solution is a bona fide polarization measurement but, in the context of our previous discussion, we prefer to keep to a more general definition of polarimetry, that is the measurement of any arbitrary polarization state. The most flexible polarimeters are known as Stokes polarimeters because they measure all elements of the Stokes vector. Even if only the three normalized elements S_1/S_0, S_2/S_0, and S_3/S_0 are returned, we can still determine the DOP from this via Eq. 6.26, as well as the polarization state of the fraction of the light that is polarized. If the instrument is capable of measuring the unnormalized Stokes vector, meaning that S_0 is obtained, we have additional information on the light intensity. Many polarimeters operate at a fixed wavelength. In the case where multiple wavelengths or a range of wavelengths can be measured either in a wavelength-scanning mode or in broadband operation, one then obtains spectra of the Stokes vector elements and a spectrum of the DOP. This is the most powerful type of measurement, known as spectro-polarimetry, as it enables the complete characterization of a light source.

Although the hardware details of such instruments are beyond the scope of the current discussion, it is worth mentioning that some are designed for high speed, while others are designed for high accuracy when small changes in polarization need to be detected [9–16]. Furthermore, some polarimeters operate with the rotation of vibrating elements, while others have no moving parts [17, 18]. Such instruments are available in a very wide range of spectral regions. The basic operating principles of the underlying components that enable polarization analysis come down to a few fundamental light-material interactions that will be discussed in the following chapter. Then in Chap. 8 we will see an application of polarimetry to IR absorption spectroscopy.

6.5 Summary and Next Steps

In this chapter we have seen how light is an electromagnetic phenomenon that consists of a time- and space-varying electric and magnetic field that are perpendicular to each other, and both perpendicular to the direction of propagation. In the subsequent chapters, we will consider only the electric portion of the light, as its interaction with molecules and materials is most significant and forms the basis of our understanding of many types of optical spectroscopy. We have discussed how light may be polarized in a variety of ways, with elliptical polarization as the most general description, and linearly- and circularly-polarized states being special cases. When the light is completely polarized in some form (DOP of unity), it can be described by a 2-element Jones vector. Alternatively, a 4-element Stokes vector can be used. The Stokes representation has several advantages, the most significant of which is that it can be used to describe unpolarized (DOP of zero) and partially polarized light (with $0 < \mathrm{DOP} < 1$). The complete characterization of a light source, therefore, includes a description of its frequency (or spectrum), polarization state (characterized by the azimuth and ellipticity of the polarization ellipse, and possibly its handedness), and intensity. Instruments that can perform all of these measurements are known as spectro-polarimeters. In the following chapter, we will talk about how light interacts with materials and introduce the two fundamental properties of polarizability and susceptibility.

6.6 Exercises

1. The speed of light in a vacuum is given by

$$c = \frac{1}{\sqrt{\mu_0 \varepsilon_0}} \approx 2.998 \times 10^8 \ \mathrm{m} \cdot \mathrm{s}^{-1}$$

Considering that $\mu_0 = 1.257 \times 10^{-6}$ H·m^{-1}, and $\varepsilon_0 = 8.854 \times 10^{-12}$ F·m^{-1}, work out the units to confirm that the above speed is in m·s^{-1}.

---

2. a. Convert the complex number $4.7 - 2.3i$ into polar form. Do this by hand and then verify your result in Python.
   b. A complex number has a magnitude of 8.6 and a phase angle of 23°. Convert this into a representation on the complex Cartesian plane, $a + ib$. Do this by hand, and then verify your result in Python.

---

3. Illustrate that the normalized form of

$$\mathbf{E} = \begin{bmatrix} Ae^{i\pi/2} \\ A \end{bmatrix}$$

is equivalent to

$$\mathbf{E} = \frac{1}{\sqrt{2}} \begin{bmatrix} 1 \\ -i \end{bmatrix}.$$

The normalized form of $\mathbf{E}$ is generally obtained so that $|E_x|^2 + |E_y|^2 = 1$. Do this by hand and in Python.

---

4. The intensity of a light source in air can be evaluated using

$$I = \frac{c\varepsilon_0}{2} |\mathbf{E}|^2$$

where c is the speed of light, ε_0 is the vacuum permittivity, and $|\mathbf{E}|$ is the magnitude of the electric field vector. What is the intensity (including appropriate units) of vertically-polarized light with an electric field amplitude of 3.6 V·m^{-1}? As you work through this, think about the units for the quantities too.

---

5. a. Starting with a randomly-polarized light source, explain how you could create something that is circularly polarized. How many optical elements would you need? Describe them, and their placement in the path of the beam. If any of the components require a specific orientation, describe that too.
   b. It is possible that you have created a perfectly circularly-polarized beam, but that you are unsure if it is right or left circularly polarized. Explain how this scenario can arise. What are some different approaches to resolving this ambiguity?

---

6. Why can Jones vectors not be used to describe light that is only partially polarized?

---

7. Imagine that you want to create a polarimeter that can measure the complete Stokes vector, operating at a single frequency in the visible range, like that produced by a HeNe laser at $\lambda = 632.8$ nm.

   a. Describe the simplest design possible. For example, one that has the fewest parts. Describe how the 4 Stokes vector elements are obtained from this design.
   b. Describe the lowest cost design. What is needed for this setup? How are the Stokes vector elements obtained?
   c. Can you think of a design with no moving parts? How would that one work?

---

8. The Stokes vector

$$\mathbf{S} = \begin{bmatrix} S_0 \\ S_1 \\ S_2 \\ S_3 \end{bmatrix} = \begin{bmatrix} I_{\text{total}} \\ I_0 - I_{90} \\ I_{45} - I_{135} \\ I_{\text{rcp}} - I_{\text{lcp}} \end{bmatrix}$$

may be obtained from elements of the Jones vector using either one of the relations

$$\mathbf{S} = \begin{bmatrix} E_x E_x^* + E_y E_y^* \\ E_x E_x^* - E_y E_y^* \\ E_x E_y^* + E_y E_x^* \\ i(E_x E_y^* - E_y E_x^*) \end{bmatrix} = \begin{bmatrix} |E_x|^2 + |E_y|^2 \\ |E_x|^2 - |E_y|^2 \\ 2\mathbb{R}\,\mathrm{e}\{E_x E_y^*\} \\ -2\mathbb{I}\,\mathrm{m}\{E_x E_y^*\} \end{bmatrix}.$$

   a. Show that the two expressions for S_2 and S_3 are identical
   b. What is odd about the fact that $i = \sqrt{-1}$ appears in one of the expressions for S_3? Why is there no need to panic?

---

9. Some specific design of a Stokes polarimeter has determined that $S_1/S_0 = 0.83$, $S_2/S_0 = 0.45$, and $S_3/S_0 = -0.20$.

   a. Without getting into any detailed calculations, provide a rough description of the polarization state. To what extent can you approximately sketch the polarization ellipse?
   b. Now use the expressions in this Chapter to determine the degree of polarization, handedness (if significantly elliptical), azimuth (if significantly far from circular), and ellipticity (if significantly far from linear) of the polarization state. Compare with the results of your initial quick assessment.

References

1. J.C. Maxwell, Trans. Cambridge Phil. Soc. **10**, 155 (1855)

2. J.C. Maxwell, Philos. Mag. **161**, (1861)
3. D.S. Kliger, J.W. Lewis, C.E. Randal, *Polarized Light in Optics and Spectroscopy* (Academic Press Inc., 1990)
4. R.C. Jones, J. Opt. Soc. Am. **38**, 671 (1948)
5. H. Poincaré, *Théorie Mathématique de la Lumière* (Georges Carré, Paris, 1892)
6. M. Born, E. Wolf, *Principles of Optics*, 4th edn. (Pergamon Press Ltd., Oxford, 1970)
7. B. Kahr, Chirality **33**, 758 (2021)
8. B. Kahr, Chirality **12**, 714 (2000)
9. M.W. Wang, Y.F. Chao, K.C. Leou, F.H. Tsai, T.L. Lin, S.S. Chen, Y.W. Liu, Jap. J. Appl. Phys. **43**(2), 827 (2004)
10. W. Guan, G.A. Jones, Y.W. Liu, T.H. Shen, J. Appl. Phys. **103**(4), 043104 (2008)
11. S.A. Hall, M.A. Hoyle, J.S. Post, D.K. Hore, Anal. Chem. **85**, 7613 (2013)
12. P.S. Hauge, F.H. Dill, Opt. Commun. **14**, 431 (1975)
13. D.E. Aspnes, J. Opt. Soc. Am. A **21**, 403 (2004)
14. D.E. Aspnes, P.S. Hauge, J. Opt. Soc. Am. **66**, 949 (1976)
15. M. Shinki, M.T. Ivanov, J.S. Post, S. Vagle, J.T. Cullen, D.K. Hore, Appl. Opt. **51**, 4113 (2012)
16. D.H. Goldstein, Appl. Opt. **31**, 6676 (1992)
17. M. Ivanov, T. Eiju, in *Optoelectronics, Materials, and Devices for Communications*, vol. 4580, ed. by T.P. Lee, Q. Wang. International Society for Optics and Photonics (SPIE, 2001), vol. 4580, pp. 664–672
18. I. Naydenova, L. Nikolova, T. Todorov, F. Andruzzi, S. Hvilsted, J. Mod. Opt. **44**, 1643 (1997)

Chapter 7
Interaction Between Materials and Electric Fields

7.1 Overview

In the previous chapter we developed a sense of how the electric portion of an electromagnetic field behaves as it propagates through space. We now consider what happens when light travels through a material. The manner in which light is altered as it reflects from or is transmitted into a material is the basis of all spectroscopy, as we hope to use measurements of the properties of light and relate them to a better understanding of the properties of molecules and materials. We will start with an atomic/molecular picture that will lead us to a simple model of how individual molecules are perturbed by the electric field. We will then refine this using results from quantum mechanics before scaling up our solution to describe large collections of molecules that comprise materials. We then arrive at a definition of the complex refractive index. Using this fundamental optical constant, we will examine a formalism for arriving at the reflectance and transmittance in a variety of common situations. This will then be built into an algebraic system whereby Jones and Mueller matrices describe light-material interactions that modify the Jones or Stokes vectors of the incident light. Finally, these concepts are revisited for thin film materials and multi-layer systems.

7.2 Molecular Properties

We now build up a simple model for the polarizability, a fundamental property of atoms and molecules. We will start with a simple classical model, from which we will derive the oscillator strength that governs the interaction between the molecule and an external field. We will then replace this with a result from quantum mechanics.

D. K. Hore, *Molecular Orientation Analysis Using Polarized Vibrational Spectroscopy*,
https://doi.org/10.1007/978-3-032-04874-5_7

7.2.1 Polarizability

A dipole is a set of separated charges. The magnitude of the dipole can be characterized by the dipole moment

$$\boldsymbol{\mu} = q\mathbf{d} \tag{7.1}$$

where q is the charge in units of Coulombs, and $\mathbf{d}$ is the distance in metres. The dipole moment μ therefore has SI units of C·m in the MKS system. Chemists often like to use units of Debye (D) as 1 D $\approx 3.34 \times 10^{-21}$ C·m has a more convenient size for the relatively small dipole moments typical of molecules. Even when molecules don't have a permanent or static dipole (i.e. $|\boldsymbol{\mu}_0| = 0$), a dipole can be induced by an electric field. As shown in Fig. 7.1, although the electrons cannot migrate far from the nucleus in molecules and dielectric materials, they are still repelled by the negative end of the electric field, so a dipole is induced. The induced dipole moment can be written as

$$\boldsymbol{\mu}_{\text{induced}} = \alpha\mathbf{E} \tag{7.2}$$

where $\mathbf{E}$ is the field strength, and α is the polarizability. (Formally, μ (and therefore μ_{induced}) and $\mathbf{E}$ are vectors, but we will come back to this point when we discuss the dimensions of α shortly. For now we write α instead of $\boldsymbol{\alpha}$ as if the polarizability was a scalar property.) Overall, this relationship shows that the larger the field, the larger the induced dipole moment. Additionally, we can understand that some molecules are easier to polarize than others. A molecule with a lower polarizability resists distortion of its electron density in the presence of an applied field, (what chemists traditionally have called *hard*) and has a smaller value of α. More polarizable (a.k.a. *soft* molecules) will develop a larger dipole moment with the same applied field strength. So the total dipole moment becomes

$$\boldsymbol{\mu}_{\text{induced}} = \boldsymbol{\mu}_0 + \alpha\mathbf{E}. \tag{7.3}$$

Overall, this is called the *linear* regime, as the magnitude of the induced dipole moment is proportional to the magnitude of the applied field strength. When the field strength is large, nonlinear effects may be observed. In other words, the induced

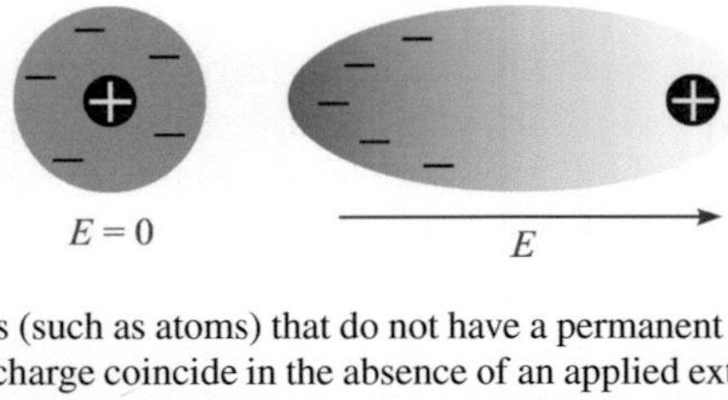

Fig. 7.1 For any species (such as atoms) that do not have a permanent dipole moment, the centres of positive and negative charge coincide in the absence of an applied external field. However, when those same species are put in an electric field, such as the electric part of an electromagnetic wave, the positive and negative charges are displaced, and a dipole is induced

dipole is no longer linearly proportional to the applied field. We can then write the total dipole moment as

$$\boldsymbol{\mu}_{\text{induced}} = \boldsymbol{\mu}_0 + \alpha\mathbf{E} + \alpha^{(2)}\mathbf{E}^2 + \alpha^{(3)}\mathbf{E}^3 + \cdots + \alpha^{(n)}\mathbf{E}^n. \tag{7.4}$$

As n increases, the magnitude of the nth order polarizability $\alpha^{(n)}$ decreases, so these effects are generally not observed unless $\mathbf{E}$ is very large. (We will later come back to discuss what it means to raise a vector like $\mathbf{E}$ to an exponent n.)

7.2.2 *Lorentz Model for Polarizability*

The Lorentz oscillator is a classical model that is widely used to provide the spectral lineshape that accounts for the frequency dependence of the optical constants in the absence of any inhomogeneous broadening. The basic components of the description are shown in Fig. 7.2, consisting of an object of mass m attached to a rigid wall by a spring with a force constant k_H. If we were to pull the spring and let go, it would oscillate about its equilibrium position $\mathbf{x} = 0$ with a restoring force given by Hooke's Law

$$\mathbf{F}_{\text{spring}} = -k_\text{H}\mathbf{x} = \omega_0^2 m\mathbf{x} \tag{7.5}$$

where we have also used the fact that the resonance frequency ω_0 is given by $\sqrt{k_\text{H}/m}$. Keep in mind that we have defined $\mathbf{x}$ as the *displacement* from the equilibrium position as this simplifies the math. If we further assume that the object has a charge q, then a light wave travelling along $\hat{z}$ with an electric field component $\mathbf{E}$ along $\hat{x}$ exerts a force on the charged mass that is given by

$$\mathbf{F}_{\text{driving}} = -q\mathbf{E}. \tag{7.6}$$

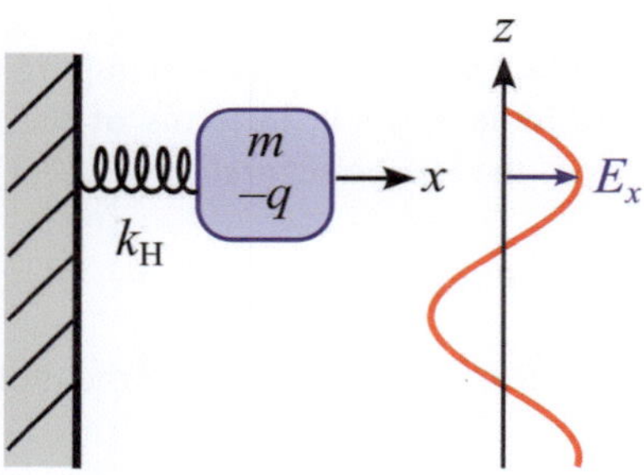

Fig. 7.2 The Lorentz model of atomic/molecular polarizability is a classical model based on a charged mass attached to a wall with a spring that can be driven to vibrate by the electric part of an electromagnetic field. Note that the mass is moving along the $\hat{\mathbf{x}}$ direction and the light wave is propagating along $\hat{\mathbf{z}}$ so the charge interacts with the x-component of the field

We now have a driven harmonic oscillator. Finally we include a damping or frictional force that resists the motion, so the amplitude of the oscillations will decay over time. This force is proportional to the velocity of the object, which in turn can be written as the derivative of the displacement with respect to time

$$\mathbf{F}_{\text{damping}} = -m\gamma \frac{\partial \mathbf{x}}{\partial t} \tag{7.7}$$

where γ is the damping constant.

Combining these contributions, we can arrive at the net force

$$\mathbf{F}_{\text{net}} = \mathbf{F}_{\text{spring}} + \mathbf{F}_{\text{driving}} + \mathbf{F}_{\text{damping}}. \tag{7.8}$$

Making use of Newton's second law, that the force is a product of the mass and acceleration, the equation of motion can be expressed as

$$m\frac{\partial^2 x}{\partial t^2} = -m\omega_0^2 x - qE - m\gamma\frac{\partial x}{\partial t}. \tag{7.9}$$

Since the electric field has a time dependence given by

$$\mathbf{E}(\hat{z}, t) = E_{\text{max}} e^{i(\mathbf{k}\cdot\hat{\mathbf{z}} - \omega t)}$$

we look for a solution to the displacement of the form

$$\mathbf{x}(t) = x_{\text{max}} e^{i(\mathbf{k}\cdot\hat{\mathbf{z}} - \omega t)}.$$

Solving this second-order differential equation results in

$$\mathbf{x}(t) = -\left(\frac{q\mathbf{E}(t)}{m}\right)\frac{1}{\omega_0^2 - \omega^2 - i\omega\gamma/m}. \tag{7.10}$$

For convenience, we can define a mass-weighted damping constant $\Gamma = \gamma/m$. As we can write the induced dipole moment in terms of the polarizability and field (Eq. 7.2), and also in terms of the charge separation (Eq. 7.1)

$$\boldsymbol{\mu}_{\text{induced}} = \alpha\mathbf{E} = -q\mathbf{x}$$

we can therefore formulate our results as a model of the molecular polarizability

$$\alpha = -\frac{q\mathbf{x}}{\mathbf{E}} = \left(\frac{q^2}{m}\right)\frac{1}{\omega_0^2 - \omega^2 - i\omega\Gamma}.$$

Note that if the frequencies ω and ω_0 and the linewidth Γ in the above expression are expressed in units of s^{-1} (Hz), charge q in C, and mass m in kg, then we end up with a polarizability in standard SI units of $C^2 \cdot s^2 \cdot kg^{-1}$ as expected. Taking a small step away from this classical picture that is tied to this specific mass m and charge q, we can generalize the expression by a single amplitude a that we will refer to as the *oscillator strength*, where currently $a = q^2/m$. Our expression is useful in a practical sense as it provides the frequency dependence of the polarizability

$$\alpha(\omega) = \frac{a}{\omega_0^2 - \omega^2 - i\omega\Gamma}. \tag{7.11}$$

Often, particularly when describing molecular vibrations in the mid-infrared, we wish to use wavenumbers in units of cm^{-1} for ω, ω_0 and Γ. In this case, a would have units of $C^2 \cdot s^2 \cdot kg^{-1} \cdot cm^{-2}$.

We can also see that, due to the presence of $i = \sqrt{-1}$ in the denominator, α is complex-valued. We examine the frequency dependence of α as suggested by Eq. 7.11. To arrive at an analytical expression that expresses this as $\alpha = \mathbb{R}\mathrm{e}\{\alpha\} + i\mathbb{I}\mathrm{m}\{\alpha\}$ we can rationalize the denominator. This is left as an end-of-chapter assignment. It is worth pointing out, though, as we are interested in numerical solutions to problems, it is seldom necessary to come up with explicit expressions for the real and imaginary components of a quantity to model in Python, since we can store `alpha` as a complex variable, and then use `alpha.real` and `alpha.imag` to access the components of interest. For demonstration, using values of $a = 1000\ C^2 \cdot s^2 \cdot kg^{-1} \cdot cm^{-2}$, $\omega_0 = 2875\ cm^{-1}$, and $\Gamma = 15\ cm^{-1}$, the plots of these quantities are shown in Fig. 7.3a for the real component and Fig. 7.3b for the imaginary component. The value where $\omega = \omega_0$ is shown with a dashed black vertical line. Additionally $\omega = \omega_0 \pm \Gamma/2$ are plotted with red lines. For historical reasons, the regions where $\mathbb{R}\mathrm{e}\{\alpha\}$ increases with increasing frequency are referred to as normal dispersion, and the small region in the vicinity of ω_0 where α decreases with increasing frequency is sometimes still referred to as anomalous dispersion. However, from considerations that were used to assemble the Lorentz model, you can see that there is nothing actually anomalous about this behaviour. The imaginary component has a simpler shape, with a characteristic peak at ω_0. It is also worth pointing out that $\mathbb{I}\mathrm{m}\{\alpha\}$ decays more quickly as $\omega > \omega_0$ or $\omega < 0$ compared to the more gradual decay of $\mathbb{R}\mathrm{e}\{\alpha\}$ away from resonance. In Fig. 7.3b, it can be observed that the half maximum of $\mathbb{I}\mathrm{m}\{\alpha\}$ (maximum and half maximum drawn with horizontal green lines) has a full width (full width at half maximum, or FWHM) of Γ.

For comparison, the magnitude $|\alpha|$ is plotted in Fig. 7.3c, along with the corresponding maximum and half maximum values (green). It is instructive to look at the phase of α, that we call ϕ_α from the definition $\alpha = |\alpha|e^{i\phi_\alpha}$. Recall that in Python this can be obtained from `phi = arctan2(alpha.imag, alpha.real)`, where we use the quadrant-preserving version of the arctangent function; this is plotted in Fig. 7.3d. Here we can see that when $\omega \ll \omega_0$, $\alpha(t)$ keeps up with $\mathbf{E}(t)$. In other words, the electrons perturbed by the field from the light wave are oscillating in phase with $\mathbf{E}$. However, as $\omega \to \omega_0$, there is a growing phase lag. The field emitted from the time-varying dipole moment is still at the same frequency as the driving

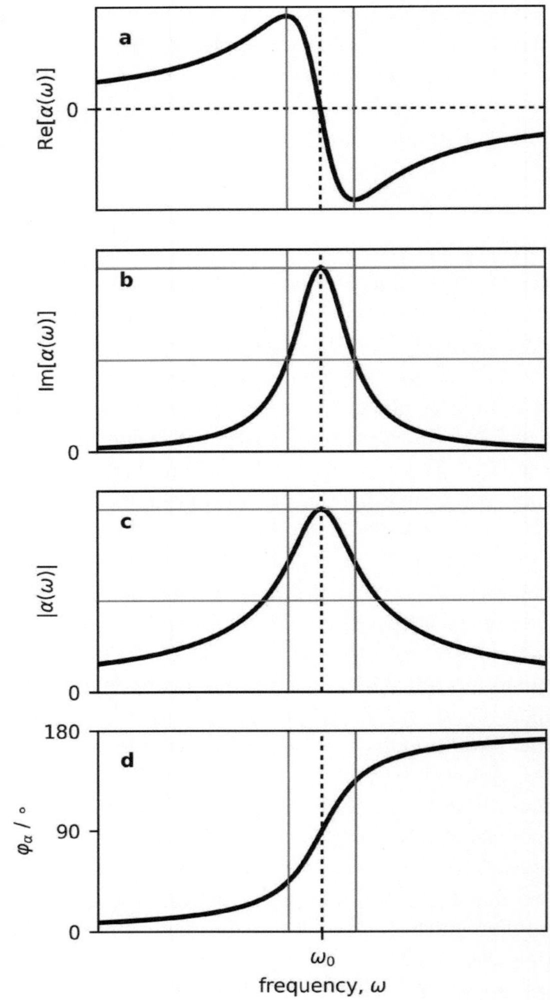

Fig. 7.3 Plots of the **a** real, **b** imaginary, **c** magnitude and **d** phase of the polarizability α as a function of frequency. In all plots, the value of ω_0 is indicated with the dashed black line, and values of $\omega = \omega_0 \pm \frac{1}{2}\Gamma$ are shown with red lines for reference. For both the imaginary component and the magnitude, the green lines highlight the respective maximum and half maximum values. This is so that the relationship between these values and Γ can be visualized

field, but is delayed in time. On resonance (when $\omega = \omega_0$), α lags behind $\mathbf{E}$ by exactly 90°. At higher frequencies, the phase shift keeps increasing until at $\omega \gg \omega_0$ ϕ_α plateaus at 180°. Altogether, one says that the phase of the polarizability changes by 180° when passing through a resonance. The magnitude of Γ affects the steepness of this transition.

Code used to generate Fig. 7.3

Note that, even though we have specified values of a, ω_0 and Γ above to facilitate re-creation of the plots, in the graphic I decided to remove the horizontal axis ticks except to indicate ω_0, This is because we wish to draw attention to the shape of these plots, which a function of the relative magnitudes of a and Γ only.

```
import numpy as np
from matplotlib import pyplot

# setup
pyplot.rc('font', size=8)
fig = pyplot.figure(figsize=(3.25, 6))
omega0 = 2875.
omega = np.linspace(omega0-50, omega0+50, 200)
Gamma = 15.
a = 1.0e3

# expression that gets used for everything
alpha = a/(omega0**2 - omega**2 - 1j*Gamma*omega)

# real part
ax1 = fig.add_subplot(411)
ax1.plot(omega, alpha.real, 'k-', lw=2)
ax1.axhline(0, lw=1, ls='--', dashes=[2,2], color='k')
ax1.set_ylabel(r'Re[$\alpha(\omega)$]')
ax1.label = 'a'

# imaginary part
ax2 = fig.add_subplot(412)
max = np.max(alpha.imag)
ax2.plot(omega, alpha.imag, 'k-', lw=2)
ax2.set_ylabel(r'Im[$\alpha(\omega)$]')
ax2.axhline(max, color='g', lw=0.75)
ax2.axhline(0.5*max, color='g', lw=0.75)
ax2.set_ylim(0, 1.1*max)
ax2.label = 'b'

# magnitude
ax3 = fig.add_subplot(413)
maxMag = np.max(abs(alpha))
ax3.plot(omega, np.abs(alpha), 'k-', lw=2)
ax3.set_ylabel(r'$|\alpha(\omega)|$')
ax3.axhline(maxMag, color='g', lw=0.75)
ax3.axhline(0.5*maxMag, color='g', lw=0.75)
ax3.set_ylim(0, 1.1*maxMag)
ax3.label = 'c'

# phase
ax4 = fig.add_subplot(414)
ax4.plot(omega, np.degrees(np.arctan2(alpha.imag, alpha.real)), 'k-', lw=2)
ax4.set_xlabel('frequency, $\omega$')
ax4.set_ylabel(r'$\varphi_\alpha$ / $\circ$')
ax4.label = 'd'

# pretty
for ax in [ax1, ax2, ax3, ax4]:
    ax.set_xticks([])
    ax.set_yticks([0])
    ax.axvline(omega0, lw=1, ls='--', dashes=[2,2], color='k')
```

```
        ax.axvline(omega0 - 0.5*Gamma, color='r', lw=0.75)
        ax.axvline(omega0 + 0.5*Gamma, color='r', lw=0.75)
        ax.set_xlim(np.min(omega), np.max(omega))
        ax.text(0.05, 0.82, ax.label, fontweight='bold', transform=ax.transAxes
            )

    ax4.set_xticks([omega0])
    ax4.set_xticklabels([r'$\omega_0$'])
    ax4.set_yticks([0, 90, 180])
    fig.set_tight_layout(True)
    pyplot.savefig('alphaShape.png', dpi=600)
```

Our primary interest will be in the vibrational polarizability. As we expect, molecules will have multiple resonances, with each of the $3N - 6$ normal modes represented by a Lorentz oscillator. The frequency-dependent polarizability can then be expressed as

$$\begin{aligned}\alpha(\omega) &= \sum_j \alpha_j \\ &= \sum_j \frac{a_j}{\omega_j^2 - \omega^2 - i\omega\Gamma_j}\end{aligned} \tag{7.12}$$

where each mode has an amplitude a_j, width Γ_j, and resonance frequency ω_j. (In the case of a single oscillator, we referred to that frequency as ω_0 in Eq. 7.11. However, now with j resonances, it is simpler to write ω_j than $\omega_{0,j}$). From our earlier description of a single Lorentz oscillator, it was perhaps difficult to see the advantage of $\mathbb{I}\text{m}\{\alpha\}$ over $\mathbb{R}\text{e}\{\alpha\}$ as they both had relatively simple line shapes. In a congested spectral region, however, when there are overlapping modes with different values of A_j and Γ_j, the difference becomes more stark. An example with only four resonances is shown in Fig. 7.4. Here we can see that it becomes difficult to identify the resonance frequencies and even the number of modes in $\mathbb{R}\text{e}\{\alpha\}(\omega)$, but the distinction remains clear in $\mathbb{I}\text{m}\{\alpha\}(\omega)$, even when some modes are partially buried and appear as shoulders on other peaks.

The code used to generate Fig. 7.4 is nearly identical to that for the case of a single mode shown above, with the exception that we have first defined a set of 4 resonance frequency ω_0 values (that we subsequently label as ω_n), 4 widths Γ_n, and 4 amplitudes A_n using

```
omega0 = [2852., 2875., 2930., 2945.]
Gamma = [15., 10., 15., 15.]
a = [0.8e3, 1.0e3, 1.2e3, 1.0e3]
```

and we then assemble the polarizability using Eq. 7.12 by means of

```
alpha = 0.
for i in range(len(omega0)):
    alpha += a[i]/(omega0[i]**2 - omega**2 - 1j*Gamma[i]*
        omega)
```

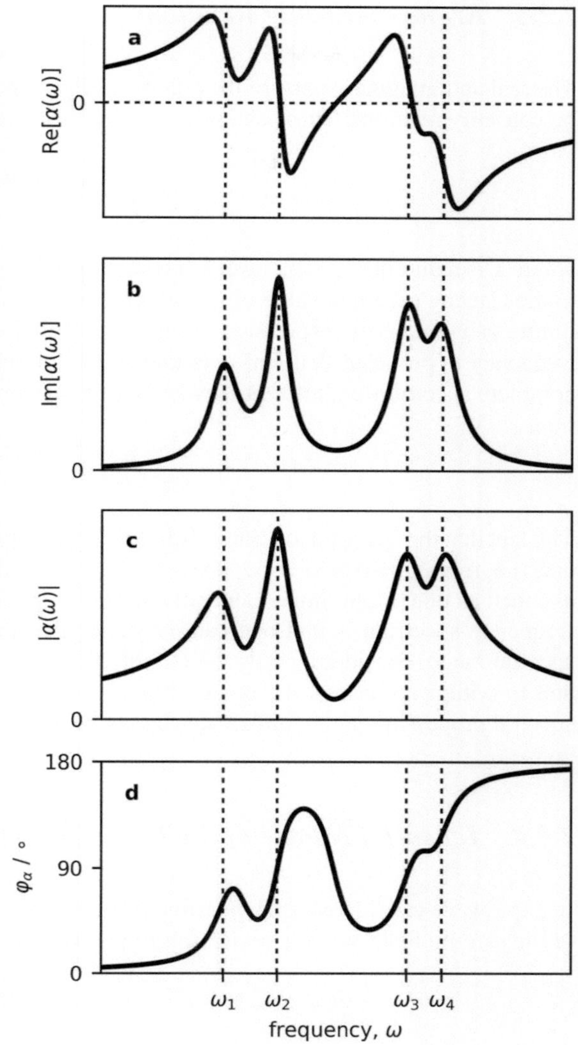

Fig. 7.4 When there are several vibrational modes present, each one may be considered as it if it was its own Lorentz oscillator, with its characteristic a, ω, and Γ values. In this case, the **a** real component of α has a rather complicated lineshape, where overlapping modes are difficult to distinguish. However, the **b** imaginary spectrum of α enables the modes to be resolved more easily, with overlapping features appearing as characteristic shoulders. For completeness, and to enable comparison with a single mode spectrum shown in Fig. 7.3, the **c** magnitude and **d** phase spectra are also shown

where the += operator provides a convenient way of incrementing the `alpha` array after it has been initially defined using `alpha = 0`. Note that there is no need to set the initial value of `alpha` to an array; we could have done that with something like `alpha = np.zeros(len(omega))` but this is unnecessary as any NumPy array can be added to a float, just as a float can be added to any array resulting in another array of the same length.

7.2.3 Kramers–Kronig Relations

The real and imaginary parts of the polarizability are not independent, but are linked by causality relations. These can be expressed as

$$\mathbb{R}\mathrm{e}\{\alpha\}(\omega) = \frac{2}{\pi}\mathscr{P}\int_0^\infty \frac{x \cdot \mathbb{I}\mathrm{m}\{\alpha\}(x)}{x^2-\omega^2}\mathrm{d}x \tag{7.13}$$

where x is a dummy variable over which the integration is performed, and $\mathscr{P}$ refers to the Cauchy principal value of the integral, essentially avoiding the small discontinuity at $x = \omega$. This expression implies that $\mathbb{R}\mathrm{e}\{\alpha\}$ can be obtained at any single frequency ω provided that $\mathbb{I}\mathrm{m}\{\alpha\}$ is known for all values of ω, i.e. we require the complete spectrum of $\mathbb{I}\mathrm{m}\{\alpha\}$. Likewise, a relationship exists in the opposite direction,

$$\mathbb{I}\mathrm{m}\{\alpha\}(\omega) = -\frac{2\omega}{\pi}\mathscr{P}\int_0^\infty \frac{\mathbb{R}\mathrm{e}\{\alpha\}(x)}{x^2-\omega^2}\mathrm{d}x. \tag{7.14}$$

The fact that the spectrum of either $\mathbb{R}\mathrm{e}\{\alpha\}$ or $\mathbb{I}\mathrm{m}\{\alpha\}$ cannot be known for *all* frequencies (i.e. from $\omega = 0$ to $\omega = \infty$) does not limit the applicability of these expressions as much as one might think. Generally, the consequence of working with a limited frequency spectrum is that one can determine the quantity of interest within that spectral range, considering only the contributions of modes within that same range, and to within a constant offset. Additional details concerning practical means of spectral processing in the mid-IR range may be found in the literature [1–3].

7.2.4 Tensorial Nature of the Polarizability

In Eq. 7.4 we considered an expansion of the dipole moment in successive powers of the electric field; we now revisit this with a slightly different notation, writing

$$\boldsymbol{\mu} = \boldsymbol{\mu}_0 + \boldsymbol{\alpha}\mathbf{E} + \boldsymbol{\alpha}^{(2)} : \mathbf{EE} + \boldsymbol{\alpha}^{(3)} \vdots \mathbf{EEE} + \cdots \tag{7.15}$$

and consider the *shape* of the relevant quantities in the above expression, as we know that $\boldsymbol{\mu}$ and $\mathbf{E}$ are vectors. We also have to reconcile what is meant by things like **EE** and **EEE** in the context of vector "multiplication". The double and triple vertical dots in Eq. 7.15 are a clue that something needs special handling. In Cartesian coordinates, these vectors have 3 elements.

$$\boldsymbol{\mu} = \begin{bmatrix} \mu_a \\ \mu_b \\ \mu_c \end{bmatrix}, \quad \mathbf{E} = \begin{bmatrix} E_x \\ E_y \\ E_z \end{bmatrix}.$$

Throughout our discussion from this point forward, we will use a, b, and c as the molecular frame and x, y, and z as the laboratory frame Cartesian coordinates and corresponding unit vectors. We are inching our way towards a discussion where the relationship between the molecular and laboratory frame will be the primary topic of interest. This is after all the goal of measuring spectra with polarized light—to determine the orientation of the molecules with respect to the sample/laboratory coordinates. Just before we get there, the intention of the current discussion is to point out the relationship between the polarizability terms, dipole moment, and electric field. To facilitate this discussion, we will momentarily consider $\boldsymbol{\mu}$ and $\mathbf{E}$ in the same coordinate system. This can done by thinking about $\boldsymbol{\mu}$ and $\mathbf{E}$ both in the laboratory frame (as would be the case in a single-molecule spectroscopy) or $\boldsymbol{\mu}$ and $\mathbf{E}$ both in the molecular frame (although we don't generate electric fields from our light sources in that way, as the polarization (of the light!) is always defined with respect to the laboratory frame, typically that of the table on which the optics are mounted). Nevertheless, we will proceed with the second option and momentarily consider the electric field vector as

$$\mathbf{E} = \begin{bmatrix} E_a \\ E_b \\ E_c \end{bmatrix}.$$

If α was simply a number (albeit complex-valued, as we understand from our study of the Lorentz oscillator), then $\boldsymbol{\mu}_{\text{ind}} = \alpha\mathbf{E}$ would amount to

$$\begin{bmatrix} \mu_a \\ \mu_b \\ \mu_c \end{bmatrix} = \alpha \begin{bmatrix} E_a \\ E_b \\ E_c \end{bmatrix} \tag{7.16}$$

which is simply a short-form notation for

$$\begin{aligned} \mu_a &= \alpha E_a \\ \mu_b &= \alpha E_b \\ \mu_c &= \alpha E_c \end{aligned} \tag{7.17}$$

where all of the 7 quantities appearing in Eq. 7.17 are scalars. The issue is that, while we could think of some species for which this could be a reasonable approximation, molecules are typically more polarizable in one direction than another. We say that the polarizability is *anisotropic* to account for the fact that, based on the details of the chemical bonds between the atoms, it is easier to distort the electron density in certain directions. For example, species that are conjugated are very polarizable along the path of conjugation that often involves double bonds and aromatic rings, and less polarizable perpendicular to that direction in the molecule. This is well described by considering that $\boldsymbol{\alpha}$ is a matrix, i.e. a tensor of rank 2

$$\boldsymbol{\alpha} = \begin{bmatrix} \alpha_{aa} & \alpha_{ab} & \alpha_{ac} \\ \alpha_{ba} & \alpha_{bb} & \alpha_{bc} \\ \alpha_{ca} & \alpha_{cb} & \alpha_{cc} \end{bmatrix} \tag{7.18}$$

which makes sense if we multiply out $\boldsymbol{\mu}_{\text{ind}} = \boldsymbol{\alpha}\mathbf{E}$ to get

$$\begin{aligned} \mu_a &= \alpha_{aa}E_a + \alpha_{ab}E_b + \alpha_{ac}E_c \\ \mu_a &= \alpha_{ba}E_a + \alpha_{bb}E_b + \alpha_{bc}E_c \\ \mu_a &= \alpha_{ca}E_a + \alpha_{cb}E_b + \alpha_{cc}E_c. \end{aligned} \tag{7.19}$$

We can now also understand the definition of a single element of the $\boldsymbol{\alpha}$ tensor. For example, α_{cb} provides the contribution to the induced dipole moment along c from the component of the electric field along b. It is important to realize that this is only part of the c component of $\boldsymbol{\mu}_{\text{ind}}$, as the other elements of $\boldsymbol{\alpha}$ need to be considered if they are non-zero.

In the case of the first-order polarizability $\boldsymbol{\alpha}^{(1)}$ or what we've been more casually calling $\boldsymbol{\alpha}$ it is clear that multiplying a vector by a matrix on the left results in another vector, so the general rules of basic linear algebra apply. But what about the higher order terms that involve $\boldsymbol{\alpha}^{(n)}$? Taking $\boldsymbol{\alpha}^{(2)}$ as an example, this tensor of rank 3 has three dimensions, and we can picture it as a 27-element Rubik's cube with dimensions of $3 \times 3 \times 3$. We can provide a similar interpretation for any given element. For example, $\alpha^{(2)}_{acb}$ provides the contribution to the induced dipole moment along a from an electric field polarized along c and another field polarized along b. We say *contribution* since this is not the entirety of the a-component of $\boldsymbol{\mu}_{\text{ind}}$, as is clear from the matrix multiplication written out explicitly above.

A word about notation: sometimes vectors are written with arrows on top like $\vec{E}$ and sometimes they are written in bold like $\mathbf{E}$. And for tensors of rank 2, one convention is to write $\overset{\leftrightarrow}{\alpha}$ or $\overset{\rightrightarrows}{\alpha}$ and another is to simply use boldface $\boldsymbol{\alpha}$ and let the dimensions (rank of the tensor) be clear from the context. We will use boldface for all vectors and tensors. Although this creates some ambiguity, it avoids having to write things like $\overset{\rightrightarrows}{\alpha}^{(2)}$, plus the rank is clear when we are additionally labelling with superscripts. Keep in mind that individual elements of a tensor are always scalars (no bold), and are recognizable by their number of indices, like $\alpha^{(2)}_{acb}$ (in addition to having our superscripted notation).

A summary of these quantities and their dimensions appears in Table 7.1, including the fact that scalars such as ω and Γ can be labelled as tensors of rank 0, and vectors such as $\boldsymbol{\mu}$ and $\mathbf{E}$ are tensors of rank 1. This generalizes the entire notation into the realm of tensor algebra. Although we will occasionally return to these concepts, practically speaking, we tend to work with expressions that involve specific elements of the tensors, so most equations are expressed with scalar quantities. Still, it is helpful to keep the dimensions in mind, as that sets up the correct number of subscripts and subsequent operations that we will perform in the upcoming chapters. Note that in Python, scalars are floating point numbers (or floats for short), including complex-valued data types. Vectors are represented by 1D NumPy arrays.

Table 7.1 Tensor notation and properties of common tensors in Cartesian coordinates

Tensor rank	Common name	Cartesian shape	Cartesian elements	Example
0	Scalar, number	No shape (1 × 1)	1	ω
1	Vector	3 × 1	3	$\boldsymbol{\mu}$
2	Matrix	3 × 3	9	$\boldsymbol{\alpha}$
3	(Cube)	3 × 3 × 3	27	$\boldsymbol{\alpha}^{(2)}$
4	—	3 × 3 × 3 × 3	81	$\boldsymbol{\alpha}^{(3)}$
⋮	—	⋮	⋮	⋮
n	—	n-dimensional	3^n	$\boldsymbol{\alpha}^{(n)}$

Tensors of rank 2 can be represented either by the special NumPy `matrix` data type (restricted to exactly two dimensions), or by a 2D `array` data type. Higher dimensional tensors are simply multi-dimensional NumPy arrays. This is one of the reasons why there is a move towards deprecating the `matrix` data type in favour of `array`; arrays in NumPy are scalable and mutable, and array-based operations are more general. On the other hand, SymPy is fully capable but less flexible when it comes to tensor algebra, requiring the conversion between data types if matrix–vector multiplication is also desired.

It is important not to confuse the scalar nature of the frequency ω with the notion of what happens when using a NumPy array data type to store a range of ω values. When plotting spectral properties, we have typically created a NumPy array using something like `omega = np.linspace(2800, 3000, 100)`. Even though this is an array, and it can be said that array-based operations *vectorize* code, ω is very much a scalar quantity! It's just that we wish to plot $\alpha(\omega)$, and creating an array of scalars is a convenient way to evaluate $\boldsymbol{\alpha}$ at many ω values.

Returning to our expression for the frequency dependence of the polarizability, we now express the Lorentzian amplitude as an element of a rank 2 tensor via the notation

$$\alpha_{lm}(\omega) = \sum_q \frac{a_{q,lm}}{\omega_q^2 - \omega^2 - i\omega\Gamma_q} \tag{7.20}$$

where the index q runs over all of the vibrational modes. This expression highlights that the nonresonant contribution to $\boldsymbol{\alpha}$ and each of the q resonant amplitudes can take on 9 values. In the simplest case that we consider throughout this study, each of the 9 elements is described with the same resonance frequency ω_q and width Γ_q.

We now take the final departure from the Lorentz model, keeping only the frequency dependence expressed in the denominator of the resonance contributions. To understand the true origins of the amplitudes $A_{q,lm}$, we need to turn to the insight afforded by quantum mechanics.

7.2.5 *Quantum Mechanical Description of Polarizability*

In the quantum mechanical description of molecules and materials, all information is contained in the wavefunction ψ. Every physical property m associated with an experimental observable has an associated operator $\hat{M}$. If m is an eigenvalue of $\hat{M}$, then we can obtain its value from

$$\begin{aligned} m &= \int_{-\infty}^{\infty}\int_{-\infty}^{\infty}\int_{-\infty}^{\infty} \psi^*(x, y, z)\hat{M}\psi(x, y, z)\, \mathrm{d}x\, \mathrm{d}y\, \mathrm{d}z \\ &= \int_{\Omega} \psi^* M \psi\, \mathrm{d}\Omega \\ &= \langle\psi|M|\psi\rangle \end{aligned} \tag{7.21}$$

where the second line introduces the shorthand notation in which Ω is a placeholder for the set of variables (either Cartesian or spherical polar) that describe all of the three-dimensional space. When using the Ω notation, it is also common to write multiple integrals as a single integral, where most or all of the variables are hidden in this representation. The last line introduces Dirac's notation, where the wavefunction is written as a *ket* $|\psi\rangle$, and its transposed complex conjugate (or Hermitian transpose) $\psi^{\mathrm{H}} = (\psi^{\mathrm{T}})^*$ often simply written ψ^* is designated by the *bra* $\langle\psi|$. When bras and kets are combined to form a *braket*, integration over all space is implied, so

$$\langle\psi|\psi\rangle \equiv \int_{\Omega} \psi^*\psi\, \mathrm{d}\Omega. \tag{7.22}$$

When comparing the above expression with Eq. 7.21, the difference is that the ket in Eq. 7.21 is first modified by the operator $\hat{M}$, resulting in a different wavefunction.

As an example, if we want to know the a-component of the ground state ψ_0 dipole moment, we can arrive at this through

$$\begin{aligned} \mu_a &= \int_{-\infty}^{\infty} \psi_0^* \hat{\mu}_a \psi_0\, \mathrm{d}a \\ &= \langle\psi_0|\mu_a|\psi_0\rangle. \end{aligned} \tag{7.23}$$

Likewise, if we want the b-component of the dipole moment of the first excited state ψ_1 this can be obtained from

$$\begin{aligned} \mu_b &= \int_{-\infty}^{\infty} \psi_1^* \hat{\mu}_b \psi_1\, \mathrm{d}b \\ &= \langle\psi_1|\mu_b|\psi_1\rangle. \end{aligned} \tag{7.24}$$

There is a special quantity of interest that doesn't have a direct classical analog, that is the transition dipole moment (TDM) which connects two different states. For the $0 \rightarrow 1$ (some write $1 \leftarrow 0$) transition, the a-component of the TDM is given by

$$\begin{aligned}\mu_a^{01} &= \langle\psi_1|\mu_a|\psi_0\rangle \\ &= \langle 1|\mu_a|0\rangle\end{aligned} \tag{7.25}$$

where we have introduced an even shorter notation, by writing $\langle 1|$ for $\langle\psi_1|$, and $|0\rangle$ for $|\psi_0\rangle$. We will later see how both the magnitude and the direction of the TDM are important.

Quoting the results obtained from perturbation theory [4–7], one arrives at the Kramers–Heisenberg–Dirac (KHD) formulation of the polarizability. For a simple two-level system with only a $0 \rightarrow 1$ transition,

$$\alpha_{lm}(\omega) = \frac{1}{\hbar}\langle 0|\mu_l|1\rangle \otimes \langle 1|\mu_m|0\rangle \left[\frac{1}{\omega_0 - \omega - i\Gamma/2} + \frac{1}{\omega_0 + \omega + i\Gamma/2}\right] \tag{7.26}$$

where $\hbar = h/2\pi$ is the reduced Planck constant, and the $\otimes$ symbol denotes the outer (tensor) product of two vectors.

We have now seen three ways that two vectors (tensors of rank 1) can be multiplied. The first is the dot product or inner product, where $\mathbf{u} \cdot \mathbf{v}$ produces a scalar (tensor of rank 0), with a NumPy syntax of `dot(u, v)`. The second is the vector or cross product $\mathbf{u} \times \mathbf{v}$ that produces a vector (tensor of rank 1) with the same number of elements as $\mathbf{u}$ and $\mathbf{v}$, using `cross(u, v)`. And now there is the outer or tensor product $\mathbf{u} \otimes \mathbf{v}$ that produces a tensor of rank 2, achieved with `outer(u, v)`. Consider the following examples

```
import numpy as np
a = np.array([1, 2, 3]
b = np.array([4, 5, 6])

np.dot(a, b)
>>> 56

np.cross(a, b)
>>> array([-3,  6, -3])

np.outer(a, b)
>>> array([[ 4,  5,  6],
           [ 8, 10, 12],
           [12, 15, 18]])
```

and note that we have opted to use straightforward NumPy arrays with the appearance of row vectors rather than column vectors. This is the preferred style whenever possible, for simplicity in the code and output. In this case, however, `a.T` simply produces `a` again, so one must keep the dimensions in mind when writing NumPy code to perform linear algebra operations.

Returning to the polarizability, the first term in Eq. 7.26 describes a resonance interaction, as it causes α_{lm} to become large when $\omega \rightarrow \omega_0$. The second term is an anti-resonance contribution, and is generally much smaller than the second term for frequencies in the neighbourhood of ω_0. We can clean this expression up a bit by writing the two frequency-dependent terms with a single denominator using

$$\frac{1}{a} + \frac{1}{b} = \frac{a+b}{ab}$$

and, assuming that $\Gamma \ll \omega_0$, thereby dropping terms in Γ^2. This results in

$$\alpha_{lm}(\omega) = \left(\frac{2\omega_0}{\hbar}\right) \frac{\langle 0|\mu_l|1\rangle \otimes \langle 1|\mu_m|0\rangle}{\omega_0^2 - \omega^2 - i\omega\Gamma}. \tag{7.27}$$

Comparing this result with the simple classical model derived as the Lorentz oscillator (Eq. 7.20), we see that the oscillator strength in the expression for the polarizability is given by the product of two transition dipole moments. In SI units, if each TDM has units of C·m, one can verify that we again end up with a polarizability in units of $C^2 \cdot s^2 \cdot kg^{-1}$.

7.3 Materials

We now move from the microscopic scale of molecules to the macroscopic dimensions of materials. We can make this jump relatively easily, as much of the nomenclature and background we have introduced can be scaled up in a straightforward manner, with only small differences in notation.

7.3.1 Susceptibility

In a material, it is not practical to consider all of the individual dipole moments, so we collectively describe all of the interactions discussed above in terms of the dipole moment per unit volume V, or *polarization*, $\mathbf{P}$.

$$\mathbf{P} = \frac{1}{V} \sum_{\text{molecules}} \boldsymbol{\mu}. \tag{7.28}$$

This is not to be confused with the orientation of the electric field vector described in the previous chapter. Unfortunately both of these quantities have the same name, polarization, in a subject where both are frequently discussed together. The summation in Eq. 7.28 is a vector sum, so we have to take the directions of the dipole moments into account in addition to their magnitudes. For most materials, the sum of the permanent dipole moments is zero, so there is no static polarization outside the realm of ferroelectrics. However, the net result can be non-zero for the induced contributions to the polarization.

$$\mathbf{P} = \varepsilon_0 \left(\boldsymbol{\chi}\mathbf{E} + \boldsymbol{\chi}^{(2)}\mathbf{E}^2 + \boldsymbol{\chi}^{(3)}\mathbf{E}^3 + \cdots + \boldsymbol{\chi}^{(n)}\mathbf{E}^n \right). \tag{7.29}$$

Here ε_0 is the familiar vacuum permittivity, and $\boldsymbol{\chi}^{(n)}$ is now the nth order susceptibility that incorporates the volume number density into its definition. Just as the polarizability $\boldsymbol{\alpha}$ described how much the electron density of an atom or molecule can be perturbed by an external field, the susceptibility $\boldsymbol{\chi}$ tells us how much the electron density of the material can be distorted. Both words are good descriptions of the same thing; polarizability historically is the microscopic quantity, and susceptibility the macroscopic quantity. We will soon see how the linear susceptibility $\boldsymbol{\chi}$, or $\boldsymbol{\chi}^{(1)}$ to be pedantic, is responsible for the absorption of light (Chap. 8) the second-order susceptibility $\boldsymbol{\chi}^{(2)}$ accounts for important nonlinear optical phenomena such as frequency doubling (Chap. 10), and the third-order susceptibility $\boldsymbol{\chi}^{(3)}$ is responsible for Raman scattering (Chap. 9). As we have discussed the tensoral properties of $\boldsymbol{\alpha}$, the shapes of the various $\boldsymbol{\chi}^{(n)}$ tensors in Eq. 7.29 may already be understood: $\boldsymbol{\chi}$ is rank 2 tensor (better known as a 3×3 matrix), $\boldsymbol{\chi}^{(2)}$ is a $3 \times 3 \times 3$ cube, and so on.

We introduced the Lorentz model as a crude but useful way to picture the polarizability of molecules. We could have developed the same picture for a material, where the charged mass on a spring is actually a collection of masses. One would then end up with an expression for the susceptibility of the material as

$$\begin{aligned} \boldsymbol{\chi}(\omega) &= \boldsymbol{\chi}_{\mathrm{NR}} + \sum_j \boldsymbol{\chi}_j \\ &= \boldsymbol{\chi}_{\mathrm{NR}} + \sum_j \frac{\mathbf{A}_j}{\omega_j^2 - \omega^2 - i\omega\Gamma} \end{aligned} \tag{7.30}$$

that is written in direct analogy with its microscopic analogue Eq. 7.12. Now $\mathbf{A}_j$ are the oscillator strengths, but defined for the collection of molecules. This time we add an offset $\boldsymbol{\chi}_{\mathrm{NR}}$ that includes contributions to the susceptibility that are not due to vibrational resonances. Apart from $\boldsymbol{\chi}_{\mathrm{NR}}$, all of the plots we have made of $\boldsymbol{\alpha}(\omega)$ (examining the real and imaginary parts, magnitude and phase) and all of our discussion about the phase around resonance, apply in the same way to $\boldsymbol{\chi}(\omega)$.

Comparing Eqs. 7.12 and 7.30, one arrives at

$$\boldsymbol{\chi}_j = \frac{N_{\mathrm{V}}}{\varepsilon_0} \langle \boldsymbol{\alpha}_j \rangle \tag{7.31}$$

where N_{V} is the molecular number density, the number of molecules per unit volume. The set of angled brackets $\langle \cdots \rangle$ imply that we are talking about an ensemble average, that is the average of the collection of the molecules that takes the orientation of each molecule into account. If we consider how we used to write this expression as a sum over molecules divided by the volume, this makes sense as we now don't have each molecule's $\boldsymbol{\alpha}$ value appearing individually in the expression. The sum over molecules divided by the volume V has been replaced with the number density, and so we must correspondingly replace the individual $\boldsymbol{\alpha}$ values with the average $\boldsymbol{\alpha}$ value. Furthermore, if we wish to express both $\boldsymbol{\alpha}$ and $\boldsymbol{\chi}$ as a collection of Lorentz oscillators, then a similar relationship can be established for each vibrational mode amplitude, i.e. the oscillator strengths, to write

$$\mathbf{A}_j = \frac{N_\mathrm{V}}{\varepsilon_0}\langle \boldsymbol{a}_j \rangle. \tag{7.32}$$

We will return to this concept several times when we describe the analysis of molecular orientation by IR absorption, Raman scattering, and nonlinear optical spectroscopy in the subsequent chapters.

7.3.2 Dielectric Constant and Refractive Index

If we briefly return to the wave equation and now consider light propagation *inside* a dielectric material, we arrive at

$$\frac{\partial^2 \mathbf{E}}{\partial z^2} = \mu \boldsymbol{\varepsilon} \frac{\partial^2 \mathbf{E}}{\partial t^2} \tag{7.33}$$

where μ is the magnetic permeability of the material (don't confuse this with the dipole moment!), and $\boldsymbol{\varepsilon}$ is the electric permittivity of the material. So, inside the material, the wave travels with speed

$$v = \frac{1}{\sqrt{\mu \boldsymbol{\varepsilon}}} \tag{7.34}$$

and this is slower than the speed in vacuum by a factor equal to

$$\frac{c}{v} = \sqrt{\frac{\mu}{\mu_0}\frac{\boldsymbol{\varepsilon}}{\varepsilon_0}}. \tag{7.35}$$

For most materials, interaction with magnetic fields is weak, and therefore $\mu \approx \mu_0$. This results in

$$\frac{c}{v} \approx \sqrt{\frac{\boldsymbol{\varepsilon}}{\varepsilon_0}} = \sqrt{\boldsymbol{\varepsilon}_\mathrm{R}} = N \tag{7.36}$$

where $\boldsymbol{\varepsilon}_\mathrm{R}$ is the relative permittivity or dielectric constant, and we define N as the refractive index. We will later see that $\boldsymbol{\varepsilon}_\mathrm{R}$ and therefore N are complex-valued. Although some texts use n or $\tilde{n}$ as the symbol for the complex refractive index, we will use n for the real part of the refractive index, as we'll discuss later.

The electric displacement **D** is the charge per unit area that would be displaced if a layer of charges are put in an electric field of strength **E**. In a vacuum

$$\mathbf{D} = \varepsilon_0 \mathbf{E} \tag{7.37}$$

and we see that the displacement and the field are related by the vacuum permittivity ε_0. In a material

$$\mathbf{D} = \boldsymbol{\varepsilon}\mathbf{E} = \varepsilon_0 \mathbf{E} + \mathbf{P} \tag{7.38}$$

where $\mathbf{P}$ is the polarization (the dipole moment per unit volume). If we substitute $\mathbf{P} = \varepsilon_0 \boldsymbol{\chi} \mathbf{E}$ into this expression we arrive at

$$\begin{aligned} \boldsymbol{\varepsilon}\mathbf{E} &= \varepsilon_0 \mathbf{E} + \varepsilon_0 \boldsymbol{\chi} \mathbf{E} \\ \boldsymbol{\varepsilon}\mathbf{E} &= \varepsilon_0 (1 + \boldsymbol{\chi}) \mathbf{E} \\ \frac{\boldsymbol{\varepsilon}}{\varepsilon_0} &\equiv \boldsymbol{\varepsilon}_{\mathrm{R}} = N^2 = 1 + \boldsymbol{\chi}. \end{aligned} \tag{7.39}$$

As seen in Eq. 7.36 the ratio of the permittivity of a material to that of a vacuum is called the relative permittivity $\boldsymbol{\varepsilon}_{\mathrm{R}}$ or the dielectric constant. A word of caution: many books and papers simply write $\boldsymbol{\varepsilon}$ for the dielectric constant, and then one has to be careful to interpret the meaning from the context. Also note that $\boldsymbol{\varepsilon}_{\mathrm{R}}$, N, and $\boldsymbol{\chi}$ are all dimensionless quantities, and all encode the same information!

As we will see below, these values are also complex-valued, in general. So care must be taken not to confuse $\boldsymbol{\varepsilon} = N^2$ (correct) with $|N|^2$ (incorrect). The complex-valued dielectric constant is actually the *square* (not the magnitude squared!) of the complex-valued refractive index. If we define a complex number as $z = a + ib$, then $z^2 = a^2 - b^2 + 2iab$, while $|z|^2 = a^2 + b^2$. In Python, z^2 is given by `z**2` or simply `z*z`, regardless of whether z is real or complex-valued. But recall that $|z|^2$ is given by `np.abs(z)**2`.

The above expressions also provide the relationship between the real and imaginary components of the other optical constants. For example, since the dielectric constant can be expressed as $\boldsymbol{\varepsilon}_{\mathrm{R}} = 1 + \boldsymbol{\chi}$, this implies that $\mathbb{R}\mathrm{e}\{\boldsymbol{\varepsilon}_{\mathrm{R}}\} = 1 + \mathbb{R}\mathrm{e}\{\boldsymbol{\chi}\}$ and that $\mathbb{I}\mathrm{m}\{\boldsymbol{\varepsilon}_{\mathrm{R}}\} = \mathbb{I}\mathrm{m}\{\boldsymbol{\chi}\}$.

Throughout the rest of this book, we will refer to $\mathbb{R}\mathrm{e}\{N\}$ as n, or use n for the refractive index to indicate that it has no imaginary component for a particular material at the wavelength of interest. Although this is fairly standard, there is less consensus on what to call $\mathbb{I}\mathrm{m}\{N\}$. I prefer to call it κ. Most literature uses the symbol k instead, and although that is the norm, I can't bring myself to do it. That is just a bit too crazy for me, since we will constantly need to multiply both of these quantities by the wavevector. I therefore prefer to reserve k for the wavevector, wherever possible. The refractive index will come up frequently in our discussion, and we will write $N = n + i\kappa$. From the relationship between the refractive index and the dielectric constant, $(n + i\kappa)^2 = \mathbb{R}\mathrm{e}\{\boldsymbol{\varepsilon}_{\mathrm{R}}\} + i\mathbb{I}\mathrm{m}\{\boldsymbol{\varepsilon}_{\mathrm{R}}\}$ we have

$$\begin{aligned} \mathbb{R}\mathrm{e}\{\boldsymbol{\varepsilon}_{\mathrm{R}}\} &= n^2 - \kappa^2 \\ \mathbb{I}\mathrm{m}\{\boldsymbol{\varepsilon}_{\mathrm{R}}\} &= 2n\kappa. \end{aligned} \tag{7.40}$$

In air, we have seen that $\mathbf{E}(z,t) = E_{\max} e^{i(\mathbf{k}_0 \cdot \hat{z} - \omega t)}$, where the wavevector in air (essentially that in a vacuum) is defined as $|\mathbf{k}_0| = 2\pi/\lambda_0$. In a material, this expression is written with the wavevector in the material $|\mathbf{k}| = 2\pi/\lambda$

$$\mathbf{E}(z,t) = E_{\max} e^{i(\mathbf{k} \cdot \hat{z} - \omega t)}$$

where λ is the wavelength inside the material. Since we can write

$$|k| = \frac{2\pi}{\lambda} = \frac{2\pi N}{\lambda_0} \tag{7.41}$$

the expression for the propagation of the electric field in time and space can be expressed in a manner that explicitly uses the complex refractive index of the material,

$$\mathbf{E}(z,t) = E_{\max} e^{i\mathbf{k}_0 N \cdot \hat{z}} e^{-i\omega t}.$$

For transparent materials $\kappa = 0$ and so $N = n$ resulting in

$$\mathbf{E}(z,t) = E_{\max} e^{i\mathbf{k}_0 n \cdot \hat{z}} e^{-i\omega t}$$

i.e. the wavevector has been modified from its state in vacuum to accommodate the shorter wavelength inside the material. But for materials that absorb light, i.e. $\kappa > 0$,

$$\begin{aligned} \mathbf{E}(z,t) &= E_{\max} e^{i\mathbf{k}_0 (n+i\kappa) \cdot \hat{z}} e^{-i\omega t} \\ &= E_{\max} e^{i\mathbf{k}_0 n \cdot \hat{z}} e^{-\mathbf{k}_0 \kappa \cdot \hat{z}} e^{-i\omega t} \\ &= E_{\max} e^{i(\mathbf{k}_0 n \cdot \hat{z} - i\omega t)} e^{-\mathbf{k}_0 \cdot \kappa \hat{z}}. \end{aligned} \tag{7.42}$$

We observe that the combination of the imaginary unit $i = \sqrt{-1}$ and the real part of the refractive index causes a phase shift that contributes to the overall phase of the field. But the multiplication of i from the propagation phase with i from the imaginary part of the complex refractive index results in real-valued exponential decay. That is, the amplitude of the field starts at $E_{\max}$ but then decays along the propagation direction z as shown in Fig. 7.5.

Code to generate Fig. 7.5

Figure 7.5 could have been created in a vector illustration program such as Inkscape. In that case, setting the position of the text labels could be done more organically using the mouse, and the colour of objects could be determined by clicking on them in the typical way. However, even though many such programs have some ability for function plotting, extensive function specification is of course simpler in Python. This is therefore an opportunity to have a look at a relatively

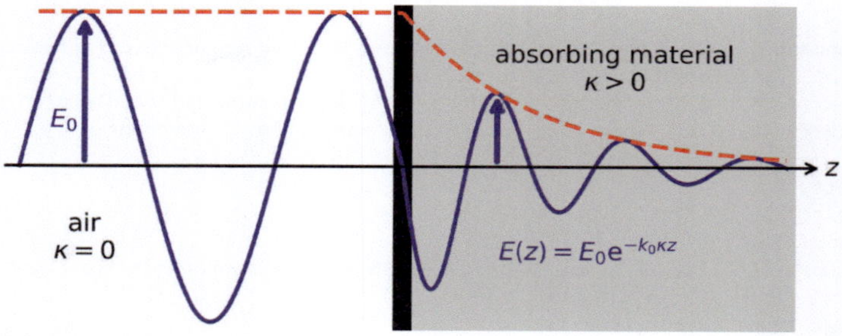

Fig. 7.5 When a wave travels from a transparent ($\kappa = 0$ so $N = n$) material to an absorbing material ($N = n + i\kappa$, where $\kappa > 0$) the wavelength decreases and the amplitude of the wave decreases exponentially while inside the material

simple code that primarily draws lines, arrows, rectangles, and text of various colours.

```
import numpy as np
from matplotlib import pyplot
import matplotlib.patches as mpatches

# setup
pyplot.rc('font', size=9)
fig = pyplot.figure(figsize=(4.5, 2))
ax = fig.add_subplot()
z1 = np.linspace(-1, 0, 500)
z2 = np.linspace(0, 1, 500)

# field in air
E1 = np.sin(-z1*1.5*2*np.pi)
ax.plot(z1, E1, 'b-', lw=1.5)
ax.plot([-0.95, 0], [1, 1], 'r--')
pos = -0.83
ax.annotate('', xy=(pos, 1), xytext=(pos, 0), arrowprops=dict(arrowstyle
     ='->', color='b', lw=2))
ax.text(-0.92, 0.25, '$E_0$', color='b')
ax.text(pos, -0.6, 'air\n$\kappa=0$', ha='center')

# material
width = 0.05
ax.add_artist(mpatches.Rectangle((-0.5*width, -1.05), width, 2.1, ec='none
     ', fc='k'))

# field in material
ax.add_artist(mpatches.Rectangle((0.5*width, -1.05), 1., 2.1, ec='none', fc
     ='0.8'))
E2 = np.sin(-z2*3*2*np.pi) * np.exp(-3*z2)
ax.plot(z2, E2, 'b-', lw=1.5)
ax.plot(z2, np.exp(-3*z2), 'r--')
pos = 0.245
```

```
ax.annotate('', xy=(pos, 0.5), xytext=(pos, 0), arrowprops=dict(arrowstyle
    ='->', color='b', lw=2))
ax.text(0.25, -0.6, r'$E(z) = E_0 {\rm e}^{-k_0 \kappa z}$', color='b')
ax.text(0.55, 0.5, 'absorbing material\n$\kappa > 0$', ha='center')

# pretty
ax.text(1.1, 0, '$z$', va='center')
ax.axis('off')
ax.annotate('', xy=(1.1, 0), xytext=(-1.05, 0), arrowprops=dict(arrowstyle
    ='->'))
fig.set_tight_layout(True)
pyplot.savefig('fieldDecay.png', dpi=600)
```

Note our use of LaTeX syntax in order to specify math type, including subscripts. We also point out that the location of objects such as lines and arrowheads is sometimes offset from the actual coordinates of the objects they annotate for the sake of clarify.

Finally, consider the implication of $\kappa > 0$ on the intensity of the electric field as it travels through the material.

$$\begin{aligned} I &\propto |E|^2 \\ &= E_{\text{max}}^2 e^{-2\kappa \mathbf{k}_0 \cdot \hat{z}} \\ &= I(z). \end{aligned} \tag{7.43}$$

As shown in Fig. 7.6, the intensity of the field decays twice as quickly as the field amplitude.

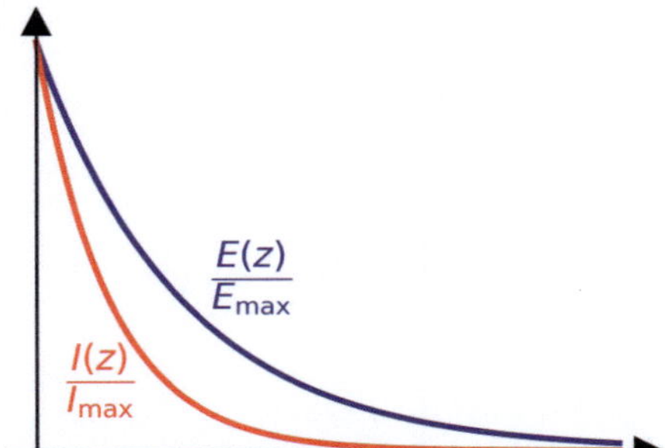

Fig. 7.6 Decay of the electric field amplitude (blue) E_{max} with distance z as an electromagnetic wave travels through an absorbing medium, in comparison to the manner in which the intensity, proportional to the magnitude squared of the field, decays (red). In order to compare the decay rates, the field is normalized to unit amplitude $E(z)/E_{\text{max}}$ as is the intensity $I(z)/I_0$

7.4 Reflection and Transmission at Interfaces

In Chap. 8 we will study the absorption of infrared light by materials when light incident from a controlled source is transmitted or reflected from a sample. One of the ways in which the experiment is inherently quantitative is that the light intensity is not measured directly; the ratio of the reflected or the transmitted intensity to the incident intensity is recorded. Therefore, to understand how those quantities can be connected to the arrangement of molecules within the sample, we will want to first take a closer look at the reflection and transmission, in particular at how the amplitude and phase of the electric field is altered in the process.

7.4.1 Electric Field Amplitudes and Phases

Figure 7.7 illustrates the general case of a wave travelling in a medium we refer to as i towards an abrupt change in the material, where the refractive index suddenly changes from N_i to N_t. Such interfaces, devoid of any surface roughness, without gradients in the refractive index, and without consideration of molecular details, are often referred to as Fresnel-type interfaces. At the interface, there is a wave reflected back into medium i, and a wave transmitted into the second medium t. The drawing indicates only the wavevectors of the incident k_i, reflected k_r, and transmitted k_t waves, with a depiction of the electric fields omitted. From here on, we will prefer the term *transmission* to *refraction*, since that conveniently enables us to reserve the subscript r for the reflected beam and t for the transmitted (refracted) beam. From conservation of linear momentum, Snell's law of reflection states that $\theta_\mathrm{r} = \theta_\mathrm{i}$, and Snell's law of refraction enables us to calculate the transmitted angle θ_t from

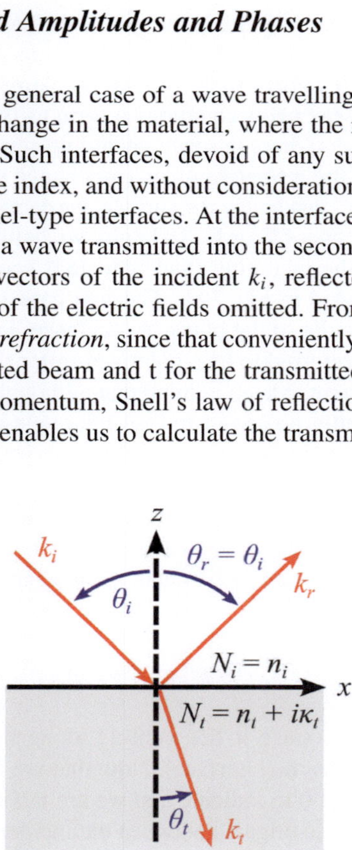

Fig. 7.7 Light incident at the boundary of two semi-infinite media with refractive indices N_i for the incident phase, and N_t for the transmitted phase. Snell's Law of reflection provides $\theta_\mathrm{r} = \theta_\mathrm{i}$, and Snell's law of refraction provides the transmitted angle N_t. The orientation of the electric fields is not shown in this figure, but they can be anywhere perpendicular to the indicated wavevectors

$$N_\mathrm{i} \sin\theta_\mathrm{i} = N_\mathrm{t} \sin\theta_\mathrm{t}. \tag{7.44}$$

In nearly all practical cases, we find ourselves interested in situations where the incident medium is transparent, and so $N_\mathrm{i} = n_i$ ($\kappa_i = 0$). If this wasn't true, the problem would be rather complicated, as the amplitude of the wave would be attenuated as it approaches the interface. If $\kappa_\mathrm{t} > 0$, the transmitted wave would decrease exponentially as the beam travels in medium t, the familiar phenomenon of absorption described by Beer's law; we would also be interested in the consequences for the reflected beam. These are the two situations of primary interest that we will discuss further in the next chapter in a dedicated study of absorption and absorbance measurements. For now, note that Fig. 7.7 did not specify the polarization of any of the waves. But drawing only the wavevectors, we know only that **E** can be anywhere perpendicular to **k**.

Our goal is to describe the magnitude and phase of the reflected and transmitted electric field relative to the incident field. There are two special cases to consider that will provide solutions to all sorts of more general problems. These are based on two linearly-polarized states. Maxwell's equations inform us that the tangential parts of **E** and **B** are continuous everywhere in space, including across the sharp interface. In other words, even though the refractive index can change sharply from $z = 0^-$ just above the surface to $z = 0^+$ just below the surface, we cannot have any discontinuity in the *tangential* components of **E** or **B**. To begin, we define the *plane of incidence* as that plane containing the incoming (and reflected) wavevector and the surface normal. This plane is shaded in Fig. 7.8a. The first case is that where the electric field vector **E** is perpendicular to the plane of incidence. From now on, this special polarization state will be referred to as s-polarized (after the German *senkrecht* for perpendicular). It is important to keep in mind that, although the s-polarized incident light appears to be vertically-polarized in Fig. 7.8a, an s-polarized field, although always linearly polarized, is not always vertically polarized. The description of the field orientation must consider the plane of incidence for the particular interface under investigation. Returning to our problem, Fig. 7.8b illustrates the particular electric and magnetic field vectors associated with each of the three beams we are discussing. Continuity across the $z = 0$ boundary between the two media enables us to write

$$\begin{gathered} E_\mathrm{i}(z = 0^+) + E_\mathrm{r}(z = 0^+) = E_\mathrm{t}(z = 0^-) \\ E_\mathrm{i}^0 + E_\mathrm{r}^0 = E_\mathrm{t}^0 \end{gathered} \tag{7.45}$$

for the electric field vector, since it lies entirely along the plane of the interface (all components tangential to the interface). Note that we have adopted the simpler notation with the superscript 0 to indicate that we are talking about the fields at the interface, and we no longer distinguish between the incident ($z = 0^+$) and transmitted ($z = 0^-$) side, since it can be understood from the subscripts i, r, and t. For the magnetic field, however, only a component of the field is tangential to the interface, so we can write

$$-B_\mathrm{i}^0 \cos\theta_\mathrm{i} + B_\mathrm{r}^0 \cos\theta_\mathrm{r} = -B_\mathrm{t}^0 \cos\theta_\mathrm{t}. \tag{7.46}$$

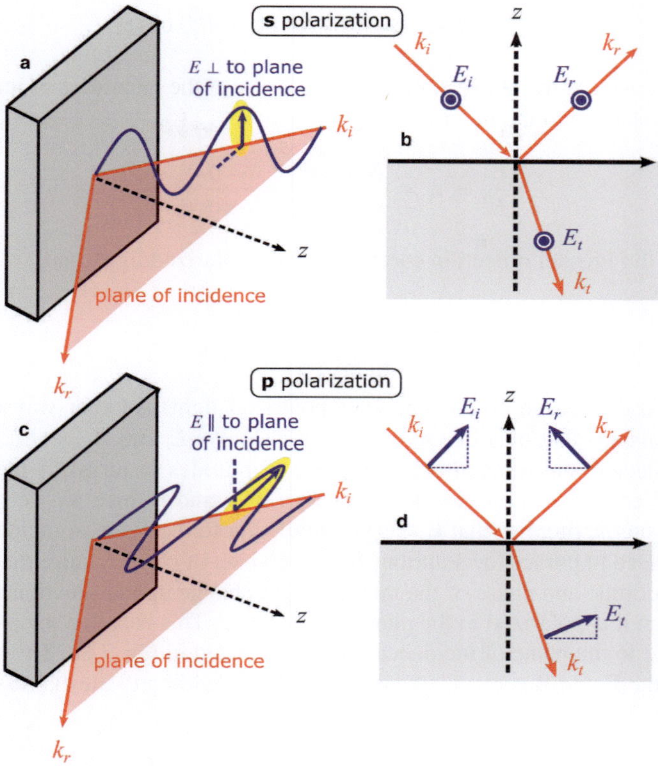

Fig. 7.8 **a** Light incident at a Fresnel-type interface with the electric field vector perpendicular to the plane of incidence, the so-called s-polarization. Notice that the plane of incidence is defined as that plane containing the incoming wavevector and the surface normal. **b** The corresponding components of the electric and magnetic fields. **c** The same geometry, but with p-polarized light defined as having the electric field vector in the plane of incidence and **d** the corresponding electric and magnetic fields with respect to the propagation vectors

Maxwell's equations provide the relationship between the electric and magnetic fields as

$$\mathbf{B} = \frac{N}{c}\mathbf{E}$$

that will enable us to re-write these expressions in terms of the electric components only, as we are ultimate interested in the amplitudes of the reflected and transmitted electric fields. This results in

$$-\frac{N_\mathrm{i}}{c}E_\mathrm{i}^0\cos\theta_\mathrm{i} + \frac{N_\mathrm{i}}{c}E_\mathrm{r}^0\cos\theta_\mathrm{r} = -\frac{N_\mathrm{t}}{c}E_\mathrm{t}^0\cos\theta_\mathrm{t} \tag{7.47}$$

which can be further simplified to

$$N_\mathrm{i}(E_\mathrm{r}^0 - E_\mathrm{i}^0)\cos\theta_\mathrm{i} = -N_\mathrm{t}(E_\mathrm{r}^0 + E_\mathrm{i}^0)\cos\theta_\mathrm{t}. \tag{7.48}$$

We can now re-write the above expressions to arrive at the quantities of interest, the ratio of the reflected to the incident field

$$\frac{E_\mathrm{r}}{E_\mathrm{i}} \equiv r_\mathrm{s} = \frac{N_\mathrm{i}\cos\theta_\mathrm{i} - N_\mathrm{t}\cos\theta_\mathrm{t}}{N_\mathrm{i}\cos\theta_\mathrm{i} + N_\mathrm{t}\cos\theta_\mathrm{t}} \tag{7.49}$$

that we call the Fresnel reflection coefficient for s-polarized light, and

$$\frac{E_\mathrm{t}}{E_\mathrm{i}} \equiv t_\mathrm{s} = \frac{2N_\mathrm{i}\cos\theta_\mathrm{i}}{N_\mathrm{i}\cos\theta_\mathrm{i} + N_\mathrm{t}\cos\theta_\mathrm{t}} \tag{7.50}$$

the Fresnel transmission coefficient for s-polarized light. Another way of stating these relationships that may be more intuitive is $E_\mathrm{r} = rE_\mathrm{i}$ and $E_\mathrm{t} = tE_\mathrm{i}$, hence we have the reflected and transmitted (complex) amplitudes in terms of the incident amplitude. Recall that these are the s versions of r and t since we derived these expressions on the premise that **E** was perpendicular to the plane of incidence.

We also need to pursue the solutions for another set of r and t, those that describe how the amplitude and phase of the incident light change upon reflection and transmission when it is polarized *in* the plane of incidence. This is called the p-polarized case (parallel to the plane of incidence), and is depicted in Fig. 7.8c. The analogous drawing of the **E** and **B** vectors appears in Fig. 7.8d, leading to the expressions

$$B_\mathrm{i}^0 + B_\mathrm{r}^0 = B_\mathrm{t}^0 \tag{7.51}$$

and

$$E_\mathrm{i}^0\cos\theta_\mathrm{i} - E_\mathrm{r}^0\cos\theta_\mathrm{r} = E_\mathrm{t}^0\cos\theta_\mathrm{t}. \tag{7.52}$$

Making the same substitutions as before, we can re-write the first equation as

$$\frac{N_\mathrm{i}}{c}E_\mathrm{i}^0 + \frac{N_\mathrm{r}}{c}E_\mathrm{r}^0 = \frac{N_\mathrm{t}}{c}E_\mathrm{t}^0 \tag{7.53}$$

to arrive at our final expressions

$$\frac{E_\mathrm{r}}{E_\mathrm{i}} \equiv r_\mathrm{p} = \frac{N_\mathrm{t}\cos\theta_\mathrm{i} - N_\mathrm{i}\cos\theta_\mathrm{t}}{N_\mathrm{t}\cos\theta_\mathrm{i} + N_\mathrm{i}\cos\theta_\mathrm{t}} \tag{7.54}$$

that we call the Fresnel reflection coefficient for p-polarized light, and

$$\frac{E_\mathrm{t}}{E_\mathrm{i}} \equiv t_\mathrm{p} = \frac{2N_\mathrm{i}\cos\theta_\mathrm{i}}{N_\mathrm{i}\cos\theta_\mathrm{t} + N_\mathrm{t}\cos\theta_\mathrm{i}} \tag{7.55}$$

the Fresnel transmission coefficient for p-polarized light. From the above expressions, we can see that these are field amplitude coefficients, but we take care to distinguish the terms *amplitude* and *magnitude*. There are several ways in which these coefficients carry phase information. The first is when r is real but negative. This means that the reflected field is phase shifted by 180° with respect to the incident field. Although we have already mentioned that we are primarily concerned with transparent incident media, consider that when N_t is complex-valued, all four Fresnel coefficients will be complex valued.

The Fresnel equations are a thing of beauty. To better appreciate their form, we will create a plot that shows the angle-dependence of the four coefficients in the case where $N_i = 1$ (air) and $N_t = 1.5$ (transparent glass in the visible region) in Fig. 7.9a. The first observation we make is that, in the case of external reflection (low index to high index) with both phases transparent, all four coefficients are real. We can therefore represent all four quantities in Fig. 7.9a on a scale that extends from -1 to 1. The magnitude $|r_s| \rightarrow |r_p| \rightarrow 1$ as $\theta_i \rightarrow 90°$. This suggests that the surface generally becomes more reflective at more oblique angles. At $\theta_i = 90°$ there is technically no reflection, and the light remains in the incident phase, so it is proper that $|E_r|/|E_i| = 1$ at that angle. Likewise, $|t_s| \rightarrow |t_p| \rightarrow 0$ as $\theta_i \rightarrow 90°$, in agreement with our intuition. One surprising feature is that $r_s = -r_p$ at $\theta_i = 0°$. According to the definition, the plane of incidence is not unique at normal incidence, since the incident wavevector is then parallel to the surface normal. In that case, s- and p-polarized states do not formally exist. One would then expect the solutions of r_s and r_p to be identical (matching in magnitude and phase) at $\theta_i = 0°$. The fact that $r_s = -r_p$ comes from the fact that the handedness of the coordinate system is inverted upon reflection, and this is accounted for in the sign of r_p. Finally, there is an amazing natural phenomena encoded in the Fresnel equations that is the zero crossing of r_p at the value $\theta_i = \arctan(n_t/n_i)$, approximately 56.3° in the case of $n_i = 1$ and $n_t = 1.5$. This special angle is known as the Brewster angle, often abbreviated θ_B, and is formally defined as the angle at which $r_p = 0$. There are many interesting phenomena and useful applications associated with the Brewster angle. The code used to generate Fig. 7.9 appears below.

Code used to generate Fig. 7.9

```
import numpy as np
from matplotlib import pyplot

thetaDeg = np.linspace(0, 90, 100)
n1 = 1.
n2 = 1.5

theta1 = np.radians(thetaDeg)
theta2 = np.arcsin(n1*np.sin(theta1)/n2)
rs = (n1*np.cos(theta1) - n2*np.cos(theta2))/(n1*np.cos(theta1) + n2*np.cos
     (theta2))
rp = (n2*np.cos(theta1) - n1*np.cos(theta2))/(n1*np.cos(theta2) + n2*np.cos
     (theta1))
ts = 2*n1*np.cos(theta1) / (n1*np.cos(theta1) + n2*np.cos(theta2))
tp = 2*n1*np.cos(theta1) / (n1*np.cos(theta2) + n2*np.cos(theta1))
```

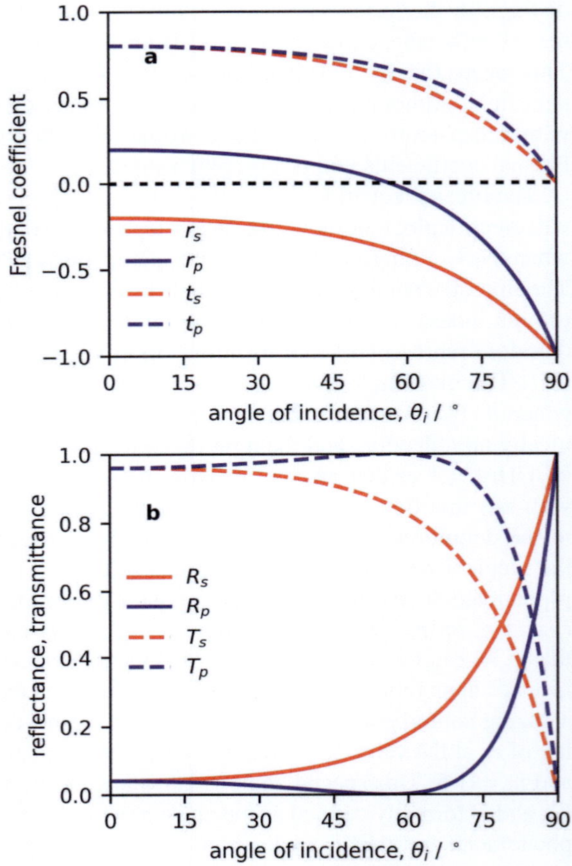

Fig. 7.9 **a** The four Fresnel coefficients that provide the reflected and transmitted field amplitude for the case of s- and p-polarized light. **b** The corresponding transmittance and reflectance curves

```
fig = pyplot.figure(figsize=(3.25, 5))
pyplot.rc('font', size=8)

ax1 = fig.add_subplot(211)
ax1.plot(thetaDeg, rs, 'r-', label='$r_s$')
ax1.plot(thetaDeg, rp, 'b-', label='$r_p$')
ax1.plot(thetaDeg, ts, 'r--', label='$t_s$')
ax1.plot(thetaDeg, tp, 'b--', label='$t_p$')
ax1.set_ylabel('Fresnel coefficient')
ax1.set_ylim(-1,1)
ax1.set_yticks(np.arange(-1, 1.01, 0.5))
ax1.axhline(0, color='k', ls='--', dashes=[2,2])
pyplot.legend(loc=0, frameon=False)
ax1.text(0.08, 0.85, 'a', fontweight='bold', transform=ax1.transAxes)

ax2 = fig.add_subplot(212)
```

```
ax2.plot(thetaDeg, abs(rs)**2, 'r-', label='$R_s$')
ax2.plot(thetaDeg, abs(rp)**2, 'b-', label='$R_p$')
ax2.plot(thetaDeg, 1 - abs(rs)**2, 'r--', label='$T_s$')
ax2.plot(thetaDeg, 1 - abs(rp)**2, 'b--', label='$T_p$')
ax2.set_ylabel('reflectance, transmittance')
ax2.set_ylim(0,1)
pyplot.legend(loc=0, frameon=False)
ax2.text(0.08, 0.81, 'b', fontweight='bold', transform=ax2.transAxes)

for ax in [ax1, ax2]:
   ax.set_xlabel(r'angle of incidence, $\theta_i$ / $^\circ$')
   ax.set_xlim(0, 90)
   ax.set_xticks(range(0, 91, 15))

fig.set_tight_layout(True)
pyplot.savefig('fresnel_external.png', dpi=600)
```

7.4.2 Snell's Laws

One of the consequences and most practical outcomes of the Fresnel equations is that they lead directly to Snell's law of reflection

$$\theta_\mathrm{i} = \theta_\mathrm{r} \tag{7.56}$$

and Snell's law of refraction

$$N_\mathrm{i} \sin \theta_\mathrm{i} = N_\mathrm{t} \sin \theta_\mathrm{t}$$

that we already encountered as Eq. 5.1, but we now label with the commonly used subscripts of 1 rather than *i* for the incident phase and 2 rather than *t* for the refracted phase

$$N_1 \sin \theta_1 = N_2 \sin \theta_2. \tag{7.57}$$

One must be careful, though, when using labels for the phases, as some problems require calculations for beams travelling in either direction. When light travels from medium 2 to medium 1, the incident phase has index N_2 and angle θ_2. Therefore, it would be wise to keep using the i (incident) and t (transmitted) labels, but we'll switch to numbers for most of the rest of the discussion in this book. Take care, you've been warned!

Note that Eq. 7.57 applies when light is approaching an isotropic medium from another isotropic medium. This means that the manner in which the molecules are arranged is random, and hence all properties of the material—including optical properties such as the refractive index and susceptibility—are the same in all directions. We are ultimately leading up to a discussion of materials in which the molecules are ordered, and hence the optical constants are anisotropic, i.e. they are not the same in all directions. Since the beam typically approaches from an isotropic medium (such as air) $N_{1x} = N_{1y} = N_{1z}$, and we simply write N_1 for all Cartesian components of the incident refractive index. However, if the refracted material is anisotropic then it is, in general, described by the set of refractive indices $N_{2x} = n_{2x} + i\kappa_{2x}$, $N_{2y} = n_{2y} + i\kappa_{2y}$ and $N_{2z} = n_{2z} + i\kappa_{2z}$. In other words, the speed at which light propagates through the material, and the amount to which the electric field amplitude is diminished, differs depending upon whether the field is parallel to the material's x, y, or z directions. We'll get into all that in due time but, for now, the point we wish to make is that, for anisotropic materials, Snell's Law of Refraction becomes

$$N_1 \sin\theta_1 = N_{2z} \sin\theta_2 \tag{7.58}$$

noting that we must use the z component of the refractive index N_{2z} in the transmitted medium.

Snell's Law of Refraction (either in its isotropic form in Eq. 7.57 or the anisotropic version in Eq. 7.58) is required in almost any task of interest to us in order to evaluate the refracted angle θ_2 given the incident angle θ_1 and the refractive indices N_1 and N_2. Although this seems straightforward as one can write

$$\theta_2 = \arcsin\left(\frac{N_1 \sin\theta_1}{N_2}\right) \tag{7.59}$$

there are some dangers lurking in the evaluation of this expression. We need to carefully consider how we would solve for the refracted angle using Python and NumPy. When performing the calculation, one needs to take care in the evaluation of the inverse sine function $\sin^{-1} z$ also known as $\arcsin z$ [8]. Although

$$\arcsin(x + iy) = a + ib \tag{7.60}$$

is defined over any set of numbers in the complex plane (i.e. $-\infty < x < \infty$ and $-\infty < y < \infty$), it is not single-valued, and it is customary to return a solution for which $-\pi/2 < a < \pi/2$. Within this range, a few characteristic values for which $a > 0$ are plotted with solid red lines in Fig. 7.10; some values for which $a < 0$ are plotted as dashed red lines. The imaginary part of the solution to Eq. 7.60 with $b > 0$ is plotted with solid blue lines, and that with $b < 0$ with dashed blue lines. Two branch cuts indicated by the dashed black lines are visible. Where $y = 0$ one is located at $-\infty < x < -1$; the other appears at $1 < x < \infty$. As these regions represent discontinuities, care must be taken when evaluating functions for which $z = x + iy$ is close to one of the branch cuts as this may result in an unstable assignment of the sign of b. One solution is to make sure that the arcsine routine is robust in the region of the branch points. Some programming languages and libraries take care of this by default, others have dedicated higher accuracy arcsine functions for this purpose [9]. To be safe, one could evaluate the arcsine in a different manner, for

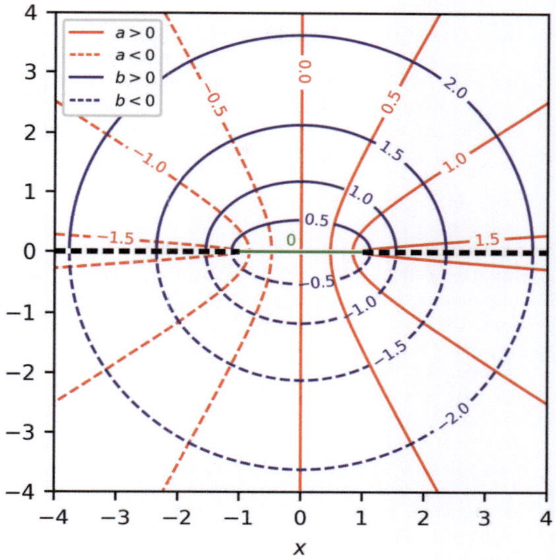

Fig. 7.10 Solutions to Eq. 7.60; regions with $a > 0$ are plotted with red solid lines, $a < 0$ with red dashed lines. Regions with $b > 0$ are plotted with blue solid lines, and $b < 0$ with blue dashed lines. Solutions for real-valued arguments of arcsine ($-1 < x < 1$ and $y = 0$) are indicated by the green line. The two branch cuts are illustrated with the black dashed lines

example, with the expression [10, 11]

$$\arcsin z = i \ln(\sqrt{1 - z^2} - iz) \tag{7.61}$$

that may provide better performance closer to the branch cuts.

Code to generate Fig. 7.10

```
import numpy as np
from matplotlib import pyplot
from matplotlib.lines import Line2D

def arcsin(z1):
    z = z1.astype(complex)
    return 1j*np.log(np.sqrt(1.0 - z*z) - 1j*z)

# setup
pyplot.rc('font', size=8)
fig = pyplot.figure(figsize=(3.25, 3.25))
ax = fig.add_subplot(aspect='equal')
extent = 4
x, y = np.meshgrid(np.linspace(-extent, extent, 500), np.linspace(-extent,
     extent, 500))

# contours
z = arcsin(x + 1j*y) # note we are not using np.arcsin, but our own version
c1 = ax.contour(z.real, levels=np.arange(-2, 2.05, 0.5), colors='r',
     linewidths=1, extent=[-extent, extent, -extent, extent])
```

```
ax.clabel(c1, inline=True, fontsize=6, manual=[(0,3), (1.5,2.5), (-1.5,2.5)
    , (2.5,1.5), (-2.5,1.5), (3,0.5), (-3,0.5)])
c2 = ax.contour(z.imag, levels=np.arange(-2, 2.05, 0.5), colors='b',
    linewidths=1, extent=[-extent, extent, -extent, extent])
ax.clabel(c2, inline=True, fontsize=6, manual=[(2.5,2.75), (1.5,1.75),
    (1,1), (0.25,0.5), (0.25,-0.5), (1,-1), (1.5,-1.75), (2.5,-2.75)])

# branch cuts
ax.plot([-extent, extent], [0, 0], 'w-', lw=2)
ax.plot([-1, 1], [0, 0], 'g-', lw=1)
ax.text(-0.25, 0.1, '0', color='g', fontsize=6)
ax.plot([-extent, -1], [0, 0], 'k--', lw=2, dashes=[2,1])
ax.plot([1, extent], [0, 0], 'k--', lw=2, dashes=[2,1])

# pretty
ax.set_xticks(range(-extent, extent+1))
ax.set_yticks(range(-extent, extent+1))
ax.set_xlabel('$x$')
ax.set_ylabel('$y$')

# legend
rSolid = Line2D([0], [0], color='r', lw=1, label=r'$a>0$')
rDashed = Line2D([0], [0], color='r', lw=1, ls='--', dashes=[2,1], label=r'
    $a<0$')
bSolid = Line2D([0], [0], color='b', lw=1, label=r'$b>0$')
bDashed = Line2D([0], [0], color='b', lw=1, ls='--', dashes=[2,1], label=r'
    $b<0$')
ax.legend(handles=[rSolid, rDashed, bSolid, bDashed], loc='upper left',
    framealpha=1, fontsize=6)

fig.set_tight_layout(True)
pyplot.savefig('branchCuts.png', dpi=600)
```

As an example, we can consider evaluating the refracted angle θ_2 when a beam is incident from a material with $n_1 = 1.5$ towards air ($n_2 = 1$) approaching at $\theta_1 = 65°$. This case of internal reflection will be described in more detail shortly, but here we draw attention to the point that simply rearranging Snell's Law to Eq. 7.59 will result in a form that cannot be evaluated, as such an expression will result in an error in many programming languages (including Python) if n_1, n_2, and θ_1 were defined as real numbers, as the argument of `arcsin` is greater than unity and hence out of range. In Python, one solution that avoids such an error is to define at least one of the variables as a complex-valued type, thereby coercing `arcsin` to provide a complex-valued result. This is fine when $\kappa > 0$ for one of the phases, as the refractive index will naturally be complex-valued. But what happens when one or both media are transparent? We could define $N_2 = n_2 + i\kappa_2 = 1 + 0i$. This satisfies the desired complex data type; however, substituting this value for N_2 into Eq. 7.59 results in $z \approx 1.359 + 0i$, exactly on one of the branch cuts. This can be avoided by setting $N_2 = 1 + 10^{-6}i$, making κ_2 sufficiently and realistically small. If κ_2 is made too small, however, machine precision limitations could return a result that is not sufficiently far from one of the branch cuts. Many of the code examples in the subsequent chapters use Eq. 7.59 for convenience, being aware of possible limitations. However, it is straightforward to make a custom Python module `optics` and then define a function that uses Eq. 7.60 as `optics.arcsin()` as an alternative to what NumPy provides in `np.arcsin()` that should work with all codes requiring evaluation of the arcsine function. Such a custom module would be a great place to store functions returning the Fresnel coefficients too!

7.4.3 Reflectance and Transmittance

We arrived at this point in our journey by considering how the electric field of a light wave changes as it travels through different materials. Although this is a fundamental consideration in the study of light-matter interaction (as the local field in the region where the molecules are located drives the molecular response we aim to detect), the fields themselves are not detected in an experiment. Rather, we measure the intensity of transmitted and/or reflected light. The transmittance T is the ratio of the transmitted intensity I_t to the incident intensity that we have previously referred to as I_i but is also typically called I_0. If the beam is incident from medium 1, then the labelling scheme we have been discussing can also refer to the incident intensity as I_1. Since the intensity is proportional to the square of the electric field amplitude, the transmittance is related to the square of the magnitude of the electric field transmission
coefficient t

$$T = \frac{I_t}{I_i} = \frac{N_2 \cos\theta_2}{N_1 \cos\theta_1}|t|^2. \tag{7.62}$$

The prefactor involving the refractive indices of the incident and transmitted media and the beam angles accounts for the fact that the beam may have a different diameter upon entering (from the light source) the sample and leaving (to the detector). As we are talking about the irradiance, or the power per unit area, this beam size correction is required. When the transmission coefficient t_s is used in Eq. 7.62, the s-polarized transmittance T_s is obtained. Likewise T_p is readily calculated by using t_p in Eq. 7.62. This expression is the one provided in most introductory optics books [12–14]. However, you may wonder what happens in the event that the refractive index is complex, and the refracted angle is also complex, as we need to end up with a real number for T. The solution depends on the input beam polarization

$$T_s = \left(\frac{N_t \cos\theta_t + N_t^* \cos\theta_t^*}{N_i \cos\theta_i + N_i^* \cos\theta_i^*}\right)|t_s|^2 \tag{7.63a}$$

$$T_p = \left(\frac{N_t \cos\theta_t^* + N_t^* \cos\theta_t}{N_i \cos\theta_i^* + N_i^* \cos\theta_i}\right)|t_p|^2 \tag{7.63b}$$

where the asterisk denotes complex conjugation. A discussion and derivation of Eq. 7.63 may be found in Ref. [15]. The calculation of transmittance inside an absorbing medium doesn't present itself very often, as the phase in which the light is detected (usually air) is most often transparent. Otherwise, the placement of the detector would make a difference, as the amplitude of the wave decreases due to absorption. You may then be curious about the meaning of something like $\cos\theta_i^*$, whether this implies $(\cos\theta_i)^*$ or $\cos(\theta_i^*)$. This is suggested as an exercise in one of the end-of-chapter problems.

Another quantity that can be measured is the reflectance R, the ratio of the reflected intensity I_r to the incident intensity I_0. As the diameter of the incident and reflected beams is necessarily the same (they travel in the same medium, usually air), R is directly obtained from the electric field reflection coefficient r

$$R = \frac{I_r}{I_i} = |r|^2. \tag{7.64}$$

In the case where all media (especially medium 2, which is typically our sample of interest) are transparent, and if we can ignore loss due to scattering, then $R + T = 1$. Since the expression for R is more convenient to evaluate, this also provides a facile route for the calculation of T. The reflectance and transmittance for our example of external reflection are shown in Fig. 7.9b where it can be observed that the reflectance generally increases with angle, while the transmittance decreases. However, at values of $\theta_i < \theta_B$, R_p initially decreases as the angle of incidence increases until the Brewster angle is reached, after which both R_s and R_p increase. We can also see that $T_p = 1$ at $\theta_i = \theta_B$. This is not a general rule, but is a consequence of $\kappa_2 = 0$.

7.4.4 Internal Reflection

We now take a second pass at the discussion, merely changing the order of the materials so the beam approaches from a material with $N_1 = 1.5$ and encounters a phase with $N_2 = 1$. This is commonly seen when light exits glass. Since the incident medium now has the larger refractive index (momentarily considering just the real part since $\kappa_1 = \kappa_2 = 0$) this configuration is known as internal reflection. All of the expressions required to handle this scenario are exactly as for external reflection, but the results obtained are surprisingly different in many important ways. This problem is complicated in a fascinating way, and there are several technologies, unique experiments, and natural phenomena that result from the behaviour of a subset of angles in an internal reflection experiment. Although it may not be apparent at first glance from the expressions for r_s, r_p, t_s and t_p provided above, when we substitute $n_1 < n_2$, there is angle above which the solutions are no longer real, even when $\kappa_1 = \kappa_2 = 0$. Since we are now dealing with complex values of these coefficients we have two options for inspecting and discussing them: we either plot their real and imaginary components or plot their magnitude and phase. As a reminder, we can easily convert between the two representations expressed as $c = a + ib = |c|e^{id}$

using Euler's identity $c = |c| \cos d + i|c| \sin d$, which provides $|c| = \sqrt{a^2 + b^2}$ and $d = \arctan(b/a)$, taking care to preserve the quadrant information of the phase angle.

The real components of the four Fresnel coefficients are plotted in Fig. 7.11a, and their imaginary parts appear in Fig. 7.11b. The first thing we note is that the Brewster angle has shifted to an angle lower than where it appears in external reflection. Its value is still provided by $\theta_B = \arctan(n_2/n_1)$ but that now evaluates to approximately 34°. The more striking observation is that there is another angle of interest, at which the real values of the Fresnel coefficients are sharply cusped, and above which the coefficients take on complex values. This is the critical angle, θ_c, that marks the onset of total internal reflection (TIR). The TIR phenomenon is most clearly seen if we instead plot the magnitude of the coefficients in Fig. 7.11c, or the actual reflectance and transmittance in Fig. 7.11e. Here we can see that, for $\theta > \theta_c$ the reflectance $R = 1$ and the transmittance is then of course zero.

Code used to generate Fig. 7.11

```
import numpy as np
from matplotlib import pyplot

thetaDeg = np.linspace(0, 90, 100)
n1 = 1.5
n2 = 1. + 0j

theta1 = np.radians(thetaDeg)
theta2 = np.arcsin(n1*np.sin(theta1)/n2)
rs = (n1*np.cos(theta1) - n2*np.cos(theta2))/(n1*np.cos(theta1) + n2*np.cos
     (theta2))
rp = (n2*np.cos(theta1) - n1*np.cos(theta2))/(n1*np.cos(theta2) + n2*np.cos
     (theta1))
ts = 2*n1*np.cos(theta1) / (n1*np.cos(theta1) + n2*np.cos(theta2))
tp = 2*n1*np.cos(theta1) / (n1*np.cos(theta2) + n2*np.cos(theta1))

fig = pyplot.figure(figsize=(4.5, 5.5))
pyplot.rc('font', size=8)

ax1 = fig.add_subplot(321)
ax1.plot(thetaDeg, rs.real, 'r-', label='Re[$r_s$]')
ax1.plot(thetaDeg, rp.real, 'b-', label='Re[$r_p$]')
ax1.plot(thetaDeg, ts.real, 'r--', label='Re[$t_s$]')
ax1.plot(thetaDeg, tp.real, 'b--', label='Re[$t_p$]')
ax1.set_ylabel('real')
ax1.set_ylim(-1,3)
ax1.axhline(0, color='0.8', zorder=-999)
pyplot.legend(loc='upper right', frameon=False, fontsize=6)
ax1.text(0.08, 0.85, 'a', fontweight='bold', transform=ax1.transAxes)

ax2 = fig.add_subplot(322)
ax2.plot(thetaDeg, rs.imag, 'r-', label='Im[$r_s$]')
ax2.plot(thetaDeg, rp.imag, 'b-', label='Im[$r_p$]')
ax2.plot(thetaDeg, ts.imag, 'r--', label='Im[$t_s$]')
ax2.plot(thetaDeg, tp.imag, 'b--', label='Im[$t_p$]')
```

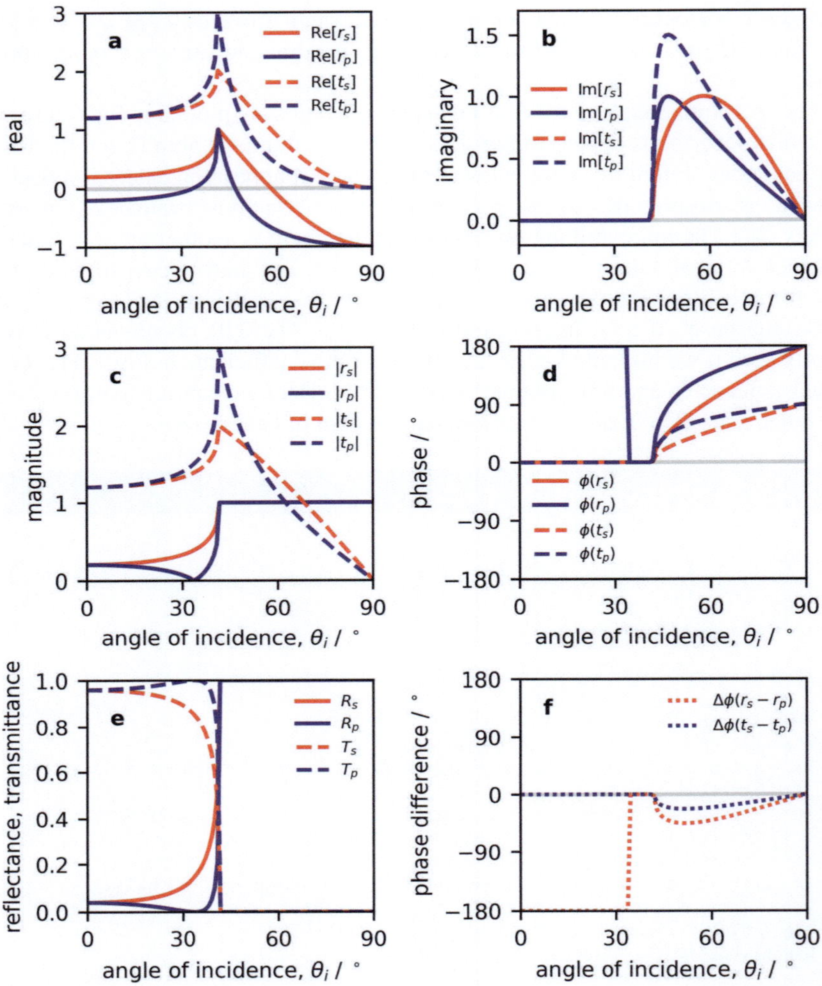

Fig. 7.11 The **a** real, **b** imaginary, **c** magnitude, and **d** phase of the Fresnel reflection (solid lines) and transmission (dashed lines) coefficients for internal reflection, where the beam approaches air ($N_2 = 1$) from glass ($N_1 = 1.5$). s-polarized coefficients are plotted in red; p-polarized versions in blue. **e** The corresponding transmittance and reflectance. **f** The difference in phase between s- and p-polarized coefficients in reflection (red) and transmission (blue)

```
ax2.set_ylabel('imaginary')
ax2.set_ylim(-0.2,1.7)
ax2.axhline(0, color='0.8', zorder=-999)
pyplot.legend(loc='center left', frameon=False, fontsize=6)
ax2.text(0.08, 0.85, 'b', fontweight='bold', transform=ax2.transAxes)

ax3 = fig.add_subplot(323)
ax3.plot(thetaDeg, abs(rs), 'r-', label='$|r_s|$')
ax3.plot(thetaDeg, abs(rp), 'b-', label='$|r_p|$')
ax3.plot(thetaDeg, abs(ts), 'r--', label='$|t_s|$')
ax3.plot(thetaDeg, abs(tp), 'b--', label='$|t_p|$')
ax3.set_ylabel('magnitude')
ax3.set_ylim(0,3)
pyplot.legend(loc='upper right', frameon=False, fontsize=6)
ax3.text(0.08, 0.85, 'c', fontweight='bold', transform=ax3.transAxes)

ax4 = fig.add_subplot(324)
ax4.plot(thetaDeg, np.degrees(np.arctan2(rs.imag, rs.real)), 'r-', label='$
    \phi(r_s)$')
ax4.plot(thetaDeg, np.degrees(np.arctan2(rp.imag, rp.real)), 'b-', label='$
    \phi(r_p)$')
ax4.plot(thetaDeg, np.degrees(np.arctan2(ts.imag, ts.real)), 'r--', label='
    $\phi(t_s)$')
ax4.plot(thetaDeg, np.degrees(np.arctan2(tp.imag, tp.real)), 'b--', label='
    $\phi(t_p)$')
ax4.set_ylabel(r'phase / $^\circ$')
ax4.set_ylim(-180, 180)
ax4.set_yticks(range(-180, 181, 90))
ax4.axhline(0, color='0.8', zorder=-999)
pyplot.legend(loc='lower left', frameon=False, fontsize=6)
ax4.text(0.08, 0.85, 'd', fontweight='bold', transform=ax4.transAxes)

ax5 = fig.add_subplot(325)
ax5.plot(thetaDeg, np.abs(rs)**2, 'r-', label='$R_s$')
ax5.plot(thetaDeg, np.abs(rp)**2, 'b-', label='$R_p$')
ax5.plot(thetaDeg, 1. - np.abs(rs)**2, 'r--', label='$T_s$')
ax5.plot(thetaDeg, 1. - np.abs(rp)**2, 'b--', label='$T_p$')
ax5.set_ylabel('reflectance, transmittance')
ax5.set_ylim(0, 1)
pyplot.legend(loc=0, frameon=False, fontsize=6)
ax5.text(0.08, 0.81, 'e', fontweight='bold', transform=ax5.transAxes)

ax6 = fig.add_subplot(326)
phase_rs = np.degrees(np.arctan2(rs.imag, rs.real))
phase_rp = np.degrees(np.arctan2(rp.imag, rp.real))
phase_ts = np.degrees(np.arctan2(ts.imag, ts.real))
phase_tp = np.degrees(np.arctan2(tp.imag, tp.real))
ax6.plot(thetaDeg, phase_rs - phase_rp, 'r--', dashes=[1,1], label=r'$\
    Delta \phi(r_s - r_p)$')
ax6.plot(thetaDeg, phase_ts - phase_tp, 'b--', dashes=[1,1], label=r'$\
    Delta \phi(t_s - t_p)$')
ax6.set_ylabel('phase difference / $^\circ$')
ax6.set_ylim(-180, 180)
ax6.set_yticks(range(-180, 181, 90))
ax6.axhline(0, color='0.8', zorder=-999)
pyplot.legend(loc='upper right', frameon=False, fontsize=6)
ax6.text(0.08, 0.85, 'f', fontweight='bold', transform=ax6.transAxes)
```

```
for ax in [ax1, ax2, ax3, ax4, ax5, ax6]:
    ax.set_xlabel(r'angle of incidence, $\theta_i$ / $^\circ$')
    ax.set_xlim(0, 90)
    ax.set_xticks(range(0, 91, 30))

fig.set_tight_layout(True)
pyplot.savefig('fresnel_internal.png', dpi=600)
```

7.5 Absorbance and Absorptance

In the next chapter, we will discuss absorption of IR radiation in detail, ultimately getting into how absorption spectra obtained using polarized light can be used to extract information about the orientation and conformation of molecules. But, just before we get into those details, we can at least extend our discussion of reflection and transmission to include the case where the refracted medium is absorbing, that is $N_2 = n_2 + i\kappa_2$ with $\kappa_2 > 0$. Since we are dealing with a single interface between two semi-infinite media (medium 1 extends from $z = -\infty$ to the interface at $z = 0^-$; medium two starts at $z = 0^+$ and extends to $z = \infty$) there will not be much sense in describing the transmitted beam. In other words, medium 2 is thick enough that the field is sufficiently attenuated after travelling a large distance that we do not detect any transmitted light. Note that this is not always the case, and we will later discuss absorbing media of finite thickness. But for now, we will not consider t_s, t_p, T_s, or T_p. Figure 7.12 presents the results for varying values of κ_2. Examining the results of s-polarized external reflection with $N_1 = 1$ and $N_2 = 1.5 + i\kappa_2$ (Fig. 7.12a), we first notice that the reflectance increases as κ_2 increases. Large values of κ_2 are typical of metals, and hence metals are frequently used to make mirrors. The same trend is seen in the p-polarized results (Fig. 7.12b), where we also notice a softening of what used to be the Brewster angle. This is the reason why, strictly speaking, the Brewster angle is a phenomenon associated with transparent media. We can also notice that, for either polarization, there is no noticeable difference in reflectance when $\kappa_2 < 0.1$. That is, the blue ($\kappa = 0$) and orange ($\kappa = 0.01$) lines in Fig. 7.12 are practically hidden underneath the green ($\kappa = 0.1$) line. This is an important consideration, as that would already be a significant attenuation coefficient for a typical organic functional group normal mode vibration in the infrared.

The case of internal reflection is even more interesting. Here we have used the same values of N_2, but consider $N_1 = 4$. This is typical of a material such as germanium in the mid-infrared, where the approximation $\kappa_1 \approx 0$ is also reasonable. The s-polarized reflectance is plotted in Fig. 7.12c, and the p-polarized case in Fig. 7.12d. Now we observe something curious: although below what used to be the critical angle (when $\kappa_2 = 0$) the reflectance increases with κ_2, the trend is reversed above this angle ($\theta_c = \arcsin(1.5/4) \approx 22°$). What we are observing is that

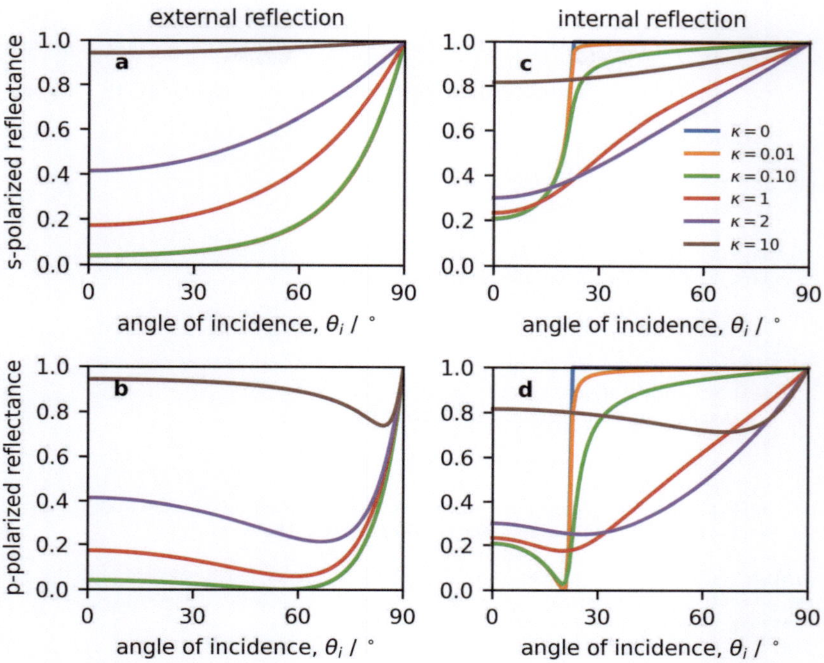

Fig. 7.12 The effect of an absorbing medium on the reflectance as a function of the angle of incidence. We first plot the case of external reflection where $N_1 = 1$ and $N_2 = 1.5 + i\kappa_2$ for **a** s-polarized reflectance and **b** p-polarized reflectance. The internal reflection cases are modelled with $N_1 = 4$ and $N_2 = 1.5 + i\kappa_2$ with specific κ_2 values shown in the inset to panel c for **c** s-polarization and **d** p-polarization

our total internal reflection is not quite so total anymore. When $\kappa_2 = 0$ we used to have all of the incident intensity reflected, and now the reflectance has been attenuated by the presence of a medium with $\kappa_2 > 0$. This phenomenon is referred to with the appropriately-named oxymoron, attenuated total (internal) reflection, formally ATIR, then ATiR, and these days simply ATR. We will return to further study the ATR phenomenon in the next chapter; here we simply point out that all of these effects are fully accounted for by the simple Fresnel equations.

Code used to generate Fig. 7.12

```
import numpy as np
from matplotlib import pyplot

def r(n1, n2):
    theta1 = np.radians(thetaDeg)
    theta2 = np.arcsin(n1*np.sin(theta1)/n2)
```

```
    rs = (n1*np.cos(theta1) - n2*np.cos(theta2))/(n1*np.cos(theta1) + n2*np
         .cos(theta2))
    rp = (n2*np.cos(theta1) - n1*np.cos(theta2))/(n1*np.cos(theta2) + n2*np
         .cos(theta1))
    return (rs, rp)

def go(n2):
    # external reflection
    rs, rp = r(1.0, n2)
    ax[0,0].plot(thetaDeg, abs(rs)**2)
    ax[1,0].plot(thetaDeg, abs(rp)**2)

    # internal reflection
    if n2.imag < 0.001:
        kappa = r'$\kappa = 0$'
    elif n2.imag < 0.9:
        kappa = r'$\kappa = %4.2f$' % n2.imag
    else:
        kappa = r'$\kappa = %d$' % int(n2.imag)
    rs, rp = r(4.0, n2)
    ax[0,1].plot(thetaDeg, np.abs(rs)**2, label=kappa, clip_on=False)
    ax[1,1].plot(thetaDeg, np.abs(rp)**2, clip_on=False)

pyplot.rc('figure', figsize=(4.5, 3.75))
pyplot.rc('font', size=8)
thetaDeg = np.linspace(0, 90, 100)
fig, ax = pyplot.subplots(2, 2)

go(1.5+0j)
go(1.5+0.01j)
go(1.5+0.1j)
go(1.5+1j)
go(1.5+2j)
go(1.5+10j)

ax[0,1].legend(loc=0, frameon=False, fontsize=6)
ax[0,0].set_ylabel('s-polarized reflectance')
ax[1,0].set_ylabel('p-polarized reflectance')
ax[0,0].set_title('external reflection', fontsize=8)
ax[0,1].set_title('internal reflection', fontsize=8)
ax[0,0].text(0.08, 0.87, 'a', fontweight='bold', transform=ax[0,0].
    transAxes)
ax[1,0].text(0.08, 0.87, 'b', fontweight='bold', transform=ax[1,0].
    transAxes)
ax[0,1].text(0.08, 0.87, 'c', fontweight='bold', transform=ax[0,1].
    transAxes)
ax[1,1].text(0.08, 0.87, 'd', fontweight='bold', transform=ax[1,1].
    transAxes)

for i in range(2):
    for k in range(2):
        ax[i,k].set_xlabel(r'angle of incidence, $\theta_i$ / $^\circ$')
        ax[i,k].set_xlabel(r'angle of incidence, $\theta_i$ / $^\circ$')
        ax[i,k].set_xlim(0, 90)
        ax[i,k].set_ylim(0, 1)
        ax[i,k].set_xticks(range(0, 91, 30))
```

```
fig.set_tight_layout(True)
pyplot.savefig('fresnel_absorbance.png', dpi=600)
```

We have seen that the intensity of the light is proportional to the square of the electric field amplitude. If we write $I_0 \propto |E_0|^2$ for the incident light intensity, and $I_\mathrm{R} \propto |E_\mathrm{R}|^2$ for the reflected intensity, we see that our definition of the reflectance follows from

$$R = \frac{I_\mathrm{R}}{I_0} = \frac{|E_\mathrm{R}|^2}{|E_0|^2} = \frac{|rE_0|^2}{|E_0|^2} = |r|^2. \tag{7.65}$$

The transmittance can be determined in a similar way, and I_T/I_0 comes into play, but with the additional consideration that the incident and transmitted fields are travelling in different media. If some of the light has been absorbed by a material with $\kappa_2 > 0$, then some of the light is unaccounted for in the transmittance and reflectance and $R + T < 1$. We define another quantity, the *absorptance*

$$A' = \frac{I_\mathrm{A}}{I_0} \tag{7.66}$$

that enables us to write $R + T + A' = 1$ for any system. As a word of caution, not many people discuss the absorptance so it is difficult to decide on what symbol to use for this quantity. In cases where it is described, the most common notation is A, but we will use A' in order to distinguish it from the *absorbance*.

Although it is possible to measure T and/or R, this is seldom done in practice for applications such as those considered here. Instead, the beam spectrum I_0 is not measured directly, but the transmitted (or reflected) intensity from the sample I_t (or I_r) and a judiciously-chosen reference sample I_t^ref (or I_r^ref) are used to calculate the absorbance, A. Since there are many possible choices of a reference sample, there is no precise definition of absorbance, and it is therefore necessary to clarify what this quantity represents each time it is introduced. For example, suppose you are using ATR-IR to study phospholipids at the interface between the prism (IRE) and the bulk aqueous phase. The reference sample can be a dry prism (i.e. the IRE–air interface) or the same aqueous phase, but without the lipids (something close to the prism–water interface). The latter choice makes sense as it might be better at revealing the lipid vibrational modes that interfere with water (although this is not a rigorous separation), but the point is that either choice is valid, but will result in drastically different values of the absorbance. Such experimental considerations will be revisited in Chap. 8 when we discuss IR absorption spectroscopy. In general, we define the transmission absorbance A_t and reflection absorbance A_r for these two quantities

$$A_t = -\log_{10}\left[\frac{I_t}{I_t^{ref}}\right] \tag{7.67a}$$

$$A_r = -\log_{10}\left[\frac{I_r}{I_r^{ref}}\right]. \tag{7.67b}$$

We intentionally use this pedantic nomenclature in order to avoid confusion with the absorptance A'. Readers should be aware that most works use the symbol A for all three quantities! Note that Eq. 7.67 are equivalent to

$$A_t = -\log_{10}\left[\frac{T}{T^{ref}}\right] \tag{7.68a}$$

$$A_r = -\log_{10}\left[\frac{R}{R^{ref}}\right] \tag{7.68b}$$

since the ratios T/T^{ref} and R/R^{ref} are independent I_0, thereby eliminating the need to measure the beam spectrum when comparing measured ratios to those calculated assuming $I_0 = 1$. This discussion will be continued in detail in the next chapter devoted to IR absorption spectroscopy. But I wanted to introduce the basic concept of absorbance and absorptance here to illustrate that these quantities are readily calculated if you know the reflectance or transmittance for a set of samples (with one used as a reference). Some end-of-chapter problems present opportunities to calculate such absorption spectra.

7.6 Interaction of Polarized Light with Materials: Jones and Mueller Matrices

In Chap. 6 we introduced a Jones vector description of polarized light, and a Stokes vector description that can be used to describe any form of light, even if it is only partially polarized. Although the characterization of the polarization state of light (polarimetry) is a measurement science in its own right, the primary motivation for us is to understand how materials have caused changes in the polarization, and then ultimately to relate this to the organization of the molecules that comprise those materials. In addition to the samples we study, there are polarization-altering elements that form important parts of the spectrometers used in the measurements. Of those components, the two basic types are polarizers and waveplates, also known as retarders. Such elements can be described in the form of simple matrices. The Jones matrix $\mathbf{M}$ is the general transformation that relates the electric field components of any input field $\mathbf{E}_{in}$ to those of the transmitted or reflected output field $\mathbf{E}_{out}$,

$$\mathbf{E}_{out} = \mathbf{M} \cdot \mathbf{E}_{in}. \tag{7.69}$$

Similarly, the Mueller matrix **M** provides the analogous transformation for the Stokes vector

$$\mathbf{S}_{\text{out}} = \mathbf{M} \cdot \mathbf{S}_{\text{in}}. \tag{7.70}$$

The length of the vectors **E** and **S** inform us of the dimensions of the rank-2 tensors **M**. The Jones vector describes the orthogonal linear components of the field perpendicular to the propagation direction, and so **M** is a 2×2 matrix with 4 complex-valued elements; the Mueller matrix **M** is a 4×4 matrix with 16 real-valued elements.

7.6.1 Polarizers

Polarizers, almost always linear polarizers, are devices that accept any incident polarization state, and produce plane-polarized light, usually upon transmission at normal incidence, although there are other possibilities. In simple cases, the azimuth of the resulting linear polarized light can be controlled simply by rotating the polarizer about its normal axis. This is a prime example of how the elements of a Jones matrix are often easy to determine by inspection. For example, since

$$\begin{gathered}\mathbf{E}_{\text{out}} = \mathbf{M} \cdot \mathbf{E}_{\text{in}} \\ \begin{bmatrix} E_x \\ E_y \end{bmatrix}_{\text{out}} = \begin{bmatrix} M_{11} & M_{12} \\ M_{21} & M_{22} \end{bmatrix} \begin{bmatrix} E_x \\ E_y \end{bmatrix}_{\text{in}}\end{gathered} \tag{7.71}$$

if we were to need the Jones matrix of a polarizer with a horizontal transmission axis, we seek the elements of **M** for which

$$\begin{bmatrix} 1 \\ 0 \end{bmatrix} = \begin{bmatrix} M_{11} & M_{12} \\ M_{21} & M_{22} \end{bmatrix} \begin{bmatrix} E_x \\ E_y \end{bmatrix}_{\text{in}}. \tag{7.72}$$

In other words regardless of $\mathbf{E}_{\text{in}}$, we want to end up with all of the light horizontal, blocking any component of the input state with a projection along the vertical direction. By inspection, we realize that the solution has $M_{11} = 1$ and $M_{12} = M_{21} = M_{22} = 0$ and so we conclude that

$$\mathbf{M}_{\text{polarizer}}(0^\circ) = \begin{bmatrix} 1 & 0 \\ 0 & 0 \end{bmatrix}$$

where we have added the azimuthal angle of the transmission axis, 0° indicating that it is horizontal. This is an ideal polarizer as it blocks all of the vertical component, and transmits 100% of the horizontal component of **E**, without imparting any phase shift. A similar analysis would lead us to the conclusion that the Jones matrix of a polarizer that transmits vertically-polarized light would be

$$\mathbf{M}_{\text{polarizer}}(90^\circ) = \begin{bmatrix} 0 & 0 \\ 0 & 1 \end{bmatrix}.$$

In principle, the same sort of analysis can be carried out with Stokes vectors to provide the results of a horizontal and vertical transmission Mueller matrix

$$\mathbf{M}_{\text{polarizer}}(0^\circ) = \frac{1}{2}\begin{bmatrix} 1 & 1 & 0 & 0 \\ 1 & 1 & 0 & 0 \\ 0 & 0 & 0 & 0 \\ 0 & 0 & 0 & 0 \end{bmatrix} \text{ and } \mathbf{M}_{\text{polarizer}}(90^\circ) = \frac{1}{2}\begin{bmatrix} 1 & -1 & 0 & 0 \\ -1 & 1 & 0 & 0 \\ 0 & 0 & 0 & 0 \\ 0 & 0 & 0 & 0 \end{bmatrix}.$$

As an example, consider a 45° linearly-polarized beam passing through an ideal polarizer with a vertical transmission axis. The Jones vector of the output field is given by

$$\begin{aligned} \mathbf{E}_{\text{out}} &= \mathbf{M}_{\text{polarizer}}(90^\circ)\, \mathbf{E}_{\text{in}}(45^\circ) \\ &= \begin{bmatrix} 0 & 0 \\ 0 & 1 \end{bmatrix} \left(\frac{1}{\sqrt{2}} \begin{bmatrix} 1 \\ 1 \end{bmatrix} \right) \\ &= \frac{1}{\sqrt{2}} \begin{bmatrix} 0 \\ 1 \end{bmatrix}. \end{aligned}$$

The same problem solved using Stokes vectors would look like

$$\begin{aligned} \mathbf{S}_{\text{out}} &= \frac{1}{2}\begin{bmatrix} 1 & -1 & 0 & 0 \\ -1 & 1 & 0 & 0 \\ 0 & 0 & 0 & 0 \\ 0 & 0 & 0 & 0 \end{bmatrix} \begin{bmatrix} 1 \\ 0 \\ 1 \\ 0 \end{bmatrix} \\ &= \frac{1}{2}\begin{bmatrix} 1 \\ -1 \\ 0 \\ 0 \end{bmatrix}. \end{aligned}$$

It's a good idea to verify if the solution makes sense according to the definition of the Stokes vector provided in Eq. 6.13, and whether you calculate the same transmittance value using this approach. Although NumPy arrays work perfectly and are the workhorse for these types of analysis, the linear algebra features in SymPy are a great tool for deriving and comparing expressions for this type of work.

7.6.2 Retarders

A retarder, or waveplate, is a material or device that shifts the relative phase between orthogonal components of **E**. An ideal retarder does so without modifying its amplitude components. Let us first consider a retarder whose slow axis is horizontal, that is, horizontally-polarized light passing through the retarder is delayed more than vertically-polarized light. A simple retarder is a birefringent material, in this case oriented such that its refractive index is larger along the horizontal axis. Retarders must be specified according to the phase shift they impart and the orientation of their slow axis. In general, the Jones matrix for a retarder with horizontal (0°) slow axis and retardation $\Delta\phi$ is

$$\mathbf{M}_{\text{ret}}(\Delta\phi, 0°) = \begin{bmatrix} e^{i\Delta\phi} & 0 \\ 0 & 1 \end{bmatrix} = \begin{bmatrix} 1 & 0 \\ 0 & e^{-i\Delta\phi} \end{bmatrix} = \begin{bmatrix} e^{\frac{1}{2}i\Delta\phi} & 0 \\ 0 & e^{-\frac{1}{2}i\Delta\phi} \end{bmatrix}.$$

Of the three versions provided above, the first is the most compact. The second form recognizes that, since phase is a relative quantity, it makes no difference if we delay the horizontal component of the field by $\Delta\phi$ rad or if we advance the vertical component by $\Delta\phi$ rad (an effective "delay" by $-\Delta\phi$ rad). For the same reason, if neither of those choices sit well with you, then you can go with the third option: split the difference and delay the horizontal component by $\frac{1}{2}\Delta\phi$ rad and advance the vertical component by $\frac{1}{2}\Delta\phi$ rad. It doesn't matter: either way the x-component will emerge phase shifted by $\Delta\phi$ rad with respect to the y-component. Fortunately for the indecisive, there are not so many choices available in the Mueller matrix, and the same operation is achieved with

$$\mathbf{M}_{\text{ret}}(\Delta\phi, 0°) = \begin{bmatrix} 1 & 0 & 0 & 0 \\ 0 & 1 & 0 & 0 \\ 0 & 0 & \cos\Delta\phi & -\sin\Delta\phi \\ 0 & 0 & \sin\Delta\phi & \cos\Delta\phi \end{bmatrix}.$$

7.6.3 Matrix Descriptions of Reflection and Transmission at Interfaces

When light interacts with materials at normal incidence, it is conventional to consider the direction of propagation as z, and then the Jones vector specifies the x- and y-components of the field. However, it is also possible to use p- and s-polarized states as basis vectors, so the light transmitted across an interface can be obtained via

$$\begin{bmatrix} E_{\text{p}} \\ E_{\text{s}} \end{bmatrix}_{\text{out}} = \begin{bmatrix} t_{\text{p}} & 0 \\ 0 & t_{\text{s}} \end{bmatrix} \begin{bmatrix} E_{\text{p}} \\ E_{\text{s}} \end{bmatrix}_{\text{in}} \tag{7.73}$$

if the material is isotropic (no mixing of s- and p-states). Similarly, the fields upon reflection are described as

$$\begin{bmatrix} E_{\text{p}} \\ E_{\text{s}} \end{bmatrix}_{\text{out}} = \begin{bmatrix} r_{\text{p}} & 0 \\ 0 & r_{\text{s}} \end{bmatrix} \begin{bmatrix} E_{\text{p}} \\ E_{\text{s}} \end{bmatrix}_{\text{in}}. \tag{7.74}$$

In the case of reflection, care must be taken in the sign of r_p as a result of the change in handedness of the coordinate system described earlier. Some authors use a different sign in the definition of r_p, and then change the sign in the Jones matrix above. Whatever the convention, we can conveniently switch from (E_x, E_y) to (E_p, E_s) basis vectors for the description of the polarization.

Although it is relatively straightforward to describe the corresponding Mueller matrices, this is a good time to mention that, as expected, any Jones matrix $\mathbf{M}_J$ can be converted into a Mueller matrix $\mathbf{M}_M$. This transformation [16] is given by

$$\mathbf{M}_M = \mathbf{C} \cdot (\mathbf{M}_J \otimes \mathbf{M}_J^*) \cdot \mathbf{C}^{-1} \tag{7.75}$$

where

$$\mathbf{C} = \frac{1}{\sqrt{2}} \begin{bmatrix} 1 & 0 & 0 & 1 \\ 1 & 0 & 0 & -1 \\ 0 & 1 & 1 & 0 \\ 0 & -i & i & 0 \end{bmatrix}.$$

Here the symbol $\otimes$ is the Kronecker product, a special type of outer product. We have recently seen how the outer product of two vectors (tensors of rank 1) can be used to form a matrix (tensor of rank 2). In Eq. 7.75, we build a 4×4 matrix (tensor of rank 2) from two 2×2 matrices (tensors of rank 2). To be explicit, if

$$\mathbf{A} = \begin{bmatrix} a_{11} & a_{12} \\ a_{21} & a_{22} \end{bmatrix}$$

and

$$\mathbf{B} = \begin{bmatrix} b_{11} & b_{12} \\ b_{21} & b_{22} \end{bmatrix}$$

then

$$\mathbf{A} \otimes \mathbf{B} = \begin{bmatrix} a_{11}\mathbf{B} & a_{12}\mathbf{B} \\ a_{21}\mathbf{B} & a_{22}\mathbf{B} \end{bmatrix}$$
$$= \begin{bmatrix} a_{11}b_{11} & a_{11}b_{12} & a_{12}b_{11} & a_{12}b_{21} \\ a_{11}b_{21} & a_{11}b_{22} & a_{12}b_{21} & a_{12}b_{22} \\ a_{21}b_{11} & a_{21}b_{12} & a_{22}b_{11} & a_{22}b_{21} \\ a_{21}b_{21} & a_{21}b_{22} & a_{22}b_{21} & a_{22}b_{22} \end{bmatrix}.$$

In Python, this can be achieved using the NumPy `kron` function. As an example

```
A = np.array([[1,2],[3,4]])
>>> A
array([[1, 2],
       [3, 4]])
B = np.array([[5,6],[7,8]])
>>> B
array([[5, 6],
```

```
       [7, 8]])
>>> np.kron(A, B)
array([[ 5,  6, 10, 12],
       [ 7,  8, 14, 16],
       [15, 18, 20, 24],
       [21, 24, 28, 32]])
```

Note that this same procedure whereby the Kronecker product and the **A** operators are used can be evaluated easily in SymPy

```
import sympy as sp
rp, rs = sp.symbols('r_p r_s', complex=True)
Jones = sp.Matrix([[rp, 0], [0, rs]])

A = 1/sp.sqrt(2) * sp.Matrix([[1, 0, 0, 1], [1, 0, 0, -1],
     [0, 1, 1, 0], [0, -sp.I, sp.I, 0]])

Mueller = A * sp.KroneckerProduct(Jones, sp.conjugate(
    Jones)) * sp.Inverse(A)
```

to produce

$$\mathbf{M}_{\text{reflection}} = \frac{1}{2}\begin{bmatrix} |r_\mathrm{p}|^2 + |r_\mathrm{s}|^2 & |r_\mathrm{p}|^2 - |r_\mathrm{s}|^2 & 0 & 0 \\ |r_\mathrm{p}|^2 - |r_\mathrm{s}|^2 & |r_\mathrm{p}|^2 + |r_\mathrm{s}|^2 & 0 & 0 \\ 0 & 0 & 2\mathbb{R}\mathrm{e}\{r_\mathrm{p}r_\mathrm{s}^*\} & 2\mathbb{I}\mathrm{m}\{r_\mathrm{p}r_\mathrm{s}^*\} \\ 0 & 0 & -2\mathbb{I}\mathrm{m}\{r_\mathrm{p}r_\mathrm{s}^*\} & 2\mathbb{R}\mathrm{e}\{r_\mathrm{p}r_\mathrm{s}^*\} \end{bmatrix} \tag{7.76}$$

if you were in need of the form of the Mueller matrix for reflection from an isotropic surface. The actual output of `sp.pprint(Mueller)` is not as tidy as what is displayed in Eq. 7.76, but could be made to look nicer with various additional tricks to recognize things like $r_\mathrm{p}r_\mathrm{p}^* = |r_\mathrm{p}|^2$.

7.6.4 *Rotated Polarization-Altering Elements*

Based on the above examples, it is probably straightforward to imagine what the Jones and Mueller matrices of the retarders would look like if the slow axis was vertical instead of horizontal, that is with the fast axis now vertical. However, when using any Jones or Mueller matrix, one often needs a form that is rotated by an arbitrary angle with respect to the horizontal. This is easily achieved via the transformation

$$\mathbf{M}(\theta) = \mathbf{R}(-\theta) \cdot \mathbf{M}(0^\circ) \cdot \mathbf{R}(\theta) \tag{7.77}$$

where the rotation operator **R** has a Jones version

$$\mathbf{R}(\theta) = \begin{bmatrix} \cos\theta & \sin\theta \\ -\sin\theta & \cos\theta \end{bmatrix} \tag{7.78}$$

and a Mueller version

$$\mathbf{R}(\theta) = \begin{bmatrix} 1 & 0 & 0 & 0 \\ 0 & \cos 2\theta & \sin 2\theta & 0 \\ 0 & -\sin 2\theta & \cos 2\theta & 0 \\ 0 & 0 & 0 & 1 \end{bmatrix}. \tag{7.79}$$

For retarders, we will therefore use the general notation

$$\mathbf{M}_{\text{ret}}(\Delta\phi, \theta) = \mathbf{R}(-\theta) \cdot \mathbf{M}_{\text{ret}}(\Delta\phi, 0^\circ) \cdot \mathbf{R}(\theta). \tag{7.80}$$

One way to think about this transformation is a rotation of the laboratory frame by an angle θ so that the Jones/Mueller matrix of the 0° optic or material now applies, and we restore the lab to its proper orientation with $\mathbf{R}(-\theta)$ after we are done. For a more precise definition, this is an identity that follows from a more formal method of transforming tensors of arbitrary rank. We will discuss those methods further when we encounter them in subsequent chapters.

7.6.5 The Effect of Multiple Polarization-Altering Elements

The real power of using Jones and Mueller matrices is that relatively complex systems can easily be modelled by successively building up the response of the constituent effects. For a series of n optical elements, it is then necessary to know (or create) only the appropriate matrices for each element, and then operate on the incident Jones vector with them in turn as

$$\mathbf{E}_{\text{out}} = \mathbf{M}_n \cdot \mathbf{M}_{(n-1)} \cdots \mathbf{M}_2 \cdot \mathbf{M}_1 \cdot \mathbf{E}_{\text{in}} \tag{7.81}$$

or

$$\mathbf{S}_{\text{out}} = \mathbf{M}_n \cdot \mathbf{M}_{(n-1)} \cdots \mathbf{M}_2 \cdot \mathbf{M}_1 \cdot \mathbf{S}_{\text{in}} \tag{7.82}$$

in the case of Stokes vectors. The method is therefore readily lends itself to implementation in an algorithm to be run by a computer, and we will extensively make use of such operations in our Python codes, primarily using NumPy.

7.7 Light Propagation in Stratified Systems

Up until this point, we have considered light propagating in a semi-infinite phase (mostly air) approaching the boundary (located at $z = 0$) with another semi-infinite phase. The second phase is most often the material of interest, usually absorbing with $N_2 = n_2 + i\kappa_2$. This was also the case in our recent discussion of internal reflection, where medium 1 is still transparent, only with a larger refractive index (for example $N_1 = 4$ in the case of Ge in the mid-infrared). In many real samples, however, we must consider that the sample has a finite thickness. This may be encountered when studying monolayers deposited on a substrate, or a thin film that has been prepared by spin-coating, or even a multi-layer system that has self-assembled or been fabricated. We will now see how we can extend the formalism developed so far to determine the reflectance and transmittance for such systems. Along the way, we will develop expressions for the electric field amplitude and phase as a function of distance in the material. Those complex-valued field amplitudes will be of use when we later consider local field effects.

7.7.1 *Geometric Approach*

There have been many methods and formalisms proposed for the reflection and transmission coefficients for thin films, and systems composed of multiple layers of thin films [12, 17]. Some of these have been popularized due to their convenience, especially before computers were widely available and analytical approaches were favoured. Some of those are still interesting, even though, by default, we typically seek to implement numerical solutions for comparison with experimental data. We begin with the consideration of what is historically referred to as a 3-phase (or single layer) system, where a material of thickness d and complex refractive index $N_2 = n_2 + i\kappa_2$ is sandwiched between two bulk phases. The phase in which the beams are initially incident will be referred to as material 1, with refractive index n_1 ($N_1 \approx n_1$ as we consider transparent materials for which $\kappa_1 \approx 0$). In cases where the transmitted beam is detected, people typically choose medium 3 to be transparent as well (refractive index n_3). However, in reflection experiments there is no such requirement, so we cover the more general case where the refractive index of that medium is written as $N_3 = n_3 + i\kappa_3$.

Referring to Fig. 7.13, we can see that, if the film thickness d is small compared to the diameter of the beam, the detected field in the reflected direction is a linear superposition of

$$E_\mathrm{r} = E_{\mathrm{r}1} + E_{\mathrm{r}2} + E_{\mathrm{r}3} + \cdots \tag{7.83}$$

which we assume to be perfectly overlapped in space. The first term in the series arises from a single reflection from the interface between media 1 and 2, and so $E_{\mathrm{r}1} = r_{12}E_\mathrm{i}$, where r_{12} refers to the simple Fresnel reflection coefficient defined in

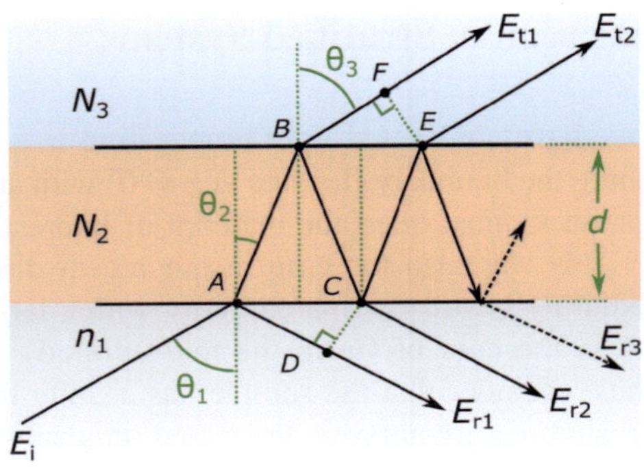

Fig. 7.13 The reflection $\tilde{r}$ and transmission $\tilde{t}$ coeeicients include multiple reflections inside a thin film can be arrived at from simple geometric considerations by determining the appropriate combination of simple Fresnel reflection r and transmission t coefficients that account for each of the substituent reflected and transmitted beams that are superimposed to form the net reflected and transmitted fields

the previous section, making the appropriate choice of r_s or r_p depending on the polarization of the beam. We previously referred to such coefficients simply as r, but our new notation r_{12} simply implies $r(N_1, N_2, \theta_1)$, i.e. that the incident medium is material 1, and the beam is refracted into medium 2. This overcomes the challenge we described when switching from the i and t descriptors of the incident and transmitted materials, especially critical to sort out as we now account for beams crossing the boundaries in both directions. The second beam originates from the incident field first penetrating into the film, propagating through to the interface between materials 2 and 3, reflecting from that surface, propagating back up, and being transmitted from material 2 (the film) into material 1. This path may be described as

$$E_{r2} = E_i t_{12} r_{23} e^{i\Delta\phi} t_{21} \tag{7.84}$$

where t is the s- or p-polarized Fresnel transmission coefficient defined earlier, and $e^{i\Delta\phi}$ accounts for the phase shift between beams 1 and 2 as a result of the double pass through the film. Referring to the notation defined in Fig. 7.13, $\Delta\phi$ can be calculated from the optical path length difference. We need to evaluate the difference between two passes in the film via $\overline{AB}$ and $\overline{BC}$ versus the distance $\overline{AD}$ in the incident medium to arrive at

$$\begin{aligned}
\Delta\phi &= \frac{2\pi}{\lambda}\left[N_2\left(\overline{AB}\right) + N_2\left(\overline{BC}\right) - n_1\left(\overline{AD}\right)\right] \\
&= \frac{2\pi}{\lambda}\left(N_2\frac{d}{\cos\theta_2} + N_2\frac{d}{\cos\theta_2} - 2n_1 d\sin\theta_1\tan\theta_2\right) \\
&= \frac{4\pi}{\lambda}N_2\cos\theta_2.
\end{aligned} \tag{7.85}$$

A similar set of considerations enables us to write an expression for the third reflected beam

$$E_{r3} = E_{\rm i} t_{12} r_{23} e^{i\Delta\phi} (r_{21} r_{23} e^{i\Delta\phi}) t_{21}. \tag{7.86}$$

Here we use parentheses to help identify a component that can later be factored when we write a similar expression for an arbitrary term E_{rn} in this infinite series. We need to transmit the incident field into the film, make two reflections inside the film for each subsequent beam, and then refract back into the incident medium to meet up with the other beams. We can therefore write the general expression

$$E_{\rm rn} = E_i t_{12} r_{23} e^{i\Delta\phi} (r_{21} r_{23} e^{i\Delta\phi})^{n-2} t_{21}. \tag{7.87}$$

Our end goal is to arrive at the overall reflection coefficient for this system, that we will refer to as $\tilde{r}$ in order to distinguish it from the simple two-phase Fresnel reflection coefficient r. We therefore need to evaluate

$$\tilde{r} = \frac{E_{\rm r}}{E_{\rm i}} = \frac{E_{\rm r1}}{E_{\rm i}} + \frac{\sum\limits_{n=2}^{\infty} E_{\rm rn}}{E_{\rm i}}. \tag{7.88}$$

We can compare this with the well-known expression for a converging infinite geometric series

$$\sum_{n=0}^{\infty} ax^n = \frac{a}{1-x} \tag{7.89}$$

that is valid when $-1 < x < 1$. In our case, $|x| = |r_{21} r_{23} e^{i\Delta\phi}| < 1$ since each value of $|r| < 1$. So we can use Eq. 7.89 to write the reflection coefficient as

$$\tilde{r} = r_{12} + \frac{t_{12} r_{23} t_{21} e^{i\Delta\phi}}{1 - r_{21} r_{23} e^{i\Delta\phi}}. \tag{7.90}$$

This expression is used to obtain $\tilde{r}_{\rm s}$ and $\tilde{r}_{\rm p}$ depending on whether $r_{\rm s}$ and $t_{\rm s}$, or $r_{\rm p}$ and $t_{\rm p}$ are used throughout Eq. 7.90. There are slightly more compact ways of writing Eq. 7.90 that make use of some identities, but we'll keep it as-is since the current form better reveals the underlying considerations upon which it was constructed.

Now we need the set of transmission coefficients $\tilde{t}_{\rm s}$ and $\tilde{t}_{\rm p}$. We again turn to Fig. 7.13 to see that the overall transmission coefficient $\tilde{t}$ can be obtained using a similar set of considerations, adding up the transmitted rays in medium 3 shown on the top of Fig. 7.13. This results in

$$\begin{aligned} E_{t1} &= E_i t_{12} t_{23} e^{i\Delta\phi/2} \\ E_{t2} &= E_i t_{12} t_{23} e^{i\Delta\phi/2} (r_{23} r_{21} e^{i\Delta\phi}) \\ &\vdots \\ E_{tn} &= E_i t_{12} t_{23} e^{i\Delta\phi/2} (r_{23} r_{21} e^{i\Delta\phi})^{n-1} \end{aligned} \tag{7.91}$$

where the phase difference is given by

$$\begin{aligned} \Delta\phi &= \frac{2\pi}{\lambda} \left[N_2 \left(\overline{BC}\right) + N_2 \left(\overline{CE}\right) - N_3 \left(\overline{BF}\right) \right] \\ &= \frac{2\pi}{\lambda} \left(\frac{2dN_2}{\cos\theta_2} - 2dN_3 \sin\theta_3 \tan\theta_2 \right) \\ &= \frac{4\pi}{\lambda} N_2 \cos\theta_2 \end{aligned} \tag{7.92}$$

to arrive at the same solution for $\Delta\phi$ as was determined for the reflected field, even though we now consider medium 3 in place of medium 1. Comparing with the form of Eq. 7.89, we see that this is also a converging series, and we end up with

$$\tilde{t} = \frac{t_{12} t_{23} e^{i\Delta\phi/2}}{1 - r_{23} r_{21} e^{i\Delta\phi}}. \tag{7.93}$$

The reflectance and transmittance can now be obtained in the usual way, very similar to Eqs. 7.64 and 7.62 except that we are now marking these quantities with a tilde to emphasize that they came from considerations of multiple beam interference. Also note that Eqs. 7.64 and 7.62 were derived for a two-phase system, so T_p and T_s contained N_1 and N_2, but $\tilde{T}_p$ and $\tilde{T}_s$ in Eq. 7.94 contain N_1 and N_3. Information about the optical constants n_2 and κ_2 of the film are also embedded in those expressions, through $\tilde{t}_p$ and $\tilde{t}_s$, and the corresponding reflection coefficients.

$$\tilde{R}_p = |\tilde{r}_p|^2 \tag{7.94a}$$

$$\tilde{R}_s = |\tilde{r}_s|^2 \tag{7.94b}$$

$$\tilde{T}_p = \left(\frac{N_3 \cos\theta_N + N_3^* \cos\theta_3^*}{N_1 \cos\theta_1 + N_1^* \cos\theta_1^*} \right) |\tilde{t}_p|^2 \tag{7.94c}$$

$$\tilde{T}_s = \left(\frac{N_3 \cos\theta_N^* + N_3^* \cos\theta_N}{N_1 \cos\theta_1^* + N_1^* \cos\theta_1} \right) |\tilde{t}_s|^2. \tag{7.94d}$$

The method we have illustrated above can be extended to systems that have more than one layer in much the same way. However, the expressions get rather complicated considering the many options for transmitted and reflected beams at each layer. In such cases, people generally prefer to use a matrix-based approach as described below.

7.7.2 Matrix-Based Approach

An alternative approach describes the general case of light incident from a transparent semi-infinite phase (such as air, $N_1 = n_1$, $\kappa_1 = 0$) with angle θ_1 with respect to the surface normal $\hat{z}$ onto a stratified system consisting of an arbitrary number of thin layers. Each layer j of thickness d_j has refractive index $N_j = n_j + i\kappa_j$. Finally the beam encounters the last semi-infinite phase with $N_N = n_N + i\kappa_N$. The exact solution to Maxwell's equations can been formulated in a compact extensible form [18–20] where each of the layers (from $j = 2$ to $j = N - 1$) is accounted for by a matrix $\mathbf{M}_j$ that relates the tangential component U of the electric field, and V of the magnetic fields from the first surface ($z = z_1 = 0$) to the last surface ($z = z_N$)

$$\begin{bmatrix} U_1 \\ V_1 \end{bmatrix} = \mathbf{M}_2\mathbf{M}_3\mathbf{M}_4 \cdots \mathbf{M}_{N-1} \begin{bmatrix} U_{N-1} \\ V_{N-1} \end{bmatrix} \tag{7.95}$$

where

$$\mathbf{M}_{\mathrm{s},j} = \begin{bmatrix} \cos\beta_j & -\dfrac{i}{p_j}\sin\beta_j \\ -ip_j\sin\beta_j & \cos\beta_j \end{bmatrix} \tag{7.96}$$

in the case of an s-polarized field with $p_j = N_j \cos\theta_j$ if the layer is isotropic, or $p_j = N_{y,j}\cos\theta_j$ if the medium is anisotropic, thereby probing the y-component of the refractive index. Here θ_j is the refracted angle in medium j for s-polarized light ($\mathbf{E}$ along $\hat{y}$, perpendicular to the plane of incidence). For p-polarized light (components of $\mathbf{E}$ along $\hat{x}$ and $\hat{z}$, parallel to the plane of incidence), one has

$$\mathbf{M}_{\mathrm{p},j} = \begin{bmatrix} \cos\beta_j & -\dfrac{i}{q_j}\sin\beta_j \\ -iq_j\sin\beta_j & \cos\beta_j \end{bmatrix} \tag{7.97}$$

where $q_j = \cos\theta_j/N_j$ for isotropic layers, or $q_j = \cos\theta_j/N_{x,j}$ if one is using a different refractive index N_x along the x-direction. Keep in mind that, when using anisotropic refractive indices, the appropriate form of Snell's Law of refraction (Eq. 7.58) should be used where $N_{z,j}$ takes the place of the isotropic refractive index N_j. The term accounting for the phase of the fields between the layers is defined as

$$\beta_j = \frac{2\pi d_j p_j}{\lambda}. \tag{7.98}$$

We can then define a single matrix $\mathbf{M}$ that represents the entire system comprising of all layers

$$\mathbf{M} = \mathbf{M}_2\mathbf{M}_3\mathbf{M}_4 \cdots \mathbf{M}_{N-1} \tag{7.99}$$

so that

$$\begin{bmatrix} U_1 \\ V_1 \end{bmatrix} = \mathbf{M} \begin{bmatrix} U_{N-1} \\ V_{N-1} \end{bmatrix} \tag{7.100}$$

which can be used to arrive at the reflection and transmission coefficients

$$\tilde{r}_{\mathrm{p}} = \frac{(M_{11} + M_{p,12}q_N)q_1 - (M_{p,21} + M_{22}q_N)}{(M_{11} + M_{p,12}q_N)q_1 + (M_{p,21} + M_{22}q_N)} \tag{7.101a}$$

$$\tilde{r}_{\mathrm{s}} = \frac{(M_{11} + M_{s,12}p_N)p_1 - (M_{s,21} + M_{22}p_N)}{(M_{11} + M_{s,12}p_N)p_1 + (M_{s,21} + M_{22}p_N)} \tag{7.101b}$$

$$\tilde{t}_{\mathrm{p}} = \left(\frac{N_1}{N_N}\right) \frac{2q_1}{(M_{11} + M_{p,12}q_N)q_1 + (M_{p,21} + M_{22}q_N)} \tag{7.101c}$$

$$\tilde{t}_{\mathrm{s}} = \frac{2p_1}{(M_{11} + M_{s,12}p_N)p_1 + (M_{s,21} + M_{22}p_N)} \tag{7.101d}$$

where p_1 and q_1 are in the first layer, p_N and q_N are in the last layer, and M_{ij} are elements of the overall s- or p-version of the $\mathbf{M}$ matrix that have the same diagonal elements M_{11} and M_{22}. Note that we have put a tilde ($\tilde{r}_{\mathrm{p}}$ etc.) on these coefficients to distinguish them from the versions of the Fresnel coefficients that we originally derived (Eqs. 7.54, 7.49, 7.55 and 7.50) for two semi-infinite phases. The distinction isn't that important, but I'd like to keep using this notation as it will be convenient for us in later chapters. If the reflectance and transmittance are desired, those quantities can now be obtained from the usual expressions (see Eqs. 7.64, 7.62 and 7.94)

$$\tilde{R}_{\mathrm{p}} = |\tilde{r}_{\mathrm{p}}|^2 \tag{7.102a}$$

$$\tilde{R}_{\mathrm{s}} = |\tilde{r}_{\mathrm{s}}|^2 \tag{7.102b}$$

$$\tilde{T}_{\mathrm{p}} = \left(\frac{N_N \cos\theta_N + N_N^* \cos\theta_3^*}{N_1 \cos\theta_1 + N_1^* \cos\theta_1^*}\right) |\tilde{t}_{\mathrm{p}}|^2 \tag{7.102c}$$

$$\tilde{T}_{\mathrm{s}} = \left(\frac{N_N \cos\theta_N^* + N_N^* \cos\theta_N}{N_1 \cos\theta_1^* + N_1^* \cos\theta_1}\right) |\tilde{t}_{\mathrm{s}}|^2 \tag{7.102d}$$

with the small difference (from Eq. 7.94) that we use the subscripted index N instead of 3 for the last medium, since we can have an arbitrary number of layers inside the stack.

7.8 Electric Field Profiles

In the previous section, we developed all the machinery needed to calculate the s- and p-polarized transmittance and reflectance for very general situations. Those quantities are primarily useful as they can be calculated for any type of sample (including those modelled with an arbitrary number of layers), and then again for a reference sample, and the ratio can be used to generate the absorbance spectra.

When we discuss absorption spectra and their interpretation in detail in Chap. 8 we will see that there are some situations in which an ideal reference sample is not available to permit the type of quantitative analysis that is desired. In such cases, it is necessary to calculate the mean square electric fields at an arbitrary location within the material, i.e. within a specific point of our multi-layer stack. This is so that we can evaluate the *local* electric fields in the regions where the molecules exist. More on that point in Chap. 8, but this is a good place to get into the details of the methods, since we are half-way there already when we have formulated all of the constituent **M** matrices.

We now need to invert **M** inside each layer. Following the method proposed by Hansen [18] and extended to anisotropic layers by Yamamoto and Ishida [21], one obtains the characteristic **N** matrices

$$\mathbf{N}_{s,k}(z) = \left[\mathbf{M}_{s,k}(z)\right]^{-1} = \begin{bmatrix} \cos\beta_k(z) & \dfrac{i}{p_k}\sin\beta_k(z) \\ -ip_k\sin\beta_k(z) & \cos\beta_k(z) \end{bmatrix} \tag{7.103}$$

with

$$\beta_k(z) = \frac{2\pi}{\lambda} p_k(z - z_{k-1})$$

for s-polarized light and

$$\mathbf{N}_{p,k}(z) = \left[\mathbf{M}_{p,k}(z)\right]^{-1} = \begin{bmatrix} \cos\beta_k(z) & \dfrac{i}{q_k}\sin\beta_k(z) \\ -iq_k\sin\beta_k(z) & \cos\beta_k(z) \end{bmatrix} \tag{7.104}$$

for p-polarized light, where z_{k-1} is the thickness of phase $k-1$. These **N** matrices may be used to obtain the Cartesian components of the electric field at any location z in material k according to

$$\begin{bmatrix} U_k(z) \\ V_k(z) \end{bmatrix} = N_k(z) \prod_{j=k}^{2} N_j \begin{bmatrix} U_{N-1} \\ V_{N-1} \end{bmatrix} \tag{7.105a}$$

$$W_k(z) = \frac{N_1 \sin(\theta_1)}{N_k^2} U_k(z). \tag{7.105b}$$

Finally the quantities of interest arrive via

$$E_x(z) = U_k(z) \tag{7.106a}$$

$$E_y(z) = V_k(z) \tag{7.106b}$$

$$E_z(z) = W_k(z). \tag{7.106c}$$

As a result of these steps, explicit formations of this result (for example, for 2- or 3-phase systems), or approximate equations for $\langle|\mathbf{E}|^2\rangle$, have historically been of interest. Nevertheless, this formalism indicates that, so long as the refractive index N and thickness d are known for each material, the Cartesian components of the electric field profiles may readily be determined. Technically, these expressions do provide solutions for the electric fields at any arbitrary point z; however, the solutions are dependent on the magnitude and phase of the incident field. It is therefore more universal to describe the solutions with respect to the incident fields, i.e. with respect to the components of the field at $z = -\infty$. Finally, although we express the results in terms of Cartesian components, the incident fields are usually written in terms of the s- and p-polarized states. Here s is along $\hat{y}$, and we take projections of p along the $\hat{x}$ and $\hat{z}$ directions. Putting it all together, the local field *corrections*, the ratio of the fields at any point z with respect to the s- or p-polarized incident fields can be expressed as

$$L_{xx}(z) = \frac{E_x(z)}{E_\mathrm{p}(z = -\infty)} \tag{7.107a}$$

$$L_{yy}(z) = \frac{E_y(z)}{E_\mathrm{s}(z = -\infty)} \tag{7.107b}$$

$$L_{zz}(z) = \frac{E_z(z)}{E_\mathrm{p}(z = -\infty)}. \tag{7.107c}$$

The majority of reports concerned with IR absorption of molecules at surfaces treat the system as a 3-phase problem. We will later (in Chap. 10) see that three-phase systems are central to the treatment of vibrational SFG data, even when the system is comprised of what appears to be only two adjacent bulk phases. Typically, external reflection geometries consider air/vacuum–(surface layer)–substrate. ATR experiments often consider IRE–(surface layer)–bulk phase (either air or water). Although the above formalism is especially well-suited to numerical calculations using NumPy arrays, where there are only 2 or 3 phases of interest, the analytical solutions are also easy to come by. We consider the common scenario of a three-phase system and arrive at expressions to the distant-dependent fields in the incident medium ($z < 0$)

$$L_{xx}(z) = \left[1 - |\tilde{r}_\mathrm{p}| \exp\left(i\phi_\mathrm{p} - i\frac{4\pi z n_1 \cos\theta_1}{\lambda}\right)\right] \cos\theta_1 \tag{7.108a}$$

$$L_{yy}(z) = 1 + |\tilde{r}_\mathrm{s}| \exp\left(i\phi_\mathrm{s} - i\frac{4\pi z n_1 \cos\theta_1}{\lambda}\right) \tag{7.108b}$$

$$L_{zz}(z) = \left[1 + |\tilde{r}_\mathrm{p}| \exp\left(i\phi_\mathrm{p} - i\frac{4\pi z n_1 \cos\theta_1}{\lambda}\right)\right] \sin\theta_1 \tag{7.108c}$$

where ϕ is the phase of $\tilde{r}$ that can be obtained from arctan $(\mathbb{I}\mathrm{m}\{\tilde{r}\}/\mathbb{R}\mathrm{e}\{\tilde{r}\})$. In the film (medium 2, where $0 < z < d$), we have

$$L_{xx}(z) = (1 - \tilde{r}_\mathrm{p}) \cos\left(\frac{2\pi z N_2 \cos\theta_2}{\lambda}\right) \cos\theta_1$$
$$+ i \frac{n_1 \cos\theta_2}{N_2}(1 + \tilde{r}_\mathrm{p}) \sin\left(\frac{2\pi z N_2 \cos\theta_2}{\lambda}\right) \quad (7.109a)$$

$$L_{yy}(z) = (1 + \tilde{r}_\mathrm{s}) \cos\left(\frac{2\pi z N_2 \cos\theta_2}{\lambda}\right)$$
$$+ i \frac{n_1 \cos\theta_1}{N_2 \cos\theta_2}(1 - \tilde{r}_\mathrm{s}) \sin\left(\frac{2\pi z N_2 \cos\theta_2}{\lambda}\right) \quad (7.109b)$$

$$L_{zz}(z) = \left(\frac{n_1}{N_2}\right)^2 (1 + \tilde{r}_\mathrm{p}) \cos\left(\frac{2\pi z N_2 \cos\theta_2}{\lambda}\right) \sin\theta_1$$
$$+ i \frac{n_1 \cos\theta_1}{N_2 \cos\theta_2}(1 - \tilde{r}_\mathrm{p}) \sin\left(\frac{2\pi z N_2 \cos\theta_2}{\lambda}\right) \sin\theta_1, \quad (7.109c)$$

and in the transmitted phase (medium 3, where $z > d$), the fields are given by

$$L_{zz}(z) = \tilde{t}_\mathrm{p} \exp\left[-\frac{2\pi \mathbb{I}\mathrm{m}\{N_3 \cos\theta_3\}}{\lambda}(z - d)\right] \cos\theta_3 \quad (7.110a)$$

$$L_{yy}(z) = \tilde{t}_\mathrm{s} \exp\left[-\frac{2\pi \mathbb{I}\mathrm{m}\{N_3 \cos\theta_3\}}{\lambda}(z - d)\right] \quad (7.110b)$$

$$L_{zz}(z) = \frac{n_1}{N_3} \tilde{t}_\mathrm{p} \exp\left[-\frac{2\pi \mathbb{I}\mathrm{m}\{N_3 \cos\theta_3\}}{\lambda}(z - d)\right] \sin\theta_1. \quad (7.110c)$$

Note that we can collapse these expressions to describe a two-phase system by setting the thickness d of the middle layer to zero, and also replacing all references to N_3 with N_2.

We now consider an example of a three-phase system consisting of a 1 μm thick isotropic film with $N_2 = 1.45 + 0.2i$ sandwiched between air ($N_1 = 1$) and water ($N_3 = 1.30 + 0.1i$) with a 1650 cm^{-1} beam approaching from the air at an angle of 50°. The various components of $|\mathbf{L}|$ are plotted as a function of propagation distance in solid lines, and the phase of the local field correction is plotted with dashed lines. Regions in air are drawn with a white background, those inside the absorbing organic layer are shaded grey, and those in the substrate are shaded yellow. One can observe that the tangential components L_{xx} and L_{yy} are continuous across the air–film and film–water boundary, but that L_{zz} is discontinuous. The extent of the discontinuity is related to the ratio of the refractive indices of the neighbouring phases. This is the reason why there is a large step in L_{zz} about $z = 0$, due to the contrast between the refractive index of air and the organic layer. By the same reasoning, the difference between the refractive indices of the organic layer and glass is not nearly as extreme, hence there is a smaller discontinuity at the $z = d$ interface.

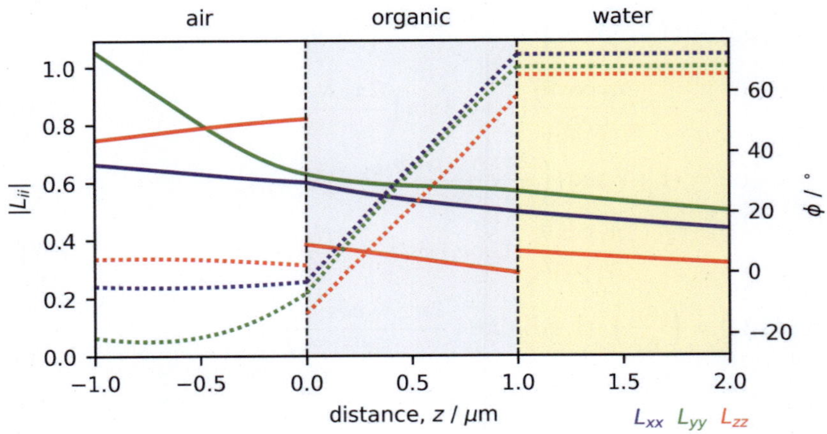

Fig. 7.14 Plots of the magnitude (solid lines) and phase (dashed lines) of L_{xx} (blue), L_{yy} (green) and L_{zz} (red) as a 1650 cm^{-1} beam approaches a 1 μm thick organic layer situated at $z = 0$ at an angle of 50°. The field profiles are shown starting 1 μm away from the surface (negative values of z), inside the organic layer ($0 < z < 1$ μm) and up to 2 μm into the glass substrate. Since the glass substrate is weakly absorbing with $\kappa_3 = 0.1$ there is a slow exponential decay of the field amplitudes so that all components will have $|L_{ii}| \to 0$ as $z \to \infty$

Code used to generate Fig. 7.14

```
import numpy as np
from optics import rtilde, arcsin
from matplotlib import pyplot

def plot(z, Ex, Ey, Ez):
    ax.plot(z, np.abs(Ex), 'b-')
    ax.plot(z, np.abs(Ey), 'g-')
    ax.plot(z, np.abs(Ez), 'r-')

    phix = np.degrees(np.arctan2(Ex.imag, Ex.real))
    phiy = np.degrees(np.arctan2(Ey.imag, Ey.real))
    phiz = np.degrees(np.arctan2(Ez.imag, Ez.real))
    ax2.plot(z, phix, color='b', ls='--', dashes=[1,1])
    ax2.plot(z, phiy, color='g', ls='--', dashes=[1,1])
    ax2.plot(z, phiz, color='r', ls='--', dashes=[1,1])

# setup
pyplot.rc('font', size=8)
fig = pyplot.figure(figsize=(4.5, 2.5))
N1 = 1.
N2 = 1.45 + 0.2j
N3 = 1.3 + 0.1j
wl = 10000./1650. # microns
d = 1. # therefore also in microns
theta = np.radians(50.)
ax = fig.add_subplot()
```

```
ax2 = ax.twinx()

# Fresnel coefficients that take thin film interference into account
rs, rp, ts, tp = rtilde(N1, N2, N3, d, wl, theta)
phis = np.arctan2(rs.imag, rs.real)
phip = np.arctan2(rp.imag, rp.real)

# region 1
z1 = np.linspace(-d, 0, 100)
arg = 1j*4*np.pi*z1*N1*np.cos(theta)/wl
Lxx = (1. - np.abs(rp) * np.exp( 1j*phip - arg )) * np.cos(theta)
Lyy = 1. + np.abs(rs) * np.exp( 1j*phis - arg )
Lzz = (1. + np.abs(rp) * np.exp( 1j*phip - arg )) * np.sin(theta)
plot(z1, Lxx, Lyy, Lzz)

# region 2
z2 = np.linspace(0, d, 100)
theta2 = arcsin(N1*np.sin(theta)/N2)
arg = 2*np.pi*z2*N2*np.cos(theta2)/wl
Lxx = (1. - rp) * np.cos(arg) * np.cos(theta) + 1j*N1*np.cos(theta2)/N2 *
      (1 + rp) * np.sin(arg)
Lyy = (1. + rs) * np.cos(arg) + 1j*N1*np.cos(theta)/(N2*np.cos(theta2)) *
      (1 - rs) * np.sin(arg)
Lzz = (N1/N2)**2 * (1 + rp) * np.cos(arg) * np.sin(theta) + 1j*N1*np.cos(
      theta)/(N2*np.cos(theta2)) * (1 - rp) * np.sin(arg) * np.sin(theta)
plot(z2, Lxx, Lyy, Lzz)

# region 3
z3 = np.linspace(d, 2*d)
theta3 = arcsin(N1*np.sin(theta)/N3)
arg = -2*np.pi * (N3*np.cos(theta3)).imag/wl * (z3 - d)
Lxx = tp * np.exp(arg) * np.cos(theta3)
Lyy = ts * np.exp(arg)
Lzz = tp * np.exp(arg) * np.sin(theta3)
plot(z3, Lxx, Lyy, Lzz)

# more pretty things
fig.text(0.75, 0.05, r'$L_{xx}$', color='b')
fig.text(0.80, 0.05, r'$L_{yy}$', color='g')
fig.text(0.85, 0.05, r'$L_{zz}$', color='r')
up = ax.get_ylim()[1]
ax.text(0.5*np.min(z1), up*1.05, 'air', ha='center')
ax.text(0.5*d, up*1.05, 'organic', ha='center')
ax.text(0.5*(d + np.max(z3)), up*1.05, 'water', ha='center')
ax.set_ylabel(r'$| L_{ii} |$')
ax2.set_ylabel(r'$\phi$ / $^\circ$')
ax.set_xlim(np.min(z1), np.max(z3))
ax.set_ylim(bottom=0)
ax.axvspan(0, d, color='0.95')
ax.axvspan(d, np.max(z3), color='lemonchiffon')
ax.axvline(0, color='k', lw=0.75, ls='--')
ax.axvline(d, color='k', lw=0.75, ls='--')
ax.set_xlabel(r'distance, $z$ / $\mu$m')
fig.set_tight_layout(True)
pyplot.savefig('compareFields.png', dpi=600)
```

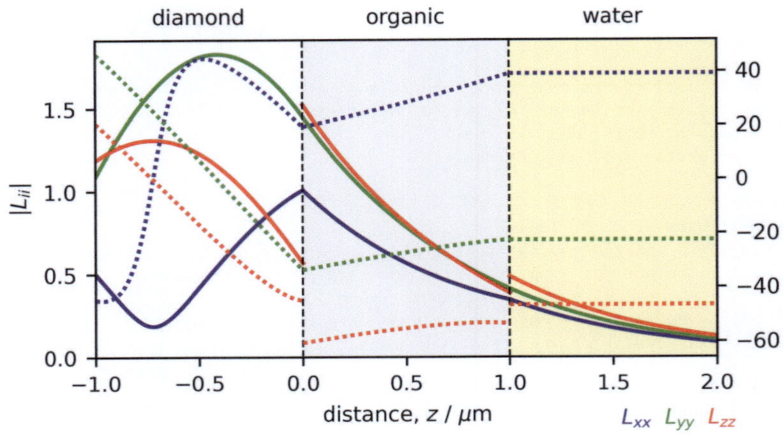

Fig. 7.15 Plots of the magnitude and phase of the local field factors when the incident medium is diamond $N_1 = 2.4$ and all other parameters are the same as in Fig. 7.14

If we now consider an internal reflection experiment as encountered in a typical ATR-IR configuration where the incident phase is a high-index crystal transparent in the mid-IR, we use the same code, simply replacing $N_1 = 1$ with $N_1 = 2.4$ to represent diamond. The result is shown in Fig. 7.15. It is interesting to observe that even though the material of interest (phase 2) has the same optical constants $N_2 = n_2 + i\kappa_2$ as we modelled for the case of external reflection, the electric field profiles are substantially different from those shown in Fig. 7.14. As we will see when we study absorption spectroscopy in the next chapter, these fields must be explicitly accounted for in the event that it is not possible to compare the signal to that of a reference sample that provides an "experimental" determination of the **L** factors. Also note that this illustrates that components of **L** can be substantially larger than unity, indicating that the electric field at a particular point is larger than that of the incident field. This does not violate conservation of energy, as the magnitude of the overall electric field vector is reduced. However, it is desirable to take advantage of such local field enhancements when probing molecules with small absorption coefficients.

7.9 The Generation of Colours from the Transmission or Reflection of White Light

7.9.1 Additive and Subtractive Mixing of Colours

It is interesting to consider the particularly simple form of the Fresnel equations at normal incidence, where the incident material is air ($N_1 = n_1 = 1$) and the refracted material has $N_2 = n + i\kappa$. At normal incidence there is no s- or p-polarization, since the plane of incidence is not defined when the incident wavevector and the surface normal are parallel. In this case, we have

$$R = \frac{(n-1)^2 + \kappa^2}{(n+1)^2 + \kappa^2} \tag{7.111}$$

$$T(z = 0^+) = \frac{4n}{(n+1)^2 + \kappa^2} \tag{7.112}$$

where the notation $z = 0^+$ indicates that we are just inside the absorbing medium, before the beam propagates any appreciable distance in the material. If the beam traverses the second medium to a distance d (with the path length $z = d$ since we are at normal incidence) we have an additional attenuation resulting in the final expression

$$T(z = d) = \frac{4n}{(n+1)^2 + \kappa^2} \exp\left[-\frac{4\pi\kappa d}{\lambda}\right]. \tag{7.113}$$

Just as the reflectance and the transmittance are the fractions of the incident intensity that are reflected and transmitted, recall that the absorptance A (sometimes also referred to by the slightly outdated term absorptivity) is the fraction of the incident intensity that is absorbed. At $z = 0^+$, at the surface immediately on the material side of the interface,

$$A(z = 0^+) = 1 - T(z = 0^+) - R = 0$$

which verifies the relationship $T(z = 0) + R = 1$. At a distance $z = d$ we have

$$\begin{aligned} A &= 1 - T(z = d) - R \\ &= \frac{4n}{(n+1)^2 + \kappa^2} \left(1 - \exp\left[-\frac{4\pi\kappa d}{\lambda}\right]\right) \end{aligned} \tag{7.114}$$

where $4\pi\kappa/\lambda$ is the absorption coefficient appearing in Beer's law, to be discussed in Chap. 8. Setting $d = 0$ results in $A = 0$ as we have already demonstrated. The other limiting case of $d = \infty$ results in $A_\infty = T(z = 0^+)$. In other words, all of the light transmitted into the medium is absorbed.

Typically the goal in characterizing the optical constants of materials is to make measurements of R, T, or A, perhaps at various angles of incidence and polarization, and using the resulting data to determine n and κ. Likewise, given the dispersion $n(\lambda)$ and $\kappa(\lambda)$, one can predict the reflection and transmission spectra in the corresponding wavelength range.

The colour of an object, based on the observation of transmitted or reflected light, is related to this spectral data, albeit in an indirect way. Predicting the colour from the spectra requires information related to the wavelength sensitivity of the eye, and knowledge of how the mixing of colours is perceived by the brain. For example, a light source may appear green due to the excitation of green receptors or by the mixing of yellow and blue light. Similarly, when both red and blue receptors are activated but green is not present, the stimulation is perceived as magenta. The combination of multiple light sources is often referred to as additive mixing of colours. One of the best ways to experiment with this phenomenon is to set up mounts so that red, blue, and green light sources may be held in a fixed orientation as the colours are projected onto a wall or white screen. By varying the degree to which two or all three circles of light are overlapped, different colours can be created.

This is in contrast to the subtractive mixing employed in the creation of pigments for paint. Here the idea is that objects are illuminated with broadband (typically white) light sources, and certain wavelengths are selectively transmitted or reflected according to the optical constants of the pigments. A common explanation is that the dyes absorb certain colours, and the ones that are not absorbed (or absorbed the least) are transmitted and reflected. This explanation is straightforward in the case of transmitted light (essentially following Beer's law), but does not intuitively agree with our expressions for reflectance. The contradiction may be posed in the following way: Imagine an object that absorbs in the red and therefore transmits in the blue. A thin sheet of blue paper or a transparent blue glass filter (a filter for red!) are examples. If this object is held in front of a white light bulb, the transmitted light appears blue. The absorption spectrum, as one might expect, has a peak absorbance in the red, and comparatively little absorbance in the blue. But in reflection, this object also appears blue. The casual explanation is that the red components of the light are absorbed while the blue portions are reflected. If we consider Eq. 7.111, one might be tempted to substitute κ_{high} for red, and κ_{low} or even $\kappa \approx 0$ for blue, while noting that there is almost negligible change in the refractive index over these wavelengths. If one does this calculation, however, the results are surprising, as summarized in Fig. 7.16.

Here we plot $R(n, \kappa)$, the reflectance as a function of the real and imaginary parts of the material's refractive index, at normal incidence from air. Although very large regions $1 < n < 3$ and $0 < \kappa < 3$ are explored in the top row, the bottom row displays a zoomed-in version of a more realistic range $1.33 < n < 1.34$ and $0 < \kappa < 0.01$. Here it is clearly seen that changing the value of κ alone has a negligible effect on the reflectance, and therefore cannot be responsible for the blue light that is visible when $\kappa_{\text{red}} \approx 0.01$ and $\kappa_{\text{blue}} \approx 0$. In fact, close inspection of Fig. 7.16 reveals that the vertical stripes are slightly curved in the direction opposite to what we might expect, showing that for constant n, a material with higher κ is more reflective than a material (or wavelength) with lower κ. The answer to this caper lies in the variation of the real part of the refractive index in the vicinity of the absorption band. Even though the relative change in n is many orders of magnitude smaller than the relative change in κ for these blue and red wavelengths, Eq. 7.111 (and Fig. 7.16) reveals that small changes in n are much more likely to result in large changes in the reflectance compared to even large variations in κ. Furthermore, higher n results in higher reflectance. So the explanation that red absorbers reflect blue light is in itself not entirely satisfactory. The underlying reason comes from the associated change in the refractive index: the strong absorption band in the red

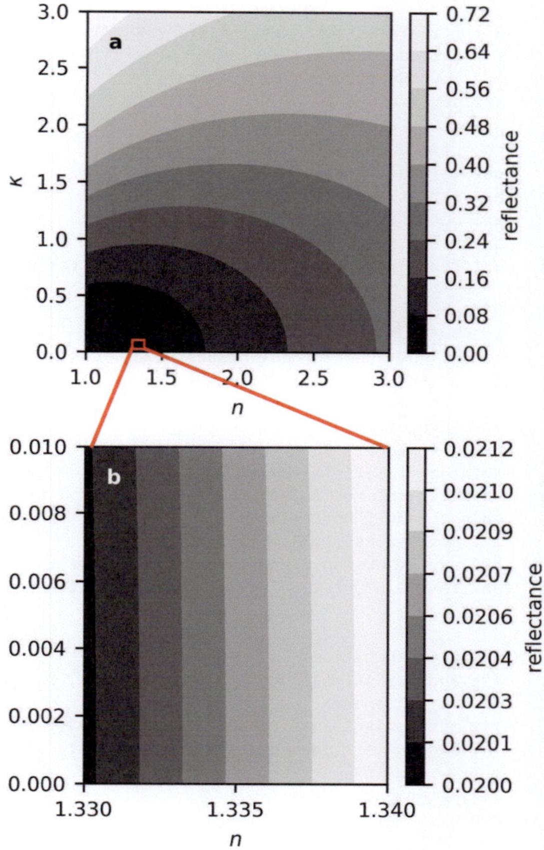

Fig. 7.16 **a** An illustration of the relationship between the reflectance and the real (n) and imaginary (κ) parts of the complex refractive index. **b** A zoomed-in version around the region of the refractive index of common dielectric materials with typical absorption coefficients of organic compounds

results in a $n_{\text{blue}} > n_{\text{red}}$. We note that similar expressions may be derived for multi-layer deposition of dyes, and dyes combined with transparent layers. Non-normal incidence can also be incorporated simply by using the full form of the reflectance and transmittance expressions in the matrix-based approach introduced earlier.

Code used to generate Fig. 7.16

```
pyplot import matplotlib.patches as mpatches import matplotlib.lines
as lines

pyplot.rc('font', size=8)
fig = pyplot.figure(figsize=(3.25, 5))

# contour - wide range
ax1 = fig.add_subplot(211)
n, k = np.meshgrid(np.linspace(1, 3, 200), np.linspace(0, 3, 200))
```

```
R = ((1-n)**2 + k**2)/((1+n)**2 + k**2)
c = ax1.contourf(R, levels=10, origin='lower', extent=[1, 3, 0, 3], cmap=
    pyplot.cm.Greys_r)
ax1.set_xlabel('$n$')
ax1.set_ylabel(r'$\kappa$')
cb = fig.colorbar(c)
cb.set_label('reflectance')
ax1.set_xticks(np.arange(1., 3.1, 0.5))
ax1.set_yticks(np.arange(0., 3.1, 0.5))
ax1.set_ylim(0, 3)
ax1.text(0.07, 0.89, 'a', transform=ax1.transAxes, fontweight='bold')

# contour - zoomed in
ax2 = fig.add_subplot(212)
n, k = np.meshgrid(np.linspace(1.33, 1.34, 200), np.linspace(0, 0.01, 200))
R = ((1-n)**2 + k**2)/((1+n)**2 + k**2)
c = ax2.contourf(R, origin='lower', extent=[1.33, 1.34, 0, 0.01], cmap=
    pyplot.cm.Greys_r)
ax2.set_xlabel('$n$')
ax2.set_ylabel(r'$\kappa$')
cb = fig.colorbar(c, format='%6.4f')
cb.set_label('reflectance')
ax2.set_xticks([1.33, 1.335, 1.34])
ax2.set_yticks(np.arange(0., 0.011, 0.002))
ax2.text(0.07, 0.89, 'b', transform=ax2.transAxes, fontweight='bold', color
    ='w')

# fancy annotations
fig.add_artist(mpatches.Rectangle((0.294, 0.59), 0.315-0.295, 0.01, ec='r',
    fc='none'))
fig.add_artist(lines.Line2D([0.227, 0.293], [0.48, 0.59], color='r'))
fig.add_artist(lines.Line2D([0.315, 0.72], [0.59, 0.48], color='r'))

fig.set_tight_layout(True)
pyplot.savefig('nkappaR.png', dpi=600)
```

7.9.2 Structural Colouration

Thin film interference phenomena critically depend on 4 properties: the refractive indices of all phases, the thickness of thin layer, the angle of incidence (that in turn determines the refracted angles in the other phases), and the wavelength. If we use the formalism we have developed in this chapter, and set the incident angle $\theta_1 = 0°$ (making all refracted angles 0°) we can study the effect of wavelength for a given layer thickness. Of course, we can do this for any desired angle, but normal incidence is convenient in that $R_p = R_s$. If we picture a source of white light (like a lamp, or the sun) that is equally bright across the visible region between 400–700 nm, we then can avoid having to model the *incident* spectrum since we are assuming unit intensity across all colours. This light then impinges on a free-standing thin film with $N_2 = 1.4$, i.e. there is air on both sides ($N_1 = N_3 = 1$).

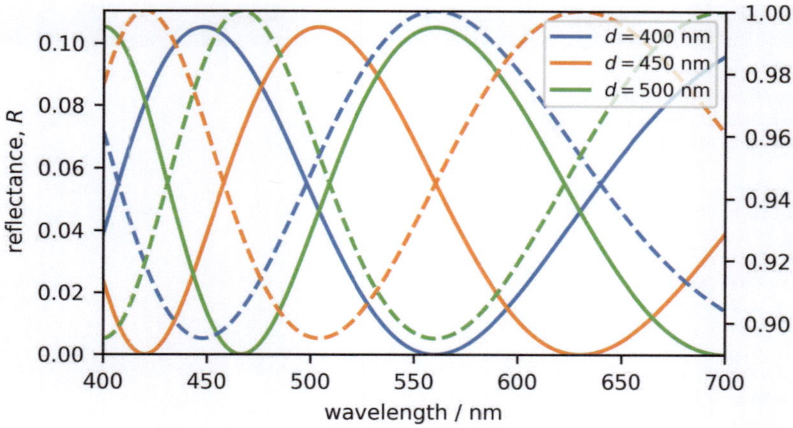

Fig. 7.17 Normal incidence reflectance (solid lines) and transmittance (dashed lines) spectra from 400–700 nm for a free-standing transparent film with refractive index 1.4. The plots correspond to film thicknesses of 400 nm (blue), 450 nm (orange), and 500 nm (green)

The reflectance is plotted in Fig. 7.17 in solid lines for three different film thicknesses. In the first case (blue curve), the film is 400 nm thick, and we can see that approximately 10% of the incident light is reflected for $\lambda = 400$ nm and again at $\lambda > 700$ nm, but none of the 560 nm is reflected back towards the observer. In other words, we have dramatically altered the spectrum of the incident light in this reflection, even though the material is transparent ($\kappa_2 = 0$); none of the green light is reflected. If the sandwiched structure instead has a width of 450 nm, (orange line), the same unit-intensity incident white light spectrum now has none of its components near 630 nm reflected; the reflected spectrum is devoid of red. This is a dramatic change in the colour by simply by changing the thickness by a mere 50 nm. The corresponding transmittance is plotted using dashed lines in Fig. 7.17, where it is observed that minima in the reflectance correspond to maxima in the transmittance, and vice versa. This makes sense from a conservation of energy principle, as we are modelling transparent materials. Another way to think about this is that the ratio between the reflectance and transmittance is modulated as a function of wavelength, angle, and/or film thickness. Transmission diffraction gratings are based on this principle, as are interference filters. This is the reason why the desired properties of the transmitted or reflected light are achieved only for the angle of incidence (typically normal incidence) for which the device was designed. There is recent interest in developing photonic crystals based on structural colouration in order to fine tune their optical properties [22–24].

Code used to generate Fig. 7.17

```
import numpy as np
from AXE227 import fields2
from matplotlib import pyplot

# setup
pyplot.rc('font', size=8)
fig = pyplot.figure(figsize=(4.5, 2.5))
ax1 = fig.add_subplot()
ax2 = ax1.twinx()
N1 = 1.0 + 0j
N2 = 1.4 + 0j
N3 = 1.0 + 0j
theta1 = np.radians(0.)
wavelength = np.linspace(400, 700, 1000)

def go(d):
    Rp, Rs, Tp, Ts = fields2(N1, N2, N3, d, wavelength, 0, theta1)[:4]
    ax1.plot(wavelength, Rp, label=r'$d=$%d nm' % int(d))
    ax2.plot(wavelength, Tp, ls='--')

go(400.)
go(450.)
go(500.)

# pretty
fig.legend(loc=(0.65, 0.73), fontsize=7)
ax1.set_ylim(bottom=0)
ax2.set_ylim(top=1)
ax1.set_ylabel('reflectance, $R$')
ax2.set_ylabel('transmittance, $T$')
ax1.set_xlim(400, 700)
ax1.set_xlabel('wavelength / nm')
fig.set_tight_layout(True)
pyplot.savefig('visibleReflectanceInterference.png', dpi=600)
```

In addition to such optical engineering applications, nature uses this phenomenon to create many colours. Some of the most striking examples are blue. Figure 7.18 shows peacock feathers, a *Morpho* butterfly, and *Pollia condensata* berries [25]. The blue colour in each of these does not result from blue pigment (i.e. no κ-based phenomena is altering the transmission or reflection). Rather, microscopic biological structures are arranged in order to create precisely the right thickness to produce that blue, upon reflection of white light.

7.10 Summary and Next Steps

Chapter 6 described properties of light, and now we have seen how light interacts with molecules and materials. One of the key concepts of this chapter are the Fresnel reflection coefficients r_s and r_p and the corresponding transmission coefficients

Fig. 7.18 **a** Peacock feather. Photograph by Alex Duarte. **b** *Morpho didius* butterfly. Photograph of museum specimen by Didier Descouens; Creative Commons Attribution-Share Alike 4.0 International license. **c** *Pollia condensata* berries. Photograph by Juliano Costa; Creative Commons Attribution-Share Alike 3.0 Unported license

t_s and t_p. These are immediately useful to obtain the reflectance and transmittance at a single boundary. However, we have also seen that those same 4 coefficients can be used to build up a much more complex system. Even when one additional phase is added in the form of a thin film, the resulting reflectance and transmittance can have a rich wavelength, angle, thickness, and refractive index dependence. An understanding and quantitative evaluation of such systems can be attained from the simple reflection and transmission coefficients. In more complex cases, when there is more than one layer in which thin film interference effects need to be considered, a matrix approach can be used. The beautify of this formalism is that, if it is implemented in a Python code (have you made your personal `optics` module yet?), that code can then be re-used by many different programs for a wide range of calculations. Finally, we have seen that the electric field profiles can be modelled as the beam propagates through the material. In some cases all that is desired is to know the values of the mean-squared field at a particular interface. In other cases, the complete depth-dependence is of interest, as we can then integrate the field over the region at which our targeted molecules interact with the light.

From here on, there are three largely separate discussions to be had. The next chapter is devoted to IR absorption spectroscopy. We have done most of the heavy lifting for that already, as we have seen how to determine the absorbance in a wide range of different experimental geometries. Still, there are some points specific to IR absorption that will be discussed in Chap. 8, before tackling the topic of how to get information about the orientation distribution of molecules from polarized IR measurements. Chapter 9 then turns to Raman spectroscopy, which is a departure from almost everything we have discussed to this point as Raman scattering is an incoherent process. Nevertheless, after we develop the basic tools for orientation analysis in Chap. 8, they can be extended to the more information-rich Raman spectra. Finally, Chap. 10 combines the salient features of IR and Raman orientation analysis in the study of surfaces and buried interfaces as vibrational sum-frequency generation is discussed. That will be an opportunity to continue the discussion of surface electric fields that we started here.

7.11 Exercises

1. Is $\cos(\theta^*) = (\cos\theta)^*$? The asterisk represents complex conjugation. How are either/both of these quantities related to $\cos\theta$? Illustrate your answer not only for a specific case, but for the general case of any angle θ.

---

2. A beam of light travels from air into a polymer, and is then transmitted into air again. Consider $N_{\text{air}} = 1$, incident angle θ, and $N_{\text{polymer}} = n + i\kappa$.

   a. According to Snell's Law,

   $$N_1 \sin\theta_1 = N_2 \sin\theta_2$$

   the refracted angle θ_2 inside the polymer is complex-valued. What is the value of θ_2?

   b. What about the refracted (exit) angle (call it θ_3) at the polymer–air interface? Is it real or complex? What is the value? I recommend to work this out by hand. If you are surprised by your result, plug in some values for the refractive indices and verify your result using a small Python script.

---

3. The Lorentz model results in the polarizability

   $$\alpha(\omega) = \sum_j^N \frac{a_j}{\omega_j^2 - \omega^2 - i\omega\Gamma_j}$$

   where N is the total number of modes.

   a. Re-write this expression in a form that better illustrates the real and imaginary parts, i.e. as $\alpha(\omega) = \mathbb{Re}\{\alpha(\omega)\} + i\mathbb{Im}\{\alpha(\omega)\}$.

   b. Taking full advantage of what NumPy and Matplotlib have to offer, plot $\mathbb{Re}\{\alpha(\omega)\}$ and $\mathbb{Im}\{\alpha(\omega)\}$ to convince yourself of the shape of these functions. For simplicity, consider $N = 1$, $a_1 = 1$, $\omega_1 = 2875\ \text{cm}^{-1}$, and $\Gamma_1 = 15\ \text{cm}^{-1}$. Do this in two ways: First plot the two functions that you have derived in part a. Second, define the complex-valued $\alpha(\omega)$ directly as it appears above, and let Python (numerically) evaluate the real and imaginary parts using the built-in methods of the complex data type. Use some good design principles when overlaying the two sets of plots on the same axes, so you can best distinguish them.

---

4. Now let's scale things up to the level of a material. The linear susceptibility is given by

$$\chi(\omega) = \sum_{j}^{N} \frac{A_j}{\omega_j^2 - \omega^2 - i\omega\Gamma_j}$$

Plot the dispersion of the real $n(\omega)$ and imaginary $\kappa(\omega)$ parts of the refractive index $N = n + i\kappa$ for a single oscillator ($N = 1$) with resonance frequency 2875 cm^{-1} and width $\Gamma = 15\ \mathrm{cm}^{-1}$. Recall that the complex refractive index is given by $N(\omega) = \sqrt{1 + \chi(\omega)}$. Compare two cases: the first of a "weak" oscillator with $A = 10^3$, and then a "strong" oscillator with $A = 10^5$. On your graphs, mark the position of ω_0 (in Matplotlib `axvline` is handy for this). What do you observe between the two cases?

---

5. a. Plot graphs of the Fresnel reflection coefficients r_s and r_p as a function of angle of incidence for the case of external reflection from air $n_1 = 1$ to glass $n_2 = 1.5$. Assume that both materials are transparent ($\kappa = 0$) for this unspecified wavelength. Although the code for this has been provided, it is recommended that you try to plot these directly from the Fresnel equations.
   b. In a separate panel, plot the reflectance values R_s and R_p.
   c. Compare your plots to the ones shown in this Chapter. If they are not the same, think about why.
   d. Repeat this calculation of the angle-dependence of these 4 quantities for the case of internal reflection, where the beam travelling in glass approaches the glass–air interface. If you encounter any difficulties here, you should be suspicious of the `np.arcsin()` function. There are several ways to tame this function; one of them has already been presented. Explore some simple options that produce the expected results.

---

6. A Fresnel rhomb is referred to as an achromatic retarder, since the phase shift between s- and p-components it imparts has only a very small (sometimes negligible) wavelength dependence. Commercial Fresnel rhombs are typically available for visible wavelengths, and impart a phase shift of 90° after two internal reflections.

   Suppose you wanted to make a similar type of rhomb for the mid-infrared out of ZnSe, but that resulted in a phase shift of ±45° after two reflections (i.e. $\Delta\phi = \pm 22.5°$ after each reflection). Assume that $n_{\mathrm{ZnSe}} = 2.4$ throughout the region of interest. ZnSe is nearly transparent in the mid-IR, so you can assume that $\kappa_{\mathrm{ZnSe}} = 0$.

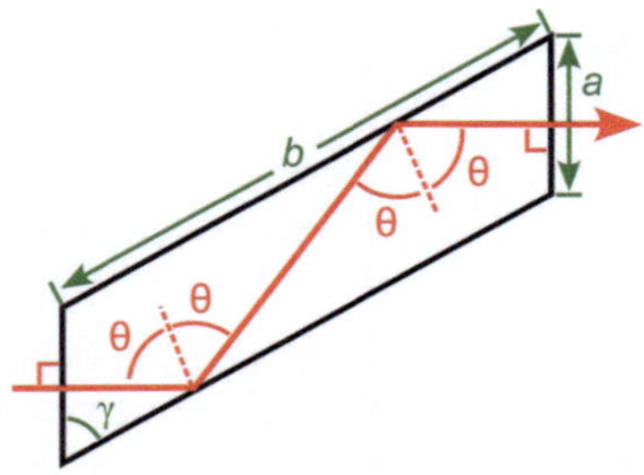

a. Describe the precise shape that this crystal needs to be cut. Referring to the drawing above, the most critical thing is to specify the angle γ. If there are multiple solutions, which one is most practical?
b. Does the dimension a matter? What consideration goes into deciding how large a should be? The dimension b then depends on your desired values of a and γ.
c. Achromatic retarders are often not truly achromatic. In reality, what determines the exact wavelength range over which this device can be expected to provide the desired retardation?
d. What could the practical utility be of having a rhomb that offers 45° phase shift in comparison to the typical 90° retardation?

---

7. Green light ($\lambda = 532$ nm) approaches a silver ($N = 0.054 + 3.429i$ at) mirror at $\theta_i = 45°$. Calculate the reflectance, and describe the polarization of the reflected light when the incident light is

   a. p-polarized
   b. s-polarized
   c. right-hand circularly polarized

---

8. The same green light approaches the same silver mirror as described above. But this time the angle of incidence is 15° (i.e. close to normal incidence) and the light is linearly polarized, with its azimuth 45° from the horizontal.

   a. What is the polarization state of the reflected light?
   b. What if there was an array of 4 mirrors rather than just the one, with the angle of incidence at 15° with respect to each mirror. What is the final polarization state now?
   c. If you wanted the light to be linearly polarized at 45° after the last (4th) mirror, but there was no room to install any polarization optics except for before the first mirror, is there another way to solve this problem? I'm specifically

thinking about the case where you could create a special polarization state such that the state after 4 reflections is the desired LP state. What would this incident state look like? How could you determine this precisely? Is there a simpler way to achieve this result?

---

9. The Jones and Müller matrices presented in this chapter are for ideal optical elements.

   a. An ideal polarizer produces perfectly polarized light and doesn't impart any phase shift. What do the matrices for a non-ideal polarizer look like? For example, some polarizers have an extinction ratio of 10,000:1.
   b. An ideal waveplate doesn't absorb any light. If it did, the performance would still be okay as long as it didn't *selectively* absorb light parallel or perpendicular to its optical axis. If this were the case, what would the corresponding matrices look like?
   c. Can you think of any other phenomena that result in non-ideal effects? How are those handled in the corresponding operators?

---

10. We have seen that any rotated optical element may be described by

$$\mathbf{M}(\theta) = \mathbf{R}(-\theta) \cdot \mathbf{M}(0°) \cdot \mathbf{R}(\theta)$$

where $\mathbf{R}(\theta)$ is the rotation matrix that is appropriate depending on whether $\mathbf{M}$ is a Jones or Mueller matrix. In the following chapters, we will see a more general way of transforming tensors of arbitrary rank. The above expression is a special case for transforming tensors of rank 2. Make a note of this point, and come back to reconcile this result.

---

11. In general, Jones matrices do not commute. That is, if you had a Jones matrix **A** and another one **B** the result **AB** and **BA** are not the same. However, one often encounters materials that have optical properties described by two different Jones matrices. As an example, a sample may exhibit linear birefringence (LB, like a waveplate) and circular birefringence (CB, resulting in optical rotation). Even though the Jones matrices for LB and CB are well-defined, we can see why neither the matrix product $\mathbf{J}_{\mathrm{LB}}\mathbf{J}_{\mathrm{CB}}$ nor $\mathbf{J}_{\mathrm{CB}}\mathbf{J}_{\mathrm{LB}}$ are satisfactory—it is not as though the sample is responsible for LB along half its thickness and CB along the second half! We need a new Jones matrix that accounts for both optical effects. This is precisely what Jones **N**-matrices are for.

    a. If this topic is of interest, work out the answer using additional information about Jones **N** matrices in Ref. [16].
    b. How can the same thing be achieved if the resulting Müller matrix is desired?

12. Consider a film of plastic wrap, with a thickness of 5.7 μm. At normal incidence, $T = 0.64$ at the probe wavelength. What is the value of κ at this wavelength? If additional information is required to solve for κ, state what is required and make some assumptions so you can arrive at a solution. *Hint:* consider what $\mathbf{E}_{\text{out}}$ is as a result of

- crossing the air–film interface
- travelling through the thickness of the film
- crossing the film–air interface

and then determine the intensity of the transmitted light from $\mathbf{E}_{\text{out}}$.

13. a. It has just started to rain, but it is still sunny out. If your window is covered in a thin layer of water, has it become brighter or darker inside your room, compared to when your window was dry? Use $N_{\text{glass}} = 1.5$ and $N_{\text{water}} = 1.33$ and assume the rays of sunlight are approaching normal to the glass ($\theta_{\text{i}} = 0°$).
    b. How about the other way around? You are in a shower with a glass door, and the bathroom light is on outside. When you start the shower and the glass becomes wet, does it appear brighter or darker?

14. The figure below illustrates six different geometries for studying the same film. In each case, consider the film thickness to be $d = 5$ nm with $N_2 = 1.32 + 0.36i$. They are (a) transmission, (b) external reflection, internal reflection (c) below and (d) above the critical angle with the medium on the other side of the film air. When water is present adjacent to the organic film, we can again consider the cases of internal reflection (e) below and (f) above the critical angle.

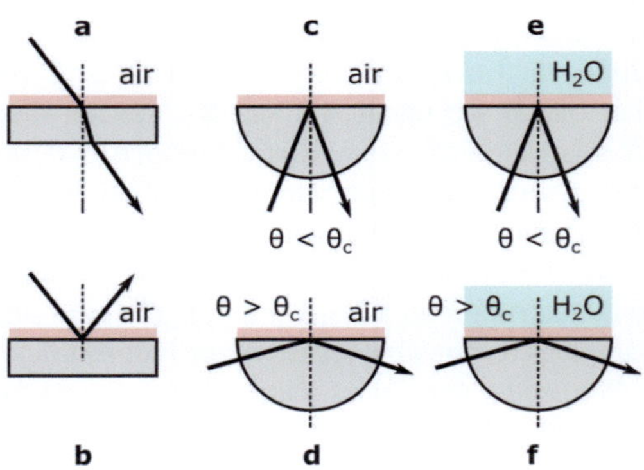

For each case, make plots showing

- the s- and p-polarized transmittance and reflectance;
- the difference in the transmittance (if applicable) and reflectance (if applicable) with and without the thin film present (i.e. compared to what would be observed at the medium 1–medium 3 interface);

- the relative mean-squared electric fields in terms of Cartesian components $\langle L_{xx}^2 \rangle$, $\langle L_{yy}^2 \rangle$, and $\langle L_{zz}^2 \rangle$;
- absorbance;
- absorptance

each as a function of angle of incidence over the range $0° < \theta_1 < 90°$. For the solid substrates, consider that they are made of germanium ($N = 4$). For water, use $N = 1.30 + 0.10i$.

---

15. You wish to perform ATR-FTIR experiments studying protein adsorption at the solid–liquid interface. You plan to use a 45° germanium prism with 100 nm of Si deposited and a 50-nm oxide layer (SiO_2) that is exposed to a protein solution. We don't formally discuss IR absorption until the next chapter, but all of the key concepts that get us most of the way there (to the polarized reflectance) have now been covered.

    a. Using online resources such as `refractiveindex.info` look up the complex refractive index values of all the materials (including your protein) at a single IR wavenumber, say 2875 cm^{-1}. You may find that it is not as easy as it should be to get a complex refractive index, even for common materials, in the mid-infrared. When it comes to the protein, bovine serum albumin has been fairly well characterized, but you may have to look elsewhere to find its optical constants. If you are just interested in getting your code running, feel free to make some substitutions.
    b. Write some code that calculates the s- and p-polarized reflectance with and without the protein adsorbed. You can assume an adsorbed protein thickness, or make a simple model that has the thickness changing over time.
    c. Rather than working at a single wavelength, this problem can be extended to calculate the reflectance spectrum over a range, like 400–4000 cm^{-1}. Notice that `refractiveindex.info` often includes either dispersion relationships, or n and κ data that you can download. You can write a function that calls `scipy.interpolate.interp1d` or something similar to look up the refractive index at the specified wavenumber. As a stepping stone to creating the full spectrum, you could implement a code that executes the previous step (single frequency reflectance) for an arbitrary wavenumber.
    d. Assuming that you have no control over the thickness of any of these layers, is there a specific angle of incidence (rather than 45°) that would offer you more contrast? In other words, is it easier to detect changes in protein layer concentration for a particular beam angle?

---

16. Are *Pollia condensata* berries edible?

---

17. We have seen that the polarizability may be expressed as

$$\boldsymbol{\alpha} = \begin{bmatrix} \alpha_{aa} & \alpha_{ab} & \alpha_{ac} \\ \alpha_{ba} & \alpha_{bb} & \alpha_{bc} \\ \alpha_{ca} & \alpha_{cb} & \alpha_{cc} \end{bmatrix}.$$

   a. Create such a quantity in both NumPy (I recommend using the `array` rather than the simpler but deprecated `matrix` data type). Multiply this by an electric field vector such as

$$\mathbf{E} = \begin{bmatrix} E_x \\ E_y \\ E_z \end{bmatrix}$$

   to confirm that you've set things up properly.

   b. Now do this using symbolic expressions in SymPy. This is also an opportunity to practice your tensor creation and labelling, in case you don't want to always have to type all the elements in manually. Being able to work with such expressions will be extremely helpful in the following chapters.

References

1. K. Yamamoto, H. Ishida, Vib. Spectrosc. **15**, 27 (1997)
2. K. Yamamoto, A. Masui, H. Ishida, Appl. Opt. **33**, 6285 (1994)
3. T.G. Mayerhöfer, V. Ivanovski, J. Popp, Spectrochim. Acta, Part A **270**, 120799 (2022)
4. J.L. McHale, *Molecular Spectroscopy*, 1st edn. (Pearson Education, New York, 2008)
5. S. Mukamel, *Principles of Nonlinear Optical Spectroscopy* (Oxford University Press, New York, 1995)
6. P. Atkins, R. Friedman, *Molecular Quantum Mechanics*, 5th edn. (Oxford University Press, Oxford, UK, 2011)
7. D.A. Long, *The Raman Effect: A Unified Treatment of The Theory of Raman Scattering by Molecules* (Wiley, New Jersey, USA, 2002)
8. A.D. Rich, D.J. Jeffrey, SIGSAM Bull. **30**, 25 (1996)
9. T.E. Hull, T.F. Fairgrieve, P.T.P. Tang, A.C.M. Trans, Math. Softw. **23**, 299 (1997)
10. G.B. Arfken, H.J. Weber, F.E. Harris, *Mathematical Methods for Physicists: A Comprehensive Guide*, 7th edn. (Academic Press, Waltham, MA, 2013)
11. E.M. Stein, R. Shakarchi, *Complex Analysis, Princeton Lectures in Analysis*, vol. 2 (Princeton University Press, Princeton, NJ, 2003)
12. E. Hecht, A. Zajac, *Optics* (Addision-Wesley, Reading, MA, 1979)
13. A. Nussbaum, R.A. Phillips, *Contemporary Optics for Scientists and Engineers* (Prentice-Hall Inc, Englewood Cliffs, New Jersey, 1976)
14. G.R. Fowles, *Introduction to Modern Optics*, 2nd edn. (Dover Publications, New York, 1989)
15. H. Weber, J. Mod. Opt. **61**, 1219 (2014)
16. D.S. Kliger, J.W. Lewis, C.E. Randal, *Polarized Light in Optics and Spectroscopy* (Academic Press Inc., 1990)
17. M. Born, E. Wolf, *Principles of Optics*, 4th edn. (Pergamon Press Ltd., Oxford, 1970)
18. W.N. Hansen, J. Opt. Soc. Am. **58**, 380 (1968)

19. R. Dietel, A. Büchtemann, Vib. Spectrosc. **34**, 205 (2004)
20. P.H. Axelsen, M.J. Citra, Prog. Biophys. Molec. Biol. **66**, 227 (1996)
21. K. Yamamoto, H. Ishida, Appl. Spectrosc. **48**, 775 (1994)
22. L.R.P. Areias, J.P.S. Farinha, A.C.S. Appl, Mater. Interfaces **16**, 1587 (2024)
23. C. Bittner, G. Bleyer, N. Nees, M. Carneiro, L. Pflug, M. Stingl, N. Vogel, Adv. Mater. Interfaces **11**, 2300986 (2024)
24. Y. Xia, S. Gao, H. He, X. Wang, Z. Zeng, B. Chen, N. Zou, M. Cao, S. Wang, J. Phys. Chem. C **124**, 16083 (2020)
25. S. Vignolini, P.J. Rudall, A.V. Rowland, A. Reed, E. Moyroud, R.B. Faden, J.J. Baumberg, B.J. Glover, U. Steiner, Proc. Nat. Acad. Sci. USA **109**, 15712 (2012)

Chapter 8
Infrared Absorption Spectroscopy

8.1 Absorption Spectroscopy

From the standpoint of linear optics, we have seen that a feature-rich description of the optical properties of molecules is contained in the polarizability, $\boldsymbol{\alpha}$. The macroscopic version of this information, providing a description of materials, is expressed in terms of the susceptibility $\boldsymbol{\chi}$, or the dielectric constant $\boldsymbol{\varepsilon}_{\mathrm{R}}$, or the refractive index N. All four quantities are complex valued; we have also seen that equivalent information is contained in the real and imaginary parts although, when given the choice, $\mathbb{I}\mathrm{m}\{x\}$ is preferable over $\mathbb{R}\mathrm{e}\{x\}$ and over $|x| = \sqrt{\mathbb{R}\mathrm{e}\{x\}^2 + \mathbb{I}\mathrm{m}\{x\}^2}$ in most circumstances due to what can be easily and accurately obtained from the respective lineshape functions for the resonances in the mid-infrared. In this chapter, we will investigate how structural information can be obtained from polarized absorption measurements in the mid-infrared by virtue of these quantities.

Consider an experiment as shown in Fig. 8.1a where light is passed through a solution in a cuvette with path length d. This is a typical experimental configuration in other spectral regions, such as the UV–visible range, but not common in the mid-infrared. Nevertheless it will be instructive for us to first consider this simple geometry as we explore some concepts. There is a familiar but ill-defined quantity called the absorbance, A. Unlike the absorptance $\tilde{A} = 1 - T - R$, the absorbance does not have any standard definition. One thing is for sure, $A \neq -\log_{10} T$. We know this because the transmittance $T = I_{\mathrm{sample}}/I_0$ and we want A to be as closely related to $\kappa = \mathbb{I}\mathrm{m}\{N\}$ as possible. For this type of transmission experiment only, i.e. in this particular geometry, and when an appropriate reference sample is used

$$A_{\mathrm{t}} = \bar{\alpha} d, \tag{8.1}$$

where $\bar{\alpha}$ is the absorption coefficient in units of reciprocal length (historically cm^{-1}), and the path length is then specified in the inverse units (i.e. cm) so that the absorbance is dimensionless. Here the subscript t is applied in A_{t} to signify that this quantity has been determined in the transmission configuration. Such notation is not really necessary, as modelling absorbance to compare with experimental data

D. K. Hore, *Molecular Orientation Analysis Using Polarized Vibrational Spectroscopy*,
https://doi.org/10.1007/978-3-032-04874-5_8

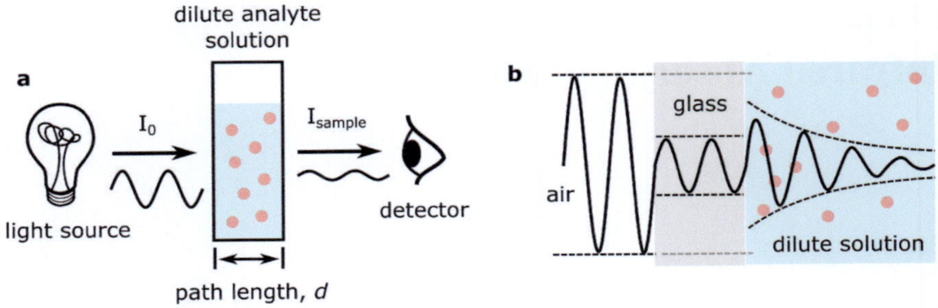

Fig. 8.1 **a** Geometry of a typical absorption experiment, where the analyte is in solution in a solvent that is transparent to the wavelengths of interest. **b** The electric field amplitude is decreased upon transmission when travelling from air to an optically denser medium (even in the absence of absorption). This is because some of the light is reflected at the interface. The transmitted amplitude at the boundary may be determined from the Fresnel transmission coefficient

relies on many details that need to be specified, as we will soon encounter. The absorbance is useful because $\bar{\alpha}$ is characteristic of a particular material, and this definition separates the effects of the material property from those of merely changing the path length d in a thicker or thinner sample. However, for solutions, it is more convenient to characterize $\bar{\alpha}$ per unit concentration, expressed as the molar extinction coefficient, ϵ, so

$$A_{\mathrm{T}} = \epsilon c d, \tag{8.2}$$

where c is the concentration. In this way, it is not necessary to re-characterize $\bar{\alpha}$ when only the concentration changes, as ϵ is characteristic of the solute in a particular solvent. This is approximately correct in dilute solutions where solute–solute interactions can be neglected, and each solute molecule is surrounded by molecules of solvent.

An issue with the transmittance is that it is heavily influenced by reflection at the air–glass and glass–solution boundaries of the cuvette, as shown in Fig. 8.1b. This is addressed by preparing a reference sample, commonly referred to as a blank, as illustrated in Fig. 8.2a. If the intensity of the light transmitted by the analyte in solution is I_{sample} and that transmitted by the blank is I_{ref}, then

$$A_{\mathrm{T}} = -\log_{10}\frac{I_{\mathrm{sample}}}{I_{\mathrm{ref}}}. \tag{8.3}$$

If the cuvette used for the blank is made of the same glass material as that used for the sample (a matched set of cuvettes), or if they are indeed the same cuvette, and the solvent is the same in both cases, the above expression for the absorbance can be simplified. We further assume that the glass and solvent are transparent in the wavelength region of interest. An important additional assumption is that, in the case of dilute samples, the real part of the analyte solution refractive index is essentially that of the pure solvent. We can then write

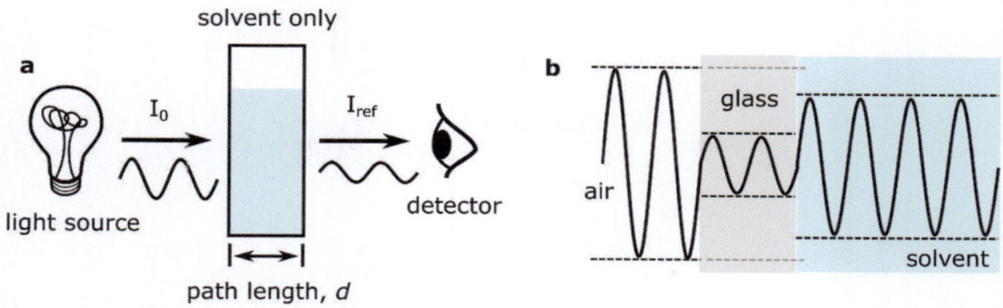

Fig. 8.2 **a** In order to measure an absorbance that is most closely related to κ of the sample, a reference or "blank" is used where the cuvette and solvent have the same refractive indices as in the sample measurement (Fig. 8.1). **b** Transmitted electric field amplitudes of the light wave in air, the glass wall of the cuvette, and in the transparent solvent

$$E_{\text{out}}^{\text{ref}} = t_4 t_3 e^{ik_0 nd} t_2 t_1 E_{\text{in}}, \tag{8.4}$$

where t_1 is the Fresnel transmission coefficient for the air–glass boundary, t_2 for the glass–solvent boundary, t_3 for the solvent–glass, and t_4 for the glass–air interface. Expressions useful in the determination of these transmission coefficients were discussed in Chap. 7 . Keep in mind that they generally depend on the beam polarization (except at normal incidence as we are briefly considering here), and the refractive index of both adjacent phases. So $t_1 = t_1(N_1, N_2, \theta_1)$. When the beam travels in the solvent between the two inside walls of the cuvette, its phase is shifted by $k_0 nd$ as indicated in Eq. 8.4. Also note that, in the regions where the molecules of interest exist, the electric field is modified from its incident value, and it continues to change as it travels through the absorbing medium. In this simple example, we are calling the field correction factors t, but a more general expression is described by $\mathbf{L}$ in Eqs. 7.108–7.110 , where the specific Fresnel coefficients necessary are automatically determined. Taking the magnitude square of this expression allows us to evaluate

$$I_{\text{ref}} = |t_4 t_3 t_2 t_1|^2 I_0. \tag{8.5}$$

We can now write the analogous expressions for the analyte solution

$$E_{\text{out}}^{\text{sample}} = t_4 t_3' e^{ik_0(n+i\kappa)d} t_2' t_1 E_{\text{in}}, \tag{8.6}$$

where t_2' is the Fresnel transmission coefficient for the glass–solution interface and t_3' is for the solution–glass interface. If we consider that the sample is dilute enough that the real part of the analyte solution refractive index is that of the blank, then $t_2 \approx t_2'$ and $t_3 \approx t_3'$ and we can take the square of the magnitude of this expression to obtain

$$I_{\text{sample}} = |t_4 t_3 t_2 t_1|^2 e^{-2k_0 \kappa d} I_0. \tag{8.7}$$

We are now in a position to evaluate the absorption coefficient

$$\begin{aligned}\bar{\alpha} &= \frac{A_t}{d} = \frac{1}{d}\log_{10}\left(\frac{I_{\mathrm{ref}}}{I_{\mathrm{sample}}}\right) \\ &= \frac{1}{d}\log_{10} e^{2k_0\kappa d} \\ &= \frac{2k_0\kappa}{\ln 10} \\ &= \frac{k_0}{n\ln 10}\mathbb{I}\mathrm{m}\{\chi\}\end{aligned} \tag{8.8}$$

where we have used the relationship $\mathbb{I}\mathrm{m}\{\chi\} = 2n\kappa$ that results from a combination of $\mathbb{I}\mathrm{m}\{\epsilon_\mathrm{R}\} = 2n\kappa$ (Eq. 7.40) and $\mathbb{I}\mathrm{m}\{\epsilon_\mathrm{R}\} = \mathbb{I}\mathrm{m}\{\chi\}$ as described in Chap. 7. We can now make the connection between $\bar{\alpha}$ (or ϵc in the case of a solution) and N, $\boldsymbol{\varepsilon}_\mathrm{R}$, and $\boldsymbol{\chi}$.

With these expressions, and the relationships between the quantities that we have established, it is instructive to consider the consequence of the oscillator strength (the numerator in our Lorentz oscillator) on the absorbance and $\mathbb{I}\mathrm{m}\{\chi\}$. For weak oscillators (left column of Fig. 8.3), the absorption coefficient (proportional to κ) is also proportional to $\mathbb{I}\mathrm{m}\{\chi\}$. Figure 8.3a shows that, even though the real part of the refractive index exhibits its characteristic shape as it passes through resonance, the extent of the change in n is negligible. This is why we can essentially ignore the effect of n that appears in the denominator of the last line of Eq. 8.8. So the absorbance (when measured and defined according to the experimental geometry we have described, and using a suitable reference sample) often provides an experimental route to accessing $\mathbb{I}\mathrm{m}\{\chi\}$. The reverse is also true: if χ is known, perhaps by modelling or fitting an absorption band to a Lorentz profile, then the spectrum of χ can easily be expressed as spectra of the real and imaginary components of the refractive index for use in the transmission and reflection calculations [1]. The case of a strong oscillator is shown in the right column of Fig. 8.3. There it can be seen that the change in refractive index is significant (Fig. 8.3b). There is also a shift between the resonance frequency and the peak of the κ spectrum (Fig. 8.3d), creating a difference between spectra of the absorption coefficient and $\mathbb{I}\mathrm{m}\{\chi\}$ (Fig. 8.3f).

Code for Fig. 8.3

This code is designed to accommodate two different oscillator strengths A, with each set of conditions appearing in one column. So, we already know that there will be six subplots and six inset labels. Here a slightly different approach is taken, just to switch it up a bit, where the subplots are defined inside of the `go` function that is called with the oscillator strength and column number (1 or 2).

```
import numpy as np
from matplotlib import pyplot
```

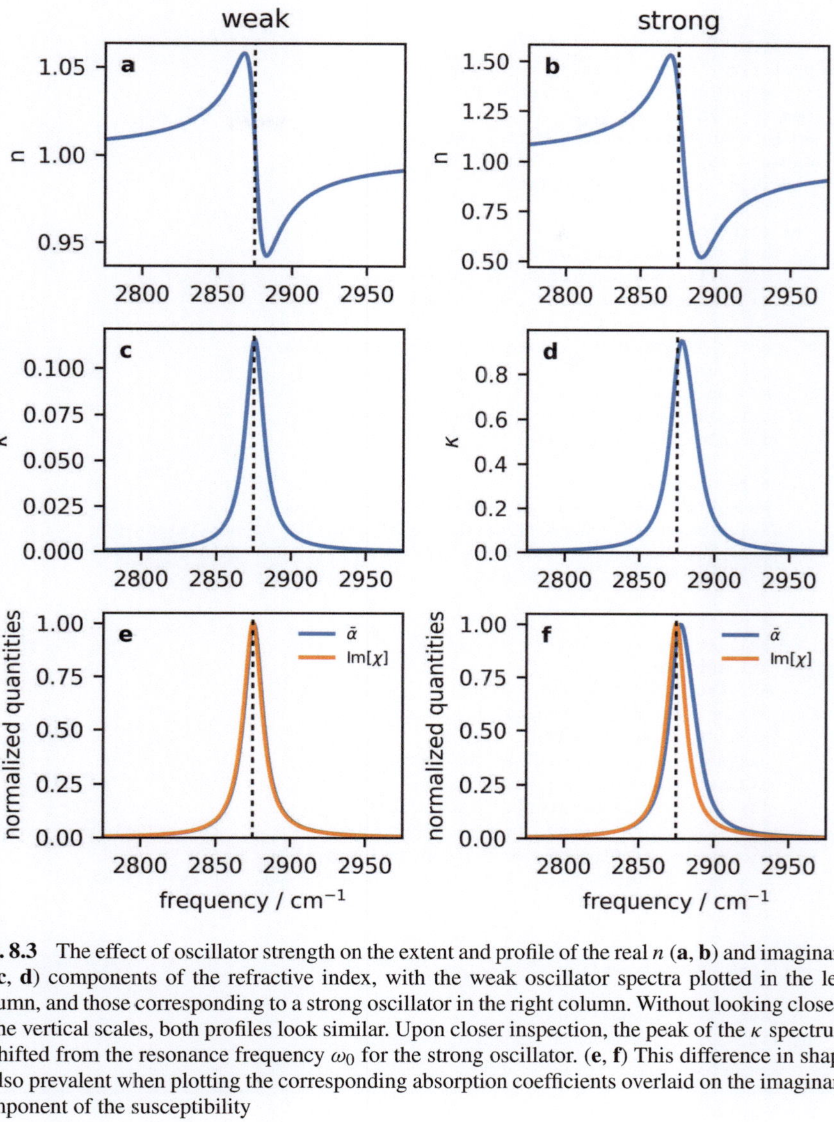

Fig. 8.3 The effect of oscillator strength on the extent and profile of the real n (**a**, **b**) and imaginary κ (**c**, **d**) components of the refractive index, with the weak oscillator spectra plotted in the left column, and those corresponding to a strong oscillator in the right column. Without looking closely at the vertical scales, both profiles look similar. Upon closer inspection, the peak of the κ spectrum is shifted from the resonance frequency ω_0 for the strong oscillator. (**e**, **f**) This difference in shape is also prevalent when plotting the corresponding absorption coefficients overlaid on the imaginary component of the susceptibility

```
# setup
pyplot.rc('font', size=8)
fig = pyplot.figure(figsize=(4.5, 5))
omega0 = 2875.
omega = np.linspace(omega0-100, omega0+100, 200)
Gamma = 15.
labels = 'a b c d e f'.split()

def go(A, column, title):
    ax1 = fig.add_subplot(3, 2, column)
    ax2 = fig.add_subplot(3, 2, column+2)
    ax3 = fig.add_subplot(3, 2, column+4)

    chi = A/(omega0**2 - omega**2 - 1j*Gamma*omega)
    epsilor_R = 1. + chi
    N = np.sqrt(epsilor_R)
    ax1.plot(omega, N.real, label='N = %d' % int(A))
    ax2.plot(omega, N.imag)

    # for comparison
    abscoeff = 4*np.pi*omega / np.log(10) * N.imag
    ax3.plot(omega, abscoeff/max(abscoeff), label=r'$\bar{\alpha}$')
    ax3.plot(omega, chi.imag / max(chi.imag), label=r'Im[$\chi$]')
    ax3.legend(loc=0, fontsize=6, frameon=False)

    # pretty
    ax1.set_title(title)
    ax1.set_ylabel('n')
    ax2.set_ylim(bottom=0)
    ax2.set_ylabel(r'$\kappa$')
    ax3.set_ylim(bottom=0)
    ax3.set_ylabel('normalized quantities')
    ax3.set_xlabel(r'frequency / cm$^{-1}$')
    for i, ax in enumerate([ax1, ax2, ax3]):
        ax.axvline(omega0, lw=1, ls='--', dashes=[2,2], color='k')
        ax.set_xlim(np.min(omega), np.max(omega))
        ax.text(0.05, 0.87, labels[2*i + column - 1], fontweight='bold',
             transform=ax.transAxes)

go(1.0e4, 1, 'weak')
go(1.0e5, 2, 'strong')

fig.set_tight_layout(True)
pyplot.savefig('weakStrong.png', dpi=600)
```

8.2 Experimental Geometries for Measuring Absorption Spectra from Solid or Liquid Samples

As we mentioned, although absorption spectroscopy using a cuvette and a blank (reference sample) is common in the UV and visible spectral regions, it is challenging to perform in the infrared. For one, very small path lengths (on the order of micrometres) are necessary owing to the large values of κ in the mid-infrared, in order for sufficient light to be transmitted to the detector. In other words, many materials are strongly absorbing in the mid-infrared. For measurements of solutions, some solvents have limited transparency in specific IR regions or "windows", but it is generally not easy to find solvents that are transparent in the infrared over the spectral ranges of interest. Here we explore some of the common ways in which IR absorption experiments are performed.

8.2.1 *Transmission Measurements with Diluted Samples*

Although IR absorption experiments are usually not performed with solvent, the concept of having a low concentration of absorbing analytes can be extended to solid samples. For powders, one option to overcome the path length issue is to mix the analyte with a transparent matrix such as KBr and press the mixture into a pellet. The transmittance of the reference sample can be obtained using a pellet that contains only KBr. We can then again apply $A_{\mathrm{T}} = -\log_{10}(I_{\mathrm{sample}}/I_{\mathrm{ref}})$ and isolate κ exactly as described above, using the same assumptions for the two sets of Fresnel transmission coefficients. This time, the output fields involve only two transmission coefficients instead of four, but the idea is the same. We must keep in mind, however, that there are circumstances where we cannot dilute the solid samples, or don't want to. This occurs in the case of non-powder samples, or when we wish to have the local fields due to the presence of neighbouring analyte molecules to be taken into consideration, i.e. studying a prepared sample that is more representative of a true bulk sample.

8.2.2 *External Reflection Measurements*

In an external reflection experiment, light of intensity I_0 is incident upon a sample at an angle θ_1 as shown in Fig. 8.4. The intensity of the reflected light $I_{\mathrm{r}} < I_0$. This means that the reflectance $R = I_{\mathrm{r}}/I_0 < 1$. This experiment is relatively easy to set up, I_{r} is easy to measure, and it is possible (although it may not be convenient) to measure I_0 if desired. However, a reference measurement is typically used in place of I_0. One can define the absorbance for such an experiment as

$$A_{\mathrm{R}} = -\log_{10}\left(\frac{I_{\mathrm{sample}}}{I_{\mathrm{ref}}}\right) = -\log_{10}\left(\frac{R_{\mathrm{sample}}}{R_{\mathrm{ref}}}\right). \tag{8.9}$$

In reflection geometries, the appearance of the resulting absorbance spectrum can vary dramatically depending on the choice of the reference sample. In one of the end-of-chapter problems in Chap. 7 , the same sample properties (in terms of n and κ) were investigated under different experimental geometries. In some cases, nega-

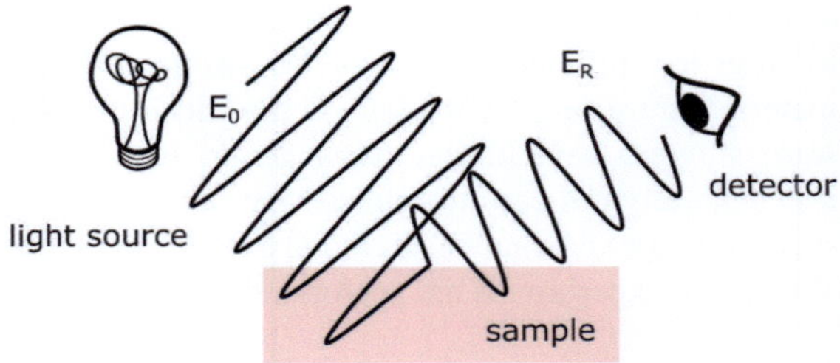

Fig. 8.4 Experimental geometry for external reflection measurements. The incident beam approaches the sample from air, and the reflected intensity is measured

tive absorbance bands are observed. This does not mean that the sample generates light! It simply indicates that more light is reflected from the sample (R_{sample} is still less than unity) than from the reference (and $R_{\text{ref}} < 1$ too). Furthermore, in addition to the sign of A_{R}, the appearance of an absorption band can be either upwards pointing or downwards pointing. This is merely a consequence of the interplay between n and κ of the sample and the corresponding values of the reference at the wavelength and angle of incidence of interest.

8.2.3 *Internal Reflection Measurements*

Attenuated total internal reflection (ATR) is a popular measurement modality that offers several benefits compared to the external reflection geometry. We will not go into those details here, and instead refer interested readers to the many excellent descriptions of ATR-IR spectroscopy in the literature [2–7]. The basic idea is that light is directed towards a sample that is in contact with a prism, referred to as an internal reflecting element (IRE) as shown in Fig. 8.5. Ideally the prism has a high refractive index and is transparent over the wavelength range of interest. If the sample is also transparent, and has a refractive index lower than the prism (i.e. $n_2 < n_1$) then, if the angle of incidence exceeds the critical angle, we have total internal reflection. That is, $R_s = R_p = 1$. This is certainly the case when no sample is in contact with the prism, i.e. when material 2 is air as $n_2 \approx 1$ and $\kappa_2 \approx 0$.

The point of this experiment, however, is to measure the absorbance spectra of absorbing samples, i.e. $\kappa_2 > 0$. In this case, even though $n_2 < n_1$ and the angle exceeds the critical angle (strictly defined only when $\kappa_2 = 0$, but people usually ignore that detail), the internal reflection is no longer total. In other words, $R < 1$ and so less light reaches the detector when the sample is present. This enables another variation of an absorbance measurement using the same definition as in the external reflection measurement described above.

Figure 8.6 plots R_s and R_p as a function of angle when the IRE has $n_1 = 2.4$, and the sample has $n_2 = 1.5$ for three different values of κ_2. We can see that when $\kappa_2 = 0$, both R_s and R_p sharply rise to unity at the critical angle. However, when κ_2

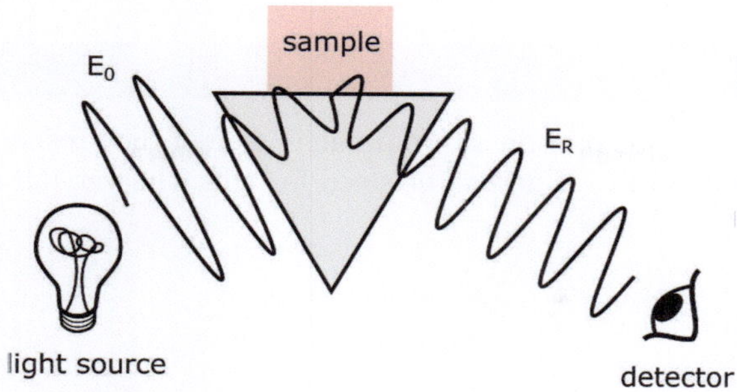

Fig. 8.5 In an internal reflection experiment, the incident beam approaches from a material that has a higher refractive index than the sample. This is achieved using an internal reflecting element (IRE), typically in the shape of a prism. For practical reasons, materials for IREs are selected so that n_{IRE} is large (so that it can be larger than n_{sample} for a wide variety of samples) and for which $\kappa_{\mathrm{IRE}} \approx 0$

increases, the light transmitted into the sample is significant and the reflectance is noticeably attenuated.

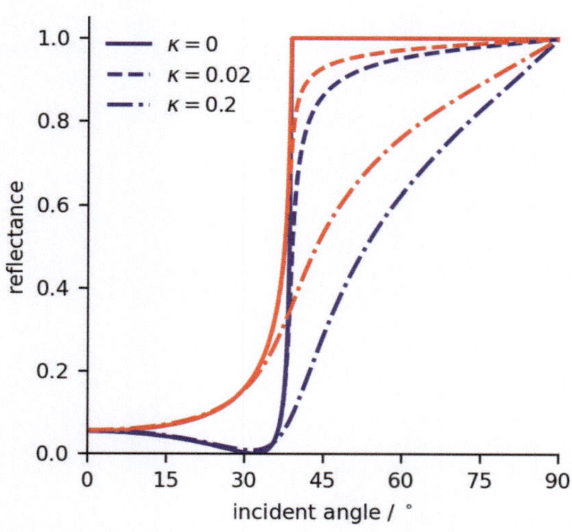

Fig. 8.6 p- (blue) and s-polarized (red) reflectance spectra when the incident medium is transparent with a refractive index of 2.4, and the refracted medium has $n = 1.5$ and varying values of κ

Code for Fig. 8.6

Note that this code uses an alternate definition of the Fresnel reflection coefficients r_s and r_p. There is no need for this; the version appearing in Eqs. 7.49 and 7.54

$$r_s = \frac{N_1 \cos\theta_1 - N_2 \cos\theta_2}{N_1 \cos\theta_1 + N_2 \cos\theta_2}$$

$$r_p = \frac{N_2 \cos\theta_1 - N_1 \cos\theta_2}{N_2 \cos\theta_1 + N_1 \cos\theta_2}$$

suffices for all cases. But this is an opportunity to illustrate something new.

```
import numpy as np
from matplotlib import pyplot

def reflectance(n1, n2, ls, label):
    theta1 = np.radians(thetaDeg)
    theta2 = np.arcsin(n1*np.sin(theta1)/n2)
    rp = -np.tan(theta1 - theta2)/np.tan(theta1 + theta2)
    rs = -np.sin(theta1 - theta2)/np.sin(theta1 + theta2)
    ax.plot(thetaDeg, np.abs(rp)**2, color='b', ls=ls, label=label)
    ax.plot(thetaDeg, np.abs(rs)**2, color='r', ls=ls, label='_nolabel')

thetaDeg = np.linspace(0., 90., 300)
pyplot.rc('font', size=8)
fig = pyplot.figure(figsize=(3.25, 3))
ax = fig.add_subplot(111)

reflectance(2.4, 1.5+0j, '-', r'$\kappa = 0$')
reflectance(2.4, 1.5+0.02j, '--', r'$\kappa = 0.02$')
reflectance(2.4, 1.5+0.2j, '-.', r'$\kappa = 0.2$')

ax.set_ylim(0, 1.05)
ax.set_xlim(0, 90)
ax.set_xticks(range(0, 91, 15))
ax.set_ylabel('reflectance')
ax.set_xlabel('incident angle / $^\circ$')
ax.spines['right'].set_visible(False)
ax.spines['top'].set_visible(False)
ax.xaxis.set_ticks_position('bottom')
ax.yaxis.set_ticks_position('left')
fig.set_tight_layout(True)
pyplot.legend(loc=0, frameon=False)
pyplot.savefig('ATRkappa.png', dpi=300)
```

Since the goal of our analysis, to characterize the absorbance, ultimately involves a ratio between two spectra, we can ignore modelling the parts where the beam travels in air or inside the IRE, and concentrate on the IRE–sample and IRE–reference interface. In other words, the fact that only a fraction of the beam is transmitted from the source into the IRE does not matter, as the associated transmission coefficients are common to the sample and reference sample intensities that are reflected towards the detector. Here we have

$$E_{\text{out}}^{\text{sample}} = r_{\text{IRE-sample}} E_{\text{in}}. \tag{8.10}$$

In this experiment, we don't remove the IRE to measure I_0, so we can't determine the reflectance. But a reference sample is used, for example, air, to provide

$$E_{\text{out}}^{\text{ref}} = r_{\text{IRE-air}} E_{\text{in}} \tag{8.11}$$

and we can then determine the absorbance

$$A = -\log_{10}\left(\frac{I_{\text{sample}}}{I_{\text{ref}}}\right) = -\log_{10}\left|\frac{r_{\text{IRE-sample}}}{r_{\text{IRE-air}}}\right|^2. \tag{8.12}$$

Throughout most of the discussion in this chapter, we've been avoiding the practical question about the input polarization. The simple (and most accurate) approach is to place a polarizer before the sample or, in the case of an ATR-IR experiment, just before the beam reaches the IRE. In that case, one simply uses either r_s or r_p in place of r, and R_s or R_p in place of R as appropriate. If one can ensure that the incident beam is unpolarized, one can use

$$R = \frac{1}{2}(R_s + R_p)$$

although care must be taken as this is seldom the case, especially when using an FTIR that has multiple mirrors comprising the interferometer directing the beam to the sample compartment [1]. Some in-spectrometer ATR-IR accessories also contain several mirrors, providing an opportunity to partially polarize the incident light. Although these seem like formidable complications, the silver lining is that, if you know the polarization of E_{in}, and you know $N = n + i\kappa$ of the sample and the angle of incidence (and n of the IRE), you can easily do the Jones vector modelling. The point here has been to arrive at κ, but if we can work backwards, there is a route to extracting what we want from the measured spectra. This ATR geometry has much in common with the external reflection experiment described previously, in that it is challenging to isolate the real and complex parts of the sample refractive index. Many schemes have been proposed for this [6, 8–17], but the general point for now is that these oblique-angle reflection experiments differ from the reference-matched normal incidence transmission experiments in their inability to isolate κ_2 in a simple way.

8.2.4 Ellipsometry

The above methods are convenient, and widely employed despite their inherent limitations. However, a small variation provides a powerful technique for measuring the complex refractive index and other properties of samples in a scheme known as ellipsometry. The general idea is based on the fact that, even for an isotropic sample, if incident light has an s- and p- component, i.e. linearly polarized at 45° or circularly polarized, then because $r_s \neq r_p$, the polarization of the reflected light will be altered. Analysis of the polarization then provides a route to determining n and κ.

There are many different schemes by which ellipsometers can be constructed [18–26]. They generally fall under two categories: nulling ellipsometers and photometric ellipsometers. In a nulling configuration, one or more elements is adjusted in order to minimize the intensity of the light at the detector, effectively determining the null condition. This is most common for a single-wavelength ellipsometer, but we will consider it here first. Within this category, there are a few possible configurations. A classical one is the polarization-compensator-sample-analyser layout, the so-called PCSA scheme as shown in Fig. 8.7. Here light incident at some angle θ_i first encounters a polarizer with azimuth θ_P, and then a compensator with retardation $\Delta\phi$ and azimuth θ_C. After reflecting from the sample, the light passes through a second polarizer (termed the analyser) oriented so that its transmission axis is at θ_A.

Using the Jones matrix formalism we have introduced in Chap. 7, we can write an expression for the output field at the detector

$$\mathbf{E}_{\text{out}} = \mathbf{M}(\theta_A) \cdot \begin{bmatrix} r_p & 0 \\ 0 & r_s \end{bmatrix} \cdot \mathbf{M}(\Delta\phi, \theta_C) \cdot \mathbf{M}(\theta_P) \cdot \mathbf{E}_{\text{in}}, \tag{8.13}$$

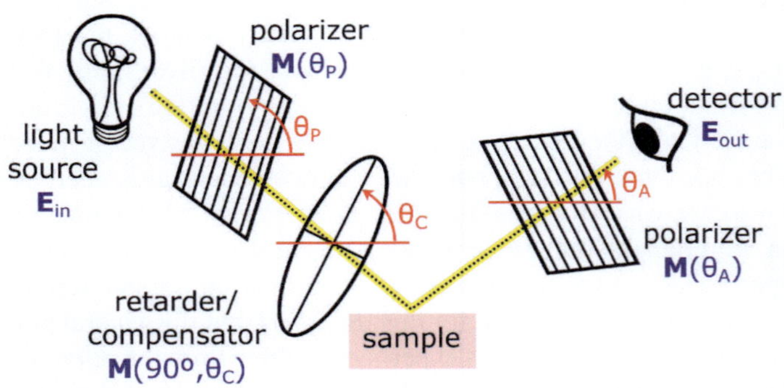

Fig. 8.7 The so-called PCSA ellipsometer configuration where the light source first passes through a polarizer (P), then a compensator (or retarder, C), then reflects of the sample (S), and is finally transmitted through a second polarizer (the analyser, A) before reaching the detector

where $\mathbf{M}(\theta_\mathrm{P})$ is the Jones matrix of a polarizer with its transmission axis set at θ_P, and $\mathbf{M}(\Delta\phi, \theta_\mathrm{C})$ represents the compensator that induces a phase shift of $\Delta\phi$ and slow axis oriented at θ_C. In a basic nulling operation, one first fixes $\theta_\mathrm{C} = 45°$ and then simultaneously or sequentially rotates θ_P and θ_A while monitoring $I_\mathrm{out} \propto |E_\mathrm{out}|^2$. The goal is to find the values of θ_P and θ_A for which $I_\mathrm{out} = 0$. The beauty of this approach can be realized by first defining the quantity

$$\rho \equiv \frac{r_\mathrm{p}}{r_\mathrm{s}} = \tan\psi e^{i\Delta} \tag{8.14}$$

so that $\tan\psi$ is the magnitude of the ratio of the two Fresnel reflection coefficients, and Δ is their difference in phase. When $I_\mathrm{out} = 0$, $\psi = \theta_\mathrm{A}$, and $\Delta = 2\theta_\mathrm{P} + 90°$. So this experiment provides direct access to the complex-valued ratio between r_p and r_s. How is this related to the optical constants? First, we must acknowledge that there are several scenarios in which the relationship between ρ and the complex refractive index is not straightforward. This includes the important application area of thin films, that we will return to later, but also the case of rough samples, refractive index gradients, etc. If we momentarily consider samples that are thick enough to have a specular reflection from their front face only, i.e. this is a Fresnel-type interface between air, and a thick isotropic material, then [19]

$$n + i\kappa = \sin\theta_\mathrm{i}\sqrt{1 + \tan^2\theta_\mathrm{i}\left(\frac{1-\rho}{1+\rho}\right)^2} \tag{8.15}$$

where θ_i is the angle of incidence.

Figure 8.8 illustrates the intensity at the detector when rotating the polarizer from $-90° \leq \theta_\mathrm{P} \leq 90°$ and the analyser from $-90° \leq \theta_\mathrm{A} \leq 90°$ for a sample with $N = 1.53 + 0.12i$. Considering that we are modelling a simple air–sample interface, where the sample is infinitely thick, this corresponds to $\Psi = 24.55°$ and $\Delta = 13.64°$. One can see two nulls within this range of polarizer and analyser angles. Either of these can be used to evaluate Ψ and Δ and arrive at $N = n + i\kappa$.

Code to generate Fig. 8.8

The following code makes use of the definitions of r_s and r_p contained in `optics.py` (see below); this more convenient (and reduces the chance of making errors) rather than writing the function definitions out each time. Also note that we are iterating through values of the polarizer angle θ_P and analyser angle θ_A using a `for` loop, rather than a slick NumPy `meshgrid`. Although this approach is slower (and slightly more cumbersome to implement) it enables us to easily perform some linear algebra inside the two nested loops, with vectors and matrices (using the NumPy `array` data type) constructed from the

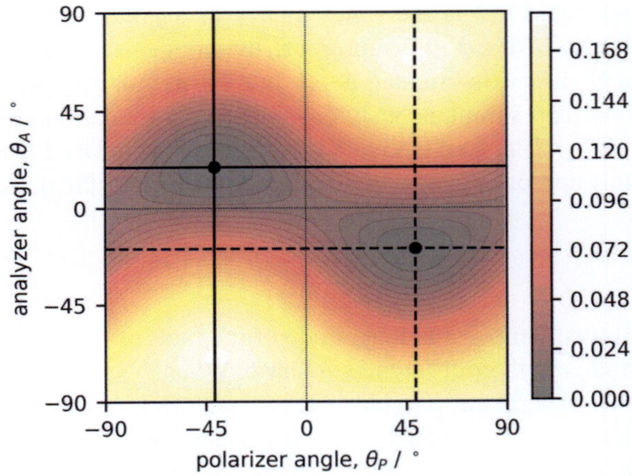

Fig. 8.8 Map of the intensity as a function of the polarizer θ_P and analyser angle θ_A for an absorbing sample ($\kappa \neq 0$). The intensity minima correspond to $\Psi = 19.09°$ and $\Delta = 8.18°$ that may then be used together with Eqs. 8.14 and 8.15 to determine $N = 1.53 + 0.12i$

two scalar quantities `thetaP` and `thetaA`. There is a way around this, "vectorizing" the entire operation with a multidimensional array, but I find that it is not worth the effort, and sacrifices some clarity. In summary, there is no shame in occasionally writing `for` loops, especially when they make use of nested NumPy operations!

```
import numpy as np
import optics
from matplotlib import pyplot

def label(solP, solA):
    # first solution
    ax.axvline(solP, color='k', lw=1)
    ax.axhline(solA, color='k', lw=1)
    ax.plot(solP, solA, 'ko', ms=4, mfc='k')
    P1 = solP
    A1 = solA

    # second solution
    A2 = -solA
    if P1 > 0:
        P2 = P1 - 90
    else:
        P2 = P1 + 90
    ax.axvline(P2, color='k', lw=1, ls='--')
    ax.axhline(A2, color='k', lw=1, ls='--')
    ax.plot(P2, A2, 'ko', ms=4, mfc='k')

    # (measured) Psi and Delta
    if A1 > 0:
        A = A1
        P = P1
    else:
        A = A2
        P = P2
```

```
    Delta = 2*P + 90.
    print(A, Delta)

# setup
thetaPdeg = np.linspace(-90., 90., 100)
thetaAdeg = np.linspace(-90., 90., 100)
thetaiDeg = 70.
C = optics.wp(np.pi/2., np.pi/4.) # retardation, azimuth
n, kappa = 1.53, 0.12
S = optics.r(1.0, complex(n, kappa), np.radians(thetaiDeg))

# calculate signal
data = []
for thetaP in np.radians(thetaPdeg):
    row = []
    for thetaA in np.radians(thetaAdeg):
        Ein = optics.lp(thetaP)
        A = optics.polarizer(thetaA)
        Eout = A @ S @ C @ Ein
        signal = optics.intensity(Eout)
        row.append(signal)
    data.append(row)
data = np.array(data).T
i = np.unravel_index(np.argmin(data), data.shape)
solA = thetaAdeg[i[0]]
solP = thetaPdeg[i[1]]

# visualize data
pyplot.rc('font', size=8)
fig = pyplot.figure(figsize=(3.5, 3.5))
ax = fig.add_subplot(111, aspect=1)
cax = ax.contourf(data, 30, origin='lower', extent=[-90, 90, -90, 90],
      cmap=pyplot.cm.hot, alpha=0.5)
pyplot.colorbar(cax, shrink=0.67)

# pretty
ax.set_xlim(-90, 90)
ax.set_ylim(-90, 90)
ax.set_xlabel(r'polarizer angle, $\theta_{\rm P}$ / $^\circ$')
ax.set_ylabel(r'analyzer angle, $\theta_{\rm A}$ / $^\circ$')
ax.axvline(0, color='k', dashes=[1,1], lw=0.5)
ax.axhline(0, color='k', dashes=[1,1], lw=0.5)
ax.set_xticks(range(-90, 91, 45))
ax.set_yticks(range(-90, 91, 45))

# locate the solution
label(solP, solA)

fig.set_tight_layout(True)
pyplot.savefig('pcsaSimulation.png', dpi=600)
```

Contents of Optics Python Module

We have defined some common operations here so that we can easily reuse them, and make our subsequent codes shorter and easier to read.

```
import numpy as np

# linearly-polarized light
def lp(theta):
   return np.array([np.cos(theta), np.sin(theta)])

# rotation matrix
def rot(theta):
   return np.array([[np.cos(theta), np.sin(theta)], [-np.sin(theta), np.cos
        (theta)]])

# rotated waveplate
def wp(retard, theta):
   wp0 = np.array([[np.exp(1j*0.5*retard), 0.], [0., np.exp(-1j*0.5*retard)
        ]])
   return rot(-theta) @ wp0 @ rot(theta)

# rotated polarizer
def polarizer(theta):
   return rot(-theta) @ np.array([[1.,0.],[0.,0.]]) @ rot(theta)

# intensity
def intensity(E):
   return float((np.conjugate(E).T @ E).real)

# stokes vector from jones vector
def stokes(E, norm=False):
   Ex = E[0]
   Ey = E[1]
   S0 = np.conjugate(Ex)*Ex + np.conjugate(Ey)*Ey
   S1 = np.conjugate(Ex)*Ex - np.conjugate(Ey)*Ey
   S2 = np.conjugate(Ex)*Ey + np.conjugate(Ey)*Ex
   S3 = 1j*(np.conjugate(Ex)*Ey - np.conjugate(Ey)*Ex)
   if norm == True:
      return np.array([1., (S1/S0).real, (S2/S0).real, (S3/S0).real])
   else:
      return np.array([S0, S1, S2, S3])

# jones reflection matrix
def r(n1, n2, rad):
   n1 = complex(n1)
   n2 = complex(n2)
   theta1 = rad
   theta2 = np.arcsin(n1*np.sin(theta1)/n2)
   rs = (n1*np.cos(theta1) - n2*np.cos(theta2))/(n1*np.cos(theta1) + n2*np.
        cos(theta2))
   rp = (n2*np.cos(theta1) - n1*np.cos(theta2))/(n2*np.cos(theta1) + n1*np.
        cos(theta2))
   return np.array([[rp, 0.], [0., rs]])

# jones transmission matrix
```

```
def t(n1, n2, rad):
   n1 = complex(n1)
   n2 = complex(n2)
   theta1 = rad
   theta2 = np.arcsin(n1*np.sin(theta1)/n2)
   ts = 2.0*np.sin(theta2)*np.cos(theta1)/np.sin(theta1 + theta2)
   tp = 2.0*np.sin(theta2)*np.cos(theta1)/(np.sin(theta1 + theta2)*np.cos(
        theta1-theta2))
   return np.array([[tp, 0.], [0., ts]])

def arcsin(z1):
    z = z1.astype(complex)
    return 1j*np.log(np.sqrt(1.0 - z*z) - 1j*z)

def rtilde(N1, N2, N3, d, wl, theta1):
    def rs(n1, n2, theta1):
        theta2 = arcsin(n1*np.sin(theta1)/n2)
        # costheta2 = sqrt(1. - (n1*sin(theta1)/n2)**2)
        return (n1*np.cos(theta1) - n2*np.cos(theta2))/(n1*np.cos(theta1) +
              n2*np.cos(theta2))

    def rp(n1, n2, theta1):
        theta2 = arcsin(n1*np.sin(theta1)/n2)
        # costheta2 = sqrt(1. - (n1*sin(theta1)/n2)**2)
        return (n2*np.cos(theta1) - n1*np.cos(theta2))/(n2*np.cos(theta1) +
              n1*np.cos(theta2))

    def ts(n1, n2, theta1):
        theta2 = arcsin(n1*np.sin(theta1)/n2)
        # costheta2 = sqrt(1. - (n1*sin(theta1)/n2)**2)
        return 2*n1*np.cos(theta1)/(n1*np.cos(theta1) + n2*np.cos(theta2))

    def tp(n1, n2, theta1):
        theta2 = arcsin(n1*np.sin(theta1)/n2)
        # costheta2 = sqrt(1. - (n1*sin(theta1)/n2)**2)
        return 2*n1*np.cos(theta1)/(n2*np.cos(theta1) + n1*np.cos(theta2))

    theta2 = arcsin(N1*np.sin(theta1)/N2)
    p = np.exp(1j*2*np.pi*N2*d*np.cos(theta2)/wl)

    # reflection coefficient, s-polarized
    r12 = rs(N1, N2, theta1)
    t12 = ts(N1, N2, theta1)
    t21 = ts(N2, N1, theta2)
    r23 = rs(N2, N3, theta2)
    r21 = rs(N2, N1, theta2)
    rtilde_s = r12 + t12*t21*r23*p*p / (1 - r21*r23*p*p)

    # transmission coefficient, s-polarized
    t23 = ts(N2, N3, theta2)
    ttitlde_s = t12*t23*p / (1 - r23*r21*p*p)

    # reflection coefficient, p-polarized
    r12 = rp(N1, N2, theta1)
    t12 = tp(N1, N2, theta1)
    t21 = tp(N2, N1, theta2)
    r23 = rp(N2, N3, theta2)
```

```
    r21 = rp(N2, N1, theta2)
    rtilde_p = r12 + t12*t21*r23*p*p / (1 - r21*r23*p*p)

    # transmission coefficient, p-polarized
    t23 = tp(N2, N3, theta2)
    ttitlde_p = t12*t23*p / (1 - r23*r21*p*p)

    return(rtilde_s, rtilde_p, ttitlde_s, ttitlde_p)


def fields(N1, Np, N2, Npp, N3, d, wl, theta1):
    rs, rp, ts, tp = rtilde(N1, N2, N3, d, wl, theta1)

    # top sur21ce
    Ex0 = np.cos(theta1) * (1 - rp)
    Ey0 = 1 + rs
    # Ez0p = sin(theta1)*(1 + rp)*(N1/N2)**2
    Ez0m = np.sin(theta1)*(1 + rp)
    Ez0 = Ez0m * (N1/Np)**2

    # bottom sur21ce
    theta3 = arcsin(N1*np.sin(theta1)/N3)
    Exd = np.cos(theta3)*tp
    Eyd = ts
    Ezdp = N1/N3 * np.sin(theta1)*tp
    # Ezdm = Ezdp * (N3/N2)**2
    Ezd = Ezdp * (N3/Npp)**2

    # return(Ex0, Ey0, Ez0, Exd, Eyd, Ezd)
    return (abs(Ex0)**2, abs(Ey0)**2, abs(Ez0)**2, abs(Exd)**2, abs(Eyd)
        **2, abs(Ezd)**2)
```

In comparison with the other reflection and transmission-based techniques we have discussed in this chapter, the advantages of ellipsometry include

- A simple setup, with only one or two additional optics beyond what is required for a polarized reflectance measurement.
- The fact that n and κ can be obtained independently, without a need for Kramers–Kronig analysis and the issues associated with such a transform.
- No calibration of the intensity is required. In the nulling scheme, any polarization bias of the detector is also not important.
- The optical constants are obtained without the need of a reference sample.
- Strong oscillators are difficult to characterize by any other method.

These advantages are just the tip of the iceberg in terms of what type of information ellipsometry can provide. Additional benefits include

- For thin samples (where the thickness d is on the order of λ), one can get d, n, and κ! A common application of ellipsometry is one where the material is known and well characterized (n and κ are known), so the film thickness d is the only parameter of interest from the Ψ and Δ measurement.
- Analysis of multi-layer films, such as coatings and laminates.

- Analysis of ordered materials such as liquid crystals.
- The study of depolarization effects.
- Spectroscopic information, i.e. Ψ and Δ as a function of wavelength.
- Imaging ellipsometers where Ψ and Δ are spatially resolved. This is especially interesting for heterogeneous samples.

A particularly powerful variant of the instrument is a Mueller matrix ellipsometer that can measure all 16 elements of the sample's 4×4 Mueller matrix [27–31]. In this way, one can characterize samples with complex anisotropy, and samples that partially depolarize the light. An example of an instrument that can measure the full Mueller matrix, as a function of wavelength throughout the mid-infrared range is shown in Fig. 8.9. The instrument consists of an FTIR, followed by a wire-grid polarizer, photoelastic modulator acting as a time-varying compensator, the sample interrogated by internal reflection through a ZnSe prism, a second photoelastic modulator, and a second wire-grid polarizer as the analyser [32]. The use of dual compensators facilitates the measurement of all of unique Mueller matrix elements.

Simulated data from this instrument is shown in Fig. 8.10, where the horizontal axis displays the detector signal as a function of the FTIR interferometer scanning mirror position. At each mirror position, the vertical trace shows the detector signal as a function of time. The time dependence of this waveform is a result of the

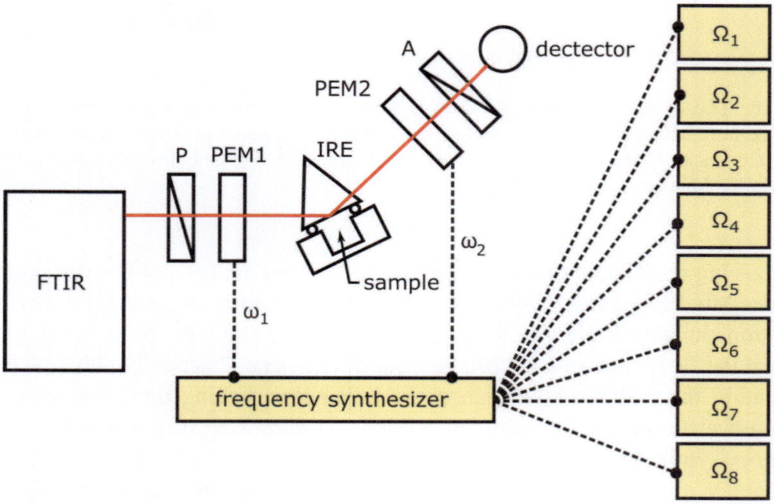

Fig. 8.9 One possible configuration of an infrared Mueller matrix ellipsometer based on an FTIR. The polarization state generator consists of a polarizer–photo elastic modulator (PEM) combination. After the light is reflected off the sample, a second PEM–polarizer pair constitutes the polarization state analyser, followed by an IR-sensitive detector. In order to determine all elements of the output Stokes vector, the detector signal is demodulated at nine frequencies of interest. Repeating this analysis at different azimuthal angles of the polarizer–PEM pairs enables all 16 elements of the sample's Mueller matrix to be determined

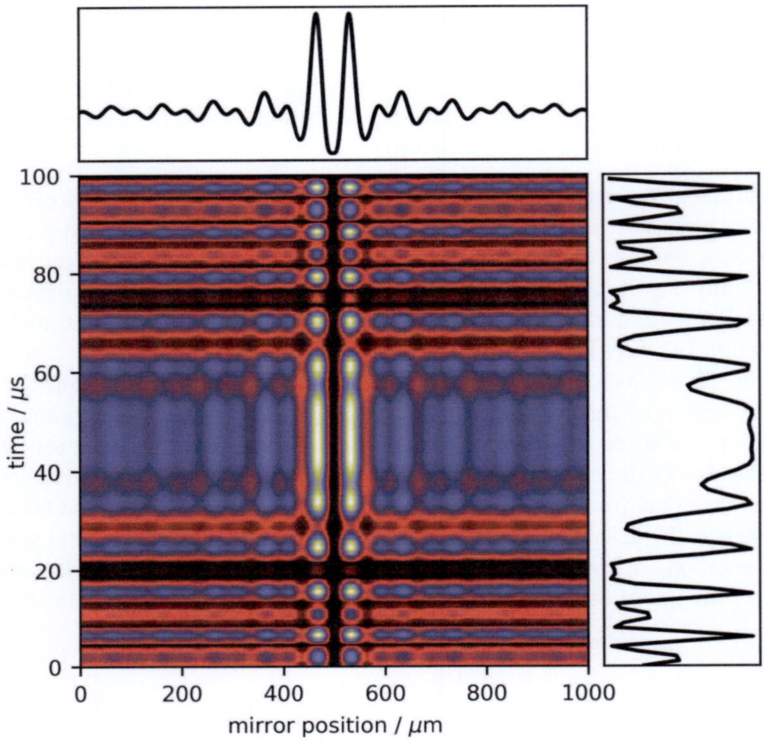

Fig. 8.10 Two-dimensional interferogram where the FTIR interferometer mirror position appears along the horizontal and the component of the signal demodulated by a lock-in amplifier at one of the frequencies of interest is along the vertical direction. A double Fourier transformation then results in the Mueller matrix elements

time-varying retardation of the compensators. Fourier analysis of the vertical slices provides Mueller matrix elements for each mirror position. A second Fourier transformation along the horizontal axis converts the signals into the frequency domain to arrive at the IR spectra of each Mueller matrix element. Practically, this is achieved by running the FTIR in step-scan mode rather than having the scanner continuously move the mirror in a smooth motion. At each step in the interferogram, a frequency synthesizer (see Fig. 8.9) is used to generate reference frequencies so that the Mueller matrix elements can be obtained using lock-in amplifiers rather than waveform analysis. Further details about the data analysis are provided in Ref. [32]. The results of such an experiment are shown in Fig. 8.11 where water is the sample that is placed in contact with the ZnSe prism [32]. Spectra of 15 elements of the Mueller matrix are shown (black traces), as they are normalized to M_{11}. For comparison, Mueller matrix spectra calculated from literature refractive index data for water [33] are plotted in grey, and the residuals are shown in blue. Although this demonstration was specifically about infrared Mueller matrix determination, the use

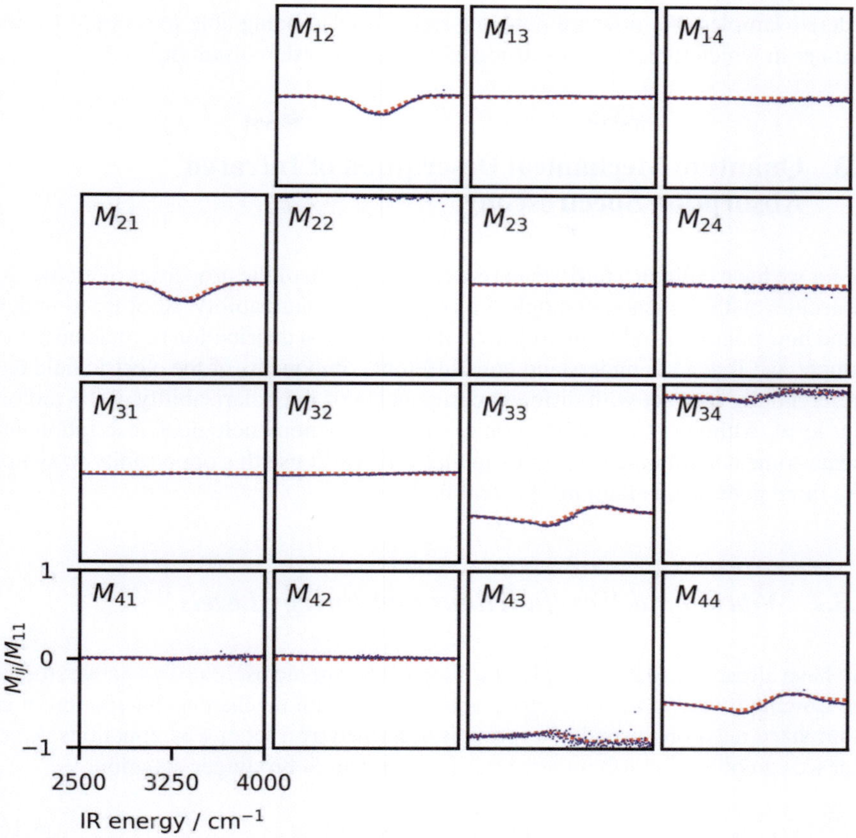

Fig. 8.11 Elements of the Mueller matrix, normalized with respect to M_{11} for water, as measured in an internal reflection experiment, as a function of the IR wavenumber in the OH stretching region. Since the sample is isotropic, elements of the Mueller matrix redundantly encode information about n and κ. Data from Ref. [32]

of photoelastic modulators in the infrared is broadly applicable to structure analysis, particularly for anisotropic samples. This field is known as polarization modulator IR absorption spectroscopy or PM-IRAS (including reflection absorption versions as PM-IRRAS) [34–40].

In summary, we have seen several methods for determining the optical constants of materials, ranging from a simple transmission-based approach when a suitable reference sample is available, to ellipsometry schemes where changes in the polarization of transmitted or reflected light can be used to work back to the n and κ values of the sample that caused those changes. While our demonstrations have been limited to isotropic samples so far, all of these approaches are equally applicable to ordered samples. We now are now one step closer to being able to account for the manner in which molecules are arranged in such an ordered sample.

8.3 Quantum Mechanical Description of Infrared Absorption Spectroscopy

So far we have built up a fairly feature-rich description of the properties of atoms and molecules with a phenomenological model of the polarizability. In order to understand how polarized light can inform on the orientation distribution of molecules in a sample, we'll need to understand how different components of the electric field can preferentially interact with different components of the polarizability at the molecular level. Although we could continue with our phenomenological description and create some anharmonic oscillators in that way, we'll use this opportunity to switch to a more general quantum mechanical framework.

8.3.1 Vibrational Wavefunctions and Energy Levels

We have already seen an example of a classical harmonic oscillator when we studied the Lorentz model in the previous chapter. In quantum mechanics, the motion of an atom-sized mass on the end of a spring is described by a set of wavefunctions $\psi_n(x)$ that we can obtain by solving the time-independent Schrödinger equation:

$$\hat{H}\psi_n(x) = E_n\psi_n(x). \tag{8.16}$$

Here the total energy (Hamiltonian) operator is given by

$$\hat{H} = -\frac{\hbar^2}{2m}\frac{\mathrm{d}^2}{\mathrm{d}x^2} + V(x), \tag{8.17}$$

where the first term is the kinetic energy operator, and the second term is an expression of the potential energy. For a harmonic potential

$$V(x) = \frac{1}{2}k_{\mathrm{H}}x^2 \tag{8.18}$$

where k_{H} is the (Hooke's law) force constant. This second-order differential equation has solutions that may be written as

$$\psi_n(x) = A_n H_n(x\sqrt{b}) \exp\left[-\frac{bx^2}{2}\right] \tag{8.19}$$

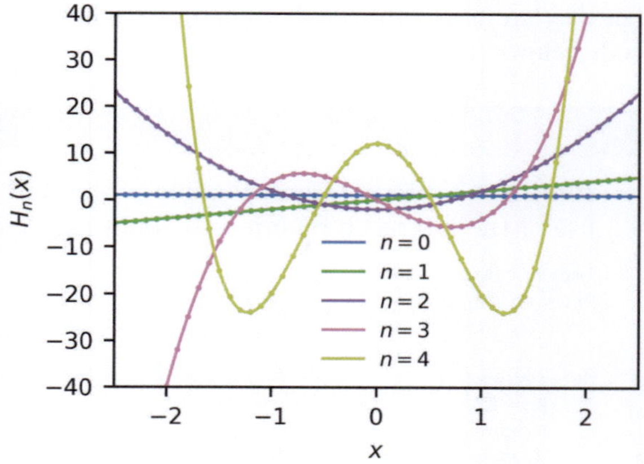

Fig. 8.12 Plots of the first five Hermite polynomials. The solid lines were generated using the built-in `scipy.special.hermite` functions, while the equivalent dotted versions were generated from the recursion relation in Eq. 8.22

where A_n is a normalization constant chosen so that the integral of ψ_n multiplied by its complex conjugate over all space is unity, resulting in

$$A_n = \frac{1}{\sqrt{2^n n!}} \left(\frac{a}{\pi}\right)^{1/4} \tag{8.20}$$

and $b = \sqrt{k_{\mathrm{H}} m / \hbar^2}$. The resulting wavefunctions are written in terms of the Hermite polynomials

$$H_n(x) = (-1)^n e^{x^2} \frac{\mathrm{d}^n}{\mathrm{d}x^n} e^{-x^2} \tag{8.21}$$

the first five of which are plotted in Fig. 8.12. Note that there are two different sets of Hermite polynomials with slightly different coefficients. The ones we use here are sometimes referred to as the physicist's Hermite polynomials. In order to illustrate how this rather complicated expression in Eq. 8.21 results in simple functions, the first five Hermite polynomials are listed here:

$$\begin{aligned}
H_0(x) &= 1 \\
H_1(x) &= 2x \\
H_2(x) &= 4x^2 - 2 \\
H_3(x) &= 8x^3 - 12x \\
H_4(x) &= 16x^4 - 48x^2 + 12.
\end{aligned}$$

It is straightforward to generate these using the recursion relation

$$H_n(x) = 2x H_{n-1}(x) - 2(n-1) H_{n-2}(x) \tag{8.22}$$

but they are also available as `hermite` from `scipy.special` as shown in the code below:

Code to generate Fig. 8.12

Plotting the Hermite polynomials is especially quick when one makes use of the `hermite` functions included in `scipy.special`.

```
import numpy as np
from scipy.special import hermite
from matplotlib import pyplot

def recursion(n):
    H = []
    H.append(np.ones(len(x)))
    H.append(2*x)
    for k in range(2, n+1):
        H.append(2*x*H[k-1] - 2*(k-1)*H[k-2])
    return H[n]

# setup
pyplot.rc('font', size=8)
fig = pyplot.figure(figsize=(3.25, 2.5))
ax = fig.add_subplot()
x = np.linspace(-2.5, 2.5, 500)

# plot the functions - two ways
for n in range(5):
    hp = hermite(n)
    p = ax.plot(x, hp(x), lw=1, label=r'$n=%d$' % n)
    ax.plot(x[::10], recursion(n)[::10], marker='o', mew=0, mfc=p[0].
         get_color(), ms=2, lw=0)

# pretty
ax.set_xlim(-2.5, 2.5)
ax.set_ylim(-40, 40)
ax.set_xlabel('$x$')
ax.set_ylabel('$H_n(x)$')
ax.legend(loc=0, frameon=False, fontsize=7)
fig.set_tight_layout(True)
pyplot.savefig('hermite.png', dpi=600)
```

The function we have called `hp` provides the *coefficients* for the nth-order Hermite polynomial. We then generate the actual polynomials by evaluating over a `linspace`; this is shown with solid lines. Out of curiosity, a simple “from scratch” Hermite polynomial generator was implemented using the recursion relation defined in Eq. 8.22. Those values are plotted with points of the same colour, overlaid on the lines.

Another Matplotlib trick to plot a few different things (lines and points in this example) in the same colour, for the cases where the colour may not be defined in advance, or is left to some acceptable default values (as is the case here), is to grab the automatically assigned colour of one plot object so that it can be reused. Here `p` is used to get the properties of the lines being plotted, and then `p[0].get_color()` retrieves the colour of the lines, for use in plotting the points using the second `plot` command. It is also attractive (and easier to see) when the density of points along `x` is reduced. Of course, we could create a second, sparser `linspace` for that, but I have opted to just evaluate and plot every 10th value of the original linspace.

Following this rather long list of substitutions, the first few vibrational wavefunctions look like this:

$$\psi_0(x) = \left(\frac{a}{\pi}\right)^{1/4} e^{-ax^2/2} \tag{8.23a}$$

$$\psi_1(x) = \left(\frac{4a^3}{\pi}\right)^{1/4} x e^{-ax^2/2} \tag{8.23b}$$

$$\psi_2(x) = \left(\frac{a}{4\pi}\right)^{1/4} (2ax^2 - 1)e^{-ax^2/2}. \tag{8.23c}$$

To be well behaved, we require $|\psi_n(x)|$ to be finite as $x \to \pm\infty$. This condition results in quantization of the energy levels. The solutions to Eq. 8.16 are restricted to

$$E_n = h\nu\left(n + \frac{1}{2}\right) \tag{8.24}$$

where $n = 0, 1, 2, 3, \ldots$. At this point, the obligatory plots are $\psi_n(x)$ and $|\psi_n(x)|^2$ as the magnitude squared is related to the probability density for finding the oscillator with a displacement x from its equilibrium position. These are shown in Fig. 8.13, with the traces offset by the energy level. Note that these wavefunctions are real valued, so $\psi^*(x)\psi(x) = |\psi(x)|^2$ amounts to simply squaring the wavefunction, and $\psi(x)$ itself can be plotted directly as its phase is restricted to positive and negative real values in Fig. 8.13a. In these plots, the classically forbidden regions are those where the potential is greater than $V(x)$, shaded in grey. However, as quantum mechanical systems are governed by wave-like behaviour, it is possible to have displacements from the equilibrium position ($x = 0$) in those forbidden regions, a phenomenon known as tunnelling.

Code to generate Fig. 8.13

In this code, the Hermite polynomials from `scipy.special` are used again, but this time, rather than plotting them directly, they are used to construct the vibrational wavefunctions simply by including some factors. The first is the amplitude factor that we have defined as `A()`

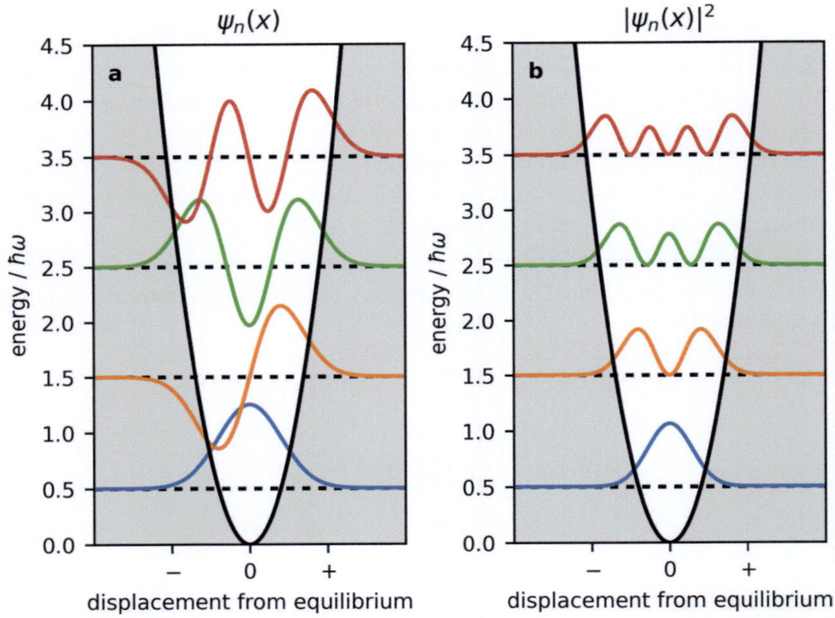

Fig. 8.13 **a** The first four vibrational wavefunctions $\psi_n(x)$ plotted vertically offset by their corresponding energy levels E_n. **b** $|\psi_n(x)|^2$ is plotted for the same set of functions

and the second is an exponential factor, ultimately producing the wavefunctions as described in Eq. 8.19.

The first of the two subplots displays the wavefunctions $\psi_n(x)$ themselves, and the second plots $|\psi_n|^2$. The command `fill_between(x, 0.5*x**2)` was used to shade the region where the potential energy exceeds $V(x)$ defined in Eq. 8.18.

```
import numpy as np
from scipy.special import hermite, factorial
from matplotlib import pyplot

def A(i, alpha):
   return (alpha/np.pi)**0.25/np.sqrt(factorial(i)*2.0**i)

# setup
pyplot.rc('font', size=8)
fig = pyplot.figure(figsize=(4.5, 3.5))
x = np.linspace(-5., 5., 1000)
alpha = 1.0
xRootAlpha = x*np.sqrt(alpha)
ax1 = fig.add_subplot(121)
ax2 = fig.add_subplot(122)

for n in range(4):
```

```
        wf = A(n,alpha)*np.polyval(hermite(n), xRootAlpha)*np.exp(-0.5*alpha*x
            **2.0)
        probability = np.conjugate(wf)*wf
        energy = n + 0.5
        for ax in [ax1, ax2]:
            ax.axhline(energy, color='k', ls='--', dashes=[2,2])
        ax1.plot(x, wf + energy)
        ax2.plot(x, probability + energy)

    # pretty
    ax1.set_title(r'$\psi_n(x)$')
    ax2.set_title(r'$|\psi_n(x)|^2$')
    ax1.label = 'a'
    ax2.label = 'b'
    for ax in [ax1, ax2]:
        ax.plot(x, 0.5*x**2, 'k-')
        ax.fill_between(x, 0.5*x**2, color='0.5', alpha=0.3)
        ax.set_xlim(-5, 5)
        ax.set_ylim(0, 4.5)
        ax.set_xlabel('displacement from equilibrium')
        ax.set_xticks([-2.5, 0, 2.5])
        ax.set_xticklabels(['$-$', '0', '$+$'])
        ax.set_ylabel(r'energy / $\hbar \omega$')
        ax.text(0.05, 0.93, ax.label, fontweight='bold', transform=ax.transAxes
            )

    fig.set_tight_layout(True)
    pyplot.savefig('vibrationalWF.png', dpi=600)
```

8.3.2 Selection Rules for IR Absorption

The dipole moment is defined as

$$\begin{aligned} \mu_x^{(mn)} &= \int_{-\infty}^{\infty} \psi_n^*(x) \cdot \mu_x \cdot \psi_m(x)\mathrm{d}x \\ &= \langle n|\mu_x|m\rangle \end{aligned} \tag{8.25}$$

where the second line introduces Dirac's notation for the integral including the shorthand notation for the wavefunctions themselves. If $m = n$, then this provides the dipole moment for state n. For example, the ground state dipole moment is given by

$$\mu_x^{(0)} = \int_{-\infty}^{\infty} \psi_0^*(x) \cdot \mu_x \cdot \psi_0(x)\mathrm{d}x$$

and the first excited state dipole moment can be obtained from

$$\mu_x^{(1)} = \int_{-\infty}^{\infty} \psi_1^*(x) \cdot \mu_x \cdot \psi_1(x) \mathrm{d}x.$$

However, when $m \neq n$, Eq. 8.25 provides the transition dipole moment (or TDM for short) that connects states m and n.

As a molecule vibrates, μ_x changes with time. We expand $u_x(t)$ as a Taylor series about the equilibrium bond length x via

$$f(x) = \sum_{n=0}^{\infty} \frac{(x-a)^n}{n!} \frac{\mathrm{d}^n}{\mathrm{d}x^n} f(a)$$

and recall that we have set up our coordinate system so that the equilibrium bond length corresponds to $x = 0$. This results in

$$u_x[x(t)] = \mu_{0x} + x(t) \left[\frac{\mathrm{d}}{\mathrm{d}x} \mu_x \right]_{x=0} + \cdots \tag{8.26}$$

where μ_{0x} is the permanent or static dipole moment, that which exists in the absence of an applied field. The second term in the expansion is the induced dipole moment. Keeping only these first two terms in the expansion, we substitute them into our expression for the TDM to obtain

$$\begin{aligned} \mu_x^{(mn)} &= \int_{-\infty}^{\infty} \psi_n^*(x) \cdot \mu_x \cdot \psi_m(x) \mathrm{d}x \\ &= \int_{-\infty}^{\infty} \psi_n^*(x) \cdot \left[\mu_{0x} + x \left(\frac{\mathrm{d}\mu_x}{\mathrm{d}x} \right)_{x=0} \right] \cdot \psi_m(x) \mathrm{d}x. \end{aligned} \tag{8.27}$$

We are now ready to substitute the vibrational wavefunctions $\psi_m(x)$ and $\psi_n(x)$ from Eq. 8.19. This results in

$$\begin{aligned} \mu_x^{(mn)} &= A_m A_n \mu_{0x} \int_{-\infty}^{\infty} H_m(x\sqrt{a}) H_n(x\sqrt{a}) e^{-ax^2} \mathrm{d}x \\ &+ A_m A_n \left(\frac{\mathrm{d}\mu_x}{\mathrm{d}x} \right)_{x=0} \int_{-\infty}^{\infty} H_m(x\sqrt{a}) x H_n(x\sqrt{a}) e^{-ax^2} \mathrm{d}x \\ &= \langle m | \mu_x | n \rangle. \end{aligned} \tag{8.28}$$

The integral in the first term vanishes since the Hermite polynomials are orthogonal. The second integral requires closer inspection since the integrand involves a product of two Hermite polynomials and the displacement x. Although some symmetry arguments can be made here [41], it is interesting to graphically inspect these functions. Figure 8.14 plots

$$f(x) = H_m(x\sqrt{a}) x H_n(x\sqrt{a})$$

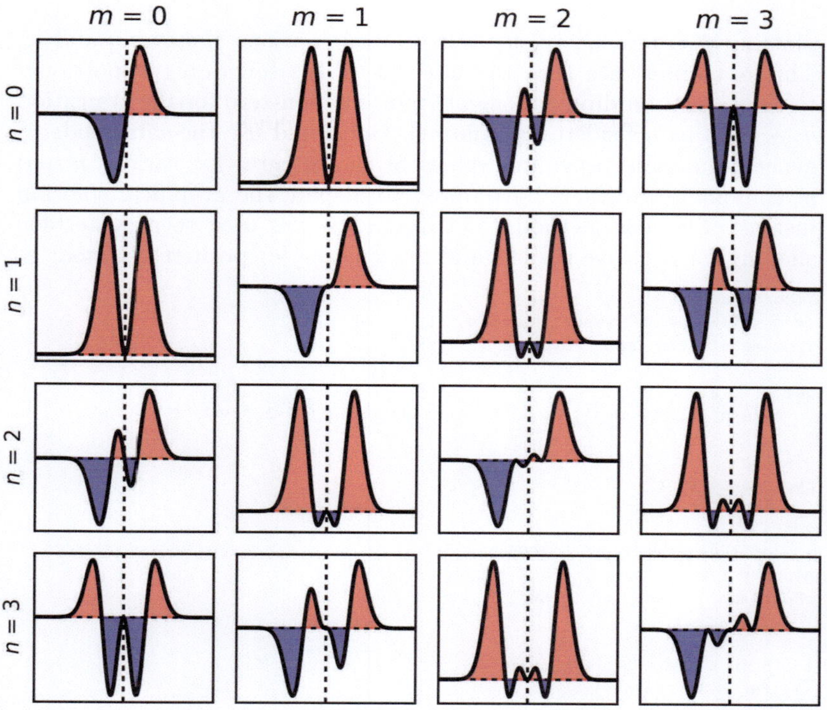

Fig. 8.14 A graphical illustration of the vibrational selection rules. The transition dipole moment connecting states n (vertical) and m (horizontal) is non-zero only when the integral in Eq. 8.28 is non-zero. Positive areas of the integrand are highlighted in red, negative areas in blue. For the integral to be non-zero the red and blue values cannot be equal, and this is observed only when $n = m \pm 1$

for various combinations of m (columns) and n (rows). For clarity, regions for which $f(x) > 0$ are highlighted in red, and regions where $f(x) < 0$ are shown in blue. The solid horizontal line indicates $f(x) = 0$, and the dashed vertical line shows the position of the equilibrium geometry at $x = 0$. Here we can see that when $m = n$, the positive and negative regions have the same area, so those integrals will vanish. Although it is more subtle to see, the same is true for $m = 1$ and $n = 3$, for example. In fact, the only non-vanishing combinations are those for which the difference between m and n is exactly ± 1. Stated more succinctly, the vibrational selection rule is $\Delta n = \pm 1$.

Code to generate Fig. 8.14

This is a very simple code that is essentially the same as the one used to make the first subplot in Fig. 8.13, except that we are now plotting $\psi_n(x) \cdot x \cdot \psi_m(x)$ for various combinations of states n and m. This is to illustrate that the integral of this function (proportional to the transition dipole moment) must be non-zero for the transition $n \rightarrow m$ to be considered as allowed. One could use the same code to generate a table providing values of the integral for various n and m combinations, but I have opted to display the graphical solution instead. The axis method `fill_between` was used to provide blue shading for negative regions and red shading for positive regions.

```
import numpy as np
from scipy.special import hermite, factorial
from matplotlib import pyplot

def A(i, alpha):
   return (alpha/np.pi)**0.25/np.sqrt(factorial(i)*2.0**i)

# setup
pyplot.rc('font', size=8)
fig = pyplot.figure(figsize=(4.5, 4))
i = 4
x = np.linspace(-5., 5., 1000)
alpha = 1.0
xRootAlpha = x*np.sqrt(alpha)

s = 1
for n in range(i):
    for m in range(i):
        ax = fig.add_subplot(i, i, s)
        lower = A(n,alpha)*np.polyval(hermite(n), xRootAlpha)*np.exp(-0.5*
             alpha*x**2.0)
        upper = A(m,alpha)*np.polyval(hermite(m), xRootAlpha)*np.exp(-0.5*
             alpha*x**2.0)
        integrand = np.conjugate(upper) * x * lower
        ax.plot(x, integrand, 'k-', lw=1.5)
        ax.fill_between(x, 0, integrand, where=integrand>0, color='r',
             alpha=0.5)
        ax.fill_between(x, 0, integrand, where=integrand<0, color='b',
             alpha=0.5)
        ax.axhline(0, color='k', lw=1, ls='--', dashes=[2,2])
        ax.axvline(0, color='k', lw=1, ls='--', dashes=[2,2])

        # pretty
        ax.set_xlim(-5, 5)
        ax.set_xticks([])
        ax.set_yticks([])
        if n == 0:
            ax.set_title('$m$ = %d' % m)
        if m == 0:
            ax.set_ylabel('$n$ = %d' % n)
        s += 1

fig.set_tight_layout(True)
pyplot.savefig('vibrationalSelection.png', dpi=600)
```

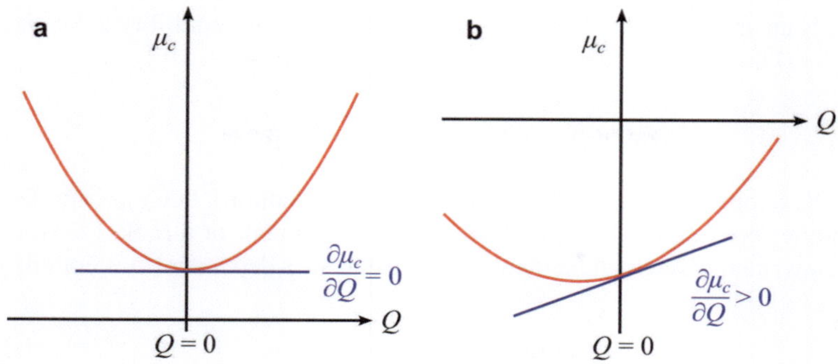

Fig. 8.15 **a** A vibrational mode for which there is a significant change in the electronic structure when the nuclear displacement changes, in such a manner so that the c-component of the dipole moment μ_c changes during the vibration, but not when the mode passes through the equilibrium geometry at $Q = 0$. Therefore this mode is not IR active. This also illustrates that the value of μ_c at $Q = 0$ does not contribute to the transition dipole moment. **b** In contrast, an IR-active mode has a non-zero slope of μ_c with respect to Q when evaluated at $Q = 0$. Also note that the fact that $\mu_c < 0$ is irrelevant to the conclusion that $(\partial \mu_c / \partial Q)_{Q=0} > 0$ for this mode. This cartoon can be compared with data obtained from electronic structure calculations of two vibrational modes of an actual molecule in Fig. 2.3

Taking a closer look at the second term in Eq. 8.28, we see that the integral we have been discussing has a prefactor of $(\mathrm{d}\mu_x/\mathrm{d})_{x=0}$. This means that the amplitude of the $0 \rightarrow 1$ transition is related to the slope of the dipole moment with respect to the vibrational coordinate (in this case simplified to x), evaluated at the position of the equilibrium geometry ($x = 0$) as shown in Fig. 8.15. In other words, if a particular vibrational mode—even for a molecule that has a large dipole moment—doesn't result in a change in the dipole moment as those atoms move about their equilibrium geometry in the course of the vibrational motion, then that mode is considered to be IR inactive. Even if IR light is incident on such a sample, and the wavelength of the light is such that the energy of the IR photons matches the energy gap for the $0 \rightarrow 1$ transition, the molecule will not be able to absorb the light in order to increase the amplitude of the vibration.

8.3.3 Connection to the Polarizability and Susceptibility

Now that we have assessed the transition dipole moment, we can get back to the molecular polarizability that we previously described in phenomenological terms. But first, a practical matter. We have seen that $\boldsymbol{\alpha}$ and $\boldsymbol{\chi}$ are rank-2 tensors with nine elements each, α_{lm} and χ_{ij}. However, $\boldsymbol{\varepsilon}_{\mathrm{R}}$, n and κ are written with a single subscript, for example, n_x or κ_y. To understand this, we need to consider how molecules are arranged inside the material, i.e. taking the distribution of their orientations into account. This will be the subject of the next section. For now we can imagine that, if there is an arrangement of molecules that causes the susceptibility to be diagonal, i.e. $\chi_{ij} = 0$ if $i \neq j$, we can then understand the notation

$$N_x = n_x + i\kappa_x = \sqrt{1 + \chi_{xx}} \tag{8.29}$$

where N_x is preferred to N_{xx} (and this is just a convention), but χ_{xx} is preferred to χ_x (another convention, but a logical one for an element of a rank-2 tensor). One can then write the diagonal elements of the frequency-dependent susceptibility as

$$\chi_{ii}(\omega) = \frac{A_{ii}}{\omega_0^2 - \omega^2 - i\omega\Gamma_j} \tag{8.30}$$

where it is understood that, if there are multiple vibrational resonances, this becomes a sum over vibrational modes, with each mode having its particular value of ω_0, A_{ii} and Γ.

Since the amplitude of the resonant term is defined as the outer product of two transition dipole moments

$$a_{lm} \propto \langle 0|\mu_l|1\rangle \otimes \langle 1|\mu_m|0\rangle, \tag{8.31}$$

when the diagonal elements of **a** such as a_{ll} are desired, they can be expressed as

$$a_{ll} \propto |\langle 1|\mu_l|0\rangle|^2 \tag{8.32}$$

as we can switch the order of the states 0 and 1 and have the same result. For isotropic (and some ordered) materials, this means that the imaginary part of the refractive index is proportional to the square of the vibrational transition dipole moment, and that (for weak oscillators with small a_{ll}) the absorbance is proportional to $\mathbb{I}\mathrm{m}\{\chi\}$ if the reference sample is chosen appropriately, as we have discussed at the beginning of this chapter.

So far, our example has been that of a simple diatomic molecule such as H–Cl, where the normal mode coordinate is along the bond axis. For a polyatomic molecule with N atoms, there are $3N - 6$ normal modes. In this case there is no simple picture of the coordinate that we have been referring to as x. Instead we have a $(3N - 6)$-dimensional normal mode coordinate Q_j for each of the j vibrational modes. We then consider that the square of the transition dipole moment is proportional to $|(\mathrm{d}\mu/\mathrm{d}Q)_{Q=0}|^2$ in order to make the connection to the oscillator strength, absorption coefficient, κ, and absorbance associated with vibrational mode j. More generally, IR-active modes are ones for which $|(\mathrm{d}\mu/\mathrm{d}Q)_{Q=0}|^2 \neq 0$ and, practically speaking, $|(\mathrm{d}\mu/\mathrm{d}Q)_{Q=0}|^2$ should be large enough to create a noticeable resonance for a particular vibrational mode to be casually labelled as IR active.

8.4 Molecular Orientation Analysis from Polarized IR Spectra

We have introduced a lot of formalism to this point, and developed some competence in using Python to write small programs for computing quantities and visualizing them. We are now in a position to use these tools to address our main objective, to characterize the molecular structure of materials. The basic idea is that, if we interrogate materials with polarized light, we can learn about the molecular organization. Ultimately this comes down to the relative orientation between the transition dipole moment vector and the electric field vector from the light beam. One thing to keep in mind: we are ultimately interested in describing the orientation of molecules, or specific parts of molecules such as individual chemical functional groups. But IR absorption spectroscopy provides information on the orientation distribution of the transition dipole moments. So, depending on the vibrational mode of interest, we may need an additional projection to relate the vector of interest within the molecular frame to the TDM.

8.4.1 *Coordinate Systems of Molecules and Materials*

For convenience, let's write the transition dipole moment that we have been abbreviating TDM as **M**. Since **E** is generated and detected in the laboratory frame, we need to project **M** from the molecular frame into the lab frame. Any quantity in one coordinate system can be projected into another (rotated) coordinate system by a transformation **D**

$$\mathbf{M}_{(abc)} \xrightarrow{\mathbf{D}} \mathbf{M}_{(xyz)}.$$

The transformation matrix **D** is aptly named the direction cosine matrix (DCM), as it consists of nine direction cosines [42–44]. These are the cosines of the angles between the unit vectors in each of the coordinate systems. For example, γ_{xc} is the angle between the laboratory-frame x-axis and the molecular-frame c-axis, so

$$\hat{\mathbf{x}} \cdot \hat{\mathbf{c}} = |\hat{\mathbf{x}}||\hat{\mathbf{c}}| \cos \gamma_{xc} = \cos \gamma_{xc}$$

as $\hat{\mathbf{x}}$ and $\hat{\mathbf{c}}$ are unit vectors. We can then write the projection operator, the direction cosine matrix, as

$$\mathbf{D} = \begin{bmatrix} \cos \gamma_{xa} & \cos \gamma_{xb} & \cos \gamma_{xc} \\ \cos \gamma_{ya} & \cos \gamma_{yb} & \cos \gamma_{yc} \\ \cos \gamma_{za} & \cos \gamma_{zb} & \cos \gamma_{zc} \end{bmatrix}. \tag{8.33}$$

The form of the DCM can be understood from considering the sequential application of three rotation matrices [42, 44]. These are the rotation about the laboratory z-axis by an angle ψ using

$$\mathbf{R}_z(\psi) = \begin{bmatrix} \cos\psi & -\sin\psi & 0 \\ \sin\psi & \cos\psi & 0 \\ 0 & 0 & 1 \end{bmatrix} \tag{8.34}$$

followed by a rotation about y by an angle θ using

$$\mathbf{R}_y(\theta) = \begin{bmatrix} \cos\theta & 0 & \sin\theta \\ 0 & 1 & 0 \\ -\sin\theta & 0 & \cos\theta \end{bmatrix} \tag{8.35}$$

and finally rotation about the new (after the previous rotations) z-axis by ϕ

$$\mathbf{R}_z(\phi) = \begin{bmatrix} \cos\phi & -\sin\phi & 0 \\ \sin\phi & \cos\phi & 0 \\ 0 & 0 & 1 \end{bmatrix} \tag{8.36}$$

so that

$$\begin{aligned} \mathbf{D}(\theta, \phi, \psi) &= \mathbf{R}_z(\phi) \cdot \mathbf{R}_y(\theta) \cdot \mathbf{R}_z(\psi) \\ &= \begin{bmatrix} D_{11} & D_{12} & D_{13} \\ D_{21} & D_{22} & D_{23} \\ D_{31} & D_{32} & D_{33} \end{bmatrix}, \end{aligned} \tag{8.37}$$

where

$$\begin{aligned} D_{11} &= \cos\phi\cos\theta\cos\psi - \sin\phi\sin\psi \\ D_{12} &= -\cos\phi\cos\theta\sin\psi - \sin\phi\cos\psi \\ D_{13} &= \sin\theta\cos\phi \\ D_{21} &= \sin\phi\cos\theta\cos\psi + \cos\phi\sin\psi \\ D_{22} &= -\sin\phi\cos\theta\sin\psi + \cos\phi\cos\psi \\ D_{23} &= \sin\theta\sin\phi \\ D_{31} &= -\cos\psi\sin\theta \\ D_{32} &= \sin\psi\sin\theta \\ D_{33} &= \cos\theta. \end{aligned}$$

These steps are illustrated in Fig. 8.16. An important property of the direction cosine matrix is that it is orthogonal, i.e. it is a square matrix whose rows and columns are orthonormal vectors. This means that

$$\mathbf{D}^{\mathrm{T}}\mathbf{D} = \mathbf{D}\mathbf{D}^{\mathrm{T}} = \mathbf{1} \tag{8.38}$$

where $\mathbf{1}$ is the 3×3 identity matrix that has ones along the diagonal and zeros in all the off-diagonal locations. From this it follows that the inverse of the direction cosine matrix is equal to its transpose

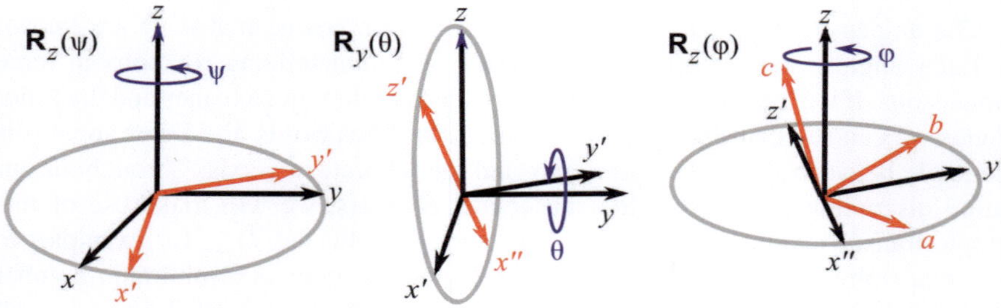

Fig. 8.16 The three successive elementary rotations that together can be used to rotate any object in three-dimensional space

$$\mathbf{D}^{-1} = \mathbf{D}^{\mathrm{T}}, \tag{8.39}$$

a very important property that we will make use of throughout the remainder of our discussion. This is especially important for symbolic calculations as the inverse is typically computationally expensive to evaluate, while forming the transpose is quick. This also means that, when using numerical evaluation (as in NumPy code), one can get an element of the inverse (through the transpose) just by switching the indices, i.e. `D[j,i]` instead of `D[i,j]`.

SymPy direction cosine matrix

Using the SymPy package to perform some symbolic algebra, we can readily construct **D** from the three constituent rotation matrices using

```
import sympy as sp
from sympy import init_printing

theta, phi, psi = sp.symbols('theta phi psi')

# direction cosine matrix
R1 = sp.Matrix([[sp.cos(psi), -sp.sin(psi), 0], [sp.sin(psi), sp.cos(psi),
     0], [0, 0, 1]])
R2 = sp.Matrix([[sp.cos(theta), 0, sp.sin(theta)], [0, 1, 0], [-sp.sin(
     theta), 0, sp.cos(theta)]])
R3 = sp.Matrix([[sp.cos(phi), -sp.sin(phi), 0], [sp.sin(phi), sp.cos(phi),
     0], [0, 0, 1]])
DCM = R3*R2*R1
```

An individual element can then easily be inspected with

```
sp.pprint(sp.simplify(DCM[0,0]))
```

The angles (θ, ϕ, ψ) that appear in the direction cosine matrix $\mathbf{D}$ are known as Euler angles, and are one of a few ways of parameterizing rotations in three dimensions. If we look at the connection between the direction cosines and the Euler angles we can see that $D_{zc} = \cos\gamma_{zc} = \cos\theta$. In other words, the Euler angle θ is the angle between the molecular c-axis and the laboratory z-axis. Throughout our future discussion in many of the subsequent chapters, we will make use of this simple transformation (compared to the more complicated D_{xa}, for example) to illustrate some concepts with simple expressions that contain a minimum number of trigonometric functions. Finally, even though not all elements of $\mathbf{D}$ depend on all Euler angles—for example $D_{11}(\theta, \phi, \psi)$, $D_{23}(\theta, \phi)$, and $D_{33}(\theta)$, we will refer to elements of the DCM as $D_{ij}(\theta, \phi, \psi)$ in general.

Now that we're going to keep track of quantities (such as dipole moments, transition dipole moments, and polarizabilities) pertaining to molecules in the molecular frame, and quantities pertaining to materials (polarization, susceptibility, refractive index, etc.) in the laboratory frame, let's fine-tune some of the notation we have introduced previously. We have written things like

$$\boldsymbol{\mu}_{\text{induced}} = \boldsymbol{\alpha}\mathbf{E}$$

but, in retrospect, all quantities need to be in the same coordinate system. It is a bit strange to consider $\mathbf{E}$ in the molecular frame, although there is nothing technically wrong with that. It's just that we will create polarized light where the azimuths of the electric field are defined with respect to the laboratory (x, y, z)-coordinates. Nevertheless, the above notation served us well enough in talking about light–molecule interactions. Now we just need to consider the coordinate systems more carefully from here on in. Another expression that we previously wrote was

$$\mathbf{P} = \varepsilon_0\boldsymbol{\chi}\mathbf{E}.$$

Now $\boldsymbol{\chi}$ and $\mathbf{E}$ are both in the lab frame, which is good. But what about the polarization? If we think about it, the dipole moments are all defined in the molecular frame, but the dipole moment per unit volume is a material (laboratory frame) quantity. Considering that

$$\mathbf{P} = \frac{1}{V}\sum_{\text{molecules}} \boldsymbol{\mu}$$

where V is the volume of the material, we can equate the two expressions to get

$$\chi_{ii} = \frac{1}{\varepsilon_0 V}\sum_{\text{molecules}} \alpha_{ii}(\theta, \phi, \psi)$$

where χ_{ii} (to discuss a particular element of $\boldsymbol{\chi}$) is a lab frame quantity, and $\boldsymbol{\alpha}$ has been projected into the lab frame. In the notation that we will adopt, there are two clues that α is no longer in its "home" coordinate system. The important one is that we use Cartesian indices i, j, k to highlight that $\boldsymbol{\alpha}$ has been projected into the lab frame. If we are being pedantic about it, we also explicitly write the dependence on the Euler angles (θ, ϕ, ψ). In other words, α_{aa}, α_{ab}, etc. are numbers that describe the molecular polarizability. The $\hat{\mathbf{a}}$, $\hat{\mathbf{b}}$, $\hat{\mathbf{c}}$ coordinates are chosen in some meaningful way for that particular molecule, and that definition must be specified and known. As an example, for elongated molecules, it is a popular choice to choose $\hat{\mathbf{c}}$ to coincide with the long axis of the molecule. In vibrational spectroscopy, the definition is often more localized. For example, $\hat{\mathbf{c}}$ might correspond to the threefold symmetry axis of a methyl group.

Enough talking—let's roll up our sleeves and do the projection. Projecting a vector from one coordinate system to another is accomplished by a straightforward use (matrix multiplication) of the direction cosine matrix. As $\boldsymbol{\alpha}$ is a rank-2 tensor, its projection requires two applications of the DCM as follows:

$$\alpha_{ij}(\theta, \phi, \psi) = \sum_{l}^{a,b,c} \sum_{m}^{a,b,c} D_{il}(\theta, \phi, \psi) \cdot D_{jm}(\theta, \phi, \psi) \cdot \alpha_{lm}. \tag{8.40}$$

As can be seen from the double summation, we need to consider all nine elements of $\boldsymbol{\alpha}$ in the molecular frame in order to get *one* element of $\boldsymbol{\alpha}$ in the laboratory frame. So this procedure would need to be repeated nine times to transform the entire tensor. In a typical Python code, this would require four nested loops. Two loops to determine all of the nine components for a single α_{ij} element, and another two outer loops for all of the α_{ij} elements. There are mathematical and programming shortcuts to this procedure, but we'll leave those aside for now.

All of this really applies to the *numerator* of $\boldsymbol{\alpha}(\omega)$, what we have been calling the amplitude $\mathbf{a}$ or oscillator strength. So if $\boldsymbol{\alpha}$ (in either coordinate system) is written as

$$\boldsymbol{\alpha}(\omega) = \frac{\mathbf{a}}{\omega_0^2 - \omega^2 - i\omega\Gamma}$$

we can write the projection as

$$a_{ij}(\theta, \phi, \psi) = \sum_{l}^{a,b,c} \sum_{m}^{a,b,c} D_{il}(\theta, \phi, \psi) \cdot D_{jm}(\theta, \phi, \psi) \cdot a_{lm}. \tag{8.41}$$

Another way of thinking about this is a situation where an IR spectrum is fit to a series of Lorentzians describing each resonant mode, and and then elements of $\mathbf{A}$ obtained by fitting are the laboratory-frame projections. In this case, we have removed the frequency dependence of the denominator, and can proceed to make a connection with the molecular frame in terms of the amplitude, only.

If we want to write out one of the elements explicitly, taking a_{xx} as an example, it would look like this:

$$\begin{aligned} a_{xx}(\theta, \phi, \psi) &= \sum_{l}^{a,b,c} \sum_{m}^{a,b,c} D_{xl}(\theta, \phi, \psi) D_{xm}(\theta, \phi, \psi) a_{lm} \\ &= D_{xa}(\theta, \phi, \psi) D_{xa}(\theta, \phi, \psi) a_{aa} \\ &\quad + 2D_{xa}(\theta, \phi, \psi) D_{xb}(\theta, \phi, \psi) a_{ab} \\ &\quad + 2D_{xa}(\theta, \phi, \psi) D_{xc}(\theta, \phi, \psi) \alpha_{ac} \\ &\quad + D_{xb}(\theta, \phi, \psi) D_{xb}(\theta, \phi, \psi) \alpha_{bb} \\ &\quad + 2D_{xb}(\theta, \phi, \psi) D_{xc}(\theta, \phi, \psi) \alpha_{bc} \\ &\quad + D_{xc}(\theta, \phi, \psi) D_{xc}(\theta, \phi, \psi) \alpha_{cc} \end{aligned} \tag{8.42}$$

where we have made use of the fact that there are only six unique elements of the 9 a_{lm} as a result of being able to permute the indices l and m far from electronic resonance.

This is a good time to mention that, if we are interested in A_{ii} (i.e. A_{ij} for which $i = j$), then one could alternatively transform the transition dipole vector $\mathbf{M}$ into the laboratory frame, and then square it. Then our example of evaluating a_{xx} would look like

$$\begin{aligned} a_{xx}(\theta, \phi, \psi) &= \left[\sum_{l}^{a,b,c} D_{xl}(\theta, \phi, \psi) M_l \right]^2 \\ &= [D_{xa}(\theta, \phi, \psi) M_a + D_{xb}(\theta, \phi, \psi) M_b + D_{xc}(\theta, \phi, \psi) M_c]^2 \\ &= D_{xa}^2(\theta, \phi, \psi)\, M_a^2 + \cdots \\ &= D_{xa}(\theta, \phi, \psi) D_{xa}(\theta, \phi, \psi)\, a_{aa} + \cdots \end{aligned} \tag{8.43}$$

and the result of Eqs. 8.42 and 8.43 ends up being the same. This may seem like a convenient shortcut, but we won't bother with this in the present discussion. We will however return to a very similar point when we study Raman spectroscopy in Chap. 9 as it will be extremely useful to have a shorter way of carrying out all of the projections in the Raman case.

Overall, this approach—directly transforming the molecular-frame property into the laboratory frame—works if we can know the contribution of each molecule individually. This is in fact the situation that one encounters when analysing data from a molecular dynamics trajectory. Each frame in the simulation contains the Cartesian (x, y, z) data from every atom in every molecule. Successive frames are snapshots of the system at future times. If one were to make a movie from those frames, the molecules would be dancing and this is the way to visualize their trajectory. Such data is also of great interest to spectroscopists, as it provides a way to relate spectroscopic observables with the orientation distribution. Let's say the simulation consists of a box of 1000 water molecules. If one loops through every molecule in the frame, then from the (x, y, z) coordinates of each oxygen and hydrogen atom, one can know the a, b, and c unit vectors for each water molecule, and the set of (θ, ϕ, ψ) for each molecule, thereby providing a means to evaluate all of the required quantities on the right side of Eq. 8.4.1. Treating data like this from molecular simulations will be described more in Chap. 11.

To analyse experimental data, however, we need a different approach as we don't have access to the Euler angles of each molecule individually. We can then use

$$\begin{aligned}\chi_{ij} &= \frac{1}{\varepsilon_0 V} \sum_{\text{molecules}} \alpha_{ij}(\theta, \phi, \psi) \\ &= \frac{1}{\varepsilon_0 V}(\text{number of molecules}) \times (\text{average value of } \alpha_{ij}) \\ &= \frac{N_\mathrm{V}}{\varepsilon_0} \langle \alpha_{ij}(\theta, \phi, \psi) \rangle\end{aligned} \tag{8.44}$$

where N_V is the number of molecules per unit volume. The angular brackets in the last line of Eq. 8.44 represent an average over all participating orientations. We say "participating" as we mean that this takes the orientation distribution into account. In other words, we are not implying that the distribution is isotropic merely by mentioning that we are averaging over *all* orientations; we mean all the ones that are present. If the molecules were in fact randomly oriented, as in our example of the box of water, Eq. 8.44 would be that of an isotropic average. But if the molecules are preferentially ordered with respect to a particular direction in the laboratory frame, then this average would reflect that. In other words, we are averaging over the actual orientation distribution of the molecules.

When the goal is to determine the structure of materials, i.e. to comment on the orientation distribution of chemical bonds, the general steps are to

1. Measure IR absorption spectra with different combinations of polarized light. This can be a transmission or reflection experiment. Reflection geometry can be in internal or external reflection schemes. Ellipsometric measurements provide even more information.
2. Have some information about $\boldsymbol{\alpha}$ in the molecular frame. There are some very special circumstances where this is not required, but that is generally not the case. This typically comes from prior knowledge. A potential source of this data is quantum chemical calculations of the transition dipole moment.
3. The last step is to piece together precisely how $\boldsymbol{\alpha}$ in the molecular frame got averaged to account for the molecular response, $\boldsymbol{\chi}$. Formally, we seek the molecular orientation distribution.

8.4.2 Orientation Distributions and Orientation Distribution Functions

The orientation distribution (OD) describes how the molecules in a material are organized. If we return to our example of the molecular dynamics simulation, we

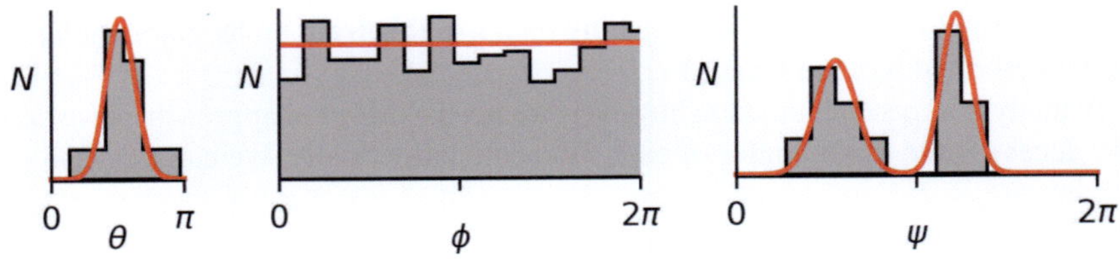

Fig. 8.17 An orientation distribution (OD) as would be described by a histogram based on sampling a large ensemble of molecules to determine the distribution of tilt (θ), azimuthal (ϕ), and twist (ψ) angles. In comparison, the results of an orientation distribution function (ODF) that approximates the OD are shown

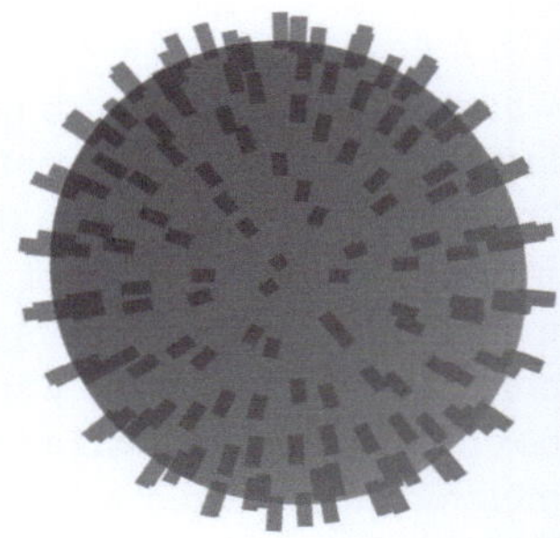

Fig. 8.18 An example of a distribution that is uniform in θ, ϕ, and ψ. Even though there is a larger population of spikes along the equator than towards the poles, there is no patch on the surface where the density of spikes is different from any other region

can go through all of the molecules in a particular frame (and even continue through multiple frames) and make histograms of the distributions of the three Euler angles. An example of such an accounting exercise is shown in Fig. 8.17 with the black line and grey shading, where the results are binned over a range of angles to better illustrate the trends. On the vertical axis we plot the number of molecules found within a particular angle range, for example, between θ and $\theta + \Delta\theta$. If we sample many molecules and choose a small bin width in plotting the histograms, the results could look like smooth functions, but it is important to keep in mind that there is no functional form here. Rather, it is the direct result of counting, and therefore a description of the true orientation distribution.

There is a minor point that we need to address with respect to the tilt angle distribution. Imagine a ball with spikes on the outside, arranged such that there is no patch on the surface where the density of spikes is different from that of any other patch. We will refer to this situation, as illustrated in Fig. 8.18, as a uniform distribution. If we plot the number of spikes found at any latitude as a function of the tilt angle (latitude), we see that the largest population exists at the equator. This makes sense, as the circle at each latitude becomes smaller towards the poles. Finally we end up at the north or south pole with a very small number of spikes (possibly a single spike). This is just a consequence of the reality of three-dimensional space, and what happens when we parameterize the rotations using Euler angles. However, as a result, it becomes challenging to inspect *perturbations* to this distribution in θ, due to the dominant shape of the isotropic distribution, that is $\sin\theta$. So the tilt angle distribution could instead be expressed as $N/\sin\theta$ versus θ. Now we have normalized the distribution with respect to what would be obtained for an isotropic distribution. Some would argue that we are obscuring rather than clarifying the tilt angle distribution with this operation. This is a matter of choice and a matter of perspective. By dividing each bin in the histogram by $\sin\theta$ (where θ is the average value of the tilt angle within that bin), we are opting to describe the tilt angle distribution with respect to an isotropic distribution, rather than the true tilt and distribution. This new representation would then appear flat for an isotropic distribution, similar to the histogram for the ϕ angle distribution in Fig. 8.17.

If we don't have access to each individual molecule (either from a molecular simulation, or from a single-molecule experiment where an ensemble can be probed one molecule at a time) then we can never know the true orientation distribution. We could instead make up a (hopefully reasonable) form of an orientation distribution function (ODF), and then use the available data to solve for the parameters in that function. One option is to assume a Gaussian distribution of tilt angles centred at θ_0 with a half-width of σ. This type of distribution can be described by the ODF

$$f(\theta) = c_{\mathrm{N}} \exp\left[-\frac{(\theta-\theta_0)^2}{2\sigma^2}\right] \tag{8.45}$$

where θ is our variable of interest, and θ_0 and σ are the parameters in this 2-parameter ODF. c_{N} is a normalization constant defined so that

$$\int_0^{\pi} f(\theta)\sin\theta \mathrm{d}\theta = 1. \tag{8.46}$$

This ensures that a molecule has a tilt angle somewhere in the range between 0 and π rad. In other words, we want to have 100% probability of finding a molecule somewhere in that complete tilt angle range.

More generally, considering an arbitrary ODF $f(\theta, \phi, \psi)$ that describes the distribution of all three Euler angles, we can write

$$\begin{aligned}\chi_{ij} &= \frac{N_{\mathrm{V}}}{\varepsilon_0}\langle\alpha_{ij}(\theta,\phi,\psi)\rangle \\ &= \frac{N_{\mathrm{V}}}{\varepsilon_0}\frac{\int_0^{2\pi}\int_0^{2\pi}\int_0^{\pi} f(\theta,\phi,\psi)\alpha_{ij}(\theta,\phi,\psi)\sin\theta \mathrm{d}\theta\mathrm{d}\phi\mathrm{d}\psi}{\int_0^{2\pi}\int_0^{2\pi}\int_0^{\pi} f(\theta,\phi,\psi)\sin\theta \mathrm{d}\theta\mathrm{d}\phi\mathrm{d}\psi}.\end{aligned} \tag{8.47}$$

The explicit denominator in the above expression takes the place of the normalization constant in the numerator, as this essentially ensures that we are working with a normalized distribution, important for the proper evaluation of χ_{ii}.

Before we get into some more realistic examples, let's pick an easy molecule for all of this. Throughout the rest of our discussion of IR absorption spectroscopy, we will simplify things by considering that

$$\mathbf{a} = \begin{bmatrix} 0 & 0 & 0 \\ 0 & 0 & 0 \\ 0 & 0 & a_{cc} \end{bmatrix} \tag{8.48}$$

so the molecule is polarizable to some degree (α_{cc}) along its molecular c-axis, but is not polarizable in any other direction. In other words, components of an electric field along $\hat{\mathbf{a}}$ will not induce any dipole moment along $\hat{\mathbf{a}}$. This is not realistic, but will simplify the math throughout the rest of this chapter to the point where we can more easily arrive at some useful general conclusions about absorption spectroscopy.

8.4.3 Isotropic Samples

We first consider an isotropic sample, i.e. one for which θ, ϕ, and ψ are uniformly distributed. For simplicity, we could consider this as $f(\theta, \phi, \psi) = 1$. The normalized form would be $f = 1/(8\pi^2)$, but we're always going to explicitly include the normalizing denominator so we'll be fine either way. I recommend getting into this habit, as it helps in cases where the ODF becomes too complicated to normalize by hand. I don't want to have to think about an analytical expression for the normalization constant if I don't need to. Ultimately—in more complex/realistic cases—the numerator and denominator will be solved numerically.

Our goal is to solve for all nine elements of A_{ij} (the numerators of χ_{ij} when a frequency-dependent denominator such as introduced in a Lorentzian lineshape (Eq. 7.11) is considered). Using our expression for the direction cosine matrix $\mathbf{D}$ and our simplified form of $\mathbf{a}$ with a_{cc} as the only non-zero term, we can carry out the projection into the lab frame, and then the integration where the population is weighted using the isotropic ODF ($f = 1$). We start by evaluating the first element χ_{xx}

$$\begin{aligned} A_{xx} &= \frac{N_\mathrm{V}}{\varepsilon_0}\langle a_{xx}(\theta, \phi, \psi)\rangle \\ &= \frac{N_\mathrm{V}}{\varepsilon_0}\langle \sum_{\ell}^{a,b,c}\sum_{m}^{a,b,c} D_{x\ell}(\theta, \phi, \psi)\, D_{xm}(\theta, \phi, \psi)\, a_{\ell m}\rangle \\ &= \frac{N_\mathrm{V}}{\varepsilon_0}\langle D_{xc}(\theta, \phi, \psi)^2 a_{cc}\rangle \\ &= \frac{N_\mathrm{V}}{\varepsilon_0}\langle \sin^2\theta\cos^2\phi\rangle\, a_{cc} \\ &= \frac{N_\mathrm{V}}{\varepsilon_0}\frac{\displaystyle\int_0^{2\pi}\int_0^{2\pi}\int_0^{\pi} \sin^2\theta\cos^2\phi\alpha_{cc}\sin\theta\mathrm{d}\theta\mathrm{d}\phi\mathrm{d}\psi}{\displaystyle\int_0^{2\pi}\int_0^{2\pi}\int_0^{\pi} f(\theta, \phi, \psi)\sin\theta\mathrm{d}\theta\mathrm{d}\phi\mathrm{d}\psi} \\ &= \frac{N_\mathrm{V}}{\varepsilon_0}\left(\frac{1}{3}a_{cc}\right). \end{aligned} \tag{8.49}$$

In the second line, we see that we need to loop through all nine elements of $a_{\ell m}$ (i.e. a_{aa}, a_{ab}, etc.) in order to evaluate their contribution to a_{xx}. The degree to which they project onto a_{xx} is given by elements of the DCM, **D**. Two applications of **D** are required for each element as we are transforming a rank-2 tensor. Finally, this is greatly simplified in the third line, as all $a_{\ell m}$ are zero except for a_{cc}. So we end up using only $D_{xc} = \sin\theta\cos\phi$, and it gets squared. The integration over the polar (tilt) angle θ has an extra $\sin\theta$ that we need to remember to include in the evaluation of that integral. As a_{cc} is a constant, ultimately we are merely integrating $\sin^3\theta\cos^2\phi$ over the full extent of all three Euler angles.

One can carry out the above operation to evaluate the remaining eight elements A_{ij} in the laboratory frame and one finds that $A_{xx} = A_{yy} = A_{zz}$, with the other 6 A_{ij} where $i \neq j$ equal to zero. This brings us one step closer to understanding how quantities like the refractive index appear with single subscripts, for example, $N_x = \sqrt{1 + \chi_{xx}}$. At least in the case of the isotropic sample we have studied so far, all non-zero elements of χ appear on the diagonal, so one index can be dropped in the subscripting $xx \to x$.

We have previously discussed how the absorbance is proportional to the imaginary component of χ, which in turn is proportional to the absorption coefficient and the imaginary part of the refractive index, κ. We now see that an unordered (isotropic) sample would produce the same absorbance spectrum regardless of whether the light was x-, y-, or z-polarized. Also, the x, y, and z susceptibilities are each proportional to one-third of a_{cc}.

8.4.4 Orientation Distributions That are Uniform in ϕ and ψ Only

In the case of an isotropic sample, we started with the projection of $\boldsymbol{\alpha}$ (or its oscillator strength numerator **a**) into the molecular frame, and arrived at a prediction of the spectra one would measure in the lab. Let's now think about this in the opposite way. Suppose one measured absorbance spectra of a sample using x-, y- and z-polarized light. (This is not easy to do for most samples in practice, but let that slide for the moment.) If the resulting absorbance was the same for the x and y cases, but a different spectrum was obtained for z (i.e. the same resonance frequencies ω_j, and the same width Γ_j for each of the j vibrational modes), this is an indication that the distribution of transition dipole moments is somehow preferentially oriented with respect to the z-axis, but uniformly distributed in the xy-plane. In other words, the distribution of ϕ and ψ angles is uniform, but there is some sort of (non-uniform) θ distribution that we seek to understand.

Let's start with the projections into the lab frame. As a reminder, **a** is a rank-2 tensor (as signified by elements with two indices ℓ and m) so we'll require two applications of the DCM. We already know how to do it, so we can get straight to work. We find that

$$\begin{aligned} a_{xx}(\theta, \phi, \psi) &= D_{xc}(\theta, \phi, \psi)\, D_{xc}(\theta, \phi, \psi)\, a_{cc} \\ &= a_{cc} \sin^2\theta \cos^2\phi \end{aligned} \tag{8.50a}$$

$$\begin{aligned} a_{yy}(\theta, \phi, \psi) &= D_{yc}(\theta, \phi, \psi)\, D_{yc}(\theta, \phi, \psi)\, a_{cc} \\ &= a_{cc} \sin^2\theta \sin^2\phi \end{aligned} \tag{8.50b}$$

$$\begin{aligned} a_{zz}(\theta, \phi, \psi) &= D_{zc}(\theta, \phi, \psi)\, D_{zc}(\theta, \phi, \psi)\, a_{cc} \\ &= a_{cc} \cos^2\theta. \end{aligned} \tag{8.50c}$$

We could evaluate the other six elements $a_{ij}(\theta, \phi, \psi)$ now too, but we'll hold off for a moment until after we address the orientation distribution.

Now let's consider how $a_{ij} \rightarrow A_{ij}$ using the molecular number density N_{V} and the average over the participating molecular orientations. In evaluating this average, we can't integrate θ yet, since $f(\theta)$ is not uniform and so-far unknown as we haven't decided on a form for the θ distribution function yet. But as we know that the OD is uniform over ϕ and ψ, let's take care of those two angles to arrive at

$$\begin{aligned} A_{xx}(\theta) &= \frac{N_{\mathrm{V}} a_{cc} \langle \sin^2\theta \rangle \displaystyle\int_0^{2\pi}\int_0^{2\pi} \cos^2\phi \mathrm{d}\phi \mathrm{d}\psi}{\varepsilon_0 \displaystyle\int_0^{2\pi}\int_0^{2\pi} \mathrm{d}\phi \mathrm{d}\phi} \\ &= \frac{N_{\mathrm{V}}}{2\varepsilon_0} a_{cc} \langle \sin^2\theta \rangle \\ &= A_{yy}(\theta) \end{aligned} \tag{8.51}$$

and

$$A_{zz}(\theta) = \frac{N}{\varepsilon_0} a_{cc} \langle \cos^2\theta \rangle. \tag{8.52}$$

After integration, if we were to consider *all* elements of **A**, we would find that the other six are equal to zero. So this particular orientation distribution, oriented with respect to z only, has three non-zero elements of $\boldsymbol{\chi}$, of which only two are unique as $\chi_{xx}(\theta) = \chi_{yy}(\theta)$. Note that, throughout this discussion, we often infer the connection between the susceptibility $\boldsymbol{\chi}$ and its "numerator" or oscillator strength **A** as described in Eq. 7.30.

There are two options for moving forward. The first way to proceed is to assume a form of $f(\theta)$ and use available experimental data (spectra of x- and z-polarized absorbance) to solve for the parameters in your ODF. For example, a simple choice would be a Dirac delta function

$$f(\theta) = \delta(\theta - \theta_0) \tag{8.53}$$

that has every molecule perfectly aligned with a tilt angle of θ_0. This is not much of a distribution per se, but it is an ODF nonetheless. Using the property of δ functions

$$\int \delta(x-a) f(x) = f(a) \tag{8.54}$$

we can now do the final (θ) integration to obtain

$$\begin{aligned} A_{xx} = A_{yy} &= \frac{N_{\rm V} a_{cc} \int_0^\pi \delta(\theta-\theta_0)\sin^2\theta \sin\theta \mathrm{d}\theta}{2\varepsilon_0 \int_0^\pi \delta(\theta-\theta_0)\sin\theta \mathrm{d}\theta} \\ &= \frac{N_{\rm V}\alpha_{cc}\sin^3\theta_0}{2\varepsilon_0\sin\theta_0} \\ &= \frac{N_{\rm V}}{\varepsilon_0}\left(\frac{1}{2}a_{cc}\sin^2\theta_0\right) \end{aligned} \tag{8.55}$$

and

$$A_{zz} = \frac{N_{\rm V}}{\varepsilon_0} a_{cc}\cos^2\theta_0. \tag{8.56}$$

It is important to recognize that there are no *variables* remaining after our integration of all three Euler angles; we can have only *parameters* left at this point. Since our ODF contains a single parameter θ_0, we have expressions (Eqs. 8.55 and 8.56) for each of the elements of $\boldsymbol{\chi}$ in terms of θ_0.

We can now use the experimental data available to us to solve for θ_0 from

$$\frac{A_{xx}}{A_{zz}} = \frac{N_{\rm V} a_{cc}\sin^2\theta_0}{2N_{\rm V} a_{cc}\cos^2\theta_0} = \frac{1}{2}\tan^2\theta_0. \tag{8.57}$$

The vacuum permittivity ε_0 appears and then also cancels in this expression but we haven't bothered to include it in order to emphasize something more important: this result, obtained from the ratio A_{xx}/A_{zz}, is independent of the molecular number density $N_{\rm V}$. Also, in this particular case (owing to the simplicity of our **a** tensor in the molecular frame) the result is even independent of a_{cc} and therefore does not depend on any molecular quantity! This is wild as it means that we have a way to comment on the molecular orientation using experimental data, knowing (almost) nothing about the properties of those molecules that are responsible for generating the measured signals. But please take a moment to consider the various assumptions that led to this expression, and the situations under which they may or may not be valid.

There are times when such extreme simplification is warranted, however a major issue is that our ODF is too simple, in that it assumes 100% of the molecules are perfected oriented at a single angle θ_0. Even if nothing is known about the nature of the orientation distribution, one typically would like there to be a *distribution* of tilt angles. Moving beyond a delta function, the next simple form of ODF that is often encountered is a Gaussian distribution of tilt angles such as that displayed in Eq. 8.45. However, even for this relatively simple functional form, we are already faced with a problem in that such an ODF has two parameters; our experimental data affords us only one unique ratio A_{xx}/A_{zz} or (equivalently) A_{yy}/A_{zz}. Ultimately, there is no way out. A popular option is to guess σ and use the experimental ratio to solve for θ_0, or assume θ_0 and comment on σ based on the A_{ii}/A_{jj} ratio. The choice will be dictated by the specific sample being studied and its circumstances. Often there is some chemical intuition to inform either θ_0 or σ and then this approach works well. Another approach commonly seen in the literature is to explore a range of values for one parameter, and study the effects on the second parameter. For example, the θ_0 that results from $\sigma = 5°$, and then $\sigma = 10°$ and so on.

We now discuss a more general approach that does not rely on an assumptions about $f(\theta)$. This is based on a representation of $f(\theta)$ in terms of a series of basis functions

$$f(\theta) = \sum_{n=0}^{\infty} \left(n + \frac{1}{2}\right) \langle P_n \rangle P_n(\cos\theta) \tag{8.58}$$

where $P_n(\cos\theta)$ are an orthogonal set of polynomials known as the Legendre polynomials and $\langle P_n \rangle$ are weighting coefficients that describe the relative contribution of each polynomial towards creating the orientation distribution function. These coefficients are commonly referred to as order parameters.

The first five polynomials are:

$$\begin{aligned}
P_0(\cos\theta) &= 1 \\
P_1(\cos\theta) &= \cos\theta \\
P_2(\cos\theta) &= \frac{3}{2}\cos^2\theta - \frac{1}{2} \\
P_3(\cos\theta) &= \frac{5}{2}\cos^3\theta - \frac{3}{2}\cos\theta \\
P_4(\cos\theta) &= \frac{35}{8}\cos^4\theta - \frac{15}{4}\cos^2\theta + \frac{3}{8}.
\end{aligned} \tag{8.59}$$

Although there are different conventions for writing these polynomials, the most typical form is that shown above, normalized such that $P_0 = 1$. As a result, although the functions are orthogonal, this form of the Legendre polynomials is not normalized, i.e.

$$\int_0^{\pi} [P_n(\cos\theta)]^2 \sin\theta d\theta \neq 1$$

so the functions are orthogonal, but not orthonormal. There are many interesting relationships among these functions [45]. One is that they can be generated using a recursion formula

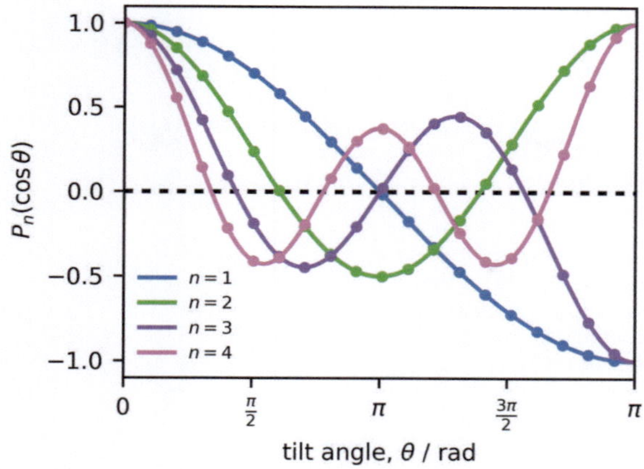

Fig. 8.19 Plots of four Legendre polynomials $P_n(\cos\theta)$. As shown in the accompanying Python code, this is also a comparison between the built-in functions from SciPy (plotted in points) and ones that I have created using the recursion formula in Eq. 8.60 (plotted in lines). As there are sometimes different definitions of functions, different normalization conventions, etc., it is a good idea to perform a check like this when using packaged expressions

$$P_{n+1} = \frac{(2n+1)\cos\theta\, P_n - n\, P_{n-1}}{n+1} \tag{8.60}$$

so starting with P_0 and P_1, any other P_n may be calculated sequentially. The last four polynomials listed in Eq. 8.59 are plotted in Fig. 8.19. Take a look at their shape, signs, zero-crossing, and the symmetry of the even and odd functions.

Code to generate Fig. 8.19

The Legendre polynomials are conveniently defined as `legendre` in `scipy.special`, so there is nothing that this code really needs to do except to plot them (points). However, since the recursion relation in Eq. 8.60 is straightforward to implement, we illustrate that too (lines).

I use this opportunity to demonstrate relabelling of the horizontal axis tick marks to values in radians. This is simply a matter of providing a list of strings to the `set_ticklabels` property of the `xaxis` object.

```
import numpy as np
from scipy.special import legendre
from matplotlib import pyplot

def Pn(n):
    P = list(np.zeros(n+1))
    P[0] = np.ones(len(theta))
    P[1] = np.cos(theta)
    for i in range(1, n):
```

```
        P[i+1] = ((2.*i + 1.)*np.cos(theta)*P[i] - i*P[i-1])/(i + 1.)
    p = ax.plot(thetaDeg, P[n], label=r'$n=%d$' % n)

    simple = np.polyval(legendre(n), np.cos(theta[::5]))
    ax.plot(thetaDeg[::5], simple, marker='o', mfc=p[0].get_color(), ms=4,
         mew=0, lw=0)

# setup
pyplot.rc('font', size=8)
thetaDeg = np.linspace(0., 180., 100)
theta = np.radians(thetaDeg)
fig = pyplot.figure(figsize=(3.25, 2.5))
ax = fig.add_subplot()

ax.axhline(0, color='k', ls='--', dashes=[2,2])
for n in range(1, 5):
    Pn(n)

ax.xaxis.set_ticks(range(0,181,45))
ax.xaxis.set_ticklabels(['0', r'$\frac{\pi}{2}$', r'$\pi$', r'$\frac{3\pi
     }{2}$', r'$\pi$'])
ax.set_xlabel(r'tilt angle, $\theta$ / rad')
ax.set_ylabel(r'$P_n(\cos \theta)$')
ax.set_xlim(0, 180)
pyplot.legend(loc=0, fontsize=6, frameon=False)

pyplot.tight_layout()
pyplot.savefig('lp.png', dpi=600)
```

The idea behind using the order parameters together with the Legendre polynomials is that $\langle P_n \rangle$ come from experimental data. So if you are able to get all of them (up to $\langle P_\infty \rangle$), you can make an ODF that describes any orientation distribution, without being confined to shapes such as Gaussian functions. This is important because the bigger problem is not that a specific ODF contains too many parameters resulting in an underdetermined system, it's that the shape of the assumed ODF may not necessarily match that of the real (unknown) OD. So a more general ODF such as that in Eq. 8.58 is appealing. Let's explore how far we can get with our current data.

Rather than using A_{xx} and A_{zz} to form the ratio A_{xx}/A_{zz}, we can instead use the spectral data to evaluate the quantity

$$\begin{aligned}\frac{A_{zz} - A_{xx}}{A_{zz} + 2A_{xx}} &= \frac{\langle a_{zz}(\theta, \phi, \psi)\rangle - \langle a_{xx}(\theta, \phi, \psi)\rangle}{\langle a_{zz}(\theta, \phi, \psi)\rangle + 2\langle a_{xx}(\theta, \phi, \psi)\rangle} \\ &= \frac{\langle \cos^2\theta\rangle - \frac{1}{2}\langle \sin^2\theta\rangle}{\langle \cos^2\theta\rangle + \langle \sin^2\theta\rangle} \\ &= \frac{3}{2}\langle \cos^2\theta\rangle - \frac{1}{2} \\ &= \langle P_2\rangle.\end{aligned} \tag{8.61}$$

In the last line we illustrate that since

$$\langle g(\theta)\rangle = \int_0^{\pi} f(\theta)g(\theta)\sin\theta \mathrm{d}\theta,$$

i.e. the average value of any function $g(\theta)$ is equal to the function multiplied (weighted) by the ODF, and integrated over the range of angles. This assumes that the orientation distribution is normalized (Eq. 8.46). Otherwise, the more general form

$$\langle g(\theta)\rangle = \frac{\int_0^{\pi} f(\theta)g(\theta)\sin\theta \mathrm{d}\theta}{\int_0^{\pi} f(\theta)\sin\theta \mathrm{d}\theta} \tag{8.62}$$

takes care of the normalization. This also points back to the reason why we use angle brackets around the P_n to denote the order parameters $\langle P_n\rangle$: the values of $\langle P_n\rangle$ are in fact the average values of the polynomials P_n, weighted according to the distribution $f(\theta)$.

One important property of the Legendre polynomials that enables them to serve as order parameters is that they each have a value of zero for an isotropic distribution, i.e. if we consider $f(\theta) = 1$,

$$\frac{1}{8\pi^2}\int_0^{\pi} P_n(\cos\theta)\sin\theta \mathrm{d}\theta = 0$$

and so $\langle P_n\rangle = 0$ for a random arrangement of molecules. However, with the exception of $P_1(\cos\theta)$ the other $P_n(\cos\theta)$ functions plotted in Fig. 8.19 clearly have additional zero-crossings. Taking $P_2(\cos\theta)$ as an example, we can find the value of its additional root from Eqs. 8.59, observing that it is located at $3\cos^2\theta = 1$ or $\theta \approx 54.7°$. This is known as the $\langle P_2\rangle$ magic angle, and is of significant consequence in orientation analysis. This means that, if an experiment measures $\langle P_2\rangle \approx 0$, we cannot tell whether the distribution is isotropic, or if all molecules are perfectly oriented at $\theta_0 = 54.7°$. This is a fundamental limitation of infrared absorption spectroscopy, and can be overcome only by the combined use of additional experiments. It is worthwhile to note that many techniques capitalize on this phenomenon as it can be used to remove the effects of anisotropy to recover some of the sought isotropic properties of the sample. This is the basis of magic angle spinning techniques in NMR spectroscopy [46].

In summary, when polarized IR spectra are acquired from a sample with uniform distributions of azimuthal (ϕ) and twist (ψ) angles, we can report the result as $\langle P_2\rangle$. Some people may be happy to receive $\langle P_2\rangle$ as a conclusion of the analysis. But you can see that, if all you can provide is $\langle P_2\rangle$ and none of the other $\langle P_n\rangle$, then it is really no different in information content from providing $\langle\cos^2\theta\rangle$ (also from Eq. 8.61). In this case, if one is willing to make a drastic assumption of a "distribution" function with infinitesimal width (Eq. 8.53), one could report on θ_0. However, this is just another way of saying that we really don't know much about the true shape of the distribution. We will shortly see that other $\langle P_n \rangle$ can come from other types of experiments (such as Raman scattering), chemical intuition, or just common sense depending on the particular sample or experiment performed.

8.5 Analysing Anisotropic Data Collected at Oblique Angles of Incidence

All of the analysis tools we have discussed in the previous section assume that we have a way of collecting absorbance spectra with light incident perpendicular to the sample surface, polarized along two of the unique the Cartesian directions. The absorbance along the x- or y-direction is then proportional to $\mathbb{I}\mathrm{m}\{\chi_{xx}\}$ (with Lorentzian amplitude A_{xx}), and that along the z-direction provides $\mathbb{I}\mathrm{m}\{\chi_{zz}\}$. This led to a straightforward way to introduce the theory of polarized absorbance and its connection to elements of $\boldsymbol{\chi}$ and $\langle P_2 \rangle$. However, there are many situations that require using transmission or reflection geometries where $\theta_1 > 0°$. Such experimental schemes have been discussed at the beginning of this chapter (transmission, external, and internal reflection), and we will now consider how to extract $\langle P_2 \rangle$ from such data. When $\theta_1 \neq 0°$, the only conceptual difference is that the components of the electric field vector necessarily contain mixed elements in the Cartesian frame. In Chap. 7 we saw that there are two special directions, p and s, both defined with respect to the plane of incidence. Recall that this is the plane that contains the incident wavevector and the surface normal. We adopt the standard coordinate system where the plane of the surface is the xy-plane, and z is the surface normal. In this coordinate system, s-polarized light has its electric field vector along y; p-polarized light has x and z components of $\mathbf{E}$. When the beam approaches the sample, the degree to which the p-polarized light projects along x versus z depends on the angle of incidence θ_1. Inside the sample, the x/z ratio depends on θ_2 that can be obtained from Snell's law of refraction. Note that for samples that are isotropic about ϕ and ψ, performing an oblique-angle transmission or reflection experiment is necessary, as the orientation described by θ is with respect to z only. If we approached such a sample at normal incidence (wavevector along z), the same absorbance spectrum would be recorded regardless of the orientation of the electric field in the xy-plane. We now take a look at such transmission and reflection cases.

8.5.1 *Transmission Experiments at Oblique Incidence*

Consider a transmission measurement performed at oblique incidence on a sample that is thick enough so that thin film interference does not have to be considered initially. Practically speaking, this requires a weak oscillator strength so that light can be transmitted through a relatively thick sample. In this way, the transmitted beams originating from multiple reflections inside the sample can all be isolated at the detector. We block all but one of those beams, so that only light that makes a single pass through the sample is recorded. When the s-polarized absorbance is measured, it can be equated to the susceptibility in an analogous way to what was described for isotropic samples (Eq. 8.8)

$$A_{\mathrm{s}} = A_y = -\log \frac{I_{\mathrm{s}}^{\mathrm{sample}}}{I_{\mathrm{s}}^{0}} = \frac{k_0 d'}{n_y \ln 10} \mathbb{I}\mathrm{m}\{\chi_{yy}\}\langle L_{yy}^2 \rangle \tag{8.63}$$

except that we have specified the y-component of the real part of the refractive index, and the χ_{yy} element of the imaginary part of the susceptibility tensor. Here d' is the oblique path length through the sample. When comparing with Eq. 8.8 we also notice that we are now including a factor $\langle L_{yy}^2 \rangle$ that represents the ratio of the mean squared electric fields in the vicinity of the absorbing molecules to that of the y-component of the mean squared field of the incident light as we have described in Chap. 7 (Eqs. 7.108–7.110). In other words, it describes the fraction of the incident s-polarized light that reaches the molecules of interest, those that absorb the light; this is often referred to as a local field correction factor. Note that we are also writing I for the intensity in Eq. 8.63, rather than T for the transmittance. The truth is, it doesn't matter whether we divide the signal (in arbitrary units) by that measured when the sample is absent (true I_0) or a reference sample that has a very similar n_2 but $\kappa_2 = 0$. One of those is required, as we need to take the ratio of signals in order to be immune to variations in the source power versus frequency (i.e. the lamp spectrum), the response curve of the detector, the transmission characteristics of any optics in the beam path, etc. In our earlier discussion at the beginning of this chapter we pointed out that, if a suitable reference sample could be found, then local field considerations could be accounted for experimentally. If no such sample exists, or if the reference sample only approximately accounts for the local fields, then we had better leave $\langle L_{yy}^2 \rangle$ in Eq. 8.63 and we might as well use I_{s}^0 to form the ratio and calculate the absorbance, for a more general discussion.

Similarly, the p-polarized absorbance is measured and related to the x- and z-components of the susceptibility and the ratio of the corresponding Cartesian components of the fields in the sample to their incident states

$$\begin{aligned} A_{\mathrm{p}} &= A_x + A_z \\ &= -\log \frac{I_{\mathrm{p}}^{\mathrm{sample}}}{I_{\mathrm{p}}^{0}} = \frac{k_0 d'}{\ln 10} \left(\frac{\mathbb{I}\mathrm{m}\{\chi_{xx}\}}{n_x} \langle L_{xx}^2 \rangle + \frac{\mathbb{I}\mathrm{m}\{\chi_{zz}\}}{n_z} \langle L_{zz}^2 \rangle \right). \end{aligned} \tag{8.64}$$

For weakly absorbing samples, there is a negligible change in the real part of the refractive index due to the vibrational resonance. One commonly assumes that the anisotropy in the real part of the refractive index is correspondingly small, and considers that $n_x = n_y \approx n_z \approx n$. This is useful as analysis of the orientation then proceeds by determining the so-called dichroic ratio

$$R_{\mathrm{D}} = \frac{A_{\mathrm{p}}}{A_{\mathrm{s}}} = \frac{A_x + A_z}{A_y} = \frac{\mathbb{I}\mathrm{m}\{\chi_{xx}\}\langle L_{xx}^2\rangle + \mathbb{I}\mathrm{m}\{\chi_{zz}\}\langle L_{zz}^2\rangle}{\mathbb{I}\mathrm{m}\{\chi_{yy}\}\langle L_{yy}^2\rangle}. \tag{8.65}$$

The convenience of Eq. 8.65 is clear from the fact that the prefactors in the absorbance expression, including the effective sample thickness d', cancel.

In this simple situation, we can make a further approximation and express the local field factors as projections of the p-polarized incident field considering the beam angle θ_2 inside the material to provide $\langle E_x^2\rangle = \langle E_{\mathrm{p}}^2\rangle \cos^2\theta_2$ and $\langle E_z^2\rangle = \langle E_{\mathrm{p}}^2\rangle \sin^2\theta_2$. Making this substitution and using Snell's law results in a revised dichroic ratio

$$R_{\mathrm{D}} = \frac{\langle E_{\mathrm{p}}^2\rangle}{\langle E_{\mathrm{s}}^2\rangle}\left[\frac{\mathbb{I}\mathrm{m}\{\chi_{xx}\}}{\mathbb{I}\mathrm{m}\{\chi_{yy}\}} + \left(\frac{\mathbb{I}\mathrm{m}\{\chi_{zz}\} - \mathbb{I}\mathrm{m}\{\chi_{xx}\}}{\mathbb{I}\mathrm{m}\{\chi_{yy}\}}\right)\frac{\sin^2\theta_1}{N_2^2}\right]. \tag{8.66}$$

The factor $\langle E_{\mathrm{p}}^2\rangle/\langle E_{\mathrm{s}}^2\rangle$ doesn't present an issue since this is the ratio of the incident fields that is known in the experiment, and can often be set to unity.

We then make use of the fact that, for our uniaxial distribution with cylindrical symmetry,

$$\begin{aligned} \mathbb{I}\mathrm{m}\{\chi_{xx}\} = \mathbb{I}\mathrm{m}\{\chi_{yy}\} &= \frac{N}{2\varepsilon_0} a_{cc}\langle \sin^2\theta\rangle \\ \mathbb{I}\mathrm{m}\{\chi_{zz}\} &= \frac{N}{\varepsilon_0} a_{cc}\langle \cos^2\theta\rangle. \end{aligned} \tag{8.67}$$

Substituting these into Eq. 8.66 results in

$$R_{\mathrm{D}} = \frac{1 - \langle\cos^2\theta\rangle + 3f\langle\cos^2\theta\rangle - f}{1 - \langle\cos^2\theta\rangle} \tag{8.68}$$

where we have defined

$$f = \frac{\sin^2\theta_1}{N_2^2}$$

for convenience. This enables us to solve for

$$\langle\cos^2\theta\rangle = \frac{R_{\mathrm{D}} - 1 + f}{R_{\mathrm{D}} - 1 + 3f}$$

that we ultimately use to express

$$\langle P_2\rangle = \frac{R_{\mathrm{D}} - 1}{R_{\mathrm{D}} - 1 + 3f} \tag{8.69}$$

and we have determined the order parameter from the experimental R_{D} value.

8.5.2 Reflection Experiments at Oblique Incidence

In a reflection geometry, we again consider the relationship between the dichroic ratio in the p/s basis and the associated quantities in the Cartesian laboratory frame, exactly as we have done for the transmission case above, and we end up with the same expression for the dichroic ratio as in Eq. 8.65. This time however, there is no simple approximation for the mean square fields. Leaving $\langle E_i^2 \rangle$ as is, we can then proceed to substitute our uniaxial projection of $\boldsymbol{\alpha}$ in place of $\boldsymbol{\chi}$ to provide

$$R_{\mathrm{D}} = \frac{\frac{1}{2}\langle L_{xx}^2 \rangle N\alpha_{cc}\langle \sin^2\theta \rangle + \langle L_{zz}^2 \rangle N\alpha_{cc}\langle \cos^2\theta \rangle}{\frac{1}{2}\langle L_{yy}^2 \rangle N\alpha_{cc}\langle \sin^2\theta \rangle}. \tag{8.70}$$

Using the relationship $\langle \sin^2\theta \rangle = 1 - \langle \cos^2\theta \rangle$ this can be rearranged to obtain

$$\langle \cos^2\theta \rangle = \frac{\langle L_{xx}^2 \rangle - R_{\mathrm{D}}\langle L_{yy}^2 \rangle}{\langle L_{xx}^2 \rangle - R_{\mathrm{D}}\langle L_{yy}^2 \rangle - 2\langle L_{zz}^2 \rangle} \tag{8.71}$$

that ultimately leads to

$$\langle P_2 \rangle = \frac{\langle L_{xx}^2 \rangle - R_{\mathrm{D}}\langle L_{yy}^2 \rangle + \langle L_{zz}^2 \rangle}{\langle L_{xx}^2 \rangle - R_{\mathrm{D}}\langle L_{yy}^2 \rangle - 2\langle L_{zz}^2 \rangle}. \tag{8.72}$$

We can therefore see that the accuracy to which $\langle P_2 \rangle$ can be obtained from experimental data is, in addition to the care required to obtain R_{D} in the experiment, closely tied to how accurately the local field corrections can be modelled, including knowledge of the sample thickness [47–63]. There are several limiting cases of interest, including monolayers or any samples whose thickness is much smaller than the evanescent wave penetration depth, and very thick samples [64, 65]. Those interested in such topics are referred to detailed studies in the literature [2, 58, 60, 61]. One of the intriguing points about expressions such as Eq. 8.72 is that, even though calculation of the local field components is required, $\langle P_2 \rangle$ can be accurately evaluated without considering anisotropy in the optical constants that enter into the mean squared fields [61, 66]. This seems like a contradiction, and it to some extent it is. But this demonstrated fact enables experimentalists to focus on those quantities that are most sensitive to obtaining $\langle P_2 \rangle$, potentially bypassing others that are more difficult to obtain accurately.

There are additional situations in which the local fields cannot be accounted for by experimental means, and therefore require explicit calculation. One case is that of a thin film studied in the previous chapter, where we have seen that the amplitude and phase of the transmitted and reflected fields depend on the film thickness as well as the optical constants. Typically the absorbance is determined from the ratio of the reflectance (or transmittance) of the sample (thin film on a substrate) to that of the bare substrate. In this case, we cannot avoid having to model the effects of the film thickness on the electric field inside the thin film. Another way of saying this is that, if the s- and p-polarized reflectance depends on the film thickness, that in itself is not an issue. But if the dichroic ratio depends on the film thickness, then that ratio cannot be related to the molecular orientation or $\langle P_2 \rangle$ without modelling the electric fields. Analysis of molecular orientation for such systems can be performed from absorption spectral collected in a transmission [67–75] or reflection geometry. Reflection experiments have the additional advantage that dielectric [76–82], semiconductor [34, 83–86], and metallic [39, 40, 87–96] substrates may be used.

Another similar situation occurs in the case of attenuated total internal reflection (ATR) IR spectroscopy of adsorbed molecular layers [58–60], where the electric field profiles inside the sample cannot be effectively "cancelled" in the measurement of the reference sample. Such ultra-thin samples can also be studied by external reflection and transmission experiments in theory, but that is very challenging due to the small (molecular length-scale!) path lengths and very low absorbance values. Internal reflection experiments can achieve high local field strengths at surfaces [2, 3], ideal for such samples.

8.6 Orientation Analysis Based on the Refractive Index

So far in our description of how oriented samples result in optical anisotropy, we have focussed on the projection of $\boldsymbol{\alpha}$ in the molecular frame into $\boldsymbol{\chi}$ in the sample/laboratory frame, and we have discussed how measurement of elements of the $\boldsymbol{\chi}$ tensor can yield $\langle P_2 \rangle$. If a suitable reference sample (with the same real part of the refractive index, but transparent) and the absorbance of the sample is low, the absorbance can be interpreted in terms of $\mathbb{I}\mathrm{m}\{\chi_{ii}\}$ elements directly. However, when any of those ideal conditions are absent, we can tackle the situation a bit differently by making the connection to the experimental absorbance in other ways.

8.6.1 Connecting the Susceptibility to the Refractive Index to Determine the Absorbance

Imagine that we model χ_{xx}, χ_{yy}, and χ_{zz} as a function of θ to establish what the susceptibility tensor elements would look like for a given value of $\langle P_2 \rangle$, as has been demonstrated using the expressions developed in the previous section of this chapter. From there, the relation $N^2 = 1 + \boldsymbol{\chi}$ can be used to obtain

$$
\begin{aligned}
n_x = n_y &= \mathbb{R}\mathrm{e}\{\sqrt{1+\chi_{xx}}\} \\
n_z &= \mathbb{R}\mathrm{e}\{\sqrt{1+\chi_{zz}}\} \\
\kappa_x = \kappa_y &= \mathbb{I}\mathrm{m}\{\sqrt{1+\chi_{xx}}\} \\
\kappa_z &= \mathbb{I}\mathrm{m}\{\sqrt{1+\chi_{zz}}\}
\end{aligned}
\tag{8.73}
$$

or, if desired, the κ versions of the above expressions can be used to model a Lorentzian profile from which the anisotropic n values are obtained by Kramers–Kronig transformation.

From here, we can extend the models of the Fresnel coefficients for the single interface (r_s, r_p, t_s, t_p) or the versions used for thin film and multi-layer systems ($\tilde{r}_s$, $\tilde{r}_p$, $\tilde{t}_s$, $\tilde{t}_p$) to include the effects of the anisotropic refractive index. This is straightforward as all expressions utilizing the material refractive index $N = n + i\kappa$ for s-polarization would now use $N_y = n_y + i\kappa_y$, and $N_x = n_x + i\kappa_x$ is substituted for the previously isotropic $N = n + i\kappa$ in all the expressions for p-polarization, and we use the anisotropic form of Snell's law of refraction (Eq. 7.58) when determining the refracted angle θ_2. With this change, we can determine $\tilde{R}_s$ and $\tilde{R}_p$ (or the corresponding polarized transmittance values) for the sample, using the anisotropic optical constants. The reference sample is typically isotropic, so the conventional expressions are used to calculate $\tilde{R}_s$ and $\tilde{R}_p$ (or transmittance). From this, the s- and p-polarized absorbances A_s and A_p can be calculated and compared to experimental A_s and A_p data. The value of $\langle P_2 \rangle$ in the model can then be adjusted until the best agreement between experiment and model is found. Alternatively, the dichroic ratio (Eq. 8.65) can be assessed, comparing model and experimental versions of R_D to isolate $\langle P_2 \rangle$. This method has been used primarily in external reflection experiments at the air–water interface [80, 81, 97, 98]. One of the flexibilities of this method is that it enables a wide variety of reference samples to be used, since the reflectance (or transmittance) of the sample and reference sample are modelled separately, each with parameters that most closely match the experimental conditions. The drawback of this approach, compared to the next method we will study, is that the $\langle P_2 \rangle$ determination is again very sensitive to the optical constants. This may seem like a strong point, since we wish to exploit this orientation sensitivity. However, there is a way to include the optical constants in a manner that makes the dichroic ratio less sensitive to errors in their modelling or approximation.

8.6.2 *Modelling the Orientation-Dependence of the Refractive Index Directly*

Although it is not straightforward, there are experimental means available to determine the anisotropic refractive indices of the sample along all three Cartesian directions, which amounts to measuring $N_x = n_x + i\kappa_x$, $N_y = n_y + i\kappa_y$ and $N_z = n_z + i\kappa_z$ [61, 99]. For the simpler uniaxial distributions we have been considering,

$N_x = N_y$. One option involves measuring the directional (x and z) components of κ, and then using the Kramers–Kronig relations to construct $n_x = n_y$ and n_z. Either way, these values can then be compared with a model for the optical anisotropy such as [80, 98, 100]

$$\begin{aligned}\kappa_{\max,x} = \kappa_{\max,y} &= \frac{\kappa_{\max}}{3}\left(1 - \langle P_2\rangle\right)\\ \kappa_{\max,z} &= \kappa_{\max}\left(\langle P_2\rangle + \frac{1-\langle P_2\rangle}{3}\right)\end{aligned} \tag{8.74}$$

and then using a Lorenztian model of the absorption band profile with κ_x and κ_z as the amplitudes

$$\begin{aligned}\kappa_x(\omega) = \kappa_y(\omega) &= \frac{\kappa_{\max,x}}{\omega_0^2 - \omega^2 - i\omega\Gamma}\\ \kappa_z(\omega) &= \frac{\kappa_{\max,z}}{\omega_0^2 - \omega^2 - i\omega\Gamma}\end{aligned} \tag{8.75}$$

with an experimentally reasonable value of the width Γ. From this, n_x and n_z can again be determined using a Kramers–Kronig relation

$$\begin{aligned}n_x(\omega) = n_y(\omega) &= n_{\infty,x} + \frac{2}{\pi}\mathscr{P}\int_0^\infty \frac{\omega'\kappa_x(\omega')}{\omega'^2 - \omega^2}\,\mathrm{d}\omega'\\ n_z(\omega) &= n_{\infty,z} + \frac{2}{\pi}\mathscr{P}\int_0^\infty \frac{\omega'\kappa_z(\omega')}{\omega'^2 - \omega^2}\,\mathrm{d}\omega'\end{aligned} \tag{8.76}$$

very similar to what was shown for the polarizability in Chap. 7 , with n_∞ being the high-frequency limit of the refractive index. Another option is to use the anisotropic κ values together with an isotropic value of n [61]. This scheme then proceeds by directly comparing the measured (or at least experimentally determined, including possible Kramers–Kronig steps) values of n_x and n_z, and/or κ_x and κ_z with the modelled version in order to arrive at the best fit $\langle P_2\rangle$. Although straightforward, the primary disadvantage of this approach is that everything rides on the optical constants. This is an issue because the anisotropic optical constants are not easy to obtain by experiments, and the orientation is not very sensitive to the real part of the refractive index, as can be gathered from the fact that an isotropic approximation to the real part is often used.

8.7 Summary

In this chapter we have reviewed some general concepts about absorption spectroscopy, and have then focused on IR absorption where resonances correspond to vibrational modes of molecules. We have seen that IR absorption spectra can be acquired in transmission or reflection mode. Reflection experiments can be external, where the incident medium (typically air) has a refractive index less than the sample, or internal, where the beams approach through another phase of larger refractive index that is in close contact with the sample. Lastly, the case of internal reflection is further broken down into cases where beams approach the sample below and above the critical angle. Internal reflection experiments above the critical angle, termed attenuated total internal reflection (ATR) IR spectroscopy, have practically taken over routine chemical analysis due to their minimal sample preparation requirements, however care must be taken when performing quantitative analysis. In all of the above cases, it is common to report IR spectral data as absorbance, where the spectra are ratioed with that of a reference material. The selection of reference spectra may be arbitrary, and has a dramatic effect on the appearance of the resulting absorbance spectra.

One of the primary goals of this chapter was to illustrate that, if the complex refractive index of the sample is known at all wavenumbers of interest, it is possible to model the spectra in each of the scenarios described above, using nothing more than the Fresnel coefficients (and potentially the local electric fields). This is powerful, as it can account for spectral features that have remarkably different appearance, merely by changing the experimental conditions. Furthermore, it is not always possible to select a reference sample (or background) that sufficiently simplifies the analysis. In such cases, being able to model the transmittance and reflectance is especially powerful as it becomes part of the data processing pipeline.

In the latter part of the chapter, we have illustrated how the arrangement of molecules in ordered samples can be obtained by acquiring data with different beam polarizations simply by adding a polarizer to the sample compartment. Even though we have considered some trivial orientation distributions, and have made simplifications regarding elements of the transition dipole moment, the formalism presented here can be directly used to solve many research problems. The illustrated techniques can also be extended with different types of orientation distribution functions containing different parameters.

8.8 Exercises

1. Write a small program that asks users to input parameters for a diatomic molecule, and then outputs the IR resonance frequency. For example, the input would be the reduced mass m (specify the expected units, e.g. kg) and the value of the force constant (e.g. $500\,\mathrm{N{\cdot}m^{-1}}$). Have your program use the simple approximation

$$f = \frac{1}{2\pi}\sqrt{\frac{k}{m}}$$

to output

a. the wavenumber of light required to induce resonance excitation in units of cm^{-1};
b. its frequency in units of Hz;
c. its wavelength in units of nm or μm (specify).

---

2. We used a numerical approach to arrive at the IR absorption selection rules, essentially plotting the integrand as a function of displacement along the vibrational mode for different states $\psi_m(x)$ and $\psi_n(x)$. Implement a SymPy version of the selection rules code to arrive at analytical versions of the expressions for the transition dipole moment that you can evaluate for states m and n. In other words, you can evaluate the actual integrals, rather than just inspecting the integrands. Remember that you can find expressions for the Hermite polynomials built-in to SymPy (just as in the case of the numerical versions in NumPy) so there is very little code that you need to write.

---

3. Visit `refractiveindex.info` and download n and κ data for octane in the mid-infrared. You'll see several sources of octane data there, some corresponding to different spectral regions, and some containing n only. Once you have located a source that contains both n and κ as a function of wavelength (check the units, they are typically micrometres), set up a wavenumber-based interpolation so that you can write `N(2875.3)` and a complex number $n + i\kappa$ is returned. An example of something very similar was provided in Chap. 3. Consider the following questions related to absorbance measurements. After you've got that set up, you'll be able to more easily answer the following questions:

   a. How thin would an octane sample have to be in order to get at least 5% of the light through at normal incidence? (Consider 5% to be the lowest acceptable value over the entire 2800–3100 cm^{-1} region, so the signal-to-noise is not too bad.) This determines whether a transmission-geometry absorption measurement is practical.
   b. Make plots of the s- and p-polarized reflectance spectra (external reflection, beams approaching octane from air), and $R_{avg} = \frac{1}{2}(R_s + R_p)$, at $\theta_i = 70°$, assuming a thick sample.
   c. Making use of a glass "reference sample", and considering that $N_{glass} = 1.457$ (and $\kappa_{glass} = 0$) throughout this entire frequency region of interest, plot the s- and p-polarized absorbance spectra for an external reflection geometry with $\theta_i = 70°$, in addition to $A_{avg} = \frac{1}{2}(A_s + A_p)$.
   d. Now consider an ATR-IR experiment using a diamond ($N = 2.4$) prism mounted such that $\theta_i = 45°$. Plot the s- and p-polarized absorbance spectra, along with A_{avg} considering air (air–diamond interface) to be the reference sample.

e. Consider one final ATR-IR experiment, this time with a germanium crystal ($N = 4.0$) at $\theta_i = 45°$. Plot the s- and p-polarized absorption spectra, and A_{avg}.

---

4. A mid-IR beam travels inside a diamond internal reflecting element ($N_1 = 2.4$ over a wide region of the mid-IR) approaching the ZnSe–air interface. Consider that the beam is polarized at 45° that is half-way in between the s- and p-orientations of the electric field vector.

   a. If the angle of incidence is $\theta_1 = 20°$, what is the polarization state of the reflected beam? You can use a variety of methods to answer this, but I generally prefer to determine $\mathbf{E}_{\text{out}}$ using r_s and r_p and then turn this into a Stokes vector.
   b. What if the angle of incidence was $\theta_1 = 30°$? If there is a significant difference between the two angles, why is that?
   c. How would your answer to the previous parts differ if there had been water above the ZnSe, rather than air? Consider that for water $N_2 = 1.30 + 0.10i$ at the wavenumber of interest.

---

5. You perform a reflection IR experiment of sodium dodecyl sulphate adsorbed at the air–water interface.

   a. What is the ideal angle of incidence?
   b. How can the absorbance be determined for such a reflection geometry? In other words, what can be used as a reference?
   c. Approximately what absorbance do you expect? Estimate this to within the nearest power of 10.
   d. Explain why the resulting IR absorption spectrum shows an intense CH_2 symmetric stretch but only a weak CH_3 symmetric stretch.

---

6. Inside the sample compartment of your FTIR, you place a polarizer, followed by a sample that is a cube with dimensions of 5.2 cm on each side. The sample has a single resonance in the 2800–3100 cm^{-1} range of interest, located at $\omega_0 = 2950$ cm^{-1} with $\Gamma = 35$ cm^{-1}.
   Each molecule that makes up the sample has a frequency-dependent polarizability given by

$$\alpha_{\ell m}(\omega) = \frac{a_{\ell m}}{\omega_0^2 - \omega^2 - i\omega\Gamma} \tag{8.77}$$

   with

$$\mathbf{a} = \begin{bmatrix} a_{aa} & a_{ab} & a_{ac} \\ a_{ab} & a_{bb} & a_{bc} \\ a_{ac} & a_{bc} & a_{cc} \end{bmatrix} = \begin{bmatrix} 0.12 & -0.04 & 0.08 \\ -0.04 & 0.18 & -0.14 \\ 0.08 & -0.14 & 0.28 \end{bmatrix} \times 10^4\ \mathrm{C \cdot V^{-1}}.$$

The values of **a** provided above will need to be divided by 10^4 in order to use them in Eq. 8.77 if ω has units of cm^{-1}.

a. Make plots of the different $\mathbb{I}\mathrm{m}\{\alpha_{\ell m}(\omega)\}$ spectra. (Note that there are only six since $a_{\ell m} = a_{m\ell}$.) Plot them all (overlay) on a single axis, and label them with a legend.

b. Assume that the molecules within the material are ordered with a narrow distribution of tilt angles θ, but are uniformly distributed with respect to their ϕ and ψ angles. The orientation distribution function (ODF) is therefore $f(\theta, \phi, \psi) = \delta(\theta - \theta_0)$. If we write the susceptibility as

$$\chi_{ij}(\omega) = \frac{A_{ij}}{\omega_0^2 - \omega^2 - i\omega\Gamma}$$

then we can relate the numerators of $\boldsymbol{\chi}$ and $\boldsymbol{\alpha}$ in

$$\begin{aligned} A_{ij}(\theta_0) &= \frac{N_{\mathrm{V}}}{\varepsilon_0} \langle a_{ij}(\theta, \phi, \psi) \rangle \\ &= \frac{N_{\mathrm{V}}}{\varepsilon_0} \left[\frac{\int_0^{2\pi} \int_0^{2\pi} \int_0^{\pi} f(\theta, \phi, \psi)\, a_{ij}(\theta, \phi, \psi) \sin\theta \, \mathrm{d}\theta \, \mathrm{d}\phi \, \mathrm{d}\psi}{\int_0^{2\pi} \int_0^{2\pi} \int_0^{\pi} f(\theta, \phi, \psi) \, \sin\theta \, \mathrm{d}\theta \, \mathrm{d}\phi \, \mathrm{d}\psi} \right]. \end{aligned}$$

For simplicity, assume that the number density (N_{V}) and ε_0 are both unity. (I worked their magnitudes into the amplitudes already.)

Write expressions for the nine elements of A_{ij} in terms of the single ODF parameter θ_0. Do this before substituting any values of $a_{\ell m}$, so you can admire the form of the resulting expressions. *Hint:* Six of them will be zero. Remember that

$$\int f(x)\, \delta(x - a)\, \mathrm{d}x = f(a).$$

c. Compare your expressions with the ones for A_{ij} in the more simple case where all elements of **a** are zero except for a_{cc}. Even though we don't expect them to be the same (because **a** has more non-zero elements now), are there some similarities that make sense?

d. Back to Python, now put in the required numerical values of $a_{\ell m}$ and use your $A(\theta_0)$ together with a tilt angle of $\theta_0 = 22°$ to define three NumPy arrays that hold values of N_x, N_y, and N_z as a function of ω, the complex-valued refractive indices along the x-, y-, and z-directions of the sample.

You don't have to show anything here, I'm just giving you a clue that this is an intermediate step.

e. Plot spectra of T_x, T_y, and T_z where T_i refers to the transmittance when the electric field is along the i-direction. Note that this experiment can be performed with the polarizer orientation fixed, rotating the cube inside the FTIR sample compartment to collect the three spectra. This way, I_0 can easily be obtained by simply removing the sample.

f. Which direction has the lowest transmittance on resonance? Why?

---

7. In the simplest ellipsometry experiment, the two ellipsometric angles Ψ and Δ are measured when light is reflected from a thick sample (no reflection from the back surface) that is isotropic. In this case, one can solve for n and κ and thereby obtain the complex-valued refractive index. Another common application is to determine the thickness of films. Here the general idea is that the film is thin enough so that the multiple reflections are perfectly spatially overlapped in the reflected beam. We now have three phases: air, film, and substrate. If both condensed phases are isotropic, that means four parameters are unknown n_{film}, κ_{film}, $n_{\text{substrate}}$, and $\kappa_{\text{substrate}}$.

   a. If $n_{\text{substrate}}$ and $\kappa_{\text{substrate}}$ are known, show how the experimentally determined Ψ and Δ values can be used to determine the film thickness d along with n_{film} if we assume $\kappa_{\text{film}} = 0$.
   b. Work out d in the case where $\kappa_{\text{film}} \neq 0$, but its value is known.
   c. Consider the flipped case, where n_{film} is known (along with both of the substrate optical constants), but both d and κ_{film} are desired.
   d. In the general case of five unknown parameters (including d), can they all be determined if Ψ and Δ are measured at more than one angle of incidence?

---

8. Two points on the surface of a sphere of radius r have coordinates $\hat{p}_1 = (r, \theta_1, \varphi_1)$ and $\hat{p}_2 = (r, \theta_2, \varphi_2)$ with respect to the origin at the centre of the sphere.

   a. Show that the angle σ between $\hat{p}_1$ and $\hat{p}_2$ is given by

$$\sigma = \arccos\left[\cos\theta_1 \cos\theta_2 + \sin\theta_1 \sin\theta_2 \cos(\varphi_2 - \varphi_1)\right]. \tag{8.78}$$

   b. The shortest distance ℓ between these points along an arc that is on the surface of the sphere is then given by $\ell = r\sigma$, where σ is in radians. (This is a nice example of the dimensionless nature of radians.) Use this expression to find the distance ℓ between Victoria, BC (48°25'42"N, 123°21'53"W) and Stockholm (59°19'46"N, 18°4'7"E). When you convert the angles into radians, note that latitude is measured from the equator, but θ is measured from the north pole. For simplicity assume the Earth is spherical with $r = 6,370$ km.

c. How much shorter is it if you could travel below the surface of the Earth to connect these two cities?
d. Use Eq. 8.78 to determine the locations (θ, φ) where the spikes are placed in Fig. 8.18 to have a uniform surface density.

References

1. M.S. Azam, M.D. Ranson, D.K. Hore, Appl. Spectrosc. **76**, 1254 (2022)
2. N.J. Harrick, *Internal Reflection Spectroscopy (Interscience Publisher* (Wiley, New York, NY, 1967)
3. M. Milosevic, *Internal Reflection and ATR Spectroscopy* (Wiley, Hoboken, NJ, 2012)
4. T.G. Mayerhöfer, *Wave Optics in Infrared Spectroscopy: Theory, Simulation, and Modeling* (Elsevier, Amsterdam, 2024)
5. T.G. Mayerhöfer, J. Popp, Appl. Spectrosc. **2025**, 465 (79)
6. T.G. Mayerhöfer, W.D.P. Costa, J. Popp, Appl. Spectrosc. **79**, 1018 (2025)
7. R. Cheng, T.G. Mayerhöfer, J. Kiefer, Appl. Spectrosc. **79**, 852 (2025)
8. G.M. Irons, H.W. Thompson, Proc. R. Soc. Lond. A **298**, 160 (1967)
9. N.I. Stas'kov, V.I. Golovachev, S.S. Gusev, V.V. Filippov, J. Appl. Spectrosco. **32**, 198 (1980)
10. L.D. Tickanen, M.I. Tejedor-Tejdor, M.A. Anderson, Appl. Spectrosc. **46**, 1848 (1992)
11. A.E. Tshmel, V.I. Vettegren, Spectrochim. Acta **29A**, 1681 (1973)
12. K. Ohta, H. Ishida, Proc. SPIE **1575**, 604 (1992)
13. K.R. Kirov, H.E. Assender, Macromolecules **37**, 894 (2004)
14. A.F. Runge, N.C. Rasmussen, S.S. Saavendra, S.B. Mendes, J. Phys. Chem. B **109**, 424 (2005)
15. T. Buffeteau, D. Blaudez, E. Péré, B. Desbat, J. Phys. Chem. B **103**, 5020 (1999)
16. D.J. Rowe, D. Smith, J.S. Wilkinson, Sci. Rep. **7**, 7356 (2017)
17. T. Hasegawa, *Quantiative Infrared Spectroscopy for Understanding of a Condensed Matter* (Springer, Tokyo, 2017)
18. R.M.A. Azzam, N.M. Bashara, *Ellipsometry and Polarized Light* (North-Holland Publishing Company, New York, 1977)
19. H.G. Tompkins, E.A. Irene, *Handbook of Ellipsometry* (William Andrew Inc, Norwich, NY, 2005)
20. A. Rothen, Rev. Sci. Instrum. **16**, 26 (1945)
21. J. Humlíček, *Handbook of Ellipsometry* (Springer Science and Business, 2005), chap. Polarized Light and Ellipsometry, pp. 3–93
22. H.G. Tompkins, *A User's Guide to Ellipsometry* (Dover Publications, New York, 2006)
23. D.E. Aspnes, Thin Sol. Films **455–456**, 3 (2004)
24. P.S. Hauge, F.H. Dill, Opt. Commun. **14**, 431 (1975)
25. D.E. Aspnes, P.S. Hauge, J. Opt. Soc. Am. **66**, 949 (1976)
26. D.E. Aspnes, J. Opt. Soc. Am. A **21**, 403 (2004)
27. E. Compain, S. Poirier, B. Drevillon, Appl. Opt. **38**, 3490 (1999)
28. G.E. Jellison, C.O. Griffiths, D.E. Holcomb, C.M. Rouleau, Appl. Opt. **41**, 6555 (2002)
29. R. Anderson, Appl. Opt. **31**, 11 (1992)
30. G.E. Jellison, F.A. Modine, Appl. Opt. **36**, 8184 (1997)
31. G.E. Jellison, F.A. Modine, Appl. Opt. **36**, 8190 (1997)
32. L.J.K. Cross, D.K. Hore, Appl. Opt. **51**, 5100 (2012)

33. D.J. Segelstein, The Complex Refractive Index of Water. Master's thesis, University of Missouri, Kansas City (1981)
34. R.A. Ramin, G. Le Bourdon, N. Daugey, B. Bennetau, L. Vellutini, T. Buffeteau, Langmuir **27**, 6076 (2011)
35. T. Buffeteau, B. Desbat, D. Blaudez, J.M. Turlet, Appl. Opt. **54**, 1646 (2000)
36. M.B.T. Bürgi, Langmuir **21**, 1354 (2005)
37. M. Bieri, T. Burgi, J. Phys. Chem. B **109**, 22476 (2005)
38. Y. Liang, D. Mauran, R.E. Prud'homme, C. Pellerin, Appl. Spectrosc. **62**, 941 (2008)
39. V. Zamlynny, I. Zawisza, J. Lipkowski, Langmuir **19**, 132 (2003)
40. T. Elzein, A. Fahs, M. Brogly, A. Elhiri, B. Lepoittevin, P. Roger, V. Planchot, J. Adhesion **89**, 416 (2013)
41. T. Engel, *Quantum Chemistry and Spectroscopy*, 2nd edn. (Pearson Education, Boston, 2010)
42. R.N. Zare, *Angular Momentum: Understanding Spatial Aspects in Chemistry and Physics* (Wiley, New York, 1988)
43. S.S. Andrews, J. Chem. Educ. **81**, 877 (2004)
44. S. Roy, K.K. Hung, U. Stege, D.K. Hore, Appl. Spectrosc. Rev. **49**, 233 (2014)
45. G.B. Arfken, H.J. Weber, F.E. Harris, *Mathematical Methods for Physicists: A Comprehensive Guide*, 7th edn. (Academic Press, Waltham, MA, 2013)
46. M.J. Duer, *Solid-State NMR Spectroscopy: Principles and Applications*, 2nd edn. (Blackwell Science, Oxford, 2010)
47. D.J. Ahn, E.I. Franses, Thin Solid Films **244**, 971 (1994)
48. D.J. Ahn, E.I. Franses, J. Phys. Chem. **96**, 9952 (1992)
49. C. Bolterauer, H. Heller, Eur. Biophys. J. **24**, 322 (1996)
50. J.W. Brauner, R. Mendelsohn, F.G. Prendergast, Biochem. **26**, 8151 (1987)
51. K.K. Chittur, Biomaterials **19**, 357 (1998)
52. M.J. Citra, P.H. Axelsen, Biophys. J . **71**, 1796 (1996)
53. A. Erbe, R.J. Bushby, S.D. Evans, L.J.C. Jeuken, J. Phys. Chem. B **111**, 3515 (2007)
54. P.A. Flournoy, W.J. Shaffers, Spectrochim. Acta **22**, 5 (1966)
55. U.P. Fringeli, H.H. Günthard, Mol. Biol. Biochem. Biophys. **31**, 270 (1981)
56. U.P. Fringeli, J. Goette, G. Preiter, M. Siam, D. Baurecht, A.I.P. Conf, Proc. **430**, 729 (1998)
57. E. Goormaghtigh, V. Raussens, J.M. Ruysschaert, Biochim. Biophys. Acta **1422**, 105 (1999)
58. D. Marsh, Biophys. J . **72**, 2710 (1997)
59. D. Marsh, Biophys. J . **75**, 354 (1998)
60. D. Marsh, M. Müller, F.J. Schmitt, Biophys. J . **78**, 2499 (2000)
61. F. Picard, T. Buffeteau, B. Desbat, M. Auger, M. Pézolet, Biophys. J . **76**, 539 (1999)
62. R.H.A. Ras, R.A. Schoonheydt, C.T. Johnston, J. Phys. Chem. A **111**, 8787 (2007)
63. N. Tillman, A. Ulman, J.S. Schildkraut, T. Penner, J. Am. Chem. Soc. **110**, 6136 (1988)
64. N.J. Harrick, A.I. Carlson, Appl. Opt. **10**, 19 (1971)
65. N.J. Harrick, J. Opt. Soc. Am. **55**, 851 (1965)
66. M.J. Pilling, P. Gardner, S. Le Vent, Surf. Sci. **582**, 1 (2005)
67. H. Akutsu, Y. Kyogoku, H. Nakahara, K. Fukuda, Chem. Phys. Lipids **15**, 222 (1975)
68. M.D. Bazzi, R.W. Woody, Biophys. J . **48**, 957 (1985)
69. J. Breton, M. Michel-Villaz, G. Paillotin, Biochimica Biophysica Acta **314**, 42 (1973)
70. E. Charney, J. Opt. Soc. Am. **45**, 980 (1955)
71. P.A. Chollet, Thin Solid Films **52**, 343 (1978)
72. A. Cunningham, G.R. Davies, I.M. Ward, Polymer **15**, 743 (1974)
73. W.S. Park, J. Korean Phys. Soc. **37**, 331 (2000)
74. M. Furuhashi, J. Yoshinobu, Jap. J. Appl. Phys. **50**, 115701 (2011)
75. H. Nakahara, K. Fukuda, J. Colloid Interface Sci. **69**, 24 (1979)
76. D. Blaudez, J.M. Turlet, J. Dufourcq, D. Bard, T. Buffeteau, B. Desbat, J. Chem. Soc., Faraday Trans. **92**, 525 (1996)
77. J.T. Buontempo, S.A. Rice, J. Chem. Phys. **98**, 5825 (1993)
78. R.A. Dluhy, J. Phys. Chem. **90**, 1373 (1986)

79. C.R. Flach, J.W. Brauner, J.W. Taylor, R.C. Baldwin, R. Mendelsohn, Biophys. J . **67**, 402 (1994)
80. C.R. Flach, A. Gericke, R. Mendelsohn, J. Phys. Chem. B **101**, 58 (1997)
81. A. Gernicke, C.R.F., R. Mendelsohn, Biosphys. J. **73**, 492 (1997)
82. J.A. Mielczarski, R.H. Yoon, J. Phys. Chem. **93**, 2034 (1989)
83. H. Brunner, U. Mayer, M. Hoffmann, Appl. Spectrosc. **51**, 209 (1997)
84. T. Hasegawa, J. Nishijo, J. Umemura, W. Theiss, J. Phys. Chem. B **105**, 11178 (2001)
85. T. Hasegawa, S. Takeda, A. Kawaguchi, J. Umemura, Langmuir **11**, 1236 (1995)
86. T. Hasegawa, J. Umemura, T. Takenaka, J. Phys. Chem. **97**, 9009 (1993)
87. D.L. Allara, J.D. Swalen, J. Phys. Chem. **86**, 2700 (1982)
88. D.L. Allara, R.G. Nuzzo, Langmuir **1**, 52 (1985)
89. M. Boncheva, H. Vogel, Biophys. J . **73**, 1056 (1997)
90. T. Buffeteau, B. Desbat, J.M. Turlet, Appl. Spectrosc. **45**, 380 (1991)
91. M.K. Debe, Appl. Surface Sci. **14**, 1 (1982)
92. M.K. Debe, J. Appl. Phys. **55**, 3354 (1984)
93. R.G. Greenler, J. Chem. Phys. **44**, 310 (1966)
94. Y. Miura, S. Kimura, Y. Imanishi, J. Umemura, Langmuir **15**, 1155 (1999)
95. C.G. Worley, R.W. Linton, E.T. Samulski, Langmuir **11**, 3805 (1995)
96. I. Zawisza, A. Lachenwitzer, V. Vamlynny, S.L. Horswell, J.D. Goddard, J. Lipkowski, Biophys. J . **85**, 4055 (2003)
97. R. Mendelson, J.W. Brauner, A. Gericke, Annu. Rev. Phys. Chem. **46**, 305 (1995)
98. R. Mendelsohn, C.R. Flach, *Infared Reflection-Absorption Spectrometry of Monolayer Films at the Air-Water Interface* (Wiley, Toronto, 2006), pp.1028–1041
99. T. Buffeteau, B. Desbat, E. Péré, J.M. Turlet, Mikrochim. Acta **14**, 631 (1997)
100. R.D.B. Fraser, T.P. MacRae, *Conformation in Fibrous Proteins and Related Synthetic Polypeptides* (Academic Press, New York, 1973)

Chapter 9
Raman Scattering Spectroscopy

9.1 Background

Raman spectroscopy is perhaps the most intriguing of the vibrational methods, as a result of the conceptual simplicity of the technique, ease (and relative low cost) with which the experiment can be performed, broad range of applicability (little or no sample preparation in many cases), and high information content, particularly when performed with polarized light [1–12].

To introduce the phenomena of vibrational Raman scattering, imagine a liquid sample, irradiated with a bright monochromatic source. In our example, we use a laser with $\lambda_{\text{ex}} = 532$ nm or $\omega_{\text{ex}} = 18797$ cm^{-1}, where the subscript "ex" designates the incident light. As shown in Fig. 9.1, in addition to the light transmitted through the sample, and the light reflected from the sample and container surfaces, there is also scattered light. In some cases, the scattering is visible only in dim conditions, when the room lights are lowered. Most of the scattered light has the same frequency $\omega_{\text{s}} = \omega_{\text{ex}}$ as the incident light; this is known as elastic or Rayleigh scattering. However, Raman's discovery was that a small fraction of the scattered light contained frequencies that were not present in the excitation source [13]. A Raman scattering experiment therefore seeks to measure the spectrum of the scattered light in order to observe these new frequencies that are somehow created by the molecules in the sample. Since Rayleigh scattering dominates, the challenge is to observe the extremely low intensity of the new frequencies at $\omega_{\text{s}} = \omega_{\text{ex}} \pm \Delta\omega$ without being overwhelmed by the background at frequency ω_{ex}. As shown in Fig. 9.1, it is therefore desirable to avoid the directions of the forward or backward scattered light as they are swamped by the transmitted/reflected light from the laser. In practice, however, these directions are often experimentally convenient, we rely on good wavelength separation (using high-quality filters and a monochromator) to enable observation of the dim light at the new frequencies.

Before we get into the details, consider a simulated spectrum of the scattered light when a 532 nm laser source is incident on a heptane sample as shown in Fig. 9.2a.

D. K. Hore, *Molecular Orientation Analysis Using Polarized Vibrational Spectroscopy*,
https://doi.org/10.1007/978-3-032-04874-5_9

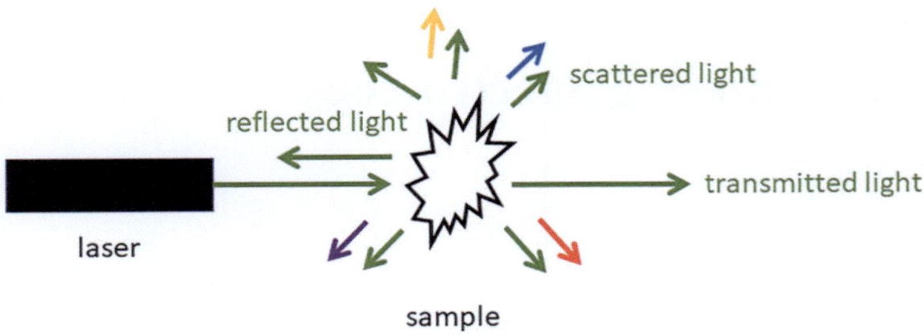

Fig. 9.1 In a basic Raman scattering experiment, the sample is illuminated with monochromatic light, and the spectrum of the scattered light is collected

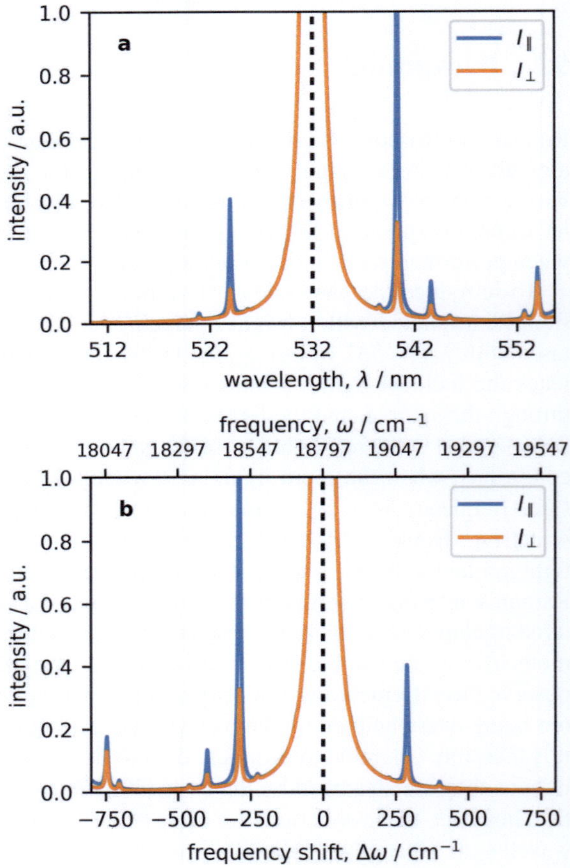

Fig. 9.2 **a** Raman spectrum of heptane, plotted as a function of wavelength in a manner analogous to how it would be observed on a spectrograph. **b** The same spectrum, plotted as a function of frequency ω_s (top axis), and then frequency shift $\Delta\omega = \omega_s - \omega_{ex}$ with respect to the excitation frequency ω_{ex}

We have cut off the strong peak at 532 nm that results from stray transmitted and/or reflected light, scattering from other objects in the light path, and Rayleigh scattering from the sample, in order to zoom in on the other spectral features. Once we do this, we notice that the frequencies of the peaks on both "sides" of the 532 nm line are the same, i.e. the there are two sets of spectra that closely resemble each other, except for their intensities. We then convert the wavelengths into frequencies, as displayed in the top axis of Fig. 9.2b. Upon closer inspection, we now see that the frequency spacing of the peaks on either side of the Rayleigh line is also the same, so the sets are in fact mirrored, with the exception of the overall greater intensity on the shorter frequency (longer wavelength) side. As a final transformation, we subtract the laser frequency ($\omega_{ex} = 18797$ cm^{-1}) from all values to obtain the bottom axis in Fig. 9.2b. This frequency shift ($\Delta\omega$) axis is what is commonly displayed as the Raman shift or simply the Raman frequency. In Fig. 9.2b, we have included the sign of $\Delta\omega$, so that frequencies lower than the Rayleigh line, the so-called Stokes shift is indicated with $\Delta\omega < 0$, and the higher frequencies (the anti-Stokes-shifted region) are displayed with $\Delta\omega > 0$. The code used to generate Fig. 9.2 will be shown closer to the end of the chapter, after we have a chance to cover a few more concepts first. It is important to point out, however, that this sign convention is not universal, and it is much more common to colloquially say $\Delta\omega$ but actually plot against the absolute value $|\Delta\omega|$. In other words, frequencies are almost never displayed with negative values and, since Stokes-shifted frequencies are more intense, it is more common to display only that region of the Raman spectrum unless there is a specific reason to require the anti-Stokes spectrum.

We now consider an energy level diagram with some transitions responsible for the spectral features. Figure 9.3a shows a set of vibrational levels in the electronic ground state and the corresponding set of vibrational levels in the first electronic excited state. Suppose that a green photon with $\lambda = 532$ nm ($\omega_{ex} = 18979$ cm^{-1}) is not energetic enough to reach the electronic excited state, for which UV light would be required as is often the case with many organic molecules that are transparent in the visible region. In such a case, we reach a virtual state as indicated by the dashed horizontal line. This is a short-lived state, and the molecule quickly returns to the ground state, thereby emitting a green photon of the same frequency $\omega_s = \omega_{ex}$. This is the Rayleigh scattering process, and it can occur from any vibrational level in the electronic ground state, so long as the transition returns to the same level, as illustrated in the transition from $v = 1$ in the electronic ground state back to the same $v = 1$ state in Fig. 9.3b. Next we consider the situation shown in Fig. 9.3c where a molecule originating in $v = 0$ returns to the electronic ground state, but ends up in $v = 1$. Here it is clear that the emitted photon with frequency ω_s (proportional to the length of the downwards pointing arrow) has $\omega_s < \omega_{ex}$ or, as the spectrometer records the data, $\lambda_s > \lambda_{ex}$ representing a Stokes-shifted photon. In the event that the molecule originates in $v = 1$ but returns to $v = 0$ after the excitation, Fig. 9.3d indicates that the anti-Stokes scattered light has $\omega_s > \omega_{ex}$ and therefore $\lambda_s < \lambda_{ex}$. The lower population of $v = 1$ compared to $v = 0$ explains the relative intensity of the anti-Stokes to Stokes bands (see Fig. 9.2), and the reason why Stokes Raman

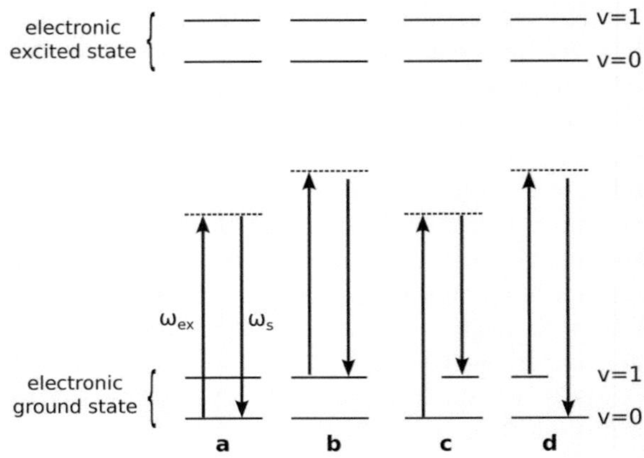

Fig. 9.3 **a** Rayleigh scattering from the lowest vibrational level in the ground electronic state, where the frequency of the scattered light ω_s is equal to that of the excitation source ω_{ex}. **b** Rayleigh scattering from $v = 1$ in the ground electronic state. **c** Stokes Raman scattering and **d** anti-Stokes Raman scattering

spectra are the default in most applications. The ratio of the intensities on either side of the Rayleigh line is given by

$$\frac{I_{\text{anti-Stokes}}}{I_{\text{Stokes}}} = \left[\frac{\omega_{\text{anti-Stokes}}}{\omega_{\text{Stokes}}}\right]^4 \exp\left[-\frac{100hc\Delta\omega}{k_B T}\right] \tag{9.1}$$

where h is the Planck constant, c is the speed of light in a vacuum, k_B is the Boltzmann constant, and T is the absolute temperature. The factor of 100 appears in Eq. 9.1 since we are specifying the Raman shift in units of cm^{-1} and the argument of the exponential function needs to be dimensionless, assuming we wish to specify all other quantities in SI base units. Equation 9.1 is what would be determined when the detection system measures the power of the scattered light. For photon-counting detectors, the expression is slightly different

$$\begin{aligned}\frac{I_{\text{anti-Stokes}}}{I_{\text{Stokes}}} &= \frac{\omega^3_{\text{anti-Stokes}}\omega_{\text{ex}}}{\omega^3_{\text{Stokes}}\omega_{\text{ex}}} \exp\left[-\frac{100hc\Delta\omega}{k_B T}\right] \\ &= \left[\frac{\omega_{\text{anti-Stokes}}}{\omega_{\text{Stokes}}}\right]^3 \exp\left[-\frac{100hc\Delta\omega}{k_B T}\right].\end{aligned} \tag{9.2}$$

One of the remarkable features of Raman spectroscopy is that, when both Stokes and anti-Stokes bands are observed for the same vibrational mode, Eq. 9.1 can be used to determine the temperature from the intensity ratio [14, 15]. However, this is practical only for low-frequency vibrations; as when $\Delta\omega$ increases the intensity ratio becomes very small, and the anti-Stokes bands become very difficult to observe.

The intensity of the scattered light I_s can be written as

$$I_s = N d\sigma_R I_0 \tag{9.3}$$

where I_0 is the intensity of the incident light, N is the molecular number density, d is the path length, and σ_R is the Raman cross section. This cross section is the ratio of the energy transferred to the Stokes modes per unit time (photons per second) to the incident intensity (photons per second per unit area). So σ_R ends up with units of area. It is convention to report all spectroscopic cross sections using cm^2, but for a dimensional analysis we can consider m^2 so that the ratio I_s/I_0 is dimensionless when N is in m^{-3} and d is specified in metres.

When the excitation frequency is far from any electronic resonance (as in the cases illustrated in Fig. 9.3), this cross section can be related to the imaginary part of the third-order polarizability [16–19]. We will come back to this point after considering a classical and quantum mechanical description of the Raman scattering process. Several reviews of the general subject of Raman scattering exist, and readers are referred to those sources for a more comprehensive treatment of the subject [20–23]. Our goal here is to introduce only those concepts necessary for a discussion of how Raman band intensities can be used to characterize the orientation of molecules in ordered materials.

9.2 Classical Model for Raman Scattering

Although our focus will be on a quantum mechanical model that describes Raman scattering, it is useful to first consider a simple classical model that can explain some of the attributes we have described. Consider the now familiar expression

$$\mu_{\text{ind}} = \alpha E \tag{9.4}$$

where the induced dipole moment μ is equal to the product of the polarizability α and the field E. We will ignore the tensorial nature of these quantities for the moment. We now invoke two simple models, one for the electric field from the excitation source at frequency ω_{ex}

$$E_{\text{ex}} = E_0 \cos(\omega_{\text{ex}} t) \tag{9.5}$$

and another for the displacement Q from the equilibrium geometry with a harmonic motion

$$Q = \cos(\omega_{\text{vib}} t) \tag{9.6}$$

where ω_{vib} is the vibrational frequency. As in our description of harmonic motion for the classical Lorentz oscillator, we can consider a simple diatomic molecule where $Q = 0$ corresponds to the equilibrium bond length. We then expand α as a Taylor series about Q to obtain

$$\alpha = \alpha_0 + \left(\frac{\partial \alpha}{\partial Q}\right)_{Q=0} Q + \cdots \tag{9.7}$$

Since we assume Q to be small, we truncate the above expression to keep only the first two terms and then our expression for the induced dipole moment becomes

$$\begin{aligned}\mu_{\text{ind}} &= \alpha E_{\text{ex}} \\ &= \left[\alpha_0 + \left(\frac{\partial \alpha}{\partial Q}\right)_{Q=0} Q\cos(\omega_{\text{vib}}t)\right] E_0 \cos(\omega_{\text{ex}}t).\end{aligned} \tag{9.8}$$

We make use of the trigonometric identity

$$\cos a \cos b = \frac{1}{2}\left[\cos(a+b) + \cos(a-b)\right]$$

to write

$$\begin{aligned}\mu(t) = {} & \alpha_0 E_0 \cos(\omega_{\text{ex}}t) + \frac{1}{2}\left(\frac{\partial \alpha}{\partial Q}\right)_{Q=0} Q\cos(\omega_{\text{ex}} + \omega_{\text{vib}})t \\ & + \frac{1}{2}\left(\frac{\partial \alpha}{\partial Q}\right)_{Q=0} Q\cos(\omega_{\text{ex}} - \omega_{\text{vib}})t.\end{aligned} \tag{9.9}$$

What we have here is an expression for the time-dependent-induced dipole moment that results from the interaction of a laser field at frequency ω_{ex} with a molecular vibration of frequency ω_{vib}. We can see that the expression consists of three terms. The first produces a dipole that oscillates at ω_{ex} and therefore accounts for Rayleigh scattering as the electromagnetic radiation emitted from this oscillating charge will be at ω_{ex}. The second term in Eq. 9.9 describes a charge distribution that oscillates at frequency $\omega_{\text{ex}} + \omega_{\text{vib}}$, and therefore radiates a field at frequency $\omega_{\text{ex}} + \omega_{\text{vib}}$, the anti-Stokes-shifted radiation. We see that Stokes-shifted frequencies $\omega_{\text{ex}} - \omega_{\text{vib}}$ are accounted for by the third term in Eq. 9.9.

Finally, we notice the factor $(\partial\alpha/\partial Q)_{Q=0}$ that appears in the Stokes and anti-Stokes terms in Eq. 9.9. This derivative of the polarizability with respect to the normal mode coordinate, evaluated at the equilibrium geometry, is termed the Raman tensor, or Raman polarizability, or derived polarizability [20] (i.e. after its derivative). Although most books and papers use the symbol α for this quantity, we will adopt the notation α^{R} to avoid confusion with the polarizability itself. The consequence of this factor when it appears in Eq. 9.9 is that, for a vibrational mode to be Raman active, we require that $\alpha^{\text{R}} \neq 0$. Figure 9.4a illustrates a situation where there is a large change in α during the vibration for the normal mode under consideration, but $\alpha^{\text{R}} = 0$ so this mode is not Raman active. In contrast, the behaviour of α with respect to Q for a Raman-active mode is shown in Fig. 9.4b. It is helpful to contrast this with the relevant quantities depicted for IR activity in Fig. 8.15. So far we have been discussing molecular vibrations as if they possess Raman activity as a binary attribute. There are certain symmetry elements that can be considered to further that discussion. In addition, there are some general rules that work in a wide variety of cases that chemists use to have some intuition about what vibrations are strong in

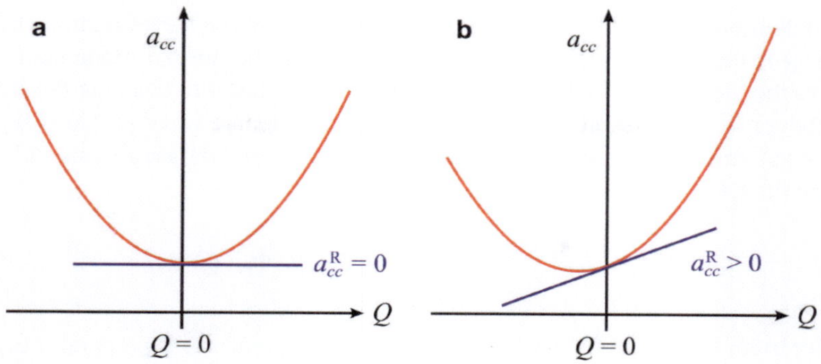

Fig. 9.4 **a** An example of a vibrational mode whose polarizability tensor element a_{cc} changes significantly during the vibration along the normal mode coordinate Q, but with a slope of zero at $Q = 0$ so therefore the Raman tensor element $a^{\mathrm{R}}_{cc} \propto \partial a_{cc}/\partial Q = 0$. **b** A Raman-active mode for which $a^{\mathrm{R}}_{cc} > 0$ as can be observed by the slope of a_{cc} vs. Q about the equilibrium geometry located at $Q = 0$. This graphical description can be compared with results calculated for actual vibrational modes of a real molecule shown in Fig. 2.3

the IR, Raman, or both. In reality, for large molecules, symmetry is often low and most of the $3N - 6$ vibrations are IR and Raman active. In such cases, it is more useful to be able to know just *how* Raman active those modes are, i.e. to understand the relative intensity of bands in the Raman spectrum.

The main conclusion from this model is that light at frequency ω_{ex} is able to induce a time-varying dipole moment at frequencies $\omega_{\mathrm{s}} = \omega_{\mathrm{ex}} \pm \omega_{\mathrm{vib}}$. This source polarization at ω_{s} induces a field E_{s} whose intensity corresponds to the Raman scattered light. Since $I_{\mathrm{s}} \propto |E_{\mathrm{s}}|^2$, the Raman cross section σ_{R} (Eq. 9.3) is proportional to $(\alpha^{\mathrm{R}})^2$. We will come back to this point in the definition of the Raman tensor after connecting this quantity to some transition dipole moments.

9.3 Quantum Mechanical Description of Raman Scattering

We have seen that the intensity of Raman scattered light is determined by transitions between vibrational states m and n and a virtual state v. In Fig. 9.3 we have considered the ground vibrational state with $m = 0$ and an excited vibrational state with $n = 1$; the virtual state was indicated by a dashed line. The associated products of transition dipole moments are given by

$$\mu_i^{(vn)}\mu_j^{(mv)} = \int_{-\infty}^{\infty} \Psi_n^*(Q)\hat{\mu}_i\Psi_v(Q)\,\mathrm{d}Q \int_{-\infty}^{\infty} \Psi_v^*(Q)\hat{\mu}_j\Psi_m(Q)\,\mathrm{d}Q \qquad (9.10)$$

using the same notation that we introduced for the IR transition dipole moment (Eq. 8.25). From our study of IR absorption spectroscopy, we know the form of the dipole moment operator $\hat{\mu}$, and we know the form of the wavefunctions $\Psi_n(Q)$ where Q is the displacement from equilibrium along the normal mode coordinate and n is the vibrational quantum number. However, in order to evaluate the expression above, we are missing information on the virtual state wavefunctions $\Psi_v(Q)$. We can get around this issue by defining the *transition polarizability* that will soon become the Raman tensor

$$\alpha_{ij}^{(mn)} = \mu_i^{(nv)} \mu_j^{(mv)} = \int_{-\infty}^{\infty} \Psi_n^*(Q) \hat{\alpha}_{ij} \Psi_m(Q)\, \mathrm{d}Q. \tag{9.11}$$

In order to make use of this, we need the form of the polarizability operator $\hat{\alpha}_{ij}$. We can use the same approach as was described for the dipole moment operator. We assume that the polarizability is a function of the atomic coordinates, and write the polarizability α as a Taylor expansion about the equilibrium geometry ($Q = 0$)

$$\hat{\alpha}_{ij} = \alpha_{ij}^0 + Q \left(\frac{\partial}{\partial Q} \alpha_{ij} \right)_{Q=0} + \cdots \tag{9.12}$$

and keep only the first two terms as the definition of $\hat{\alpha}_{ij}$.

Now we are in a position to understand the selection rule for vibrational Raman scattering by evaluating the transition polarizability

$$\begin{aligned} \alpha_{ij}^{(mn)} &= \int_{-\infty}^{\infty} \Psi_n^*(Q) \hat{\alpha}_{ij} \Psi_m(Q)\, \mathrm{d}Q \\ &= \int_{-\infty}^{\infty} \Psi_n^*(Q) \alpha_{ij}^0 \Psi_m(Q)\, \mathrm{d}Q + \int_{-\infty}^{\infty} \Psi_n^*(Q) Q \left(\frac{\partial \alpha_{ij}}{\partial Q} \right)_{Q=0} \Psi_m(Q)\, \mathrm{d}Q \\ &= \alpha_{ij}^0 \delta_{mn} + \left(\frac{\partial \alpha_{ij}}{\partial Q} \right)_{Q=0} \int_{-\infty}^{\infty} \Psi_n^*(Q) Q \Psi_m(Q)\, \mathrm{d}Q. \end{aligned} \tag{9.13}$$

We can see that the first term contains the static polarizability α_{ij}^0 multiplied by the Kronecker delta function δ_{mn} which evaluates to $\delta_{mn} = 0$ whenever $m \neq n$. Since the whole point of this is to obtain the *transition* polarizability between states m and n, we can see that the static polarizability is not relevant to the intensity of Raman bands. This is analogous to how the static dipole moment of a molecule is not related to its absorption coefficient. In the second term in Eq. 9.13, we have factored out the constant $(\partial \alpha_{ij} / \partial Q)_{Q=0}$, i.e. the Raman tensor, that appears together with the integral

$$\int_{-\infty}^{\infty} \Psi_m^*(Q) Q \Psi_n(Q)\, \mathrm{d}Q \tag{9.14}$$

which needs to be evaluated. However, these solutions are exactly the same as we explored in the case of the IR transition dipole moment, and so we can recall the result that

$$\int_{-\infty}^{\infty} \Psi_n^*(Q) Q \Psi_m(Q)\, \mathrm{d}Q = 0 \quad \text{unless} \quad n = m \pm 1$$

providing the same selection rules as was observed for IR absorption spectroscopy in Chap. 8.

Further examining Eq. 9.13, we can also understand why the relative intensities of bands in IR absorption and Raman spectra are different. The selection rules tell us what transitions make the second term in Eq. 9.13 non-zero. However, the magnitude of $\alpha_{ij}^{(mn)}$ depends on the value of the integral in Eq. 9.14 (when $n = m \pm 1$) and the magnitude of the prefactor α^{R}. For a given vibrational mode, and considering the $m = 0 \rightarrow n = 1$ transition, the numerical value of Eq. 9.14 would be the same for an IR or Raman transition, but for some of the $3N - 6$ vibrational modes of a polyatomic molecule, α^{R} will be large and therefore large Raman transitions will be observed. The magnitude of α^{R} is a different consideration, so the relative intensity of IR absorption bands may follow a different pattern from that observed in Raman spectra. Vibrational modes for which $(\partial\alpha_{ij}/\partial Q)_{Q=0}$ is significant are termed Raman active; modes for which $(\partial\mu_i/\partial Q)_{Q=0}$ is large (Fig. 8.15) are termed IR active. These considerations are not necessarily mutually exclusive, but certain molecular symmetry conditions can be considered for some general understanding. For example, centrosymmetric molecules (those containing an inversion centre) such as benzene don't have any coincident IR- and Raman-active modes. However, large molecules generally have low symmetry, and the degree to which a particular vibrational mode is Raman and/or IR active is best assessed by evaluating α_{ij}^{R} and $(\partial\mu_i/\partial Q)_{Q=0}$ directly. We will return to this point when we consider the coupling of $(\partial\alpha_{ij}/\partial Q)_{Q=0}$ and $(\partial\mu_i/\partial Q)_{Q=0}$ for a single vibrational mode in our study of SFG spectroscopy in Chap. 10.

9.4 The Raman Tensor

We briefly mentioned that the Raman cross σ_{R} section defined in Eq. 9.3 is related to the third-order response function. It can be demonstrated [16] that σ_{R} can be connected to the third-order polarizability $\boldsymbol{\alpha}^{(3)}$ and that

$$\sigma_{\mathrm{R},ij} \propto \omega_{\mathrm{ex}}\omega_{\mathrm{s}}^3 \langle \mathbb{I}\mathrm{m}\{\alpha_{ijij}^{(3)}\}\rangle \tag{9.15}$$

where the angle brackets indicate the orientational average; we will discuss this average in more detail shortly. Strictly speaking, one must consider all pathways that connect the vibrational and electronic states in the time domain in order to arrive at the Raman cross section. However, when we are sufficiently far from electronic resonance, Eq. 9.15 is approximately correct [16], and will be useful to make the connection between the various response functions that we are in the process of investigating for the different spectroscopic methods in Chaps. 8–10. Equation 9.15 introduces a contracted notation that relates the rank-2 scattering tensor on the left-hand side to a rank-4 third-order polarizability tensor on the right-hand side, albeit with repeated indices.

We can write an expression for the lineshape of a vibrational Raman response in a manner similar to what we developed for IR absorption based on q classical Lorentz oscillators

$$\alpha^{(3)}_{lmno}(\omega_s) = \sum_q \frac{a^{(3)}_{lmno,q}}{\omega_q^2 - \omega_s^2 - i\omega_s\Gamma_q} \tag{9.16}$$

where the only discernible difference is that we have written the frequency dependence that was previously ascribed to ω (the wavenumber of the mid-IR light) using ω_s (or, more generally, $\Delta\omega$ to describe the Stokes or anti-Stokes Raman frequency shift), and we are using an element of $\mathbf{a}^{(3)}$ to account for the intensity of a vibrational band in a Raman experiment. Here $\Delta\omega$ (written as ω_s in Eq. 9.16) is considered positive, regardless of whether the Stokes or anti-Stokes shift is implied. This is the more common notation, rather than writing $|\Delta\omega|$. Furthermore, due to the higher intensity, it is almost always the Stokes-shifted Raman spectrum that is recorded. Coming back to the macroscopic response from the whole material, we can write

$$\frac{I_s}{I_0} \propto \mathbb{Im}\{\chi^{(3)}_{ijkl}\} \tag{9.17}$$

where the elements of third-order susceptibility inherit the lineshape we have defined for the response of the individual molecules

$$\chi^{(3)}_{ijkl}(\omega_s) = \sum_q \frac{A^{(3)}_{ijkl,q}(\theta,\phi,\psi)}{\omega_q^2 - \omega_s^2 - i\omega_s\Gamma_q}. \tag{9.18}$$

In writing Eqs. 9.16 and 9.18 we momentarily consider a more general notation for the 81 elements of a rank-4 tensor, specified with 4 subscripted indices. However, we already know that the entire response must be able to be encoded in a 9-element rank-2 Raman tensor. This is because, far from any electronic resonance, all elements of $\mathbf{a}^{(3)}$ can be determined from

$$\mathbf{a}^{(3)} = \frac{\partial \mathbf{a}}{\partial Q} \otimes \frac{\partial \mathbf{a}}{\partial Q} \tag{9.19}$$

where $\mathbf{a}$ is the numerator of the linear (conventional) polarizability tensor $\mathbf{a}$ discussed in the previous chapter (a rank 2 quantity) and we are using Q as the generalized vibrational normal mode coordinate. The symbol $\otimes$ refers to a tensor product, where we build a rank 4 tensor from the outer product of two rank 2 tensors. For those who like to wear silly hats out in public, it is permissible to write

$$\overset{\rightrightarrows\rightrightarrows}{a}^{(3)} = \frac{\partial \overset{\rightrightarrows}{a}}{\partial Q} \otimes \frac{\partial \overset{\rightrightarrows}{a}}{\partial Q}. \tag{9.20}$$

Since we are working far from electronic resonance, recall that **a** itself is a symmetric matrix where $a_{lm} = a_{ml}$ and therefore the Raman tensor has the form:

$$\mathbf{a}^{\mathrm{R}} = \frac{\partial \mathbf{a}}{\partial Q} = \begin{bmatrix} \frac{\partial a_{aa}}{\partial Q} & \frac{\partial a_{ab}}{\partial Q} & \frac{\partial a_{ac}}{\partial Q} \\ \frac{\partial a_{ab}}{\partial Q} & \frac{\partial a_{bb}}{\partial Q} & \frac{\partial a_{bc}}{\partial Q} \\ \frac{\partial a_{ac}}{\partial Q} & \frac{\partial a_{bc}}{\partial Q} & \frac{\partial a_{cc}}{\partial Q} \end{bmatrix} = \begin{bmatrix} a^{\mathrm{R}}_{aa} & a^{\mathrm{R}}_{ab} & a^{\mathrm{R}}_{ac} \\ a^{\mathrm{R}}_{ab} & a^{\mathrm{R}}_{bb} & a^{\mathrm{R}}_{bc} \\ a^{\mathrm{R}}_{ac} & a^{\mathrm{R}}_{bc} & a^{\mathrm{R}}_{cc} \end{bmatrix}. \tag{9.21}$$

Note that others typically write $\boldsymbol{\alpha}^{\mathrm{R}} \equiv \partial\boldsymbol{\alpha}/\partial Q$ but we are using the symbol **a** rather than α since we are referring to the *amplitude* of the response function, rather than its complete form that includes the (Lorentzian) frequency dependence. I sympathize that **a** and $\boldsymbol{\alpha}$ are a bit hard to distinguish, but that's not such a bad thing!

Before returning to the inherent symmetry of $\boldsymbol{\alpha}^{(3)}$, let's take a moment to explore what a rank-4 tensor looks like. Although we can list the 81 elements in a table, and assign them indices, there is no simple way to convey the shape of a $3 \times 3 \times 3 \times 3$ quantity, as that contains one more dimension than we can depict with a typical 3D drawing. As a compromise, a partially expanded form showing all 81 elements looks like

$$\mathbf{a}^{(3)} = \begin{bmatrix} \begin{bmatrix} a^{(3)}_{aaaa} & a^{(3)}_{aaab} & a^{(3)}_{aaac} \\ a^{(3)}_{aaba} & a^{(3)}_{aabb} & a^{(3)}_{aabc} \\ a^{(3)}_{aaca} & a^{(3)}_{aacb} & a^{(3)}_{aacc} \end{bmatrix} & \begin{bmatrix} a^{(3)}_{abaa} & a^{(3)}_{abab} & a^{(3)}_{abac} \\ a^{(3)}_{abba} & a^{(3)}_{abbb} & a^{(3)}_{abbc} \\ a^{(3)}_{abca} & a^{(3)}_{abcb} & a^{(3)}_{abcc} \end{bmatrix} & \begin{bmatrix} a^{(3)}_{acaa} & a^{(3)}_{acab} & a^{(3)}_{acac} \\ a^{(3)}_{acba} & a^{(3)}_{acbb} & a^{(3)}_{acbc} \\ a^{(3)}_{acca} & a^{(3)}_{accb} & a^{(3)}_{accc} \end{bmatrix} \\ \begin{bmatrix} a^{(3)}_{baaa} & a^{(3)}_{baab} & a^{(3)}_{baac} \\ a^{(3)}_{baba} & a^{(3)}_{babb} & a^{(3)}_{babc} \\ a^{(3)}_{baca} & a^{(3)}_{bacb} & a^{(3)}_{bacc} \end{bmatrix} & \begin{bmatrix} a^{(3)}_{bbaa} & a^{(3)}_{bbab} & a^{(3)}_{bbac} \\ a^{(3)}_{bbba} & a^{(3)}_{bbbb} & a^{(3)}_{bbbc} \\ a^{(3)}_{bbca} & a^{(3)}_{bbcb} & a^{(3)}_{bbcc} \end{bmatrix} & \begin{bmatrix} a^{(3)}_{bcaa} & a^{(3)}_{bcab} & a^{(3)}_{bcac} \\ a^{(3)}_{bcba} & a^{(3)}_{bcbb} & a^{(3)}_{bcbc} \\ a^{(3)}_{bcca} & a^{(3)}_{bccb} & a^{(3)}_{bccc} \end{bmatrix} \\ \begin{bmatrix} a^{(3)}_{caaa} & a^{(3)}_{caab} & a^{(3)}_{caac} \\ a^{(3)}_{caba} & a^{(3)}_{cabb} & a^{(3)}_{cabc} \\ a^{(3)}_{caca} & a^{(3)}_{cacb} & a^{(3)}_{cacc} \end{bmatrix} & \begin{bmatrix} a^{(3)}_{cbaa} & a^{(3)}_{cbab} & a^{(3)}_{cbac} \\ a^{(3)}_{cbba} & a^{(3)}_{cbbb} & a^{(3)}_{cbbc} \\ a^{(3)}_{cbca} & a^{(3)}_{cbcb} & a^{(3)}_{cbcc} \end{bmatrix} & \begin{bmatrix} a^{(3)}_{ccaa} & a^{(3)}_{ccab} & a^{(3)}_{ccac} \\ a^{(3)}_{ccba} & a^{(3)}_{ccbb} & a^{(3)}_{ccbc} \\ a^{(3)}_{ccca} & a^{(3)}_{cccb} & a^{(3)}_{cccc} \end{bmatrix} \end{bmatrix} \tag{9.22}$$

which amounts to

$$\mathbf{a}^{(3)} = \begin{bmatrix} \begin{bmatrix} (a^{\mathrm{R}}_{aa})^2 & a^{\mathrm{R}}_{aa}a^{\mathrm{R}}_{ab} & a^{\mathrm{R}}_{aa}a^{\mathrm{R}}_{ac} \\ a^{\mathrm{R}}_{aa}a^{\mathrm{R}}_{ab} & a^{\mathrm{R}}_{aa}a^{\mathrm{R}}_{bb} & a^{\mathrm{R}}_{aa}a^{\mathrm{R}}_{bc} \\ a^{\mathrm{R}}_{aa}a^{\mathrm{R}}_{ac} & a^{\mathrm{R}}_{aa}a^{\mathrm{R}}_{bc} & a^{\mathrm{R}}_{aa}a^{\mathrm{R}}_{cc} \end{bmatrix} & \begin{bmatrix} a^{\mathrm{R}}_{aa}a^{\mathrm{R}}_{ab} & (a^{\mathrm{R}}_{ab})^2 & a^{\mathrm{R}}_{ab}a^{\mathrm{R}}_{ac} \\ (a^{\mathrm{R}}_{ab})^2 & a^{\mathrm{R}}_{ab}a^{\mathrm{R}}_{bb} & a^{\mathrm{R}}_{ab}a^{\mathrm{R}}_{bc} \\ a^{\mathrm{R}}_{ab}a^{\mathrm{R}}_{ac} & a^{\mathrm{R}}_{ab}a^{\mathrm{R}}_{bc} & a^{\mathrm{R}}_{ab}a^{\mathrm{R}}_{cc} \end{bmatrix} & \begin{bmatrix} a^{\mathrm{R}}_{aa}a^{\mathrm{R}}_{ac} & a^{\mathrm{R}}_{ab}a^{\mathrm{R}}_{ac} & (a^{\mathrm{R}}_{ac})^2 \\ a^{\mathrm{R}}_{ab}a^{\mathrm{R}}_{ac} & a^{\mathrm{R}}_{ac}a^{\mathrm{R}}_{bb} & a^{\mathrm{R}}_{ac}a^{\mathrm{R}}_{bc} \\ (a^{\mathrm{R}}_{ac})^2 & a^{\mathrm{R}}_{ac}a^{\mathrm{R}}_{bc} & a^{\mathrm{R}}_{ac}a^{\mathrm{R}}_{cc} \end{bmatrix} \\ \begin{bmatrix} a^{\mathrm{R}}_{aa}a^{\mathrm{R}}_{ab} & (a^{\mathrm{R}}_{ab})^2 & a^{\mathrm{R}}_{ab}a^{\mathrm{R}}_{ac} \\ (a^{\mathrm{R}}_{ab})^2 & a^{\mathrm{R}}_{ab}a^{\mathrm{R}}_{bb} & a^{\mathrm{R}}_{ab}a^{\mathrm{R}}_{bc} \\ a^{\mathrm{R}}_{ab}a^{\mathrm{R}}_{ac} & a^{\mathrm{R}}_{ab}a^{\mathrm{R}}_{bc} & a^{\mathrm{R}}_{ab}a^{\mathrm{R}}_{cc} \end{bmatrix} & \begin{bmatrix} a^{\mathrm{R}}_{aa}a^{\mathrm{R}}_{bb} & a^{\mathrm{R}}_{ab}a^{\mathrm{R}}_{bb} & a^{\mathrm{R}}_{ac}a^{\mathrm{R}}_{bb} \\ a^{\mathrm{R}}_{ab}a^{\mathrm{R}}_{bb} & (a^{\mathrm{R}}_{bb})^2 & a^{\mathrm{R}}_{bb}a^{\mathrm{R}}_{bc} \\ a^{\mathrm{R}}_{ac}a^{\mathrm{R}}_{bb} & a^{\mathrm{R}}_{bb}a^{\mathrm{R}}_{bc} & a^{\mathrm{R}}_{bb}a^{\mathrm{R}}_{cc} \end{bmatrix} & \begin{bmatrix} a^{\mathrm{R}}_{aa}a^{\mathrm{R}}_{bc} & a^{\mathrm{R}}_{ab}a^{\mathrm{R}}_{bc} & a^{\mathrm{R}}_{ac}a^{\mathrm{R}}_{bc} \\ a^{\mathrm{R}}_{ab}a^{\mathrm{R}}_{bc} & a^{\mathrm{R}}_{bb}a^{\mathrm{R}}_{bc} & (a^{\mathrm{R}}_{bc})^2 \\ a^{\mathrm{R}}_{ac}a^{\mathrm{R}}_{bc} & (a^{\mathrm{R}}_{bc})^2 & a^{\mathrm{R}}_{bc}a^{\mathrm{R}}_{cc} \end{bmatrix} \\ \begin{bmatrix} a^{\mathrm{R}}_{aa}a^{\mathrm{R}}_{ac} & a^{\mathrm{R}}_{ab}a^{\mathrm{R}}_{ac} & (a^{\mathrm{R}}_{ac})^2 \\ a^{\mathrm{R}}_{ab}a^{\mathrm{R}}_{ac} & a^{\mathrm{R}}_{ac}a^{\mathrm{R}}_{bb} & a^{\mathrm{R}}_{ac}a^{\mathrm{R}}_{bc} \\ (a^{\mathrm{R}}_{ac})^2 & a^{\mathrm{R}}_{ac}a^{\mathrm{R}}_{bc} & a^{\mathrm{R}}_{ac}a^{\mathrm{R}}_{cc} \end{bmatrix} & \begin{bmatrix} a^{\mathrm{R}}_{aa}a^{\mathrm{R}}_{bc} & a^{\mathrm{R}}_{ab}a^{\mathrm{R}}_{bc} & a^{\mathrm{R}}_{ac}a^{\mathrm{R}}_{bc} \\ a^{\mathrm{R}}_{ab}a^{\mathrm{R}}_{bc} & a^{\mathrm{R}}_{bb}a^{\mathrm{R}}_{bc} & (a^{\mathrm{R}}_{bc})^2 \\ a^{\mathrm{R}}_{ac}a^{\mathrm{R}}_{bc} & (a^{\mathrm{R}}_{bc})^2 & a^{\mathrm{R}}_{bc}a^{\mathrm{R}}_{cc} \end{bmatrix} & \begin{bmatrix} a^{\mathrm{R}}_{aa}a^{\mathrm{R}}_{cc} & a^{\mathrm{R}}_{ab}a^{\mathrm{R}}_{cc} & a^{\mathrm{R}}_{ac}a^{\mathrm{R}}_{cc} \\ a^{\mathrm{R}}_{ab}a^{\mathrm{R}}_{cc} & a^{\mathrm{R}}_{bb}a^{\mathrm{R}}_{cc} & a^{\mathrm{R}}_{bc}a^{\mathrm{R}}_{cc} \\ a^{\mathrm{R}}_{ac}a^{\mathrm{R}}_{cc} & a^{\mathrm{R}}_{bc}a^{\mathrm{R}}_{cc} & (a^{\mathrm{R}}_{cc})^2 \end{bmatrix} \end{bmatrix} \tag{9.23}$$

where it becomes obvious that the more practical and equivalent element-wise operation is simply

$$a^{(3)}_{lmno} = a^{\mathrm{R}}_{lm} a^{\mathrm{R}}_{no}. \tag{9.24}$$

An important conclusion is that, due to the symmetry in $\mathbf{a}^{\mathrm{R}}$ (only six of the nine elements are unique) there is an inherent restriction on the number of unique elements in $\mathbf{a}^{(3)}$ that can be obtained from Eq. 9.24. As a result, it will always be more convenient and sensible to use $\mathbf{a}^{\mathrm{R}}$, as the $\mathbf{a}^{(3)}$ notation is unnecessarily cumbersome. However, there is a caveat: $\mathbf{a}^{(3)}$ is a rank-4 tensor, while $\mathbf{a}^{\mathrm{R}}$ is a rank-2 tensor. Their *elements* can be easily interchanged using Eq. 9.24, but when transforming coordinate systems (projecting from the molecular frame to the lab frame), we will need to remember to use the operations necessary for a rank-4 tensor. We will emphasize this point again when we address orientation analysis later in this chapter, but we mention it here, as it is inextricably linked to the definition of the Raman tensor and its relationship to the Raman cross section.

We have defined the Lorentzian numerator $\mathbf{a}^{(3)}$ to be the equivalent of the "oscillator strength" (Eq. 9.16) in a Raman experiment, and so the microscopic expression providing the definition of $\boldsymbol{\alpha}^{(3)}$ and the ensemble average for the material property $\boldsymbol{\chi}^{(3)}$ (Eq. 9.18) can be combined to arrive at

$$\mathbf{A}^{(3)} = \frac{N}{\varepsilon_0} \langle \mathbf{a}^{(3)} \rangle. \tag{9.25}$$

We now consider how to project the molecular quantity $\mathbf{a}^{(3)}$ into the laboratory frame to arrive at $\mathbf{a}^{(3)}(\theta, \phi, \psi)$. We first follow the standard method as would be appropriate for a rank-4 tensor

$$\begin{aligned} a^{(3)}_{ijij}(\theta, \phi, \psi) = \sum_{l}^{a,b,c} \sum_{m}^{a,b,c} \sum_{n}^{a,b,c} \sum_{o}^{a,b,c} & D_{il}(\theta, \phi, \psi)\, D_{jm}(\theta, \phi, \psi) \\ & \times D_{in}(\theta, \phi, \psi)\, D_{jo}(\theta, \phi, \psi)\, a^{(3)}_{lmno} \end{aligned} \tag{9.26}$$

where we see that a summation over all 81 $\mathbf{a}^{(3)}$ elements is performed, since we are nesting four sums over three molecular-frame Cartesian coordinates. Note how each element $a^{(3)}_{lmno}$ is paired with the corresponding direction cosine matrix elements and so Eq. 9.26 gets the job done, albeit in a cumbersome way, applying appropriate elements of the DCM, four times (as expected for a rank-4 tensor) for each projection.

In a spontaneous Raman experiment, since we can measure only the 9-lab frame elements corresponding to $A^{(3)}_{ijij}$, this means that we require only $a^{(3)}_{ijij}(\theta, \phi, \psi)$. We can therefore simplify the notation using the Raman tensor introduced in Eq. 9.24

$$\begin{aligned} a^{(3)}_{ijij}(\theta, \phi, \psi) &= a^{\mathrm{R}}_{ij}(\theta, \phi, \psi) a^{\mathrm{R}}_{ij}(\theta, \phi, \psi) \\ &= \left[a^{\mathrm{R}}_{ij}(\theta, \phi, \psi)\right]^2 \end{aligned} \tag{9.27}$$

that leads to

$$\left[a_{ij}^{\mathrm{R}}(\theta,\phi,\psi)\right]^2 = \left[\sum_{l}^{a,b,c}\sum_{m}^{a,b,c} D_{il}(\theta,\phi,\psi)\, D_{jm}(\theta,\phi,\psi)\, a_{lm}^{\mathrm{R}}\right]^2. \tag{9.28}$$

Equation 9.28 may at first appear only slightly more compact than Eq. 9.26. But there is a significant improvement in the computation of all the projections that is available to us. The interesting thing here is that we evaluate the nine intensity elements in the laboratory frame easily and quickly if we project them all at once, rather than element-wise. Although this may seem counter-intuitive, it takes advantage of the fact that we can work with rank 2 tensors to carry out

$$\mathbf{a}^{(3)}(\theta,\phi,\psi) = \left[\mathbf{D}(\theta,\phi,\psi)\,\mathbf{a}^{\mathrm{R}}\,\mathbf{D}(\theta,\phi,\psi)^{-1}\right] \otimes \left[\mathbf{D}(\theta,\phi,\psi)\,\mathbf{a}^{\mathrm{R}}\,\mathbf{D}(\theta,\phi,\psi)^{-1}\right] \tag{9.29}$$

where $\mathbf{D}$ is the entire direction cosine matrix and $\mathbf{a}$ is the entire rank 2 tensor in the molecular frame, and so

$$a_{ijij}^{(3)}(\theta,\phi,\psi) = \left(\left[\mathbf{D}(\theta,\phi,\psi)\,\mathbf{a}^{\mathrm{R}}\,\mathbf{D}(\theta,\phi,\psi)^{-1}\right]_{ij}\right)^2 \equiv \left(a_{ij}^{\mathrm{R}}\right)^2 \tag{9.30}$$

where the subscript ij in the middle of Eq. 9.30 appears outside of the square brackets, as we are indexing an element of the rank 2 matrix that is formed from the (rank 2) matrix product of $\mathbf{D}$, $\mathbf{a}^{\mathrm{R}}$, and the inverse of $\mathbf{D}$ (that is equal to $\mathbf{D}^{\mathrm{T}}$ since $\mathbf{D}$ is orthogonal). This element can then be squared to form the single $(ijij)$ element of the rank 4 tensor $\mathbf{a}^{(3)}$. The final quantity on the right-hand side of Eq. 9.30 defines an element of the Raman tensor $\mathbf{a}^{\mathrm{R}}$.

It is important to realize that, although this compact notation works very well, we are still ultimately projecting a rank 4 tensor into the lab frame, and therefore there are four applications of the DCM applied to each element of the response in the molecular frame, regardless of whether we write out those four $\mathbf{D}$ matrices or use a formalism where two of them get squared. We also note that, if $\mathbf{a}^{(3)}$ is first constructed from the tensor product of two $\mathbf{a}^{\mathrm{R}}$ matrices, then it is a pain to transform it into the laboratory frame, as each of the 81 elements in the lab frame consists of 81 contributions from 81 $a_{lmno}^{(3)}$ in the molecular frame, so there are $81 \times 81 = 6561$ operations to be performed. However, if $\mathbf{a}^{\mathrm{R}}$ is projected into the laboratory frame first, the handy operation

$$\mathbf{a}^{\mathrm{R}}(\theta,\phi,\psi) = \mathbf{D}(\theta,\phi,\psi)\,\mathbf{a}^{\mathrm{R}}\,\mathbf{D}(\theta,\phi,\psi)^{-1}$$

appearing inside Eq. 9.30 may be used in advance, and then the two laboratory frame $\mathbf{a}^{\mathrm{R}}(\theta,\phi,\psi)$ can be coupled to arrive at $\mathbf{a}^{(3)}(\theta,\phi,\psi)$. This notation is much more compact, and easier to deal with, so we will dispense with writing the molecular response as $\alpha^{(3)}$ and from here on use the Raman tensor $\mathbf{a}^{\mathrm{R}}$. However, we will still use the notation $\mathbf{A}^{(3)}$ for the laboratory-frame amplitude, primarily to distinguish it from the response function $\mathbf{A}$ from Chap. 8, and $\mathbf{A}^{(2)}$ that we will see in Chap. 10.

9.5 Experimental Geometries and Polarization Schemes

One of the interesting things about polarization analysis in Raman experiments is that there is the possibility to use two polarizers, one before the sample that controls the incident polarization state, and one after the sample that selects the polarization state of the scattered light targeted in the analysis. In the next section, we will see how this comes about, compared to what we have discussed in the previous chapter on IR absorption spectroscopy where only a single polarizer is used. But for now, we will take a look at a typical experimental geometry, using an isotropic sample such as a solution in a cuvette as an example. This also brings up an interesting point, in that Raman spectroscopy typically uses visible or near-IR light for the excitation, and therefore doesn't have the same sample type or path length limitation as IR absorption spectroscopy. The first configuration, shown in Fig. 9.5a, has the laser incident along $\hat{x}$ and polarized along $\hat{z}$. Although the Raman scattered light may be detected in all directions it is often preferred, when practical, to collect light at 90° since that provides some spatial separation of the weak Raman scattering from the overwhelming amount of light at the incident frequency. Filters and a monochromator/spectrograph are still required, but this certainly helps. In the depicted configuration (Fig. 9.5a), the detector is placed along the $\hat{y}$-direction, with a polarizer (analyser) in front of it so that the z-polarized component of the scattered Raman light is detected.

Porto notation provides a compact way of representing all of this information as $\hat{k}_{\text{in}}(\hat{E}_{\text{in}}\hat{E}_{\text{out}})\hat{k}_{\text{out}}$ where the unit vectors are typically written as uppercase letters, so the configuration in Fig. 9.5a is written as $X(ZZ)Y$. If the analyser is rotated by 90° as shown in Fig. 9.5b, the directions of the incident and detected wavevectors remain the same, but the polarization of the detected light changes so this is now $X(ZX)Y$.

Although not as ideal, there are an increasing number of Raman setups where backscattered light is collected, i.e. a dichroic mirror acts as a beam splitter since it transmits light at the incident wavelength towards the sample, but reflects light at Stokes-shifted wavelengths towards the detector. Such a configuration is depicted in Fig. 9.5c, and is represented by the Porto notation $X(ZZ)\bar{X}$ where $\bar{X}$ indicates the negative x-direction. When such beam splitters or mirrors are used in a setup, it is important to keep in mind that the pertinent directions are with respect to the sample, rather than the placement of the laser or polarizers on the optical table. Therefore, even though the light travels along $-y$ to reach the detector, we consider its wavevector to be along $-x$ with respect to the placement of the cuvette. Likewise, when the analyser is rotated by 90° to create the setup shown in Fig. 9.5d, the Porto notation is $X(ZY)\bar{X}$. That is to say, even though the analyser's transmission axis is oriented along x in the table coordinates, we are detected the y-polarized scattered Raman light from the sample.

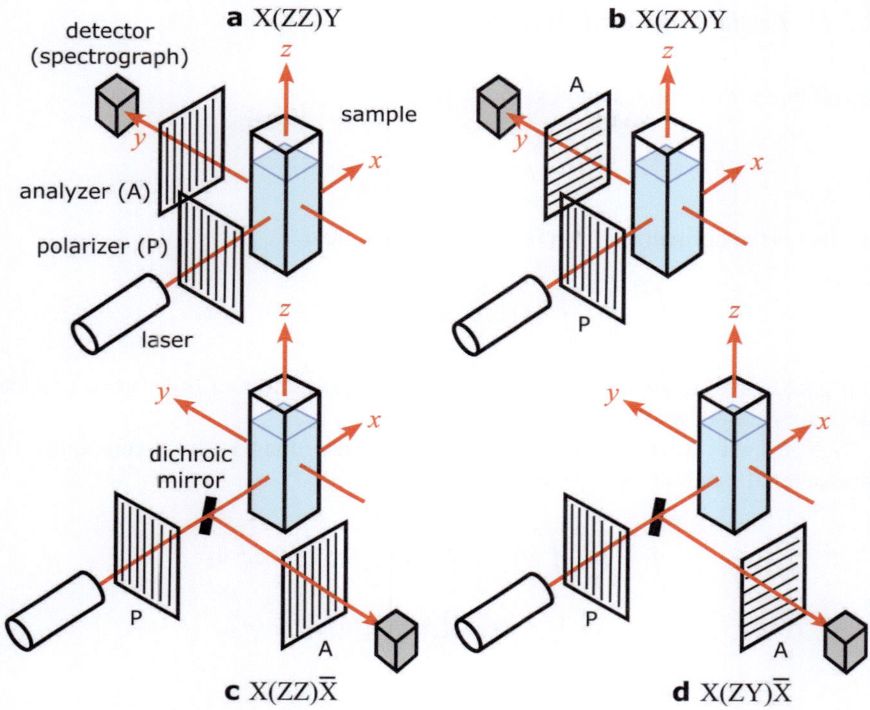

Fig. 9.5 Typical configurations for a Raman experiment, using an isotropic sample as an example. **a** A simple 90° scattering geometry where the polarizer and analyser are parallel to each other, Porto $X(ZZ)Y$. **b** The same configuration, but with the analyser rotated to create $X(ZX)Y$. **c** A back-scattering configuration that uses a beam splitter to measure $X(ZZ)\bar{X}$ and **d** $X(ZY)\bar{X}$

9.6 Extracting Molecular Information from Polarized Raman Data of Isotropic Samples

We will now follow the same treatment as was performed for the case of IR absorption spectroscopy in the previous chapter. We will again start by considering the case of an isotropic sample, and then a uniaxial distribution of molecules that is ordered with respect to their tilt angle. We will see that, in the case of Raman spectroscopy, both cases provide access to much more in-depth information, owing to the higher order response function ($\boldsymbol{\chi}^{(3)}$) that is responsible for generating the Raman signals.

9.6.1 Projections and Averages

Recall that for isotropic samples

$$f(\theta, \phi, \psi) = \frac{1}{8\pi^2}$$

and this is the normalized form of the function as then

$$\int_0^{2\pi}\int_0^{2\pi}\int_0^{\pi} f(\theta, \phi, \psi)\sin\theta \, \mathrm{d}\theta \, \mathrm{d}\phi \, \mathrm{d}\psi = 1.$$

Our goal is to evaluate the result of this uniform distribution of Euler angles on the laboratory frame $\mathbb{Im}\{\chi_{ij}^{(3)}\}$ elements.

Starting with the incident field linearly polarized along x, the x-component of the scattered radiation is given by

$$\begin{aligned}
A_{xxxx}^{(3)} &= \frac{N}{8\varepsilon_0\pi^2}\int_0^{2\pi}\int_0^{2\pi}\int_0^{\pi}\left[a_{xx}^{\mathrm{R}}(\theta, \phi, \psi)\right]^2 \sin\theta \, \mathrm{d}\theta \, \mathrm{d}\phi \, \mathrm{d}\psi \\
&= \frac{1}{5}\left[(a_{aa}^{\mathrm{R}})^2 + (a_{bb}^{\mathrm{R}})^2 + (a_{cc}^{\mathrm{R}})^2\right] + \frac{4}{15}\left[(a_{ab}^{\mathrm{R}})^2 + (a_{ac}^{\mathrm{R}})^2 + (a_{bc}^{\mathrm{R}})^2\right] \\
&\quad + \frac{2}{15}\left[a_{aa}^{\mathrm{R}}a_{bb}^{\mathrm{R}} + a_{aa}^{\mathrm{R}}a_{cc}^{\mathrm{R}} + a_{bb}^{\mathrm{R}}a_{cc}^{\mathrm{R}}\right] \\
&= A_{yyyy}^{(3)} = A_{zzzz}^{(3)}
\end{aligned} \tag{9.31}$$

where we notice that the same solution was obtained for $A_{yyyy}^{(3)}$ and $A_{zzzz}^{(3)}$ under isotropic symmetry. Such spectra would be measured with the polarizer and analyser parallel to each other, and the same result is obtained regardless of whether the fields are polarized along x-, y-, or z-directions.

If we now evaluate one of the so-called cross-polarized schemes, such as $A_{xyxy}^{(3)}$, we obtain

$$\begin{aligned}
A_{xyxy}^{(3)} &= \frac{N}{8\varepsilon_0\pi^2}\int_0^{2\pi}\int_0^{2\pi}\int_0^{\pi}\left[a_{xy}^{\mathrm{R}}(\theta, \phi, \psi)\right]^2 \sin\theta \, \mathrm{d}\theta \, \mathrm{d}\phi \, \mathrm{d}\psi \\
&= \frac{1}{15}\left[(a_{aa}^{\mathrm{R}})^2 + (a_{bb}^{\mathrm{R}})^2 + (a_{cc}^{\mathrm{R}})^2\right] + \frac{1}{5}\left[(a_{ab}^{\mathrm{R}})^2 + (a_{ac}^{\mathrm{R}})^2 + (a_{bc}^{\mathrm{R}})^2\right] \\
&\quad - \frac{1}{15}\left[a_{aa}^{\mathrm{R}}a_{bb}^{\mathrm{R}} + a_{aa}^{\mathrm{R}}a_{cc}^{\mathrm{R}} + a_{bb}^{\mathrm{R}}a_{cc}^{\mathrm{R}}\right] \\
&= A_{xzxz}^{(3)} = A_{yxyx}^{(3)} = A_{yzyz}^{(3)} = A_{zxzx}^{(3)} = A_{zyzy}^{(3)}
\end{aligned} \tag{9.32}$$

and we see that the same result is obtained for all the cross-polarized terms, i.e. $A_{ijij}^{(3)}$ where $i \neq j$.

In Raman (and other two-polarizer experiments such as fluorescence and second harmonic generation) for historical reasons the convention "in, out" is used in overall descriptions of the experiment, as we have seen in the Porto notation. However, keep in mind that for $\boldsymbol{\alpha}^{\mathrm{R}}$ and its numerator (outer product of transition dipole moments) $\mathbf{a}^{\mathrm{R}}$, we say that a^{R}_{lm} describes the induced dipole moment along $\hat{l}$ when a field is applied along $\hat{m}$ so this is "out, in". This means that an experiment to probe α^{R}_{yz} (one of the possibilities for $I_{\perp}$ shown above) would be described by Fig. 9.5d as an option, with Porto $X(ZY)\bar{X}$. In general, measurements of what we call I_{IJ} require a setup described by the general Porto notation $\_(JI)\_$ where the underscored blanks offer some flexibility in the choice of geometry. Although the Porto notation is clear, unfortunately there is no strong convention for labelling the intensity, with older texts and papers using $I_{\mathrm{in,out}}$ as well. With the exception of this Porto notation, we will choose intensity labels matching that of the tensor elements.

One can use Eqs. 9.31 and 9.32 to obtain what is known as the Raman depolarization ratio, ρ

$$\rho = \frac{I_{\perp}}{I_{\parallel}} = \frac{A^{(3)}_{xyxy}}{A^{(3)}_{xxxx}} = \frac{3\gamma^2}{45\epsilon^2 + 4\gamma^2} \tag{9.33}$$

where it is historical to present the molecular-frame tensor elements in terms of γ and ϵ (not to be confused with our other, more conventional, uses of these symbols) which are in turn given by

$$\begin{aligned}\gamma^2 = \frac{1}{2}\left(a^{\mathrm{R}}_{aa} - a^{\mathrm{R}}_{bb}\right)^2 + \frac{1}{2}\left(a^{\mathrm{R}}_{bb} - a^{\mathrm{R}}_{cc}\right)^2 + \frac{1}{2}\left(a^{\mathrm{R}}_{cc} - a^{\mathrm{R}}_{aa}\right)^2 \\ + 3\left[(a^{\mathrm{R}}_{ab})^2 + (a^{\mathrm{R}}_{bc})^2 + (a^{\mathrm{R}}_{ac})^2\right]\end{aligned} \tag{9.34}$$

and

$$\epsilon = \frac{1}{3}\left(a^{\mathrm{R}}_{aa} + a^{\mathrm{R}}_{bb} + a^{\mathrm{R}}_{cc}\right). \tag{9.35}$$

This result is very interesting to us for two reasons. The first is that it is fascinating that—for a random arrangement of molecules such as those in an aqueous solution—one obtains a different Raman spectrum depending on whether the polarizer and analyser are parallel or perpendicular to each other. (Compare this to our study of the isotropic IR absorption spectra, where all three polarizations resulted in the same IR spectra.) We can now understand that is simply a consequence of performing an isotropic average of a rank-4 tensor. The second interesting feature is that the Raman depolarization ratio may be used to obtain information on the polarizability derivatives (with respect to the normal mode coordinates) *in the molecular frame*. In the case of absorption-based spectroscopies, one would have to create a sample in which the molecules are collectively ordered to a uniform and high degree along the laboratory-frame axes to even come close to obtaining such information. But, in the case of Raman, this can be obtained directly, even in the case of (and especially in the case of) an isotropic distribution. This is yet another consequence of the manner in which elements of a rank-4 tensor are projected into the laboratory frame.

9.6.2 SymPy Example

Let's take a look at the SymPy code I used to determine these results. I'll describe the code in pieces, and then provide a complete listing at the end. We start by initializing our environment and defining the symbols we wish to use as special SymPy variables, i.e. the ones that we want to manipulate without assigning numerical values.

```
import sympy as sp
aa, ab, ac, bb, bc, cc = sp.symbols('aa ab ac bb bc cc',
    real=True) # dipole moment derivatives
theta, phi, psi = sp.symbols('theta phi psi', real=True)
```

Specifying the restriction on the variables as `real=True` isn't strictly necessary, but since we know that we will later ask SymPy to evaluate some integrals for us, I wanted to give it a head start in using any simplified forms of those integrals. To be honest, I normally start by defining variables without such restrictions, which is more general anyways, and later go back and add them if I find myself stuck, typically at the integration stage.

We are now ready to define the Raman tensor

```
alphaR = sp.Matrix([[aa, ab, ac], [ab, bb, bc], [ac, bc,
    cc]])
```

As a reminder, since all elements on the right-hand side have been defined as SymPy symbols, the object `alphaR` is already a SymPy symbol at the time of its definition.

We then define the direction cosine matrix

```
R1 = sp.Matrix([[sp.cos(psi), -sp.sin(psi), 0], [sp.sin(
    psi), sp.cos(psi), 0], [0, 0, 1]])
R2 = sp.Matrix([[sp.cos(theta), 0, sp.sin(theta)], [0, 1,
    0], [-sp.sin(theta), 0, sp.cos(theta)]])
R3 = sp.Matrix([[sp.cos(phi), -sp.sin(phi), 0], [sp.sin(
    phi), sp.cos(phi), 0], [0, 0, 1]])
DCM = R3*R2*R1
```

and carry out the coordinate frame transformation with

```
alphaRlab = DCM * alphaR * DCM.T
```

using the trick (which is actually just an identity) in Eq. 9.30. We have now taken $\mathbf{a}^{\mathrm{R}}$ in the molecular frame and transformed it to $\mathbf{a}^{\mathrm{R}}(\theta, \phi, \psi)$. Note that (unlike NumPy!) SymPy uses the $\star$ operator for matrix multiplication. As those operations are non-commutative in general, the order in which the matrices are multiplied is important. Finally, we scale up to the rank-4 quantity of interest that we have been writing as $\mathbf{a}^{(3)} = (\mathbf{a}^{\mathrm{R}})^2$. More formally, this is $\mathbf{a}^{(3)} = \mathbf{a}^{\mathrm{R}} \odot \mathbf{a}^{\mathrm{R}}$, where the $\odot$ symbol specifies that we want the element-wise multiplication of the two matrices (sometimes referred to as the Hadamard product). Keep in mind that there is some funny business going on with the contracted notation, but you know what we're talking about: this is technically the outer product $\mathbf{a}^{\mathrm{R}} \otimes \mathbf{a}^{\mathrm{R}}$, but we don't need all the extra indices that come along with scaling up the rank. Unfortunately the NumPy and SymPy notations diverge here as well, as NumPy uses * for this, but element-wise multiplication in SymPy is achieved using

```
a3 = alphaRlab.multiply_elementwise(alphaRlab)
```

but no problem; we now have everything we need.

We'll write a small function that performs the integrations to produce the isotropic average for a single element of interest

```
def isotropic(i, j):
    norm = 8*sp.pi**2
    chi3 = sp.integrate(a3[i,j] * sp.sin(theta), (theta,
        0, sp.pi), (phi, 0, 2*sp.pi), (psi, 0, 2*sp.pi))
    result = sp.simplify(chi3/norm)
    sp.pprint(result)
    return result
```

The output of this function is what I used to create Eq. 9.31 from `isotropic(0, 0)` and verified with `isotropic(1, 1)`, etc., and Eq. 9.32 from something like `isotropic(0, 1)` where arguments with off-diagonal elements of the Raman tensor are supplied. This is a classic situation where one needs to remember that list and array indexing in Python (including NumPy and SymPy) starts at zero, so when working with Cartesian coordinates, one needs to keep a mental note that x is 0, y is 1, and z is 2. Just to make sure that I'm computing the values corresponding to the elements of the tensor that I want, I sometimes define a list of strings `cartesian = ['x', 'y', 'z']` and then have my functions also print the elements of this list. That way, if I accidentally wrote `isotropic(1, 1)` when I actually should have specified `isotropic(0, 0)` to compute the xx element, I would hopefully be alerted when I see that something like `yy` was output. This is shown in the complete code listing at the bottom of this section. One could use this function to first verify that, in agreement with Eq. 9.31, all elements for which $i = j$ are the same, and that all elements for which $i \neq j$ are the same, as predicted by Eq. 9.32. After this, we can grab any representative element of those two sets and compute the depolarization ratio

```
para = isotropic(0,0)
perp = isotropic(0,1)
rho = perp/para
sp.pprint(sp.simplify(rho))
```

Even though I've asked SymPy to simplify the result before "pretty printing" it, it doesn't look recognizable as

$$\rho = \frac{\begin{array}{r}(\alpha^{\mathrm{R}}_{aa})^2 - \alpha^{\mathrm{R}}_{aa}\alpha^{\mathrm{R}}_{bb} - \alpha^{\mathrm{R}}_{aa}\alpha^{\mathrm{R}}_{cc} + 3(\alpha^{\mathrm{R}}_{ab})^2 + 3(\alpha^{\mathrm{R}}_{ac})^2 \\ +(\alpha^{\mathrm{R}}_{bb})^2 - \alpha^{\mathrm{R}}_{bb}\alpha^{\mathrm{R}}_{cc} + 3(\alpha^{\mathrm{R}}_{bc})^2 + (\alpha^{\mathrm{R}}_{cc})^2\end{array}}{\begin{array}{r}3(\alpha^{\mathrm{R}}_{aa})^2 + 2\alpha^{\mathrm{R}}_{aa}\alpha^{\mathrm{R}}_{bb} + 2\alpha^{\mathrm{R}}_{aa}\alpha^{\mathrm{R}}_{cc} + 4(\alpha^{\mathrm{R}}_{ab})^2 + 4(\alpha^{\mathrm{R}}_{ac})^2 + 3(\alpha^{\mathrm{R}}_{bb})^2 \\ +2\alpha^{\mathrm{R}}_{bb}\alpha^{\mathrm{R}}_{cc} + 4(\alpha^{\mathrm{R}}_{bc})^2 + 3(\alpha^{\mathrm{R}}_{cc})^2\end{array}}$$

when compared to Eq. 9.33. (By the way, I used help from `sp.latex` to generate this result.) There are a few ways that we could address this. One would be to figure out how to ask SymPy to make the substitutions using γ^2 (Eq. 9.34) and ϵ (Eq. 9.35). However, if our only objective is to verify whether the above result is equivalent to Eq. 9.33, save some sort of fancy rearrangement, there is a more straightforward way to achieve that. We could simply compare the two expressions by taking their ratio (ensuring that it is unity) or difference (ensuring it is zero). I typically opt for the difference, since it is less intensive to compute and simplify.

```
g2 = (aa-bb)**2/2 + (bb-cc)**2/2 + (cc-aa)**2/2 + 3*(ab**2
    + bc**2 + ac**2)
epsilon = (aa + bb + cc)/3
rhoLit = 3*g2/(45*epsilon**2 + 4*g2)
sp.pprint(sp.simplify(rho - rhoLit))
```

In ordinary Python we could check to see whether two variables held the same result by printing `rho == rhoLit` to see if the Boolean `True` was returned. This doesn't work in SymPy, because it stores the form of the expressions, rather than their evaluated result. Nevertheless, computing the difference works well. A complete listing of the code is now provided.

Isotropic depolarization ratio

```
import sympy as sp

# setup
aa, ab, ac, bb, bc, cc = sp.symbols('aa ab ac bb bc cc', real=True) #
    dipole moment derivatives
theta, phi, psi = sp.symbols('theta phi psi', real=True)
cartesian = ['x', 'y', 'z']

# create symmetric rank 2 tensor alpha - molecular frame
alphaR = sp.Matrix([[aa, ab, ac], [ab, bb, bc], [ac, bc, cc]])

# direction cosine matrix
R1 = sp.Matrix([[sp.cos(psi), -sp.sin(psi), 0], [sp.sin(psi), sp.cos(psi),
    0], [0, 0, 1]])
R2 = sp.Matrix([[sp.cos(theta), 0, sp.sin(theta)], [0, 1, 0], [-sp.sin(
    theta), 0, sp.cos(theta)]])
R3 = sp.Matrix([[sp.cos(phi), -sp.sin(phi), 0], [sp.sin(phi), sp.cos(phi),
    0], [0, 0, 1]])
DCM = R3*R2*R1

# project alpha into lab frame
alphaRlab = DCM * alphaR * DCM.T

# create the rank 4 tensor in the lab frame (wow, cool)
# alpha3Lab = sp.tensorproduct(alphaLab, alphaLab)
a3 = alphaRlab.multiply_elementwise(alphaRlab)

def isotropic(i, j):
    print('%s%s' % (cartesian[i], cartesian[j]))
```

```
        norm = 8*sp.pi**2
        chi3 = sp.integrate(a3[i,j] * sp.sin(theta), (theta, 0, sp.pi), (phi,
             0, 2*sp.pi), (psi, 0, 2*sp.pi))
        result = sp.simplify(chi3/norm)
        sp.pprint(result)
        return result

    # depolarization ratio based on these projections
    para = isotropic(0,0)
    perp = isotropic(0,1)
    rho = perp/para

    # established (literature) depolarization ratio
    g2 = (aa-bb)**2/2 + (bb-cc)**2/2 + (cc-aa)**2/2 + 3*(ab**2 + bc**2 + ac**2)
    epsilon = (aa + bb + cc)/3
    rhoLit = 3*g2/(45*epsilon**2 + 4*g2)

    # comparison
    sp.pprint(sp.simplify(rho - rhoLit))
```

9.7 Modelling Raman Spectra

We have now introduced enough concepts that we can examine the code that was used to generate the heptane Raman spectrum that was shown in Fig. 9.2. This code contains several extra components that we didn't need to discuss in our model examples, as we are now wanting to simulate the Raman spectrum of an actual molecule, heptane. The nine elements of **a** for each vibrational mode come from density functional theory (DFT) calculations. Those have been performed separately, and the values are read from a `pickle` that creates a `dipoleProperty` object where the elements a^{R}_{lm} are stored.

```
import numpy as np
from matplotlib import pyplot
import pickle

class dipoleProperty:
    def _init_(self):
        pass

def chi3(modes):
    spectrumPara = 0.
    spectrumPerp = 0.

    for m in range(len(modes)):
        info = modes[m]
```

```
        a=np.matrix([[info.alphas[0], info.alphas[1], info
            .alphas[2]], [info.alphas[3], info.alphas[4],
            info.alphas[5]],[info.alphas[6], info.alphas
            [7], info.alphas[8]]])

        ampPara = 0.2*(a[0,0]**2 + a[1,1]**2 + a[2,2]**2)
            + 4./15.*(a[0,1]**2 + a[0,2]**2 + a[1,2]**2) +
             2./15.*(a[0,0]*a[1,1] + a[0,0]*a[2,2] + a
            [1,1]*a[2,2])
        ampPerp = 1./15*(a[0,0]**2 + a[1,1]**2 + a[2,2]**2
             - a[0,0]*a[1,1] - a[0,0]*a[2,2] - a[1,1]*a
            [2,2]) + 0.2*(a[0,1]**2 + a[0,2]**2 + a
            [1,2]**2)

        prefactor = 1./(2. * info.reducedmass * f * info.
            freq)
        spectrumPara += prefactor * ampPara / ((info.freq*
            f)**2 - probe**2 - 1j*probe*Gamma)
        spectrumPerp += prefactor * ampPerp / ((info.freq*
            f)**2 - probe**2 - 1j*probe*Gamma)
    return (spectrumPara.imag, spectrumPerp.imag)

# settings
hndl = open('heptane_prop.pkl', 'rb')
modes = pickle.load(hndl, encoding='latin1')
hndl.close()
probe = np.linspace(0, 800, 2000)
f = 0.963
Gamma = 9.
pyplot.rc('font', size=8)
pyplot.rc('figure', figsize=(3.25, 5))
fig = pyplot.figure()
ax1 = fig.add_subplot(211)
ax2 = fig.add_subplot(212)
ax2b = ax2.twiny()
wl_ex = 532.
wn_ex = 1.0e7/532.
T = 298.
h = 6.626e-34
c = 2.99e8
k = 1.38e-23

# generate data
antiStokesFreq = wn_ex + probe
StokesFreq = wn_ex - probe
```

```
factor = np.exp(h*c*probe*100/(k*T))
Ipara, Iperp = chi3(modes)
antiStokesIntensityPara = wn_ex * antiStokesFreq**3 *
    Ipara
antiStokesIntensityPerp = wn_ex * antiStokesFreq**3 *
    Iperp
StokesIntensityPara = wn_ex * StokesFreq[::-1]**3 * Ipara
    * factor
StokesIntensityPerp = wn_ex * StokesFreq[::-1]**3 * Iperp
    * factor
freq = np.concatenate((StokesFreq[::-1], antiStokesFreq))
intensityPara = np.concatenate((StokesIntensityPara[::-1],
     antiStokesIntensityPara))/1.0e10
intensityPerp = np.concatenate((StokesIntensityPerp[::-1],
     antiStokesIntensityPerp))/1.0e10
rayleigh = 2.0e3/((18797 - freq)**2 + 10.)
intensityPara += rayleigh
intensityPerp += rayleigh

# frequency shift
ax2.plot(freq-wn_ex, intensityPara, label='$I_\parallel$')
ax2.plot(freq-wn_ex, intensityPerp, label='$I_\perp$')
ax2.set_xlabel(r'frequency shift, $\Delta \omega$ / cm$
    ^{-1}$')
ax2.set_ylim(0, 1)
ax2.axvline(0, color='k', ls='--', dashes=[2,2])
for ax in [ax2, ax2b]:
    ax.set_xlim(np.min(freq-wn_ex), np.max(freq-wn_ex))
labels = ax2.get_xticks().tolist()
ax2.set_ylabel('intensity / a.u.')
ax2.legend(loc='upper right')

# frequency
ax2b.set_xlabel(r'frequency, $\omega$ / cm$^{-1}$')
ax2.text(0.06, 0.88, 'b', fontweight='bold', transform=ax2
    .transAxes)
ax2b.set_xticklabels((np.array(labels)+wn_ex+0.5).astype(
    int), fontsize=7)

# wavelength
wl = 1.0e7/freq
ax1.plot(wl, intensityPara, label='$I_\parallel$')
ax1.plot(wl, intensityPerp, label='$I_\perp$')
ax1.set_xticks([512, 522, 532, 542, 552])
ax1.set_xlabel('wavelength, $\lambda$ / nm')
```

```
ax1.set_ylabel('intensity / a.u.')
ax1.set_xlim(np.min(wl), np.max(wl))
ax1.set_ylim(0, 1)
ax1.axvline(532, color='k', ls='--', dashes=[2,2])
ax1.text(0.06, 0.88, 'a', fontweight='bold', transform=ax1
    .transAxes)
ax1.legend(loc='upper right')

fig.set_tight_layout(True)
pyplot.savefig('RamanSpectrum.png', dpi=600)
```

Although the specific orientation average performed in this code corresponds to an isotropic distribution of molecules, the same overall structure could be used with a different ODF to model Raman spectra corresponding to an ordered material. The only difference is that multiple Raman spectra would be displayed corresponding to the additional polarization combination(s) of interest. Also note that this code uses Eq. 9.1 as a scaling factor between the Stokes and anti-Stokes Raman lines; an empirical lineshape for the Rayleigh peak has been used.

9.8 Extracting Structural Information from Polarized Raman Data of Ordered Samples

We will now move on to consider the next simplest orientation distribution, where ϕ and ψ are uniformly distributed, but we make no assumptions about the shape of the distribution in θ. Incidentally one can experimentally rule out the possibility that θ is also uniformly distributed by performing Raman measurements with different polarizer and analyser combinations and comparing the results with what was derived for an isotropic sample above.

In order to keep the math a bit simpler, and mainly to make the results of our expression more compact, we will make drastic assumption that in the molecular frame

$$\mathbf{a}^{\mathrm{R}} = \begin{bmatrix} 0 & 0 & 0 \\ 0 & 0 & 0 \\ 0 & 0 & a^{\mathrm{R}}_{cc} \end{bmatrix}. \tag{9.36}$$

Since we consider the azimuthal (ϕ) and twist (ψ) angles to be uniformly distributed, we can pre-integrate just those two angles

$$A_{ijij}(\theta) = \frac{N}{4\varepsilon_0 \pi^2} \int_0^{2\pi} \int_0^{2\pi} \left[a^{\mathrm{R}}_{ij}(\theta, \phi, \psi)\right]^2 \, \mathrm{d}\phi \, \mathrm{d}\psi. \tag{9.37}$$

Using our simplified form $\mathbf{a}^{\mathrm{R}}$, we perform the above integrations to determine the tilt angle dependence of all elements $A^{(3)}_{ijij}$

$$
\begin{aligned}
A^{(3)}_{xyxy}(\theta) = A^{(3)}_{yxyx} &= \frac{N}{8\varepsilon_0}(a^{\mathrm{R}}_{cc})^2[1 - 2\langle\cos^2\theta\rangle + \langle\cos^4\theta\rangle] \\
A^{(3)}_{xzxz}(\theta) = A^{(3)}_{zxzx} = A^{(3)}_{yzyz} = A^{(3)}_{zyzy} &= \frac{N}{2\varepsilon_0}(a^{\mathrm{R}}_{cc})^2[\langle\cos^2\theta\rangle - \langle\cos^4\theta\rangle] \\
A^{(3)}_{zzzz}(\theta) &= \frac{N}{\varepsilon_0}(a^{\mathrm{R}}_{cc})^2\langle\cos^4\theta\rangle \\
A^{(3)}_{xxxx}(\theta) = A^{(3)}_{yyyy} &= 3\,A^{(3)}_{xyxy}
\end{aligned}
\tag{9.38}
$$

noting that there are 10 non-zero elements of $\boldsymbol{A}^{(3)}$, of which three are unique. Before we continue, we will make a few notes about our solution and the two assumptions that went into these expressions. First, setting all elements of $\mathbf{a}^{\mathrm{R}}$ except for a^{R}_{cc} to zero is a bit extreme, and frankly unphysical. It is much more common to at least consider non-zero values for $a^{\mathrm{R}}_{aa} = a^{\mathrm{R}}_{bb}$ as well. That isn't a concern for us here, as we are not trying to compare these particular results with what has been measured in an actual experimental data set. I reiterate that our objective is how to highlight the information content in Raman data, and this simple example will be sufficient to do that. If you are working up your data, just go ahead and make $\mathbf{a}^{\mathrm{R}}$ as featured as you require for the vibrational mode(s) you are studying. It will make no difference to NumPy and SymPy whether the resulting expressions are a bit messier! Finally, we have not specified the form of the orientation distribution function $f(\theta)$, and so we write the above quantities in terms of the average values $\langle\cos^2\theta\rangle$ and $\langle\cos^4\theta\rangle$. Note that, unlike IR absorption where we had access to only $\langle\cos^2\theta\rangle$, we already see that Raman data is more sensitive to features of the orientation distribution.

We can now see that the objective is to perform polarized Raman measurements and collect spectra where the intensity of the vibrational model of interest (proportional to $\mathbb{I}\mathrm{m}\{\chi^{(3)}\}$) is obtained for

- one of $A^{(3)}_{xxxx} = 3A^{(3)}_{xyxy} = A^{(3)}_{yyyy}$
- one of $A^{(3)}_{xzxz} = A^{(3)}_{yzyz}$
- $A^{(3)}_{zzzz}$

noting the amplitudes $A^{(3)}_{ijij}$ we have been discussing come from fitting the vibrational resonances using a lineshape such as a Lorentzian or a Gaussian, or something else that is appropriate. Either way, from these three amplitudes we can form two ratios

$$
\begin{aligned}
R_1 &= \frac{A^{(3)}_{xxxx}}{A^{(3)}_{zzzz}} = \frac{3}{8}\frac{\langle\sin^4\theta\rangle}{\langle\cos^4\theta\rangle} = \frac{3}{8}\left(\frac{\langle\cos^4\theta\rangle - 2\langle\cos^2\theta\rangle + 1}{\langle\cos^4\theta\rangle}\right) \\
R_2 &= \frac{A^{(3)}_{yzyz}}{A^{(3)}_{zzzz}} = \frac{1}{2}\frac{\langle\sin^2\theta\rangle}{\langle\cos^2\theta\rangle} = \frac{1 - \langle\cos^2\theta\rangle}{2\langle\cos^2\theta\rangle}.
\end{aligned}
\tag{9.39}
$$

Being able to perform analysis using such ratios is especially important in the case of Raman spectroscopy (and SFG spectroscopy, as we will see in Chap. 10) because, unlike the case of IR absorption where the absorbance is defined based on a ratio of two intensities, the intensity of the Raman scattered light is typically expressed in arbitrary units. It is of course possible to calibrate the detector response so that SI units of intensity are obtained, but this does not simplify the situation as we typically do not know the $\mathbf{a}^{\mathrm{R}}$ tensor elements in absolute units, in addition to other experimental uncertainties. Peak fitting, for example, requires a lineshape such as a Lorentzian with a good fit value of the linewidth Γ in Eq. 9.16. For an individual band, any error in assessing Γ therefore amounts to an error in the sought element(s) for $\mathbf{a}^{\mathrm{R}}$. However, when ratios of band intensities between different polarizations for the same vibrational mode are used, the error is minimized as the same Γ (and ω_{vib}) value is used for the same vibrational mode.

Isotropic interlude

Before we continue to study an ordered sample, it is instructive to consider the values of the above two ratios in the case of an isotropic distribution. For R_1 we have

$$R_{1,\mathrm{iso}} = \frac{3}{8}\frac{\int_0^{\pi} \sin^4\theta \sin\theta \mathrm{d}\theta}{\int_0^{\pi} \cos^4\theta \sin\theta \mathrm{d}\theta} = \frac{3}{8}\frac{16/15}{2/5} = 1 \tag{9.40}$$

which is a welcome conclusion, since we expect $\langle A_{yy}^2\rangle = \langle A_{zz}^2\rangle$ for an isotropic sample. We now consider the second ratio

$$R_{2,\mathrm{iso}} = \frac{1}{2}\frac{\int_0^{\pi} \sin^2\theta \sin\theta \mathrm{d}\theta}{\int_0^{\pi} \cos^2\theta \sin\theta \mathrm{d}\theta} = \frac{1}{2}\frac{4/15}{2/5} = \frac{1}{3}. \tag{9.41}$$

That is interesting, but what further conclusion can we draw from this, since we already anticipate $\langle A_{ii}^2\rangle \neq \langle A_{ij}^2\rangle$ for $i \neq j$? We can turn to the expression for the depolarization ratio

$$\rho = \frac{3\gamma^2}{45a^2 + 4\gamma^2}$$

which, in our simplified form of $\mathbf{a}^{\mathrm{R}}$, has $\epsilon = \frac{1}{3}a_{cc}^{\mathrm{R}}$ and $\gamma^2 = (a_{cc}^{\mathrm{R}})^2$, resulting in

$$\rho_{\mathrm{iso}} = \frac{3(a_{cc}^{\mathrm{R}})^2}{\frac{45}{9}(a_{cc}^{\mathrm{R}})^2 + 4(a_{cc}^{\mathrm{R}})^2} = \frac{1}{3} \tag{9.42}$$

as we would expect.

To continue with our effort to comment on the orientation distribution of molecules, let us first review the three options we had in the case of IR absorption spectroscopy, where our polarized measurements resulted in $\langle \cos^2\theta \rangle$, or the same quantity expressed as the order parameter $\langle P_2 \rangle = \frac{3}{2}\langle \cos^2\theta \rangle - \frac{1}{2}$.

9.8.1 Assuming a Narrow Distribution of Tilt Angles

The simplest option is to assume that $\langle \cos^2\theta \rangle = \cos^2\theta_0$ and $\langle \cos^4\theta \rangle = \cos^4\theta_0$. In other words, we assume that all of the molecules are perfectly ordered at $\theta = \theta_0$. Such an orientation distribution is described by the Dirac δ-function $f(\theta) = \delta(\theta - \theta_0)$. Compared to the situation one faces with IR absorption data, this is now more constrained in that we can evaluate the result of a δ-distribution centred at θ_0 using either R_1 or R_2. From Eq. 9.39 we can see that a narrow distribution where $f(\theta) = \delta(\theta - \theta_0)$ is characterized by a particular relationship between the two ratios of band intensities

$$R_1 = \frac{3}{2}R_2^2. \tag{9.43}$$

9.8.2 Reporting Order Parameters Directly Accessed from the Raman Experiment

Some trigonometric identities have been used to express Eq. 9.39 in terms of $\langle \cos^2\theta \rangle$ and $\langle \cos^4\theta \rangle$. Those expressions can then be rearranged to obtain

$$\begin{aligned} \langle \cos^2\theta \rangle &= \frac{1}{2R_2 + 1} \\ \langle \cos^4\theta \rangle &= \frac{3(2R_2 - 1)}{(8R_1 - 3)(2R_2 + 1)}. \end{aligned} \tag{9.44}$$

We can leave it like that or further manipulate these to report the order parameters as

$$\begin{aligned} \langle P_2 \rangle &= \frac{3}{2}\langle \cos^2\theta \rangle - \frac{1}{2} \\ &= \frac{1 - R_2}{2R_2 + 1} \end{aligned} \tag{9.45}$$

and

$$\begin{aligned} \langle P_4 \rangle &= \frac{1}{8}\left(35\langle \cos^4\theta \rangle - 30\langle \cos^2\theta \rangle + 3\right) \\ &= \frac{3(2R_1R_2 - 9R_1 + 8R_2 - 1)}{(8R_1 - 3)(2R_2 + 1)}. \end{aligned} \tag{9.46}$$

This is primarily for orientation distribution purists, for whom the story is fine to end there. In fact, you will find accounts in the literature where the conclusion is to report values of these order parameters [24–29]. However this does not provide a satisfying molecular picture on its own, as we can't easily picture the distribution function from such numbers. We will next explore two options for moving forward.

9.8.3 Using Raman Data to Solve for Parameters in an Orientation Distribution Function

We can assume any shape for $f(\theta)$, and use the two ratios R_1 and R_2 to inform on any two of the parameters in our orientation distribution function. For example, if we choose a Gaussian distribution (Eq. 8.45) that is parameterized by a mean angle θ_0 and a width σ

$$f(\theta) = c_\mathrm{N} \exp\left[-\frac{(\theta-\theta_0)^2}{2\sigma^2}\right]$$

both θ_0 and σ can be obtained from knowledge of the experimentally determined $\langle P_2\rangle$ and $\langle P_4\rangle$. This is a significant advantage compared to what can be done with data from a polarized IR absorption experiment.

A useful and generalizable approach is to consider that the average value of any function can be readily determined from Eq. 8.62, and therefore we can arrive at $\langle P_2\rangle$ and $\langle P_4\rangle$ of our Gaussian function, or any chosen ODF. To be explicit,

$$\begin{aligned}\langle P_2\rangle_\mathrm{model} &= \int_0^\pi f(\theta) P_2(\cos\theta)\sin\theta\, \mathrm{d}\theta \\ &= \frac{\displaystyle\int_0^\pi \exp\left[-\frac{(\theta-\theta_0)^2}{2\sigma^2}\right]\cdot\left(\frac{3}{2}\cos^2\theta - \frac{1}{2}\right)\sin\theta\, \mathrm{d}\theta}{\displaystyle\int_0^\pi \exp\left[-\frac{(\theta-\theta_0)^2}{2\sigma^2}\right]\sin\theta\, \mathrm{d}\theta}\end{aligned} \qquad (9.47)$$

and

$$\begin{aligned}\langle P_4\rangle_\mathrm{model} &= \int_0^\pi f(\theta) P_4(\cos\theta)\sin\theta\, \mathrm{d}\theta \\ &= \frac{\displaystyle\int_0^\pi \exp\left[-\frac{(\theta-\theta_0)^2}{2\sigma^2}\right]\cdot\left(\frac{35}{8}\cos^4\theta - \frac{30}{8}\cos^2\theta + \frac{3}{8}\right)\sin\theta\, \mathrm{d}\theta}{\displaystyle\int_0^\pi \exp\left[-\frac{(\theta-\theta_0)^2}{2\sigma^2}\right]\sin\theta\, \mathrm{d}\theta}\end{aligned} \qquad (9.48)$$

where the denominators are in place for the evaluation of the normalization constant c_N in the ODF. One can establish best fit values of θ_0 and σ by minimizing the error function

$$F(\theta_0, \sigma) = \left(\langle P_2\rangle_{\text{model}} - \langle P_2\rangle_{\text{expt}}\right)^2 + \left(\langle P_4\rangle_{\text{model}} - \langle P_4\rangle_{\text{expt}}\right)^2 \tag{9.49}$$

based on the sum of the squares of the residuals.

Although this is a powerful approach, it has the inherent disadvantage that the form of the distribution function (whose parameters obtained from fits to the Raman data) may not be a reasonable representation of the distribution in the actual sample. There are some steps that can be taken to avoid this risk. For example, if the system is amenable to molecular dynamics simulations, the results of the simulations can be used to get a sense of the approximate shape of the orientation distribution. The actual distribution can then be obtained by solving for the parameters in a more appropriately shaped function. More on this approach will be described in Chap. 11. Then there is the possibility that some physical or chemical intuition can be used to propose something reasonable, either based on the natural distribution of molecules in the material, or the distribution that was induced by some sort of physical perturbation such as rubbing or stretching. In some cases, however, no such insight is available. The next approach seeks to mitigate the risks associated with incorrect assumptions in the choice of a distribution function.

9.8.4 Constructing the Most Probable Orientation Distribution

We can extend the previous method to be more flexible, constructing the orientation distribution function from $\langle P_2\rangle$ and $\langle P_4\rangle$ directly. This wasn't really a viable option when only IR data is available; we'll see that the additional information of $\langle P_4\rangle$ makes all the difference here. A completely general form of the orientation distribution can be constructed from a linear combination of Legendre polynomials (Eq. 8.59), restating the ODF that was first presented in Eq. 8.58

$$f(\theta) = \sum_{n=0}^{\infty} \frac{2n+1}{2} \langle P_n\rangle P_n(\cos\theta)$$

where the notation $P_n(\cos\theta)$ specifies that we are evaluating the Legendre polynomials as a function of $\cos\theta$. However we have to reconcile the fact that, even though we are excited about our Raman data, we do still have only two of the terms in this infinite series. On the one hand, the leading terms are more important than those for large values of n. However, even if we consider the terms up to $n = 4$, the Raman data have not provided any clue about $\langle P_1\rangle$ and $\langle P_3\rangle$. For some types of samples, it may be reasonable to assume that only even order parameters have non-zero values. If we consider a stretched polymer film as an example, molecules may be highly oriented along the stretching axis, but would not have a preference for their alignment parallel or anti-parallel to this direction. Looking at the forms of the Legendre polynomials plotted in Fig. 8.19, it is evident that the odd $P_n(\cos\theta)$ functions provide such asymmetry to the tilt angle distribution, when desired. However, even if we can safely assume that only even-valued $\langle P_n \rangle$ are present, we face the next problem: truncating the series at any point before $n = \infty$ may result in an ODF that is unphysical. For example, it may contain negative values, which is not allowed if it is to represent a probably density describing the molecular orientations. One solution is to fix values of $\langle P_n \rangle$ for all n provided by experiments ($n = 2$ and $n = 4$ in the case of Raman data) and use a concept from information theory [30] to arrive at an orientation distribution [27, 31–34].

In general, if p_i is the probability of observing any quantity i, then the usefulness of such an observation is inversely proportional to p_i. That is to say that rare events are of greater value than common events. Mathematically, it is better to define this using $\ln(p_i)^{-1} = -\ln p_i$ since that quantity approaches zero when $p_i \to 1$. If there are a large number N_m of possible measurements, then the expected number of occurrences of quantity i is $N_m p_i$. One can define the information entropy S as the information content per measurement, when the occurrences are averaged over m measurements.

$$
\begin{aligned}
S &= -\frac{1}{N_m} \sum_{i=1}^{m} N_m p_i \ln p_i \\
&= -\sum_{i=1}^{m} p_i \ln p_i \\
&= -\int p(x) \ln[p(x)]\, \mathrm{d}x.
\end{aligned}
\tag{9.50}
$$

In the last line we have replaced the sum with an integral to describe a continuous distribution of measurements, rather than several discrete measurements.

The information entropy associated with a particular orientation distribution function $f(\theta)$ is given by

$$
S[f(\theta)] = -\int_0^{\pi} f(\theta)\, \ln[f(\theta)]\, \sin(\theta)\, \mathrm{d}\theta. \tag{9.51}
$$

Therefore, our goal is to find a function $f(\theta)$ that maximizes the information entropy S. That will describe the most probable (MP) orientation distribution, so we call it $f_{\mathrm{MP}}(\theta)$. This function is subject to three constraints:

1. We require that $f_{\mathrm{MP}}(\theta) \geq 0$ for $0° \leq \theta \leq 180°$. That is, $f_{\mathrm{MP}}(\theta)$ must have only positive values over the entire range of the tilt angles. This is because we interpret $f_{\mathrm{MP}}(\theta)$ as the probability of finding a molecule oriented between θ and $\theta + \mathrm{d}\theta$.
2. We require that $\int_0^{\pi} f_{\mathrm{MP}}(\theta)\, \sin\theta = 1$, so that there is a 100% probably of finding the molecules over the complete range of angles. This is easily achieved through normalization.
3. $\int_0^{\pi} f_{\mathrm{MP}}(\theta)\, P_n(\cos\theta)\, \sin\theta = \langle P_n \rangle_{\mathrm{expt}}$ for all measured values of $\langle P_n \rangle$. So for a Raman experiment, this last point translates to two separate constraints, one for $\langle P_2 \rangle$ and one for $\langle P_4 \rangle$.

Incidentally, if we can determine such a $f_{\mathrm{MP}}(\theta)$, we have also determined all of the most probable values of $\langle P_n \rangle$, up to $n = \infty$ by way of Eq. 8.62. We now need a way of constructing $f_{\mathrm{MP}}(\theta)$ subject to the above constraints. A common method for doing this involves the use of Lagrange multipliers [35]. The basic idea is that if S has a maximum value that satisfies the constraints, then the gradient of S can be written as a linear combination of the gradients of the constraints [35]. This can be written as

$$\begin{aligned}\delta\{S[f(\theta)]\} &= \lambda_0 \delta\left(\int_0^{\pi} f(\theta)\, \sin\theta\, \mathrm{d}\theta - 1\right) \\ &\quad + \sum_n^{n\ \mathrm{expt}} \lambda_n \delta\left(\int_0^{\pi} f(\theta)\, P_n(\cos\theta)\, \sin\theta\, \mathrm{d}\theta - \langle P_n \rangle\right) \\ &= \delta\left(\lambda_0 \int_0^{\pi} f(\theta)\, \sin\theta\, \mathrm{d}\theta + \sum_n^{n\ \mathrm{expt}} \lambda_n \langle P_n \rangle\right)\end{aligned} \tag{9.52}$$

which is equivalent to

$$\begin{aligned}0 &= \delta \int_0^{\pi} \left(f(\theta) \ln[f(\theta)] + \lambda_0 f(\theta) + \sum_n^{n\ \mathrm{expt}} \lambda_n f(\theta) P_n(\cos\theta)\right) \sin\theta\, \mathrm{d}\theta \\ &= \int_0^{\pi} \left(\delta f(\theta) \ln[f(\theta)] + \lambda_0 \delta f(\theta) + \sum_n^{n\ \mathrm{expt}} \lambda_n P_n(\cos\theta) \delta f(\theta)\right) \sin\theta\, \mathrm{d}\theta \\ &= \int_0^{\pi} \left(\ln[f(\theta)] + 1 + \lambda_0 + \sum_n^{n\ \mathrm{expt}} \lambda_n P_n(\cos\theta)\right) \sin\theta\, \mathrm{d}\theta\, \delta f(\theta).\end{aligned}$$

This can be true if and only if

$$\ln[f(\theta)] = -\lambda_0 - 1 - \sum_n^{n\ \mathrm{expt}} \lambda_n P_n(\cos\theta). \tag{9.53}$$

Therefore we arrive at the solution for the most probable orientation distribution by taking the natural logarithm of both sides of Eq. 9.53 to obtain

$$\begin{aligned}f_{\mathrm{MP}}(\theta) &= \exp\left[-\lambda_0 - 1 - \sum_n^{n\ \mathrm{expt}} \lambda_n P_n(\cos\theta)\right] \\ &= \frac{\exp\left[-\sum_n^{n\ \mathrm{expt}} \lambda_n P_n(\cos\theta)\right]}{\mathrm{e}^{\lambda_0+1}}\end{aligned} \tag{9.54}$$

where the denominator is in place to satisfy the normalization constraint, and λ_n are chosen in accordance with the experimentally determined $\langle P_n \rangle$, i.e. to satisfy constraint number 3 on our list. Note that, since the denominator of Eq. 9.54 does not depend on θ (it is a constant), we can evaluate it in our usual way by ensuring that

$$\int_0^{\pi} f_{\mathrm{MP}}(\theta)\ \sin(\theta)\,\mathrm{d}\theta = 1$$

as is implied by the second constraint. We will now work through a practical example where bands in polarized Raman spectra are fit to obtain the amplitudes $A^{(3)}_{ijij}$, whose ratios R_1 and R_2 are used to arrive at $\langle P_2\rangle$ and $\langle P_4\rangle$, which are ultimately used to construct $f_{\mathrm{MP}}(\theta)$.

9.8.5 Example

As an example, imagine that a polarized Raman experiment has been performed, and a vibrational mode has been isolated for which the Raman tensor is dominated by α^{R}_{cc} so we can ignore its other eight elements. Spectra acquired in three different beam polarizations were collected, and fit to a Lorentzian bandshape to isolate the $A^{(3)}_{xxxx}$, $A^{(3)}_{xzxz}$, and $A^{(3)}_{zzzz}$ amplitudes. Those can then be used to produce two ratios, with values $R_1 = 1.408$ and $R_2 = 1.134$.

If we assume that all of the molecules are aligned with the same tilt angle θ_0, and therefore approximate the distribution with $f(\theta) = \delta(\theta - \theta_0)$, the use of R_1 results in $\theta_0 = 54.3°$. However, when using R_2, a different value of $\theta_0 = 56.4°$ is obtained by the same method. Although these values are not so different, care must be taken as a result of the sensitivity of the ratios to the condition that results in a truly narrow distribution as stated in Eq. 9.43.

We can now follow the steps outlined above to get further information about the molecular orientation by using the reported ratios of $R_1 = A^{(3)}_{xxxx}/A^{(3)}_{zzzz}$ and $R_1 = A^{(3)}_{xzxz}/A^{(3)}_{zzzz}$ to obtain $\langle P_2\rangle = -0.0410$ and $\langle P_4\rangle = -0.1565$ from the use of Eqs. 9.45 and 9.46. We now wish to construct the most probable orientation distribution given by

$$f_{\mathrm{MP}}(\theta) = \frac{\exp[\lambda_2 P_2(\cos\theta) + \lambda_4 P_4(\cos\theta)]}{\displaystyle\int_0^{\pi} \exp[\lambda_2 P_2(\cos\theta) + \lambda_4 P_4(\cos\theta)]\sin\theta\mathrm{d}\theta} \tag{9.55}$$

so the problem has been reduced to finding the Lagrange multipliers λ_2 and λ_4 for which

$$\begin{aligned}\int_0^{\pi} f_{\mathrm{MP}}(\theta) P_2(\cos\theta)\sin\theta\mathrm{d}\theta &= \langle P_2\rangle_{\mathrm{expt}}\\ \int_0^{\pi} f_{\mathrm{MP}}(\theta) P_4(\cos\theta)\sin\theta\mathrm{d}\theta &= \langle P_4\rangle_{\mathrm{expt}}.\end{aligned} \tag{9.56}$$

Table 9.1 First 20 values of the order parameters $\langle P_n \rangle$ determined for the most probable orientation distribution $f_{MP}(\theta)$

n	$\langle P_n \rangle$	Comments
0	1.0000	Necessarily true, as $\langle P_0 \rangle$ is independent of θ
1	0.0000	From $f_{MP}(\theta)$
2	−0.0410	From experimental Raman data
3	0.0000	From $f_{MP}(\theta)$
4	−0.1562	From experimental Raman data
5	0.0000	From $f_{MP}(\theta)$
6	0.0379	⋮
7	0.0000	
8	0.0226	
9	0.0000	
10	−0.0088	
11	0.0000	
12	−0.0023	
13	0.0000	
14	0.0013	
15	0.0000	
16	0.0002	
17	0.0000	
18	−0.0001	
19	0.0000	
20	0.0000	
⋮	⋮	
∞		

The search for these parameters is shown in Fig. 9.6a, in which we locate $\lambda_2 = -0.661$ and $\lambda_4 = -1.957$. Some of the order parameters $\langle P_n \rangle$ for this function are shown in Table 9.1. These were obtained by taking Eq. 9.55 and determining

$$\langle P_n(\cos\theta) \rangle = \frac{\int_0^\pi f_{MP}(\theta)\; P_n(\cos\theta)\; \sin\theta\; d\theta}{\int_0^\pi f_{MP}(\theta)\; \sin\theta\; d\theta}$$

for any desired value of n, as prescribed by Eq. 8.62. It is of particular interest to note that the values of $\langle P_2 \rangle$ and $\langle P_4 \rangle$ match those measured in the experiment; this is due to the constraint we have applied in determining $f_{MP}(\theta)$ that has required this to be true.

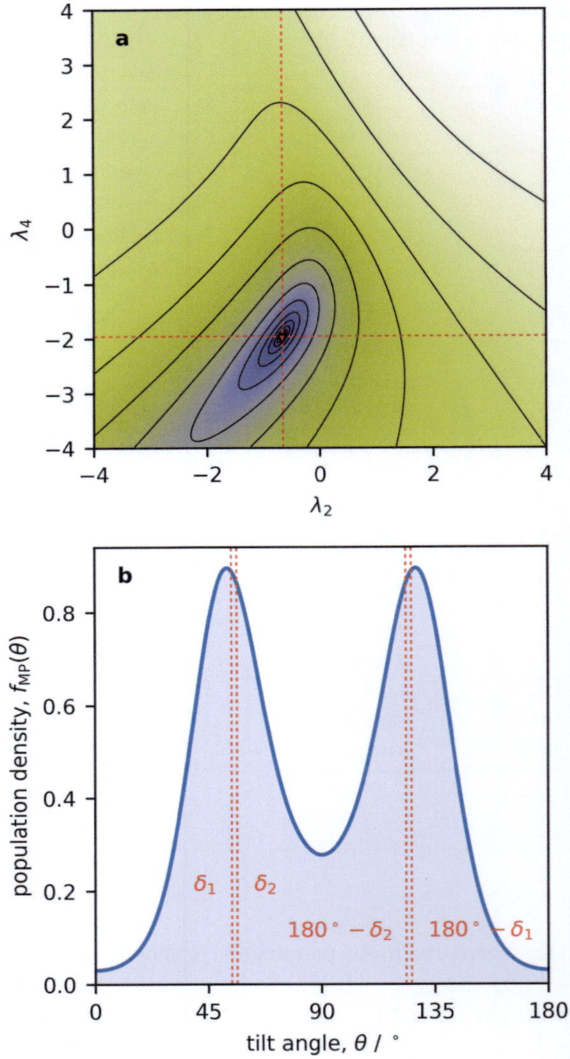

Fig. 9.6 **a** Using $\langle P_2 \rangle$ and $\langle P_4 \rangle$ values obtained from a Raman experiment to find the Lagrange multipliers λ_2 and λ_4 that satisfy the constraints in obtaining the most probable orientation distribution. **b** A plot of the most probable orientation distribution of the tilt angle. Attempts to fit the Raman data with a δ function resulted in θ_1 and θ_2 values are indicated with the dashed red lines, along with the values of $180° - \theta_1$ and $180° - \theta_2$

Another interesting observation is that $\langle P_n \rangle = 0$ for all odd values of n. If we had reason to believe that the sample didn't have any preference for the *polarity* of the orientation distribution, i.e. that alignment towards $+z$ and $-z$ are equally likely, then we would think to constrain $f_{\text{MP}}(\theta)$ so that this was the case from the start. However, we can see that this is not necessary as the most probable distribution already has this outcome. In other words, given only Raman data, $f_{\text{MP}}(\theta)$ will be symmetric about $\theta = 90°$.

More important is the fact that $\langle P_n \rangle = 0$ for all odd values of n *regardless* of the nature of the orientation distribution; this is the case as we did not specify any polarity information. This reveals that Raman spectroscopy by itself is not able to distinguish this polarity. In other words, we cannot tell if the mean tilt angle is θ_0 or $180° - \theta_0$ or both. We will later see that accessing odd values of $\langle P_n \rangle$ requires experiments based on even values of $\boldsymbol{\chi}^{(n)}$. IR absorption was a $\boldsymbol{\chi}$ or $\boldsymbol{\chi}^{(1)}$ experiment and provided access to $\langle P_2 \rangle$ only. Raman is a $\boldsymbol{\chi}^{(3)}$ experiment and results in $\langle P_2 \rangle$ and $\langle P_4 \rangle$. This also reveals that, if polarized Raman data is available, then a polarized IR experiment is unnecessary as it would not provide any new information.

Finally, even though we are showing only the first 20 values, they start becoming small, so additional values are making negligible contributions to $f_{\mathrm{MP}}(\theta)$. However, in a typical situation like this, the values of the order parameters themselves are not our primary interest. We have already made use of $\langle P_2 \rangle$ and $\langle P_4 \rangle$ to construct $f_{\mathrm{MP}}(\theta)$ and so our attention shifts to the shape of this function as shown in Fig. 9.6b. This example was inspired by a study of monolayers adsorbed at a liquid–liquid interface [36–39] but the details have been adapted to provide a more simple case study.

Code to generate Fig. 9.6

This code first calls the user-defined function `fit` to determine the values of the two Lagrange multipliers to be $\lambda_2 = -0.662$ and $\lambda_4 = -1.961$, from which $f_{\mathrm{MP}}(\theta)$ is determined.

```
import numpy as np
from matplotlib import pyplot
from scipy.optimize import brute

def fit(p):
    lambda2, lambda4 = p
    norm = np.sum(np.exp(lambda2* P2 + lambda4 * P4)*np.sin(theta)) * dtheta
    odf = np.exp(lambda2*P2 + lambda4*P4)/norm
    op2model = np.sum(P2 * odf * np.sin(theta)) * dtheta
    op4model = np.sum(P4 * odf * np.sin(theta)) * dtheta
    error = np.abs(op2model-op2expt)**2 + np.abs(op4model-op4expt)**2
    return error

# setup
pyplot.rc('font', size=8)
fig = pyplot.figure(figsize=(3.25, 6))
ax1 = fig.add_subplot(211)
ax2 = fig.add_subplot(212)
thetaDeg = np.linspace(0., 180, 200)
theta = np.radians(thetaDeg)
dtheta = theta[1] - theta[0]
P2 = 0.5*(3*np.cos(theta)**2 - 1.)
P4 = 0.125*(35*np.cos(theta)**4 - 30*np.cos(theta)**2 + 3.)

# R1, R2 = 1.4076, 1.13363
R1 = 1.408
R2 = 1.134
op2expt = (1 - R2) / (2*R2 + 1)
```

```
op4expt = 3*(2*R1*R2 - 9*R1 + 8*R2 - 1)/((2*R2+1)*(8*R1-3))
print(op2expt, op4expt)

# find the Lagrange multipliers
lambda_range = ((-4, 4), (-4, 4))
results = brute(fit, lambda_range, Ns=200, full_output=True)
lambda2, lambda4 = results[0]
print(results[0])
ax1.imshow(np.log10(results[3].T), origin='lower', cmap=pyplot.cm.
     gist_stern, extent=[lambda_range[0][0], lambda_range[0][1],
     lambda_range[1][0], lambda_range[1][1]])
ax1.contour(np.log10(results[3].T), 20, colors='k', linestyles='solid',
     linewidths=0.5, origin='lower', extent=[lambda_range[0][0],
     lambda_range[0][1], lambda_range[1][0], lambda_range[1][1]])
ax1.axhline(lambda4, color='r', ls='--', dashes=[2,2], lw=0.7)
ax1.axvline(lambda2, color='r', ls='--', dashes=[2,2], lw=0.7)
ax1.set_xlabel(r'$\lambda_2$')
ax1.set_ylabel(r'$\lambda_4$')
ax1.text(0.05, 0.92, 'a', fontweight='bold', transform=ax1.transAxes)

# visualize the distribution
norm = np.sum(np.exp(lambda2 * P2 + lambda4 * P4) * np.sin(theta)) * dtheta
odf = np.exp(lambda2 * P2 + lambda4 * P4)/norm
ax2.plot(thetaDeg, odf)
ax2.set_xlabel(r'tilt angle, $\theta$ / $^\circ$')
ax2.set_ylabel(r'population density, $f_{\rm MP}(\theta)$')
ax2.set_xticks(range(0,181,45))
ax2.fill_between(thetaDeg, 0, odf, lw=0, color='b', alpha=0.1)
ax2.set_xlim(0, 180)
ax2.set_ylim(bottom=0)
ax2.text(0.05, 0.92, 'b', fontweight='bold', transform=ax2.transAxes)

# verify the distribution
P2mod = np.sum( P2 * odf * np.sin(theta)) * dtheta
P4mod = np.sum( P4 * odf * np.sin(theta)) * dtheta
print(P2mod, P4mod)

# some extra annotations
delta1 = np.degrees(np.arctan((8*R1/3)**0.25))
delta2 = np.degrees(np.arctan(np.sqrt(2*R2)))
print(delta1, delta2)
ax2.axvline(delta1, color='r', ls='--', dashes=[2,2], lw=0.7)
ax2.axvline(180-delta1, color='r', ls='--', dashes=[2,2], lw=0.7)
ax2.axvline(delta2, color='r', ls='--', dashes=[2,2], lw=0.7)
ax2.axvline(180-delta2, color='r', ls='--', dashes=[2,2], lw=0.7)
ax2.text(39, 0.2, r'$\delta_1$', color='r')
ax2.text(63, 0.2, r'$\delta_2$', color='r')
ax2.text(76, 0.1, r'$180^\circ - \delta_2$', color='r')
ax2.text(132, 0.1, r'$180^\circ - \delta_1$', color='r')

fig.tight_layout(pad=1.5)
pyplot.savefig('methylOrange.png', dpi=600)
```

9.9 Summary

We have seen that Raman scattering provides an alternative and complementary way of studying the vibrational spectra of molecules and materials. Although Raman spectra may qualitatively appear to be very similar to IR absorption spectra, they possess a much richer information content. First, the ratio of Stokes to anti-Stokes bands of the same origin can be used to determine the local temperature. Second, for isotropic materials, the depolarization ratio provides access to the polarizability tensor elements in the molecular frame. Finally, for ordered materials, the fact that the Raman response originates from a third-order nonlinearity (through $\alpha^{(3)}$ and $\chi^{(3)}$) means that we can obtain structural information related to $\langle P_2 \rangle$ and $\langle P_4 \rangle$. Even though this appears to be just one additional order parameter compared to the $\langle P_2 \rangle$ information provided by IR absorption, we have seen that this makes a world of difference in terms of how much can be said about the orientation distribution.

9.10 Exercises

1. You want to observe the methyl symmetric stretch ($\omega_{\mathrm{vib}} = 2875\ \mathrm{cm}^{-1}$) of an alkane by Raman spectroscopy using a HeNe laser with $\lambda_{\mathrm{in}} = 633$ nm.
   a. At what wavelength (answer in nanometers) would you expect to see the Stokes-shifted Raman mode?
   b. At what wavelength would the corresponding anti-Stokes-shifted Raman mode be observed?

---

2. The Stokes/anti-Stokes intensity ratio depends on temperature, and therefore serves as a convenient way of monitoring the local temperature in samples. Assume you are using a 785 nm laser source, and record spectra in the $\pm(400 - -1800)\ \mathrm{cm}^{-1}$ range.
   a. Let's say that you can measure the entire Raman spectrum (Stokes and anti-Stokes regions) with an uncertainty of 1% in the intensity due to fluctuations from your sample. How much uncertainty in the temperature is this, assuming you are working close to ambient conditions of 20 °C?
   b. Is this situation improved (to yield more accurate temperature measurements) in a different temperature range, i.e. for warmer or colder samples?

---

3. You would like to collect Raman data of sodium dodecyl sulphate adsorbed at the air–water interface.
   a. What is the ideal type of Raman setup for this experiment?
   b. If you are performing the experiment using a Raman microscope, what are the best characteristics for the objective lens to use?

c. Collecting Raman spectra of monolayers is not easy, but do you anticipate the spectra to be of higher quality than in an IR absorption experiment? Explain your reasoning.
d. Explain why the resulting Raman spectrum shows an intense CH_2 symmetric stretch but only a weak CH_3 symmetric stretch.

---

4. If you were building a Raman setup on an optical table, you have the flexibility of placing the collection/detection optics in any location, such as 90° from the excitation source. However, when using a commercial Raman microscope, the scattered light is detected along the same direction as the excitation source ($k_{\text{out}} = -k_{\text{in}}$). For this microscope geometry

a. How many unique Raman tensor elements are there for an isotropic sample? What schemes are needed for their measurement? Use Porto notation to describe them.
b. What if the sample was ordered in the plane of the surface (about ϕ in our notation for the Euler angles), but not in the θ- or ψ-directions?
c. How about the rest of the combinations of the three Euler angles? I.e. anisotropic about each of the other two angles, or two of the three angles.
d. Think of some real situations of samples that could be ordered in such ways.
e. What is the maximum number of unique Raman tensor elements that could be measured? Does this depend on the experimental geometry? I.e. are there any limitations to the microscope configuration for certain types of samples?

---

5. For a single molecule with orientation θ with respect to z, of course, the trigonometric identity $\cos^2\theta + \sin^2\theta = 1$ applies.

a. But what about for a distribution of molecules? In this chapter, we have used the result that $\langle\cos^2\theta\rangle + \langle\sin^2\theta\rangle = 1$. Provide a proof of this relationship.
b. The consequence of the above relationship is that, if $\langle\cos^2\theta\rangle$ can be measured, $\langle\sin^2\theta\rangle$ can be known from $1 - \langle\cos^2\theta\rangle$, if desired. Now imagine that there is an experiment (that will be discussed in Chap. 10!) that could provide $\langle\cos\theta\rangle$. Can this knowledge be used to provide $\langle\sin\theta\rangle$?
c. We also made use of the identity $\langle\sin^4\theta\rangle = 1 - 2\langle\cos^2\theta\rangle + \langle\cos^4\theta\rangle$. Provide a proof of this relationship as well.
d. Is $\dfrac{\langle\sin\theta\rangle}{\langle\cos\theta\rangle} = \langle\tan\theta\rangle$?

---

6. In the case of the θ-only distributions we have considered in this chapter, everything was based on a Raman tensor with only $a^{\text{R}}_{cc} \neq 0$, i.e. only $\text{d}\alpha_{cc}/\text{d} \neq 0$.

a. Revisit the expressions we have derived if there were now two non-zero Raman tensor elements, with the addition of $a^{\text{R}}_{aa} = a^{\text{R}}_{bb}$.
b. After working out your expressions, are you still able to solve for the order parameters $\langle P_2\rangle$ and $\langle P_4\rangle$?

c. Re-write your new expressions in terms of the ratio of Raman tensor elements, using the definition $R = a^{\mathrm{R}}_{aa}/a^{\mathrm{R}}_{cc}$.

---

7. How does your analysis of Raman data to understand molecular orientation change if you know that the inherent or induced orientation has symmetry about one or more axes in the sample? For example, a distribution about ϕ that is symmetric about $\phi = 90°$ or $\phi = 180°$.

---

8. What are some advantages of assuming a trial orientation distribution function (ODF) and then using the polarized Raman data to solve for all or some of the parameters in the ODF, compared to extracting order parameters than can be used to arrive at a most probable orientation distribution?

---

9. In our discussion of IR absorption in the previous chapter, we illustrated how the two most common options for orientation analysis are to use the IR dichroic ratio to determine $\langle P_2 \rangle$ or to make the further assumption of a narrow distribution and determine θ_0 as $\langle \cos^2 \theta_0 \rangle \rightarrow \cos^2 \langle \theta_0 \rangle$.

   a. Reflect on how this situation changes with the addition of $\langle P_4 \rangle$ information in a Raman experiment.
   b. An approach based on maximum information entropy has been illustrated to arrive at the most probable orientation distribution based on experimental values of $\langle P_2 \rangle$ and $\langle P_4 \rangle$ from Raman data. Why is the same approach typically not pursed when only $\langle P_2 \rangle$ is available from IR absorption data?
   c. Can you think of any possible advantages and/or disadvantages in obtaining $\langle P_2 \rangle$ from IR data for use together with $\langle P_4 \rangle$ from Raman data? In theory, IR data is redundant since $\langle P_2 \rangle$ is also available from analysis of polarized Raman spectra. This question asks whether there are any practical considerations.

References

1. C. Sourisseau, Chem. Rev. **104**, 3851 (2004)
2. D.I. Bower, J. Poly. Sci. **10**, 2135 (1972)
3. L.G.P. Dalmolen, S.J. Picken, A.F. de Jong, W.H. de Jeu, J. Physique **46**, 1443 (1985)
4. F. Lagugné Labarthet, Annu. Rep. Prog. Chem. **103**, 326 (2007)
5. F. Lagugné Labarthet, C. Sourisseau, J. Raman Spectrosc. **27**, 491 (1996)
6. F. Lagugné Labarthet, T. Buffeteau, C. Sourisseau, J. Phys. Chem. B **102**, 5754 (1998)
7. Z. Li, R.J. Young, I.A. Kinloch, N.R. Wilson, A.J. Marsden, A.P.A. Raju, Carbon **88**, 215 (2015)
8. G. Pezzotti, J. Appl. Phys. **113**, 211301 (2013)
9. G. Pezzotti, Acta Biomater. **55**, 28 (2017)
10. M. Richard-Lacroix, C. Pellerin, Appl. Spectrosc. **67**, 409 (2013)
11. M. Richard-Lacroix, C. Pellerin, Macromolecules **46**, 5561 (2013)
12. S. Yang, S. Michielsen, Macromolecules **36**, 6484 (2003)
13. C.V. Raman, Ind. J. Phys. **2**, 387 (1927)
14. Y. Yoshikawa, S. Shigeto, Appl. Spectrosc. **74**, 1295 (2020)

15. S.P. Nikitin, C. Manka, J. Grun, J. Bowles, Rev. Sci. Instrum. **83**, 033105 (2012)
16. E.O. Potma, S. Mukamel, in *Coherent Raman Scattering Microscopy*, ed. by J.X. Cheng, X.S. Xie (CRC Press, Boca Raton, 2013), Chap. 1, pp. 3–42
17. C. Humbert, T. Noblet, Symmetry **3**, 153 (2021)
18. R.H. Stolen, J.P. Gordon, W.J. Tomlinson, H.A. Haus, J. Opt. Soc. Am. B **6**, 1159 (1989)
19. Y.R. Shen, N. Bloembergen, Phys. Rev. **137**, A1787 (1965)
20. D.A. Long, *The Raman Effect: A Unified Treatment of The Theory of Raman Scattering by Molecules* (Wiley, New Jersey, USA, 2002)
21. W.H. Weber, R. Merlin, *Raman Scattering in Materials Science* (Springer, New York, 2000)
22. R.L. McCreery, *Raman Spectroscopy for Chemical Analysis* (Wiley-Interscience, New York, 2000)
23. J.R. Ferraro, Z. Nakamoto, C.W. Brown, *Introductory Raman Spectroscopy* (Academic Press, San Diego, 2003)
24. R. Berardi, F. Spinozzi, C. Zannoni, J. Chem. Phys. **109**, 3742 (1998)
25. J. Kinosita, K., S. Kawato, A. Ikegami, Biophys. J. **20**, 289 (1977)
26. R.J. Roe, W.R. Krigbaum, J. Chem. Phys. **40**, 2608 (1964)
27. H. Schaeben, J. Appl. Cryst. **26**, 112 (1993)
28. C.D. Southern, P.D. Brimicombe, S.D. Siemianowski, S. Jaradat, N. Roberts, V. Görtz, J.W. Goodby, H.F. Gleeson, Europhys. Lett. **82**, 56001 (2008)
29. L. Svenningsson, L. Nordstierna, ACS Appl. Polym. Mater. **2**, 4809 (2020)
30. C.E. Shannon, Bell Sys. Tech. J. **27**, 379 (1948)
31. E.T. Jaynes, Phys. Rev. **108**, 171 (1957)
32. E.T. Jaynes, Phys. Rev. **106**, 620 (1957)
33. D.I. Bower, J. Polymer Sci. **19**, 93 (1981)
34. H. Schaeben, Phys. Stat. Sol. B **148**, 63 (1988)
35. G.B. Arfken, H.J. Weber, F.E. Harris, *Mathematical Methods for Physicists: A Comprehensive Guide*, 7th edn. (Academic Press, Waltham, MA, 2013)
36. T. Takenaka, T. Nakanaga, J. Phys. Chem. **80**, 475 (1976)
37. T. Nakanaga, T. Takenaka, J. Phys. Chem. **81**, 645 (1977)
38. T. Takenaka, H. Fukuzaki, J. Raman Spectrosc. **8**, 151 (1979)
39. T. Takenaka, Chem. Phys. Lett. **55**, 515 (1978)

Chapter 10
Vibrational Sum-Frequency Generation Spectroscopy

10.1 Introduction

By now we have seen how absorption-based techniques probe the susceptibility of materials, originally referred to as χ, and now more explicitly labelled as $\chi^{(1)}$ to highlight the linear relationship between the polarization (dipole moment per unit volume) and the applied field. We have also discussed how the Raman response is proportional to the imaginary part of $\chi^{(3)}$, even though the relationship between the incident field and the scattered field is also linear. Recall that recognition of the third-order response function was necessary in order to arrive at the relationships between the measured Raman spectra and the molecular orientation, and the depolarization ratio for isotropic samples. In the discussion over the past two chapters, we have progressed from $\chi^{(1)}$ to $\chi^{(3)}$, conspicuously skipping over the second-order susceptibility $\chi^{(2)}$. We now remedy this, with $\chi^{(2)}$ as the focus of the current chapter.

We have seen that the polarization may be expressed as [1–4]

$$\begin{aligned} \mathbf{P} &= \mathbf{P}^{(1)} + \mathbf{P}^{(2)} + \mathbf{P}^{(3)} + \cdots + \mathbf{P}^{(n)} \\ &= \varepsilon_0 \left(\chi^{(1)}\mathbf{E} + \chi^{(2)}\mathbf{E}\mathbf{E} + \chi^{(3)}\mathbf{E}\mathbf{E}\mathbf{E} + \chi^{(4)}\mathbf{E}^4 + \cdots + \chi^{(n)}\mathbf{E}^n \right). \end{aligned} \tag{10.1}$$

The reason that we took a detour around $\chi^{(2)}$ is that all odd-order susceptibility-based techniques that probe $\chi^{(2n+1)}$ are broadly applicable to materials characterization, as they are able to generate signals even in—and especially for—isotropic materials. If the materials happen to be ordered, then we have already discussed approaches by which polarized light can be used to obtain detailed structural information including the determination of order parameters. However, when it comes to the even-order terms $\chi^{(2n)}$, the associated techniques have more niche applications because not all materials are able to produce a resulting polarization $\mathbf{P}^{(2n)}$, at least not in the bulk of the material.

D. K. Hore, *Molecular Orientation Analysis Using Polarized Vibrational Spectroscopy*,
https://doi.org/10.1007/978-3-032-04874-5_10

10.2 Second-Order Nonlinear Optical Spectroscopy

Even though we are ultimately interested in vibrational spectroscopy, it is helpful to first consider a slightly simpler experiment, second harmonic generation (SHG), also known as frequency doubling. We will later see that, in order for any element of the $\boldsymbol{\chi}^{(2)}$ tensor to be non-zero, the material must not possess inversion symmetry. That is, $\boldsymbol{\chi}^{(2)} = 0$ at locations in the material for which $(x, y, z) \rightarrow (-x, -y, -z)$. This condition can be satisfied in a variety of ways but for now let's assume that we have such a material where at least some elements of $\boldsymbol{\chi}^{(2)} \neq 0$. Second harmonic generation is referred to as frequency doubling since this experiment uses two photons of frequency ω (most often from a single laser source) to generate $P^{(2)}(2\omega)$ and therefore re-radiates a field at frequency 2ω that is detected. We initially consider the case were both ω and 2ω are far from any resonance frequency. A typical situation would be when using a laser with frequency $\omega = 18797$ cm^{-1} (i.e. $\lambda = 532$ nm) and then detecting the signal generated at 2ω or $\lambda/2 = 266$ nm. An energy level diagram for this process is shown in Fig. 10.1. As in the case of our off-resonance Raman discussion, the dashed lines indicate short-lived virtual states. We have already gained some skills from dealing with the beastly 81-element $\boldsymbol{\chi}^{(3)}$ tensor in the last chapter; it will be more manageable to have a smaller quantity here, with $3 \times 3 \times 3 = 27$ Cartesian elements in the rank-3 $\boldsymbol{\chi}^{(2)}$ tensor. For this SHG experiment, each element $\chi^{(2)}_{ijk}$ describes the i-polarized contribution to $\mathbf{P}^{(2)}(2\omega)$ that results in the output $E_i(2\omega)$ given a j-polarized input field $E_j(\omega)$ and a k-polarized input $E_k(\omega)$. The simplest way to do this is with a single laser source, as depicted in Fig. 10.2. We previously stated that Raman experimental setups are very simple. SHG experiments now take the crown, as there is nothing to do other than shine light on the material at frequency ω and then detect the single frequency 2ω. In the case of Raman scattering, we measure the spectrum of the Stokes-shifted light. In the SHG experiment, no spectrograph is required, as we know the single frequency (2ω) that we are after. Since the generated light at 2ω may be very weak if $\chi^{(2)}_{ijk}$ is small, it is helpful to have a very bright laser source, and a filter that blocks out the laser frequency ω to aid the detector from not being saturated by this light when attempting to measure

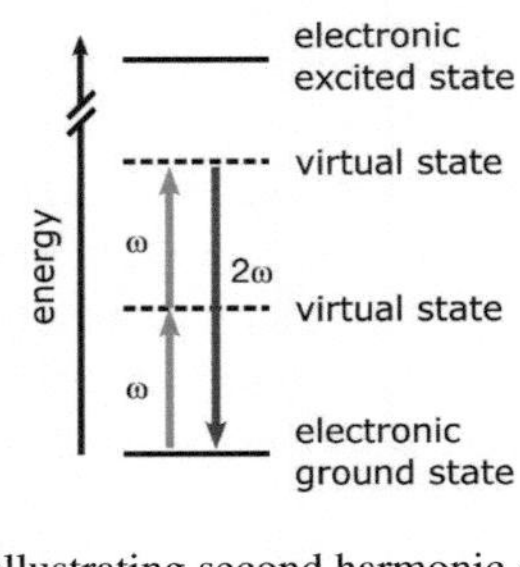

Fig. 10.1 Energy level diagram illustrating second harmonic generation when both the incoming (fundamental, at frequency ω) and the second harmonic at 2ω are far from any resonance. The dashed lines indicate virtual states

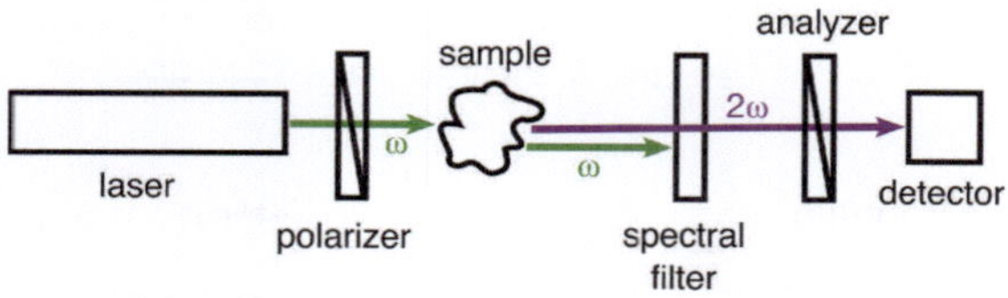

Fig. 10.2 The basic principles behind an SHG experiment where a laser at frequency ω is optionally polarized before impinging on a sample. Here we illustrate that the conversion efficiency to generate light at 2ω is typically low, so most of the transmitted light is at frequency ω from the original laser source. One therefore needs some type of spectral filtering to aid in the detection of the weak second harmonic signal

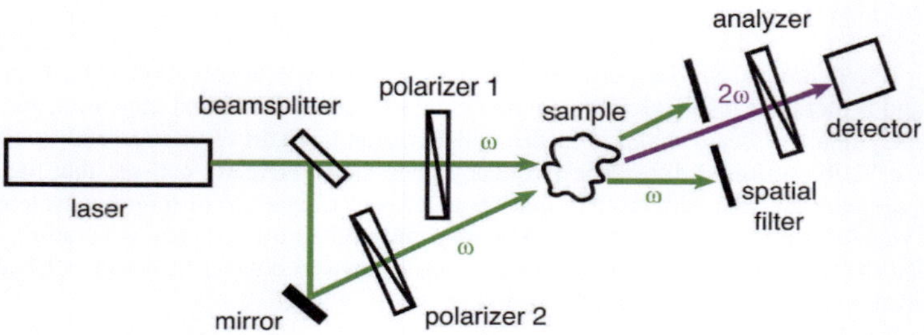

Fig. 10.3 An alternative setup for SHG spectroscopy. When compared with the setup shown in Fig. 10.2, this has the advantage that the split input beams can be polarized independently, for example, one beam s-polarized and the other p-polarized. Furthermore, the spectral filtering to isolate 2ω is greatly aided by the fact that the SHG beam can be spatially separated from the two beams at frequency ω

the dim light at frequency 2ω. In the setup pictured in Fig. 10.2, the polarizers are optional, but if they are installed, then both input photons (from the "single" input field) are j-polarized and so we have a situation described by

$$P_i^{(2)}(2\omega) = \varepsilon_0 \sum_{j}^{x,y,z} \chi_{ijj}^{(2)} E_j(\omega) E_j(\omega), \tag{10.2}$$

i.e. this is a true $[E_j(\omega)]^2$ scenario as loosely implied by the **EE** or $\mathbf{E}^2$ notation in Eq. 10.1. However, even though this is a convenient measurement to make, nothing prevents us from using two lasers, or splitting the single laser beam into two paths that cross at the sample position as shown in Fig. 10.3. Such a configuration has two advantages. First, it enables the two input photons to be independently polarized along j- and k-directions, so more elements of the $\boldsymbol{\chi}^{(2)}$ tensor may be accessed through

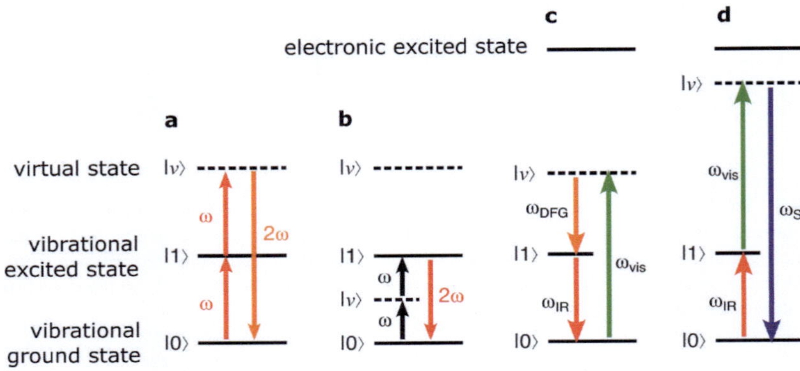

Fig. 10.4 A hypothetical SHG experiment where **a** the incident beam at frequency ω is in the mid-infrared, and is therefore resonant with a fundamental IR-active vibrational transition, and **b** where the frequency-doubled beam at 2ω is in the mid-IR so as to be in resonance with a vibrational transition. **c** Visible-infrared difference frequency generation. Here we can see that the detected light at $\omega_{\mathrm{DFG}} = \omega_{\mathrm{vis}} - \omega_{\mathrm{IR}}$ (drawn in orange) is at a lower frequency, or longer wavelength, than the input visible beam, i.e. $\omega_{\mathrm{DFG}} < \omega_{\mathrm{vis}}$. **d** Visible-infrared sum-frequency generation. Here we can see that the detected light at $\omega_{\mathrm{SFG}} = \omega_{\mathrm{vis}} + \omega_{\mathrm{IR}}$ (drawn in blue) is at a higher frequency, or shorter wavelength, than the input visible beam, i.e. $\omega_{\mathrm{SFG}} > \omega_{\mathrm{vis}}$

$$P_i^{(2)}(2\omega) = \varepsilon_0 \sum_j^{x,y,z} \sum_k^{x,y,z} \chi_{ijk}^{(2)} E_j(\omega) E_k(\omega). \tag{10.3}$$

As a bonus, we now have a spatial separation of the beams, as the field generated at 2ω travels in a unique direction. The practical consequence is that less filtering of the laser frequency ω is required.

We can now reconcile this with our interest in molecular vibrations as a means to get localized structural information about chemical functional groups that serve as intrinsic labels. So rather than a typically electronic off-resonance response, let's consider vibrational spectroscopy based on $\boldsymbol{\chi}^{(2)}$. There are four options to consider. The first is SHG with a laser source providing $\mathbf{E}(\omega)$ in the mid-IR, with the corresponding energy level diagram shown in Fig. 10.4a. The practical problem here is that if ω is in the range 400–4000 cm^{-1}, then 2ω appears at 800–8000 cm^{-1} also primarily in the mid-infrared (and bordering on the near infrared). When the generated signal is weak, it is challenging to have sensitive detectors in this spectral region. Another obvious option is to select a laser frequency ω such that 2ω is in resonance with vibrational modes of interest, as shown in Fig. 10.4b. However, if we now want 2ω to fall in the 400–4000 cm^{-1} range, we require exciting the sample with $\omega = 200$–2000 cm^{-1} light. The high-frequency end of this range is not an issue, as we will soon discuss (along with better ways to do this experiment). However, bench top laser sources in the lower frequency region are not readily available.

Furthermore, there are no sensitive, low-cost, and widely available detectors for low-light measurements in the mid-IR.

A better solution is to abandon SHG for different types of $\chi^{(2)}$ experiments. The first we consider is difference frequency generation (DFG). The energy level diagram for this process is shown in Fig. 10.4c. Here we consider an incident visible laser at frequency $\omega_{\text{vis}} = 18797\ \text{cm}^{-1}$ ($\lambda_{\text{vis}} = 532$ nm) that is far from any vibrational or electronic resonance, and hence excites the molecules in the material to a virtual state. At the same time, a laser source in the mid-infrared at ω_{IR} is applied. Since very high intensity sources are required when $\chi^{(2)}$ is small, pulsed lasers are used so that the incoming photons can be bunched up in time to create high instantaneous intensities even though the average power may be low. Ultimately, this means that the visible and IR lasers need to be overlapped on the sample, both in space and in time. In the DFG process, new photons are generated at $\omega_{\text{DFG}} = \omega_{\text{vis}} - \omega_{\text{IR}}$. If we do the math, considering that $\omega_{\text{vis}} \gg \omega_{\text{IR}}$ we see that ω_{DFG} is close in frequency to ω_{vis}. This has the great advantage that sensitive single-photon-counting detectors are readily available in the visible region. This is in fact the basis of most detection in the case of Raman spectroscopy. In fact, the analogy is deeper, as a comparison of Figs. 10.4c and 9.3c reveals that we are merely detecting the Stokes-shifted photons, just not through spontaneous Raman scattering. Since DFG is a coherent process, the generated field at $\mathbf{E}(\omega_{\text{vis}} - \omega_{\text{IR}})$ is directional. However, just as in the case of Raman scattering, this signal competes with fluorescence that can swamp the sought DFG signals at the same frequency.

This leaves a further interesting option, another $\chi^{(2)}$ process that uses the same two input beams, but generates a signal at the sum frequency [5, 6] as shown in Fig. 10.4d. From a frequency or wavelength perspective, this is analogous to an anti-Stokes Raman experiment. The anti-Stokes-shifted wavelengths are free from fluorescence, but the spontaneous anti-Stokes Raman process is usually too weak to detect, especially for high-frequency vibrations, on account of the low population in the vibrational excited state. Sum-frequency generation (SFG) does not have this limitation, as we can use a laser sources to excite the system into that state. It is important to keep in mind that this is not exactly an excitation to vibrational state $|1\rangle$ followed by a further excitation to the virtual state $|v\rangle$, as this is a concerted rather than sequential process.

These experiments can be performed in a collinear geometry with visible and IR beams combined so they approach the sample at the same angle [7–17] or a non-collinear geometry [18–27]. A non-collinear depiction of the experiment is shown in Fig. 10.5, leaving details of the sample for later discussion. Here we can see that, owing to the unique directions at which the signals are generated, six beams are present in the transmission geometry. They are the transmitted visible light at ω_{vis}, the transmitted IR beam ω_{IR} (that may be of interest for absorption monitoring), SHG at $2\omega_{\text{vis}}$, SHG at $2\omega_{\text{IR}}$ (although this is difficult to detect), the DFG beam at $\omega_{\text{vis}} - \omega_{\text{IR}}$ and the SFG beam at $\omega_{\text{vis}} + \omega_{\text{IR}}$. These six beams are also present (at a different set of angles) in the reflected direction. Finally, with the sample excited in this way (particularly at ω_{vis}), nothing prevents setting up a detection scheme to also measure the spontaneous Raman scattering (Stokes and/or anti-Stokes) or the

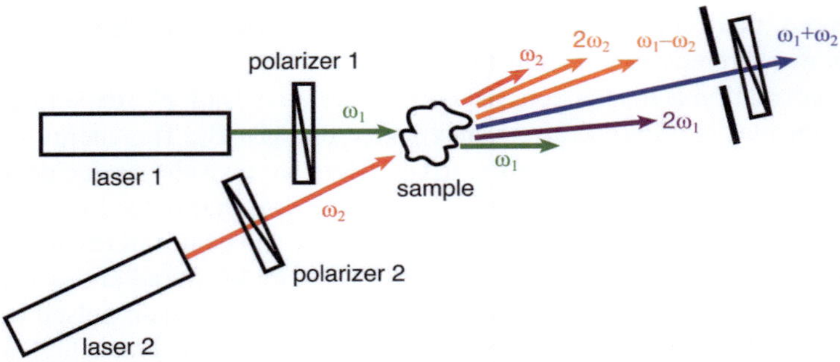

Fig. 10.5 When two lasers of different frequencies are overlapped at a sample surface, the transmitted beams contain the fundamentals (ω_1 and ω_2), the second harmonic of each fundamental ($2\omega_1$ and $2\omega_2$), and beams generated at the sum ($\omega_1 + \omega_2$) and difference ($\omega_1 - \omega_2$) frequencies. These same six beams are present in the reflected direction (not shown). Here we also illustrate that the any one of the beams (for example, that at the sum frequency) can be spatially selected and passed through an additional polarizer for analysis

fluorescence spectrum. For the remainder of this chapter, we turn our attention to vibrational sum-frequency generation.

10.3 Molecular Considerations and Nomenclature for the Hyperpolarizability

We start by examining the molecular response to $\boldsymbol{\chi}^{(2)}$, the second-order polarizability $\boldsymbol{\alpha}^{(2)}$. First, a brief discussion about the notation. $\boldsymbol{\alpha}^{(2)}$ is termed the hyperpolarizability and has historically also been given the symbol $\boldsymbol{\beta}$. However, unlike the case of Eq. 10.1 where everyone fortunately uses $\boldsymbol{\chi}^{(n)}$ for the macroscopic response, it's a bit messy in the molecular world. One commonly used convention is to write the induced dipole moment as

$$\mathbf{p} = \boldsymbol{\alpha}\mathbf{E} + \boldsymbol{\beta}\mathbf{E}^2 + \boldsymbol{\beta}^{(2)}\mathbf{E}^3 + \boldsymbol{\beta}^{(3)}\mathbf{E}^4 + \boldsymbol{\beta}^{(4)}\mathbf{E}^5 + \cdots + \boldsymbol{\beta}^{(n-1)}\mathbf{E}^n$$

where $\boldsymbol{\beta}$ (sometimes referred to as $\boldsymbol{\beta}^{(1)}$) is the (first) hyperpolarizability, $\boldsymbol{\beta}^{(2)}$ is the second hyperpolarizability, and successive terms have increasing integers in the labels, with $\boldsymbol{\beta}^{(n-1)}$ always matched to the product of n fields ($\mathbf{E}^n$ in our lazy notation). Another widely used notation is

$$\mathbf{p} = \boldsymbol{\alpha}\mathbf{E} + \boldsymbol{\beta}\mathbf{E}^2 + \boldsymbol{\gamma}\mathbf{E}^3 + \boldsymbol{\gamma}^{(2)}\mathbf{E}^4 + \boldsymbol{\gamma}^{(3)}\mathbf{E}^5 + \cdots + \boldsymbol{\gamma}^{(n-2)}\mathbf{E}^n$$

where the second hyperpolarizability is termed $\boldsymbol{\gamma}$, and all higher terms use $\boldsymbol{\gamma}^{(n)}$ that is now two orders shifted from the number of associated fields. There are a few other, less common but equally confusing, notations one occasionally finds in use. A growing number of authors in the SFG community have started to adopt the notation $\boldsymbol{\beta}^{(2)}$ for the (first) hyperpolarizability, so you are likely to encounter this written in papers after ca. 2010. In the end, a symbol is just a symbol, but this clashes with all of the above historical conventions. I prefer to write

$$\mathbf{p} = \boldsymbol{\alpha}\mathbf{E} + \boldsymbol{\alpha}^{(2)}\mathbf{E}^2 + \boldsymbol{\alpha}^{(3)}\mathbf{E}^3 + \boldsymbol{\alpha}^{(4)}\mathbf{E}^4 + \boldsymbol{\alpha}^{(5)}\mathbf{E}^5 + \cdots + \boldsymbol{\alpha}^{(n)}\mathbf{E}^n \tag{10.4}$$

for the obvious clarity in matching $\boldsymbol{\alpha}^{(n)}$ with $\mathbf{E}^n$ but also for the simplicity of associating $\boldsymbol{\alpha}^{(n)}$ as the molecular origin of the material's $\boldsymbol{\chi}^{(n)}$ response. We will therefore refer to the second-order polarizability or hyperpolarizability as $\boldsymbol{\alpha}^{(2)}$ for the remainder of our discussion.

If we consider a particular vibrational mode with resonance frequency ω_0 (so that the difference between the $|0\rangle$ and $|1\rangle$ levels depicted in Fig. 10.4d is $hc\omega_0$) we can write the resonance contribution to $\alpha^{(2)}_{lmn}$ using a Lorentz oscillator model

$$\alpha^{(2)}_{lmn}(\omega_{\text{IR}}) = \frac{a^{(2)}_{lmn}}{\omega_0 - \omega_{\text{IR}} - i\Gamma} \tag{10.5}$$

where it is more popular to use the resonance term from Eq. 7.26 rather than the full form of the expression as we had originally arrived at in Eq. 7.11. Also note that the broadening term $\Gamma/2$ appearing in Eq. 7.26 is often written as Γ as we have done in Eq. 10.5. Here, since $\boldsymbol{\alpha}^{(2)}$ and hence the corresponding oscillator strength $\mathbf{a}^{(2)}$ are rank-3 tensors, we expect that all elements $a^{(2)}_{lmn}$ are written as a product of three transition dipole moments

$$a^{(2)}_{lmn} = \langle 0|\hat{\mu}_l|v\rangle\langle v|\hat{\mu}_m|1\rangle\langle 1|\hat{\mu}_n|0\rangle. \tag{10.6}$$

From left to right, these transitions correspond to the SFG emission from the virtual state $|v\rangle$, visible excitation to the virtual state, and IR excitation from the vibrational ground state $|0\rangle$ to the vibrational excited state $|1\rangle$ as shown in Fig. 10.4d. Since the first two transitions in Eq. 10.6 both involve the virtual state, we can make use of the same simplification that we employed in our study of the Raman response, and invoke the transition polarizability that incorporates the two transition dipole moments so this becomes

$$a^{(2)}_{lmn} = \langle 0|\hat{\alpha}_{lm}|1\rangle\langle 1|\hat{\mu}_n|0\rangle. \tag{10.7}$$

If ω_{vis} is sufficiently far from any electronic resonance, $\langle 0|\hat{\alpha}_{lm}|1\rangle = \langle 1|\hat{\alpha}_{ml}|0\rangle$, i.e. the Raman tensor is symmetric about its diagonal and so we expect $a^{(2)}_{lmn} = a^{(2)}_{mln}$.

Throughout the remainder of this chapter, I'll try to be clear about $\boldsymbol{\alpha}^{(2)}$ versus $\mathbf{a}^{(2)}$. When speaking about the hyperpolarizability in general, it is tempting to write $\boldsymbol{\alpha}^{(2)}$, and I'll probably do so on occasion. However, most of our business will be conducted using $\mathbf{a}^{(2)}$, a hyperpolarizability of sorts, but technically the numerator of Eq. 10.5, so it is specific to a single vibrational mode with IR resonance frequency

ω_0, but it doesn't include the frequency dependence expressed by the Lorentzian denominator.

10.4 Second-Order Susceptibility of Bulk Materials

Just as we have done in the case of IR absorption (Chap. 8) and Raman scattering (Chap. 9), we now transition to a discussion of material properties, or a large collection of molecules where we don't keep track of the individual dipole moments. In moving from $\boldsymbol{\alpha}^{(2)}$ to $\boldsymbol{\chi}^{(2)}$ we again consider a Lorentz model for the vibrational resonance

$$\chi_{ijk}^{(2)}(\omega_{\mathrm{IR}}) = \frac{A_{lmn}^{(2)}}{\omega_0 - \omega_{\mathrm{IR}} - i\Gamma} \tag{10.8}$$

and, as we have discussed above in the case of the hyperpolarizability, it will be tempting to write $\chi^{(2)}$ throughout the remainder of this chapter (and sometimes I will), but we will be primarily concerned with the *amplitude* $\mathbf{A}^{(2)}$ that appears in Eq. 10.8.

The electric fields generated by the lasers in this experiment are defined with coordinates (describing their polarization) in the laboratory frame, and we are ultimately interested in the properties of the materials, especially the orientations of the molecules that comprise those materials. By now, our approach is familiar: We project into the lab frame and then integrate over the distribution of molecules according to

$$\begin{aligned} A_{ijk}^{(2)} &= \frac{1}{\varepsilon_0 V} \sum_{\text{molecules}} a_{ijk}^{(2)}(\theta, \phi, \psi) \\ &= \frac{N_{\mathrm{V}}}{\varepsilon_0} \langle a_{ijk}^{(2)}(\theta, \phi, \psi) \rangle \\ &= \frac{N}{\varepsilon_0} \frac{\int_0^{2\pi} \int_0^{2\pi} \int_0^{\pi} f(\theta, \phi, \psi)\, a_{ijk}^{(2)}(\theta, \phi, \psi)\, \sin\theta \,\mathrm{d}\theta}{\int_0^{2\pi} \int_0^{2\pi} \int_0^{\pi} f(\theta, \phi, \psi)\, \sin\theta \,\mathrm{d}\theta} \end{aligned} \tag{10.9}$$

where V is the volume and N_{V} is the number of molecules per unit volume. For the remainder of this chapter, we therefore consider the result of evaluating the above expression for various orientation distributions $f(\theta, \phi, \psi)$ and for different symmetries of the molecular response encoded in the hyperpolarizability $\boldsymbol{\alpha}^{(2)}$. We begin with a few bulk material considerations, and then transition to a discussion of molecules at surfaces and buried interfaces.

10.4.1 Isotropic Distribution

The most important practical aspect of SHG, DFG, and SFG, and the primary experimental motivation for these techniques can be seen from the isotropic average where $f(\theta, \phi, \psi) = 1/8\pi^2$. This is the typical situation in a liquid, or a randomly ordered polymer. Since $\mathbf{A}^{(2)}$ and $\mathbf{a}^{(2)}$ are rank-3 tensors (i.e. they have the same shape as $\boldsymbol{\chi}^{(2)}$ and $\boldsymbol{\alpha}^{(2)}$), we anticipate requiring three direction cosine matrix elements in order to carry out the molecule to lab frame projection. For this isotropic distribution we use

$$
\begin{aligned}
A_{ijk}^{(2)} = \frac{N_V}{8\pi^2\varepsilon_0} \int_0^{2\pi} \int_0^{2\pi} \int_0^{\pi} \sum_l^{a,b,c} \sum_m^{a,b,c} \sum_n^{a,b,c} D_{il}(\theta, \phi, \psi)\, D_{jm}(\theta, \phi, \psi) \\
\times D_{kn}(\theta, \phi, \psi)\, a_{lmn}^{(2)} \sin\theta \, \mathrm{d}\theta \, \mathrm{d}\phi \, \mathrm{d}\psi
\end{aligned}
\qquad (10.10)
$$

and obtain the result that, of the 27 elements of $\mathbf{A}^{(2)}$, 21 of them are zero. The only six surviving elements are those $A_{ijk}^{(2)}$ for which $i \neq j \neq k$. Among those six, only one element is unique, with

$$
\begin{aligned}
A_{xyz}^{(2)} &= A_{yzx}^{(2)} = A_{zyx}^{(2)} = -A_{xzy}^{(2)} = -A_{yxz}^{(2)} = -A_{zyx}^{(2)} \\
&= \frac{N}{6\varepsilon_0} \left(a_{abc}^{(2)} + a_{bca}^{(2)} + a_{cab}^{(2)} - a_{acb}^{(2)} - a_{bac}^{(2)} - a_{cba}^{(2)} \right) \\
&\equiv A_{\mathrm{chiral}}^{(2)}.
\end{aligned}
\qquad (10.11)
$$

Following from the discussion above, if we are far from any electronic resonance, $a_{lmn}^{(2)} = a_{mln}^{(2)}$ and so those six participating elements of $\mathbf{a}^{(2)}$ and $\mathbf{A}^{(2)}$ in Eq. 10.11 would also be zero. This is in fact a reiteration of our earlier statement that elements of $\mathbf{A}^{(2)}$ are non-zero only in the absence of inversion symmetry. Since we are modelling an isotropic environment here, a chiral phase (such as a solution of chiral molecules) is the only non-centrosymmetric isotropic case. A nice thing about SFG spectroscopy is that the chiral and achiral elements of $\mathbf{A}^{(2)}$ can be readily isolated by judicious selection of the beam polarizations [28–31]. However, as we are not considering electronic resonance any further, we will not explore this topic here. Instead, our focus will shift to the 21 achiral elements, and situations in which they are non-zero due to a break in centrosymmetry. We divide the following discussion into three parts: materials such as some crystals that are inherently non-centrosymmetric, perturbations that might alter the orientation of molecules in the bulk of materials that are initially isotropic, and surfaces.

10.4.2 Non-centrosymmetric Crystals

Non-centrosymmetric crystal classes include those with trigonal symmetry, with the example of β-barium borate $Ba(BO_2)_2$ commonly known as BBO, and those with tetragonal symmetry such as $AgGaS_2$. Such materials have large values of $\boldsymbol{\chi}^{(2)}$ as they are highly ordered in a consistent manner throughout their bulk phase. In this case, one can think about $f(\theta, \phi, \psi)$ as consisting of δ distributions about one of more of the Euler angles. There are many applications of such materials, most commonly to extend the practical frequency range of lasers, due to a limited option for lasing media that can achieve the necessary ground-excited state population inversion. In fact, DFG in $AgGaS_2$ is used to create the high-power mid-IR laser pulses for use in SFG experiments. Here, a related $\boldsymbol{\chi}^{(2)}$ process is initially used to split a visible or near-IR photon into two lower energy photons in BBO. These two photos can then interact in the DFG process. The closer the two photons are in frequency, the smaller the difference frequency, and hence the longer the wavelength of the output. There are practical limits of what can be achieved in terms of the final mid-IR tuning range, and limits to the DFG conversion efficiency. This involves consideration of a fundamental concept in nonlinear optics known as phase-matching that dictates how the electric fields generated from $\mathbf{P}^{(2)}$ add to generate the net dipole moment per unit volume. Understanding this phase-matching is typically the starting point for any course or textbook in nonlinear optics [1, 4, 5, 32]. However, we will not discuss it much further, as we will soon move away from our description of bulk materials.

10.4.3 Induced Noncentrosymmetry

An interesting application of any second-order ($\boldsymbol{\chi}^{(2)}$-based) nonlinear optical process is studying the development of centrosymmetry breaking in bulk phases. If the material where the beams are overlapped in Fig. 10.5 is centrosymmetric, such as in a bulk polymer phase, no $\boldsymbol{\chi}^{(2)}$-based response would be generated from the bulk.[1] If a DC electric field was now applied across the material causing the static dipole moments to align with the field [33, 34], this breaks the centrosymmetry and a response such as SFG can be measured.

If we make our typical assumption of a simplified molecular tensor, now applied to $\boldsymbol{\alpha}^{(2)}$, we assume that all elements are zero except for $\alpha^{(2)}_{ccc}$ and the molecules are uniformly distributed about ϕ and ψ so that $f(\phi, \psi) = 1/4\pi^2$, leaving only the tilt angle distribution to be described. Proceeding to solve for all 27 elements of $\boldsymbol{\chi}^{(2)}$, we find that

[1] Inversion symmetry is broken at the surface, and that is the subject of the next section.

$$\begin{aligned}\chi_{xxz}^{(2)}(\theta) &= \frac{N_V}{4\pi^2\varepsilon_0}\int_0^{2\pi}\int_0^{2\pi} D_{xc}(\theta,\phi,\psi)\, D_{xc}(\theta,\phi,\psi)\, D_{zc}(\theta,\phi,\psi)\, \alpha_{ccc}^{(2)}\, d\phi\, d\psi \\ &= \frac{N_V}{2\varepsilon_0}\alpha_{ccc}^{(2)}\left(\langle\cos\theta\rangle - \langle\cos^3\theta\rangle\right) \\ &= \frac{N_V}{5\varepsilon_0}\alpha_{ccc}^{(2)}\left(\langle P_1\rangle - \langle P_3\rangle\right) \\ &= \chi_{xzx}^{(2)} = \chi_{yyz}^{(2)} = \chi_{yzy}^{(2)} = \chi_{zxx}^{(2)} = \chi_{zyy}^{(2)}\end{aligned} \tag{10.12}$$

and

$$\begin{aligned}\chi_{zzz}^{(2)}(\theta) &= \frac{N_V}{4\pi^2\varepsilon_0}\int_0^{2\pi}\int_0^{2\pi} [D_{zc}(\theta,\phi,\psi)]^3\, \alpha_{ccc}^{(2)}\, d\phi\, d\psi \\ &= \frac{N_V}{\varepsilon_0}\alpha_{ccc}^{(2)}\langle\cos^3\theta\rangle \\ &= \frac{N_V}{5\varepsilon_0}\alpha_{ccc}^{(2)}\left(2\langle P_3\rangle + 3\langle P_1\rangle\right)\end{aligned} \tag{10.13}$$

and all remaining 20 elements of $\boldsymbol{\chi}^{(2)}$ are equal to zero. It is interesting to see that, while the IR absorption experiments were sensitive to $\langle\cos^2\theta\rangle$, and the Raman scattering provided information on both $\langle\cos^2\theta\rangle$ and $\langle\cos^4\theta\rangle$, the SFG signal depends on $\langle\cos\theta\rangle$ and $\langle\cos^3\theta\rangle$. Some connections between the inherent symmetries of the responsible $\boldsymbol{\chi}^{(n)}$ tensors are apparent. IR and Raman are related to a even-rank tensors (transformed by two and four direction cosine matrices, respectively), probe odd-order response functions $\boldsymbol{\chi}^{(1)}$ and $\boldsymbol{\chi}^{(3)}$, and can measure responses related to even-order parameters. SFG is related to an odd-rank even-order response function $\boldsymbol{\chi}^{(2)}$ and provides information on odd-order parameters. Since this particular situation involves $\langle\cos\theta\rangle$ and $\langle\cos^3\theta\rangle$, Eqs. 10.12 and 10.13 can also be expressed in terms of order parameters $\langle P_1\rangle$ and $\langle P_3\rangle$ based on the Legendre polynomials $P_1(\cos\theta) = \cos\theta$ and $P_3(\cos\theta) = \frac{1}{2}(5\cos^3\theta - 3\cos\theta)$ where the usual definitions

$$\begin{aligned}\langle P_1\rangle &= \langle\cos\theta\rangle \\ &= \frac{\displaystyle\int_0^{\pi} f(\theta)\cos\theta\sin\theta\, d\theta}{\displaystyle\int_0^{\pi} f(\theta)\sin\theta\, d\theta}\end{aligned} \tag{10.14a}$$

and

$$\begin{aligned}\langle P_3\rangle &= \frac{5}{2}\langle\cos^3\theta\rangle - \frac{3}{2}\langle\cos\theta\rangle \\ &= \frac{\displaystyle\frac{1}{2}\int_0^{\pi} f(\theta)(5\cos^3\theta - 3\cos\theta)\sin\theta\, d\theta}{\displaystyle\int_0^{\pi} f(\theta)\sin\theta\, d\theta}\end{aligned} \tag{10.14b}$$

have been used.

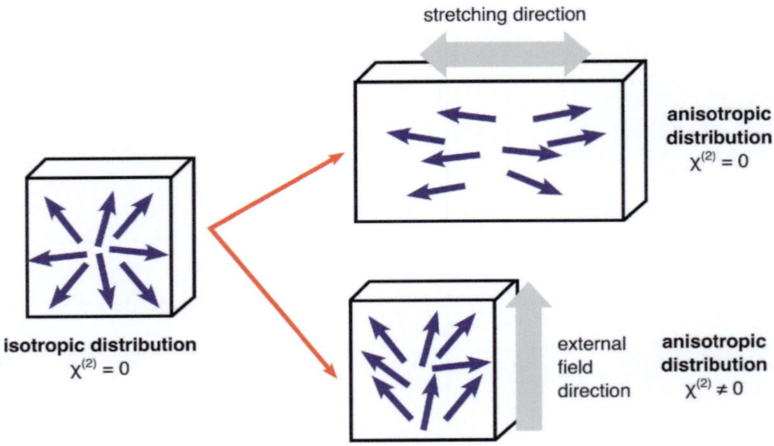

Fig. 10.6 A polymer sample where the constituent molecules are initially randomly oriented, so the orientation distribution is isotropic, and hence achiral elements of $\boldsymbol{\chi}^{(2)} = 0$. We then consider two different perturbations: if the film is stretched, the distribution becomes highly anisotropic as molecules are aligned along the stretching direction. However, since there is no preference for alignment parallel or anti-parallel to this axis, the centrosymmetry is not broken and the same elements of $\boldsymbol{\chi}^{(2)}$ are still zero. On the other hand, if a DC field is applied, molecules will align their permanent dipoles along the field, so that $\boldsymbol{\chi}^{(2)} \neq 0$ for this anisotropic arrangement

Some further understanding of this relationship can be obtained if we contrast this experiment to one where a polymer film is stretched, causing the molecules to align along the stretching direction. In the case of polarized IR absorption and polarized Raman scattering, the spectra corresponding to different elements of $\chi_{ij}^{(1)}$ and $\chi_{ijij}^{(3)}$ would show significant deviations from the isotropic state. However, as molecules are equally pulled parallel and anti-parallel to the stretching axis, this orientation distribution, even though it is highly anisotropic, is still centrosymmetric, so no SFG is generated from the bulk. On the other hand, the application of a DC field along $+x$ makes the $+x$ and $-x$ directions no longer equivalent, and so such a distribution can be monitored by IR, Raman, and SFG. These scenarios are illustrated in Fig. 10.6. We will return to this example later in our further discussion of order parameters.

10.5 Surfaces and Buried Interfaces

10.5.1 SFG Intensity in a Reflection Geometry

There are many applications of SFG spectroscopy, but a particularly niche area is to study the interface between two isotropic achiral phases. In this way, $\boldsymbol{\chi}^{(2)} = 0$ in the bulk of each phase, but is non-zero at the interface. This can be an exposed surface, such as that between a solid or liquid and vapour such as air, or a buried interface between two condensed phases. Although reflection and transmission geometries are possible, it is more common to detect the reflected SFG beam, as this lends itself to a broader class of samples that don't need to be transparent at the SFG wavelengths. Since we have already established that the SFG field originates from the polarization $\mathbf{P}^{(2)}(\omega_{\mathrm{vis}} + \omega_{\mathrm{IR}})$, and this in turn originates from the second-order polarizability, we can understand that the intensity of the SFG response (I_{SFG}) is proportional to the square of these quantities. Since $\boldsymbol{\chi}^{(2)}$ is complex valued, we consider the square of the magnitude $|\boldsymbol{\chi}^{(2)}|^2$. How this quantity is related to individual tensor elements $\chi^{(2)}_{ijk}$ is worthy of some careful consideration. But there are some macroscopic points to address first. In a reflection experiment [2, 5]

$$I_{\mathrm{SFG}} = \frac{8\pi^3 \omega_{\mathrm{SFG}}^2 \sec^2 \theta_{\mathrm{SFG}}}{\hbar c^3 n_{\mathrm{SFG}} n_{\mathrm{vis}} n_{\mathrm{IR}}} |\chi^{(2)}_{\mathrm{measured}}|^2 I_{\mathrm{vis}} I_{\mathrm{IR}} \tag{10.15}$$

where I_{vis} and I_{IR} are the intensities of the incident visible and IR beams. This is a direct consequence of the relationship originally presented in the form of Eq. 10.3

$$P_i^{(2)}(\omega_{\mathrm{SFG}}) = \varepsilon_0 \sum_j^{x,y,z} \sum_k^{x,y,z} \chi^{(2)}_{ijk} E_j(\omega_{\mathrm{vis}}) E_k(\omega_{\mathrm{IR}}). \tag{10.16}$$

In Eq. 10.15, c is the speed of light, and n are the refractive indices of all beams in the incident medium that is assumed to be transparent (we could have written N instead of n since $\kappa \approx 0$).

Another quantity that appears in Eq. 10.15 is θ_{SFG}, the angle of the reflected SFG beam with respect to the surface normal. Just as in linear optics, this comes from a consideration of conversation of momentum. If we take the visible beam as an example, we can write

$$n_1 \omega_1 \sin\theta_1 = n_2 \omega_2 \sin\theta_2 \tag{10.17}$$

where the subscript 1 refers to the incident medium (for the incoming and reflected beam) and 2 refers to the medium in which the beam is refracted. Since there is no change in frequency $\omega_2 = \omega_2$, this results in the familiar expression for Snell's law of refraction $n_1 \sin\theta_1 = n_2 \sin\theta_2$ (Eq. 5.1) and Snell's law of reflection becomes $\theta_2 = \theta_1$ (Eq. 7.56) since $n_1 = n_2$ for the same medium. We can extend this momentum conservation consideration to the SFG situation as

$$n_{\mathrm{vis}} \omega_{\mathrm{vis}} \sin\theta_{\mathrm{vis}} + n_{\mathrm{IR}} \omega_{\mathrm{IR}} \sin\theta_{\mathrm{IR}} = n_{\mathrm{SFG}} \omega_{\mathrm{SFG}} \theta_{\mathrm{SFG}} \tag{10.18}$$

where $\omega_{\mathrm{SFG}} = \omega_{\mathrm{vis}} + \omega_{\mathrm{IR}}$. We can rearrange this to write

$$\theta_{\mathrm{SFG}} = \arcsin\left(\frac{n_{\mathrm{vis}} \omega_{\mathrm{vis}} \sin\theta_{\mathrm{vis}} + n_{\mathrm{IR}} \omega_{\mathrm{IR}} \sin\theta_{\mathrm{IR}}}{n_{\mathrm{SFG}} \omega_{\mathrm{SFG}}}\right). \tag{10.19}$$

If the incident medium is air, then it is a reasonable approximation that $n_{\text{SFG}} \approx n_{\text{vis}} \approx n_{\text{IR}}$ and Eq. 10.19 simplifies to

$$\theta_{\text{SFG}} \approx \arcsin\left(\frac{\omega_{\text{vis}} \sin\theta_{\text{vis}} + \omega_{\text{IR}} \sin\theta_{\text{IR}}}{\omega_{\text{vis}} + \omega_{\text{IR}}}\right).$$

The final quantity to consider in Eq. 10.15 is what we have termed $\chi^{(2)}_{\text{measured}}$. That quantity generally consists of linear combinations of the *effective* susceptibility. Leading up to that discussion we discuss some details of the response at material surfaces.

10.5.2 The Surface Susceptibility

So far our discussion has been rather general, with the aim of treating IR, Raman, and SFG spectroscopy in a similar way. However, the second-order susceptibility and polarization described above pertains to bulk phases, just as $\boldsymbol{\chi}^{(1)}$ (previously referred to simply as $\boldsymbol{\chi}$) and $\boldsymbol{\chi}^{(3)}$ are bulk material properties. In this way, Eq. 10.1 illustrates how all $\boldsymbol{\chi}^{(n)}$ terms contribute to the dipole moment per unit volume, a bulk quantity. At the surface, we are typically interested in the dipole moment per unit area, a quantity that some authors refer to as $\mathbf{P}_{\text{S}}$. This can be obtained by integrating the bulk polarization with respect to distance (z), weighted according to the distribution $\rho(z)$. For the second-order contribution, this is

$$P^{(2)}_{\text{S},i} = \varepsilon_0 \sum_j^{x,y,z} \sum_k^{x,y,z} \int_0^\infty \rho(z)\chi^{(2)}_{ijk}(z)E_j(z)\ E_k(z)\ \text{d}z. \tag{10.20}$$

If we assume that $\chi^{(2)}_{ijk} = 0$ everywhere except for at the surface, then we can use a Dirac δ-function for the distribution $\rho(z) = \delta(z)$ positioned at $z = 0$. We then have

$$\begin{aligned} P^{(2)}_{\text{S},i} &= \varepsilon_0 \sum_j^{x,y,z} \sum_k^{x,y,z} \int_0^\infty \delta(z)\chi^{(2)}_{ijk}(z)\ E_j(z)\ E_k(z)\ \text{d}z \\ &= \varepsilon_0 \sum_j^{x,y,z} \sum_k^{x,y,z} \chi^{(2)}_{ijk}(z=0)\ E_j(z=0)\ E_k(z=0) \\ &\equiv \varepsilon_0 \sum_j^{x,y,z} \sum_k^{x,y,z} \chi^{(2)}_{\text{S},ijk} E^0_j E^0_k. \end{aligned} \tag{10.21}$$

Although some SFG experiments are designed to characterize and separate the bulk $\boldsymbol{\chi}^{(2)}_{\text{B}}$ and surface $\boldsymbol{\chi}^{(2)}_{\text{S}}$ contributions to the second-order susceptibility, for the remainder of this chapter we will be concerned with the surface susceptibility only, and therefore write $\boldsymbol{\chi}^{(2)}$ rather than $\boldsymbol{\chi}^{(2)}_{\text{S}}$. Keep in mind, however, that the units of

$\chi_{\mathrm{B}}^{(2)}$ of $\mathrm{m^2V^{-2}}$ and $\chi_{\mathrm{S}}^{(2)}$ of $\mathrm{m^2V^{-1}}$ differ. In our simplified notation, what we hereafter refer to as $\boldsymbol{\chi}^{(2)}$ therefore has units of $\mathrm{m^2V^{-1}}$. We next need to address the surface fields E_j^0 for the visible beam and E_k^0 for the IR beam that appear Eq. 10.21.

10.5.3 Surface Electric Fields

When arriving at the second-order polarization in Eq. 10.1 the relevant electric fields $\mathbf{E}(\omega_{\mathrm{vis}})$ and $\mathbf{E}(\omega_{\mathrm{IR}})$ that contribute to $\mathbf{P}^{(2)}$ are not the ones that directly come out the laser, or even the ones that have made it through all of the optics before the sample. These are the fields precisely at the location where the SFG response is generated. In our case, this is at the surface, or at the interface between the two phases. This is exactly the situation we have already encountered in Chap. 8 when we discussed how the local fields need to be explicitly accounted for in IR absorption spectroscopy except in cases where a suitable reference material is available to experimentally cancel local field corrections. Consideration of surface fields is actually a fundamental aspect of many spectroscopic techniques. When metal substrates are used, this forms the basis of the surface selection rules [35]. In the case of SFG, it is possible but rare to have reference samples that enable researchers to bypass all local field considerations, so this is typically addressed in the data analysis.

If we consider the surface to be the xy-plane located at $z = 0$, we can revisit the pictures we studied when deriving the Fresnel transmission and reflection coefficients in Chap. 7. Starting with the case of s-polarized light approaching the surface (see Fig. 10.7), just before the surface at $z = 0^+$, the net field is described by the sum of incident and reflected fields

$$\begin{aligned} E_0^+ &= E_{\mathrm{i}} + E_{\mathrm{r}} \\ &= E_{\mathrm{i}} + r_{\mathrm{s}} E_{\mathrm{i}} \\ &= (1 + r_{\mathrm{s}}) E_{\mathrm{i}} \end{aligned} \tag{10.22}$$

where r_{s} is the familiar s-polarized Fresnel reflection coefficient defined in Eq. 7.49. Since the tangential components of the field shown in Fig. 10.7a are continuous everywhere, including across the interface, $E_0^+ = E_0^- = E_0$. In other words, this is also the solution just below the surface at $z = 0^-$ and for that matter at $z = 0$. This means that we don't need to fuss about where the molecules that generated the SFG response are located. As long as they are in the vicinity of $z = 0$, the result is the same. We therefore define the y-component of the local field correction as

$$L_{yy} = \frac{E_{z=0}}{E_i} = 1 + r_{\mathrm{s}}. \tag{10.23}$$

In other words, multiplying the s-polarized incident field by L_{yy} provides the y-component of the field at the surface that contributes to $\mathbf{P}^{(2)}$. The double subscript in L_{yy} indicates that $\mathbf{L}$ is a rank-2 tensor that allows for the possibility of anisotropic local field corrections. In general, it is more flexible to write

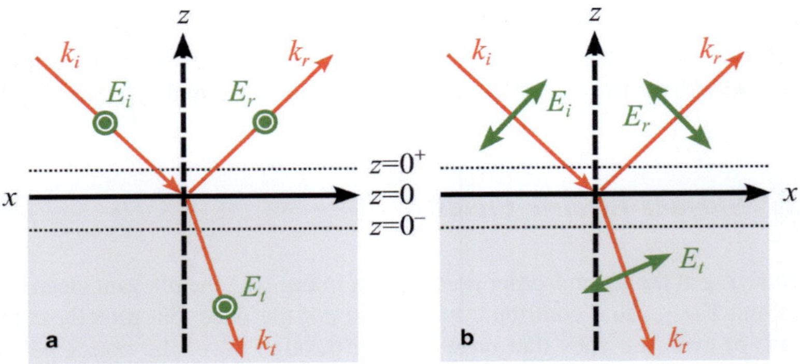

Fig. 10.7 An illustration of the fields in the vicinity of the surface when the electric field vectors are tangential to the surface, **a** as the fields along y when the incoming light is s-polarized, or **b** the x-component of the p-polarized fields

$$L_{yy} = (1 + r_\mathrm{s}) \sin\gamma \tag{10.24}$$

where γ is the angle of the incident beam polarization to the direction that would be p-polarized. This is because, if the incident beam is linearly polarized, but in a state that is mixed between s and p, we need to know what fraction ($\sin\gamma$) is along the s-direction. As the plane of incidence may be horizontal (parallel to the surface of the optical table) or vertical, it is better to specify the angle γ in this way.

In order to arrive at the x-component of the surface field, we need to consider how p-polarized light interacts with the sample, as shown in Fig. 10.7b. Since the x-component of the field is also tangent to the surface, we are in a similar situation where the solutions stepping over an infinitesimal distance along z between $z = 0^+$, $z = 0$, and $z = 0^-$ are the same, and all result in

$$\begin{aligned} E_0 &= [E_\mathrm{i} + (-E_\mathrm{r})] \cos\theta_i \cos\gamma \\ &= (1 - r_\mathrm{p}) E_\mathrm{i} \cos\theta_\mathrm{i} \cos\gamma \end{aligned} \tag{10.25}$$

where $\cos\gamma$ provides the fraction of a linearly polarized input state that projects along p. In our coordinate system, p-polarized light has field components along the x- and z-directions, the relative magnitudes of which depend on the angle of incidence. Since we are after the x-component, we consider $\cos\theta_i$ and we end up with

$$L_{xx} = \pm(1 - r_\mathrm{p}) \cos\theta_\mathrm{i} \cos\gamma. \tag{10.26}$$

The sign selection when using Eq. 10.26 depends on whether we are describing one of the incoming p-polarized (visible or IR) beams, or the reflected SFG beam. Since the incoming beams approach the surface, but the reflect beam leaves the surface, the x-components of the fields have opposite directions. We therefore use

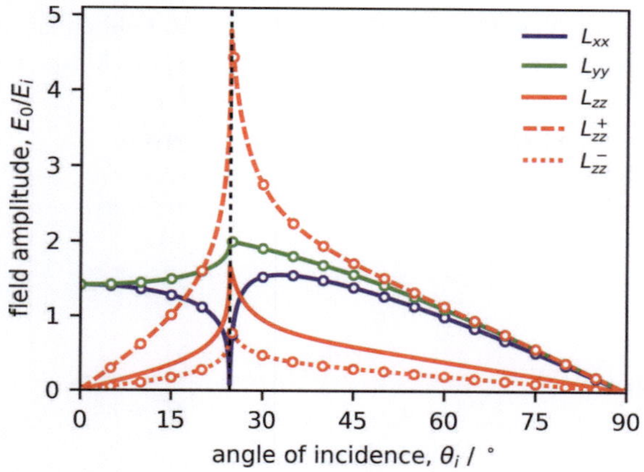

Fig. 10.8 Plots indicating the ratio of the surface amplitude to the incident field amplitude as a function of the angle of incidence. Solutions based on the simple expressions for L_{xx}, L_{yy}, and L_{zz} are plotted with lines over the complete range $0° < \theta_1 < 90°$. For the L_{zz} solution we use $N' = \frac{1}{2}(N_1 + N_2)$ in Eq. 10.28. The solution to L_{zz}^- (just inside diamond, at $z = 0^-$) is plotted with a dotted red line, and L_{zz}^+ (just into the air phase) is plotted with a dashed red line. For comparison, solutions of the full distance-dependent treatment are plotted with points. This representation enables us to see that we arrive at the same solution

$L_{xx} = (1 - r_\text{p}) \cos\theta_i \cos\gamma$ for either of the incoming beams, and $L_{xx} = -(1 - r_\text{p}) \cos\theta_\text{i} \cos\gamma$ for the reflected SFG field.

The trouble starts when we need to consider the z-component, as normal components of the fields are discontinuous and can be significantly stepped at the interface. Just above the surface (at what we've been calling $z = 0^+$),

$$E_0^+ = E_\text{i} + E_\text{r} = (1 + r_\text{p}) E_\text{i} \sin\theta_\text{i} \cos\gamma$$

so $L_{zz}^+ = (1 + r_\text{p}) \sin\theta_\text{i} \cos\gamma$. Just below the surface at $z = 0^-$ one finds

$$E_0^- = E_\text{i} + E_\text{r} = (1 + r_\text{p}) E_\text{i} \left(\frac{N_1}{N_2}\right)^2 \sin\theta_\text{i} \cos\gamma.$$

We can see that the severity of the discontinuity depends on the contrast between the refractive indices of the two adjacent bulk phases. Before continuing to address this, let's take a moment to plot the surface fields $E_x(z = 0)$, $E_y(z = 0)$, $E_z(z = 0^+)$, and $E_z(z = 0^-)$. These are shown in Fig. 10.8, plotted with lines, as a function of the angle of incidence with respect to the corresponding incoming field component for the case where $N_1 = 2.4$ and $N_2 = 1$. $E_z(\theta_1)$ is plotted in red, with the solid line displaying the values of the field normal to the surface at $z = 0^+$ and the dashed line for $z = 0^-$. The value of the critical angle is indicated by the vertical dashed black line. For comparison, solutions to the distant-dependent equations from Chap. 7 (Eqs. 7.108–7.110) are overlaid with points. This solution is also of significant interest in the analysis of ATR-IR data [36, 37], due to the field enhancement near the surface. A famous solution at $z = 0^+$ is the so-called Harrick Equations [36]:

$$L_{xx}(z = 0^+) = \frac{2\cos\theta_1\sqrt{\sin^2\theta_1 - n_{21}^2}\cos\gamma}{\sqrt{1 - n_{21}^2}\sqrt{(1 + n_{21}^2)\sin^2\theta_1 - n_{21}^2}} \tag{10.27a}$$

$$L_{yy}(z = 0^+) = \frac{2\cos\theta_1}{\sqrt{1 - n_{21}^2}}\sin\gamma \tag{10.27b}$$

$$L_{zz}(z = 0^+) = \frac{2\sin\theta_1\cos\theta_1\cos\gamma}{\sqrt{1 - n_{21}^2}\sqrt{(1 + n_{21}^2)\sin^2\theta_1 - n_{21}^2}}, \tag{10.27c}$$

where Harrick (and others in the ATR community) use the notation $n_{21} \equiv N_2/n_1$. Keep in mind that for these equations, n_1 is the refractive index of the prism, which is assumed to be transparent. Note that Eq. 10.27 is valid only for angles of incidence θ_1 above the critical angle, but our more general solution is valid at all angles, as can be seen in Fig. 10.8. In the Harrick equations, the solution at $z = 0^+$ is logical as no analyte is present inside the prism, and it is reasonable to approximate the prism–air or prism–solution interface as presenting a sharp boundary between regions of two distinct refractive indices. However, in the case of SFG, it is at times not so obvious exactly where in that infinitesimal region the signal is generated. A compromise solution is therefore

$$L_{zz} = (1 + r_\mathrm{p})\left(\frac{N_1}{N'}\right)^2\cos\theta_i\cos\gamma, \tag{10.28}$$

where setting $N' = N_1$ captures the behaviour at $z = 0^+$, $N' = N_2$ describes the situation at $z = 0^-$, and setting N' to a unique value may be used to produce something in between the two extremes. For example, $N' = \frac{1}{2}(N_1 + N_2)$ is a popular choice [6, 38]. There are more detailed theories about the behaviour of electric fields at surfaces [39–42], including models where a gradient between the $z = 0^+$ and $z = 0^-$ solutions may be mapped onto the density profile [43–47]. But for a more cursory discussion, just note that N' does not necessarily correspond to the refractive index of any particular species that is present, but accounts for the fact that the often important z-component of the field needs some special consideration.

We now have what we need to form a single element of the effective susceptibility tensor via

$$\begin{aligned}\chi^{(2)}_{ijk,\mathrm{eff}} &= L_{ii,\mathrm{SFG}}L_{jj,\mathrm{vis}}L_{kk,\mathrm{IR}}\chi^{(2)}_{ijk} \\ &= L_{ii}L_{jj}L_{kk}\chi^{(2)}_{ijk},\end{aligned} \tag{10.29}$$

where the second line introduces a simpler notation based on the SFG-vis-IR order so which L_{ii} factor is associated with which beam is implied from its position: SFG on the left, visible in the middle, and IR on the right. For example, $\chi^{(2)}_{xxz,\mathrm{eff}}$ is given by

$L_{xx,\mathrm{SFG}} L_{xx,\mathrm{vis}} L_{zz,\mathrm{IR}} \chi^{(2)}_{xxz}$ that we will now write as $L_{xx} L_{xx} L_{zz} \chi^{(2)}_{xxz}$. Since the SFG field is x-polarized in this example, the evaluation of L_{xx} would use the negative sign in Eq. 10.26 for the SFG beam, but the positive sign in front of L_{xx} for the visible beam. It is worth pointing out that another convention sometimes used in the literature is to consider only the positive sign in the formal definition of L_{xx}, but then write $\chi^{(2)}_{xxz,\mathrm{eff}} = -L_{xx} L_{xx} L_{zz} \chi^{(2)}_{xxz}$. This sounds more confusing than it is; the important part is to keep track of which convention is being followed so the correct signs are applied.

This model for **L** is developed for the case of the interface between two semi-infinite phases. For example, the interface between air and a bulk solid polymer. SFG studies are often performed on samples adsorbed at these interfaces. For example, surfactants that are isotropically distributed (and therefore not SFG active) in the aqueous phase, but are then ordered when they adsorb at the air–water interface. A common question (and one that is currently research active) is whether a description of the local fields that includes only the adjacent bulk phases (ignoring the surfactant) suffices. Another common scenario is the study of solid–liquid interfaces such as the aqueous–polymer interface. Here the beams either approach through a thin water layer (so as to minimize absorption of the IR beam) or through a thin polymer film. Thin films also create unique challenges and opportunities for the **L** factors. Rather than interrupting the flow of this discussion that will now continue towards molecular orientation analysis, we revisit these topics later in Sect. 10.8.

10.5.4 Polarization Combinations

In an SFG experiment, one can control the polarization of the incident IR and visible beams independently, and select the polarization of the detected SFG beam as shown in Fig. 10.5. There is no reason why a fancy elliptically polarized state can't be generated for the incoming visible and/or IR beams, and why the Stokes vector of the resulting SFG beam shouldn't be measured. However, for the sake of simplicity, we limit our discussion to input beams being s- or p-polarized, and the s- or p-component of the resulting SFG beam selected for detection. In this way, all of the polarization analysis can be done with simple polarizers, but it is better to use retarder–polarizer combinations for the input beams as we wish to achieve s- and p-polarized states with (ideally) equally intensities, and not sacrifice any intensity from the source lasers. Since we have three beams, this results in $2^3 = 8$ possible polarization combinations that we label SSS, SSP, SPS, SPP, PSS, PSP, PPS, and PPP using the increasing wavelength convention SFG-vis-IR. Although we have been using lowercase letters to label the s- and p-polarized fields, uppercase letters are used to label polarization combinations, often referred to as polarization schemes, since these are proxies for SFG intensities, effective susceptibility combinations, or experimental configurations, as we shall see. In the previous section, we discussed elements of the effective second-order susceptibility $\chi^{(2)}_{ijk,\mathrm{eff}}$. We now consider how these are related to what is measured in an experiment, scalar

Table 10.1 The eight possible $\chi^{(2)}_{\text{measured}}$ combinations if the two input beams are restricted to s- and p-polarized states, and the s- or p-component of the SFG field is selected for detection

$\chi^{(2)}_{\text{measured}}$	Constituent $L_{ii}L_{jj}L_{kk}\chi^{(2)}_{ijk}$ elements	Number of elements
SSS	yyy	1
SSP	yyx, yyz	2
SPS	yxy, yzy	2
SPP	yxx, yxz, yzx, yzz	4
PSS	xyy, zyy	2
PSP	xyx, xyz, zyx, zyz	4
PPS	xxy, xzy, zxy, zzy	4
PPP	xxx, xxz, xzx, xzz, zxx, zxz, zzx, zzz	8
Total number		27

quantities that we refer to as $\chi^{(2)}_{\text{measured}}$ according to the polarization scheme. One of them is especially simple, as it probes only a single element of $\boldsymbol{\chi}^{(2)}_{\text{eff}}$, that is, $\chi^{(2)}_{\text{SSS}} = L_{yy}L_{yy}L_{yy}\chi^{(2)}_{yyy}$. Keep in mind that, although we are not labelling the corresponding frequencies on the components of **L**, the three L_{yy} factors associated with $\chi^{(2)}_{yyy}$ are all different, as they are calculated using the corresponding beam angle (SFG, visible, or IR) and the appropriate frequency-dependent refractive indices. In this simple example, all of the beams were s-polarized, so their electric fields were along the laboratory y-direction. Since p-polarized fields have a component of **E** along x and z (the relative magnitudes of which depend on the beam angle), all terms in $\chi^{(2)}_{\text{measured}}$ that involve p-polarized fields incorporate multiple components of $\boldsymbol{\chi}^{(2)}_{\text{eff}}$. For example,

$$\chi^{(2)}_{\text{SSP}} = L_{yy}L_{yy}L_{xx}\chi^{(2)}_{yyx} + L_{yy}L_{yy}L_{zz}\chi^{(2)}_{yyz}.$$

Another example of $\chi^{(2)}_{\text{measured}}$ is

$$\begin{aligned}\chi^{(2)}_{\text{PSP}} = {} & L_{xx}L_{yy}L_{xx}\chi^{(2)}_{xyx} + L_{xx}L_{yy}L_{zz}\chi^{(2)}_{xyz} \\ & + L_{zz}L_{yy}L_{xx}\chi^{(2)}_{zyx} + L_{zz}L_{yy}L_{zz}\chi^{(2)}_{zyz}\end{aligned}$$

where it is additionally (and, warning, for the last time!) noted that writing $L_{xx}L_{yy}L_{xx}$ and $L_{xx}L_{yy}L_{zz}$ rather than $-L_{xx}L_{yy}L_{xx}$ and $-L_{xx}L_{yy}L_{zz}$ (as is often the case in SFG literature) signifies that the proper sign of L_{xx} for the SFG beam has been incorporated into its definition. Table 10.1 shows all the remaining combinations and the encoded $\chi^{(2)}_{ijk,\text{eff}}$ terms.

For those with previous experience collecting and analysing SFG spectra, this comprehensive list may seem surprising at first, as we typically associated the SFG response from molecules with fewer elements of the $\boldsymbol{\chi}^{(2)}$ tensor, and therefore in practice polarization schemes such as SSP often probe only a single element of $\boldsymbol{\chi}^{(2)}$ rather than a linear combination of the two elements shown in Table 10.1. This is because the next stage considers symmetries that may present at the $\boldsymbol{\chi}^{(2)}$ and/or $\boldsymbol{\alpha}^{(2)}$ level. Just before getting into that discussion, we make some general remarks about SFG spectra.

10.6 Vibrational SFG Spectra

So far we have been discussing individual elements of $\boldsymbol{\chi}^{(2)}$ without much regard for the frequency of any of the beams, apart from our calculation of the reflected SFG beam angle, and the frequency-dependent refractive index that goes into the **L** factors. Although it is common practice to refer to elements of $\boldsymbol{\chi}^{(2)}$ when describing the SFG response, we follow the convention that we have established when discussing IR and Raman spectra and instead use $A_{ijk}^{(2)}$ to signify that these are the amplitudes of the vibrationally resonant SFG response, obtained by fitting spectra to a sum of terms such as in Eq. 10.8 to arrive at

$$\chi_{ijk}^{(2)}(\omega_{\mathrm{IR}}) = \chi_{ijk,\mathrm{NR}}^{(2)} + \sum_q \frac{A_{lmn,q}^{(2)}}{\omega_q - \omega_{\mathrm{IR}} - i\Gamma_q} \tag{10.30}$$

where $\chi_{ijk,\mathrm{NR}}^{(2)}$ is a vibrationally non-resonant contribution to the response, and we are summing over q vibrational modes, each with a resonance frequency ω_q and line width Γ_q. Before considering the fitting and subsequent quantitative analysis further, let's inspect some of the qualitative features of SFG spectra and their information content.

Figure 10.9 shows the SFG spectra obtained from the interface of a solution of bovine serum albumin in contact with a polystyrene surface at pH 5 using the SSP polarization combination. As a reminder, this means that the incoming visible beam (middle index) was s-polarized, the incoming IR beam (last index) was p-polarized, and the s-component of the SFG (first index) generated from the surface was selected for detection. This was a wavelength-scanning SFG experiment performed with picosecond pulses. Initially an IR beam at $\omega_{\mathrm{IR}} = 2800\ \mathrm{cm}^{-1}$ was generated, and overlapped at the polymer–aqueous interface with a visible beam at $\lambda_{\mathrm{vis}} = 532$ nm or $\omega_{\mathrm{vis}} = 18797\ \mathrm{cm}^{-1}$. We anticipate SFG generation at $\omega_{\mathrm{SFG}} = \omega_{\mathrm{vis}} + \omega_{\mathrm{IR}} = 21597\ \mathrm{cm}^{-1}$ or $\lambda_{\mathrm{SFG}} = 463$ nm. Since we additionally know the incident beam angles of $\theta_{\mathrm{vis}} = 60°$ and $\theta_{\mathrm{IR}} = 50°$ and the refractive indices of the incident phase (polystyrene), we can use Eq. 10.19 to calculate that we expect the reflected SFG at $\theta_{\mathrm{SFG}} = 58.5°$ with respect to the surface normal. We therefore position the detector arm to this angle, and set the diffraction grating in the spectrograph so that $\lambda = 463$ nm is selected. The leftmost axis in Fig. 10.9 indicates that approximately 100 counts of SFG signal were detected. If the response is calibrated using a material with known value of $\boldsymbol{\chi}^{(2)}$, these counts can be converted into SI units of $\boldsymbol{\chi}^{(2)}$ as shown on the inner vertical axis in Fig. 10.9. Since this is an SSP experiment, the presence of this signal indicates that either $\chi_{yyx}^{(2)}$ or $\chi_{yyz}^{(2)}$ (or both) is non-zero. Once the data has been collected, typically averaging over 100 or so

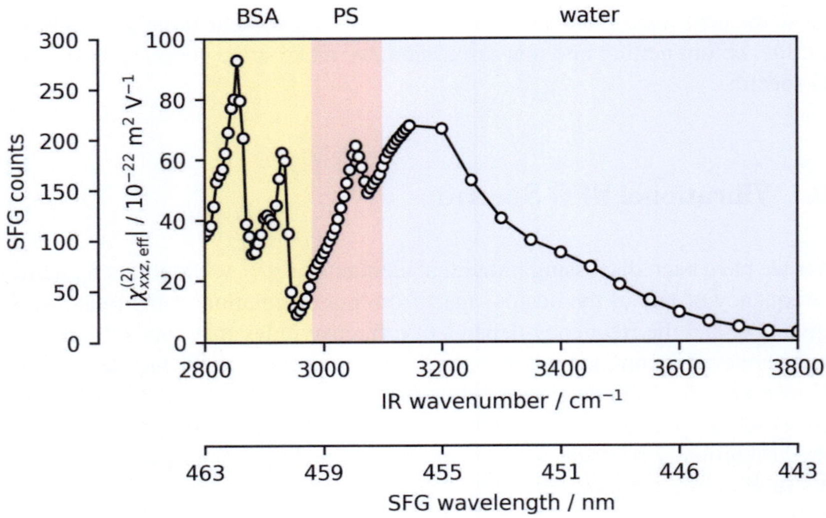

Fig. 10.9 A spectrum that illustrates SFG signal generated at the interface between polystyrene-d_2 and bovine serum albumin at pH 5. The SSP polarization scheme was used. The two vertical axes illustrate how the measured SFG signal in arbitrary units may be converted to units of the surface susceptibility using a reference material. The two horizontal axes demonstrate the detected SFG wavelength corresponding to the wavenumber of the IR beam used to generate the SFG

pulses to increase the signal-to-noise ratio, the next IR frequency is generated (for example, $\omega_{\mathrm{IR}} = 2805\ \mathrm{cm}^{-1}$), and the spectrograph is prepared to record any SFG counts at the next SFG wavelength, calculated to be $\lambda_{\mathrm{SFG}} = 459$ nm. It is interesting to note that θ_{SFG} also changes when scanning ω_{IR}, as predicted by Eq. 10.19. However, even though the detection system could be motorized in order to track this angle, this is seldom necessary. It is a good practice to use Eq. 10.19 to predict how much θ_{SFG} is predicted to change for the ω_{IR} range of interest, using knowledge of (or an estimate of) the refractive indices. Often one finds that the SFG angle varies only by a very small amount, and this can be compensated using focusing optics at the entrance of the spectrograph. If the θ_{SFG} variation is more significant, it is helpful to set the detector angle to the middle of the anticipated range, rather than optimizing it at the starting ω_{IR} value, and open the entrance slit a bit wider. This helps to ensure that the signal is not inadvertently clipped during the ω_{IR} scan.

In this example, polystyrene-d_2 was used with the polymer methylene groups deuterated in order to free up the aliphatic C–H stretching region for the observation of protein vibrational modes. Several protein C–H stretching modes are observable in the 2800–3000 cm^{-1} region. We know that any protein molecules in solution are not contributing to this response, as we have already worked out above that both $\chi^{(2)}_{yyx}$ and $\chi^{(2)}_{yyz}$ are zero for an isotropic distribution. This indicates that some proteins are adsorbed on the surface with an orientation distribution that breaks the centrosymmetry. This is not surprising, as most molecules will exhibit a preference for which side preferentially interacts with a particular surface, rendering the "up" (towards the solution phase) and "down" (towards the bulk polymer phase) directions unequally populated. Here we can see an additional benefit of SFG compared to off-resonance or electronically resonant techniques in that the transitions are local to specific chemical functional groups. It is intriguing that protein aromatic C–H stretching from residues such as phenylalanine, tyrosine, and tryptophan are almost never observed, which is purportedly related to the lack of symmetry breaking for those residues even as other parts of the protein are ordered in a polar manner at surfaces. In this example, this can be confirmed using perdeuterated polystyrene-d_8 coupled with the observation that no SFG vibrational resonances appear in the 3000–3200 cm^{-1} region. This is convenient, as aromatic C–H stretching in Fig. 10.9 may then be attributed to the phenyl rings at the polystyrene surface, providing a handle on the polymer orientation [48–57].

Finally, the remaining observation from Fig. 10.9 is the large SFG response from the water O–H stretching in the region 2900–3800 cm^{-1}. This reveals that the hydrogen bonding network, although centrosymmetric in the bulk aqueous phase, is ordered in a polar manner at the surface. The study of water organization next to surfaces is fundamental to a large range of biological, environmental, and technological applications [58–67]. When SFG spectroscopy is used, it is of additional interest to compare the shape of the water response—for example, the relative ratio of the 3200 cm^{-1} and 3400 cm^{-1} bands to that observed in bulk from IR absorption or Raman scattering. Such information can be used to comment on the orientation and coordination of interfacial water molecules.

There are three molecules of interest here: polystyrene, serum albumin, and water. SFG spectroscopy enables the interaction between all species to be probed. For example, the surface of polystyrene may be structured differently in air compared to when it is immersed in water; the corresponding interfacial water structure may be altered in the presence of the adsorbed protein. One of the exciting scientific pursuits in this field is to unravel what affects what. Is the protein adhesion driven by the interfacial hydrogen bonding environment, or is the water only along for the ride [68–70]? Similarly, is the orientation of the polymer affected by the presence of the protein? One can imagine that is possible to design a series of sequential experiments that provide answers to these questions. This above example illustrates the richness of the information that is available based on only a qualitative inspection of an SFG spectrum acquired in a single polarization scheme. Further information is available by considering polarization combinations other than SSP, most notably information that can be linked to the orientation distribution at the surface.

10.7 Interfaces with Azimuthal Symmetry About the Surface Normal

Some materials are ordered in the xy-plane of the interface, for example, a rubbed polymer that is used as an alignment layer in liquid crystal displays [71–73]. This in turn can cause adsorbates to display some degree of surface epitaxy, that is, they can be ordered in a way that is commensurate with the in-plane orientation of the substrate [71, 74–78]. Although such systems and their analysis are fascinating, we restrict ourselves to the much simpler and often encountered case where molecules are ordered at the surface (as in the case of our polystyrene example) or adsorb at the surface (water and serum albumin) in a manner that does not display any preference for the azimuthal angle ϕ about the surface normal z. In an end-of-chapter problem, you can try keeping the same 27 elements of $\boldsymbol{\alpha}^{(2)}$ as we defined at the beginning of this chapter, the 18 unique ones with $\alpha^{(2)}_{lmn} = \alpha^{(2)}_{mln}$, and can perform the integration about ϕ and evaluate each of the 27 $\boldsymbol{\chi}^{(2)}$ elements to end up with $\chi^{(2)}_{ijk}(\theta, \psi)$.

10.7.1 Molecules with No Preference for Their Orientation About One of Their Molecular Axes

We first consider the case where molecules have the maximum number (27) of non-zero but symmetric (18 unique) elements in their $\boldsymbol{\alpha}^{(2)}$ tensor, but have no preference for the way they are orientated about a particular molecular axis, typically the molecular "long" axis that is designated c. This results in

$$\begin{aligned}\chi^{(2)}_{xxz} &= \frac{N_\mathrm{A}}{4\varepsilon_0}\Big[\Big(2\alpha^{(2)}_{ccc} - 2\alpha^{(2)}_{aca} - 2\alpha^{(2)}_{bcb} + \alpha^{(2)}_{aac} + \alpha^{(2)}_{bbc}\Big)\langle\cos\theta\rangle \\ &\quad + \Big(2\alpha^{(2)}_{aca} + 2\alpha^{(2)}_{bcb} - 2\alpha^{(2)}_{ccc} + \alpha^{(2)}_{aac} + \alpha^{(2)}_{bbc}\Big)\langle\cos^3\theta\rangle\Big] \\ &= \frac{N_\mathrm{A}}{5\varepsilon_0}\Big[\Big(\alpha^{(2)}_{ccc} - \alpha^{(2)}_{aca} - \alpha^{(2)}_{bcb} + 2\alpha^{(2)}_{aac} + 2\alpha^{(2)}_{bbc}\Big)\langle P_1\rangle \\ &\quad + \frac{1}{2}\Big(2\alpha^{(2)}_{aca} + 2\alpha^{(2)}_{bcb} - 2\alpha^{(2)}_{ccc} + \alpha^{(2)}_{aac} + \alpha^{(2)}_{bbc}\Big)\langle P_3\rangle\Big] \\ &= \chi^{(2)}_{yyz}\end{aligned} \tag{10.31a}$$

along with

$$\begin{aligned}\chi^{(2)}_{xzx} &= \frac{N_\mathrm{A}}{4\varepsilon_0}\Big[\Big(2\alpha^{(2)}_{ccc} - \alpha^{(2)}_{aac} - \alpha^{(2)}_{bbc}\Big)\langle\cos\theta\rangle \\ &\quad + \Big(2\alpha^{(2)}_{aca} + \alpha^{(2)}_{aac} + 2\alpha^{(2)}_{bcb} + \alpha^{(2)}_{bbc} - 2\alpha^{(2)}_{ccc}\Big)\langle\cos^3\theta\rangle\Big]\end{aligned} \tag{10.31b}$$

$$= \frac{N_A}{10\varepsilon_0}\left[\left(2\alpha^{(2)}_{ccc} - \alpha^{(2)}_{aac} - \alpha^{(2)}_{bbc} + 3\alpha^{(2)}_{aca} + 3\alpha^{(2)}_{bcb}\right)\langle P_1\rangle\right.$$
$$\left. + \left(2\alpha^{(2)}_{aca} + \alpha^{(2)}_{aac} + 2\alpha^{(2)}_{bcb} + \alpha^{(2)}_{bbc} - 2\alpha^{(2)}_{ccc}\right)\langle P_3\rangle\right]$$
$$= \chi^{(2)}_{zxx} = \chi^{(2)}_{yzy} = \chi^{(2)}_{zyy}$$

and

$$\chi^{(2)}_{zzz} = \frac{N_A}{2\varepsilon_0}\left[\left(2\alpha^{(2)}_{aca} + \alpha^{(2)}_{aac} + 2\alpha^{(2)}_{bcb} + \alpha^{(2)}_{bbc}\right)\langle\cos\theta\rangle\right. \tag{10.31c}$$
$$\left. - \left(2\alpha^{(2)}_{aca} + 2\alpha^{(2)}_{bcb} + \alpha^{(2)}_{aac} + \alpha^{(2)}_{bbc} - 2\alpha^{(2)}_{ccc}\right)\langle\cos^3\theta\rangle\right]$$
$$= \frac{N_A}{5\varepsilon_0}\left[\left(2\alpha^{(2)}_{aca} + 2\alpha^{(2)}_{bcb} + \alpha^{(2)}_{aac} + \alpha^{(2)}_{bbc} - 3\alpha^{(2)}_{ccc}\right)\langle P_1\rangle\right.$$
$$\left. - \left(2\alpha^{(2)}_{aca} + 2\alpha^{(2)}_{bcb} + \alpha^{(2)}_{aac} + \alpha^{(2)}_{bbc} - 2\alpha^{(2)}_{ccc}\right)\langle P_3\rangle\right]$$

with the other 20 elements equal to zero. Here, and throughout the remainder of this chapter, N_A is the surface number density, the number of molecules per unit area. This goes together with our definition that includes the units of the surface (as opposed to bulk) $\boldsymbol{\chi}^{(2)}$ tensor. We find that there are seven non-zero elements of $\chi^{(2)}$, three of which are unique, and two of which are independent [47]. There are two practical consequences of this result. The first is that SFG signal is expected in only four of the eight polarization schemes that we initially listed: SSP, SPS or equivalently PSS, and PPP. Note that although the information content in SPS and PSS is equivalent (they are each a probe of $\chi^{(2)}_{xzx} = \chi^{(2)}_{zxx} = \chi^{(2)}_{yzy} = \chi^{(2)}_{zyy}$) SPS and PSS spectra generally have different intensities, on account of the different $\mathbf{L}$ factors associated with the x- versus y-projections. This means that one can select to measure either SPS or PSS, whichever is easier to set up for any reason. If both are possible, then one can select the scheme that has the largest value of $|L_{ii}L_{jj}L_{kk}|^2$ for a particular set of beam angles in order to maximize the signal-to-noise ratio. The second important result here is that SSP and SPS (or PSS) probe a single isolated element (due to the equivalences in Eqs. 10.31). PPP, on the other hand, is a different story, and presents a unique set of opportunities and challenges. One way of looking at this is from the perspective of Table 10.1. Of the eight contributions to PPP listed there, four of them are zero according to Eqs. 10.31. So PPP probes a linear combination of the three unique non-zero $\boldsymbol{\chi}^{(2)}$ elements. This result is fairly important, so it is worth writing out the expression explicitly:

$$\begin{aligned}
\chi^{(2)}_{\mathrm{PPP}} &= L_{xx}L_{xx}L_{zz}\chi^{(2)}_{xxz} + L_{xx}L_{zz}L_{xx}\chi^{(2)}_{xzx} + L_{zz}L_{xx}L_{xx}\chi^{(2)}_{zxx} \\
&\quad + L_{zz}L_{zz}L_{zz}\chi^{(2)}_{zzz} \\
&= L_{xx}L_{xx}L_{zz}\chi^{(2)}_{xxz} + (L_{xx}L_{zz}L_{xx} + L_{zz}L_{xx}L_{xx})\,\chi^{(2)}_{xzx} \\
&\quad + L_{zz}L_{zz}L_{zz}\chi^{(2)}_{zzz} \\
&= L_{xx}L_{xx}L_{zz}\chi^{(2)}_{yyz} + (L_{xx}L_{zz}L_{xx} + L_{zz}L_{xx}L_{xx})\,\chi^{(2)}_{yzy} \\
&\quad + L_{zz}L_{zz}L_{zz}\chi^{(2)}_{zzz}.
\end{aligned} \tag{10.32}$$

The first line in Eq. 10.32 follows straight from the definitions we have established, with our usual disclaimer of remembering to take care of the proper sign of L_{xx} when it applies to the SFG beam (first index). In the second line of Eq. 10.32 we make use of the fact that $\chi^{(2)}_{xzx} = \chi^{(2)}_{zxx}$ far from electronic resonance (because we imposed this symmetry on $\boldsymbol{\alpha}^{(2)}$). Finally, the last line in Eq. 10.32 highlights the fact that, since $\chi^{(2)}_{xxz} = \chi^{(2)}_{yyz}$, this contribution of the information content to the PPP response is present, in an isolated form, in the SSP spectrum. And the same goes for the xzx element, as it can be obtained from the SPS or PSS response.

As a further point in cases where $\theta_{\mathrm{vis}} \approx \theta_{\mathrm{IR}}$, we find that $|L_{xx}L_{zz}L_{xx}| \approx -|L_{zz}L_{xx}L_{xx}|$. This combined with the fact that the associated $\boldsymbol{\chi}^{(2)}$ elements are equal often results in a situation where

$$\chi^{(2)}_{\mathrm{PPP}} \approx L_{xx}L_{xx}L_{zz}\chi^{(2)}_{xxz} + L_{zz}L_{zz}L_{zz}\chi^{(2)}_{zzz}.$$

It is also interesting to note that we started with an $\boldsymbol{\alpha}^{(2)}$ tensor that had 18 unique elements but, owing to the macroscopic symmetry about ϕ and ψ, only seven elements of $\boldsymbol{\alpha}^{(2)}$ appear in these 7 $\boldsymbol{\chi}^{(2)}$ elements. Nevertheless, if the molecular response is truly this complicated—i.e. the vibrational mode doesn't have any higher symmetry that would further simplify $\boldsymbol{\alpha}^{(2)}$ by introducing some identical and/or zero elements—then any experimental efforts to analyse the molecular orientation, even though we have only θ-dependence, must have some way of accessing values for those 7 $\boldsymbol{\alpha}^{(2)}$ elements.

10.7.2 Vibrational Modes with Rotational Symmetry About a Molecular Axis

We now explore the case of a simpler molecular-frame $\boldsymbol{\alpha}^{(2)}$ tensor. Unlike our discussion of IR and Raman response functions where we kept only α_{cc} and $\alpha^{(3)}_{cccc}$, we won't simply the $\boldsymbol{\alpha}^{(2)}$ tensor quite to that extent. However, for high symmetry modes such as the methyl symmetric stretch, $\boldsymbol{\alpha}^{(2)}$ can be approximated with only three non-zero elements: $\alpha^{(2)}_{aac} = \alpha^{(2)}_{bbc}$ and $\alpha^{(2)}_{ccc}$. Note that this is equivalent to starting out with the full form of an (achiral) $\boldsymbol{\alpha}^{(2)}$ tensor and additionally integrating about ψ in the molecular frame. This essentially imposes rotational symmetry about the molecular c-axis to render the a- and b-directions indistinguishable. We can either work this result out from scratch, or start with the result shown in Eq. 10.31 and set the unwanted $\alpha^{(2)}$ elements to zero. Either way, this results in

$$\begin{aligned} A^{(2)}_{xxz} = A^{(2)}_{yyz} &= \frac{N_{\mathrm{A}}}{2\varepsilon_0}\left[\left(\alpha^{(2)}_{aac} + \alpha^{(2)}_{ccc}\right)\langle\cos\theta\rangle - \left(\alpha^{(2)}_{ccc} - \alpha^{(2)}_{aac}\right)\langle\cos^3\theta\rangle\right] \\ &= \frac{N_{\mathrm{A}}}{5\varepsilon_0}\left[\left(4\alpha^{(2)}_{aac} + \alpha^{(2)}_{ccc}\right)\langle P_1\rangle - \left(\alpha^{(2)}_{ccc} - \alpha^{(2)}_{aac}\right)\langle P_3\rangle\right] \end{aligned} \tag{10.33a}$$

$$\begin{aligned} A^{(2)}_{xzx} = A^{(2)}_{zxx} = A^{(2)}_{yzy} = A^{(2)}_{zyy} &= \frac{N_{\mathrm{A}}}{2\varepsilon_0}\left(\alpha^{(2)}_{ccc} - \alpha^{(2)}_{aac}\right)\left(\langle\cos\theta\rangle - \langle\cos^3\theta\rangle\right) \\ &= \frac{N_{\mathrm{A}}}{5\varepsilon_0}\left(\alpha^{(2)}_{ccc} - \alpha^{(2)}_{aac}\right)\left(\langle P_1\rangle - \langle P_3\rangle\right) \end{aligned} \tag{10.33b}$$

$$\begin{aligned} A^{(2)}_{zzz} &= \frac{N_{\mathrm{A}}}{\varepsilon_0}\left[\alpha^{(2)}_{aac}\langle\cos\theta\rangle + \left(\alpha^{(2)}_{ccc} - \alpha^{(2)}_{aac}\right)\langle\cos^3\theta\rangle\right] \\ &= \frac{N_{\mathrm{A}}}{5\varepsilon_0}\left[\left(2\alpha^{(2)}_{aac} + 3\alpha^{(2)}_{ccc}\right)\langle P_1\rangle + 2\left(\alpha^{(2)}_{ccc} - \alpha^{(2)}_{aac}\right)\langle P_3\rangle\right]. \end{aligned} \tag{10.33c}$$

The first row of each entry in Eq. 10.33 is written in terms of $\langle\cos\theta\rangle$ and $\langle\cos^3\theta\rangle$ since that is what would be obtained directly after projecting from the lab frame into the molecular frame. These expressions have then been re-written in terms of $\langle P_1\rangle$ and $\langle P_3\rangle$ using Eq. 10.14. Working out the relationship between the two forms is presented as an end-of-chapter problem.

As our overarching goal is to use resonant amplitudes $A^{(2)}_{ijk}$ from polarized spectra to comment on the orientation of molecules, we are faced with the typical situation of dealing with the molecular number density, in this case, the surface number density, N_{A}. Fitting experimental spectra to a lineshape such as a Lorentzian and then taking amplitude ratios such as $A^{(2)}_{xxz}/A^{(2)}_{xzx}$ (from the SSP and SPS combinations) is the canonical way of addressing this, and it simultaneously removes the need to know $\chi^{(2)}_{\text{measured}}$ in absolute units. When we followed this approach in our study of IR absorption and Raman scattering, together with our overly simplified molecular response tensors (keeping only α_{cc} and $\alpha^{(3)}_{cccc}$ in the respective cases), forming ratios also eliminated any required knowledge of the molecular-frame $\boldsymbol{\alpha}^{(n)}$ elements. We noted that, when additional $\alpha^{(n)}$ elements appear, that this is not always possible. Here we encounter our first case where there are two unique $\boldsymbol{\alpha}^{(2)}$ elements: $\alpha^{(2)}_{aac}$ (that is equal to $\alpha^{(2)}_{bbc}$) and $\alpha^{(2)}_{ccc}$. However, since there are exactly two elements, we can define a single ratio $R_{\text{surface}} = \alpha^{(2)}_{acc}/\alpha^{(2)}_{ccc}$ and then re-write Eq. 10.33 as

$$\begin{aligned} \chi^{(2)}_{xxz} &= \frac{N_{\mathrm{A}}}{2\varepsilon_0}\alpha^{(2)}_{ccc}\left[(1 + R_{\text{surface}})\langle\cos\theta\rangle - (1 - R_{\text{surface}})\langle\cos^3\theta\rangle\right] \\ &= \frac{N_{\mathrm{A}}}{5\varepsilon_0}\alpha^{(2)}_{ccc}\left[(4R_{\text{surface}} + 1)\langle P_1\rangle - (1 - R_{\text{surface}})\langle P_3\rangle\right] \end{aligned} \tag{10.34a}$$

$$\begin{aligned}\chi^{(2)}_{xzx} &= \frac{N_A}{2\varepsilon_0}\alpha^{(2)}_{ccc}\,(1 - R_{\text{surface}})\left(\langle\cos\theta\rangle - \langle\cos^3\theta\rangle\right)\\ &= \frac{N_A}{5\varepsilon_0}\alpha^{(2)}_{ccc}\,(1 - R_{\text{surface}})\left(\langle P_1\rangle - \langle P_3\rangle\right)\end{aligned} \tag{10.34b}$$

$$\begin{aligned}\chi^{(2)}_{zzz} &= \frac{N_A}{\varepsilon_0}\alpha^{(2)}_{ccc}\left[R_{\text{surface}}\langle\cos\theta\rangle + (1 - R_{\text{surface}})\langle\cos^3\theta\rangle\right]\\ &= \frac{N_A}{5\varepsilon_0}\alpha^{(2)}_{ccc}\left[(2R_{\text{surface}} + 3)\langle P_1\rangle + 2(1 - R_{\text{surface}})\langle P_3\rangle\right]\end{aligned} \tag{10.34c}$$

where the common factor of $\alpha^{(2)}_{ccc}$ in Eq. 10.34 would be eliminated along with N_A when ratios of the amplitudes are considered in a subsequent analysis step.

Before we continue with the orientation analysis, an interesting point follows from our definition of R_{surface}. Comparing with Eq. 10.7 we have

$$\begin{aligned}R_{\text{surface}} &= \frac{\alpha^{(2)}_{aac}}{\alpha^{(2)}_{ccc}} = \frac{\langle 0|\hat{\alpha}_{aa}|1\rangle\langle 1|\hat{\mu}_c|0\rangle}{\langle 0|\hat{\alpha}_{cc}|1\rangle\langle 1|\hat{\mu}_c|0\rangle}\\ &= \left.\frac{\langle 0|\hat{\alpha}_{aa}|1\rangle}{\langle 0|\hat{\alpha}_{cc}|1\rangle}\right|_{\text{surface}}\end{aligned} \tag{10.35}$$

since the c-component of the IR transition dipole cancels in this ratio, and we are left with only the ratio of the transition polarizability tensor elements. If we picture these same molecules in the bulk phase, they have

$$R_{\text{bulk}} = \left.\frac{\langle 0|\hat{\alpha}_{aa}|1\rangle}{\langle 0|\hat{\alpha}_{cc}|1\rangle}\right|_{\text{bulk}} = \frac{\alpha^{\text{R}}_{aa}}{\alpha^{\text{R}}_{cc}} \tag{10.36}$$

which looks conspicuously similar to Eq. 10.35 and can be expressed as a ratio of Raman tensor elements. In fact, the only reason why $R_{\text{surface}} \neq R_{\text{bulk}}$ is in the case where the electronic structure at the interface differs from that in the bulk [47, 79]. The interest in Eq. 10.36 stems from the fact that, if the distribution in the bulk is isotropic, then R_{bulk} can be obtained from the Raman depolarization ratio $\rho = I_\perp/I_\parallel$ through

$$\rho = \frac{3}{4 + 5\left(\dfrac{1 + 2R_{\text{bulk}}}{R_{\text{bulk}} - 1}\right)^2}. \tag{10.37}$$

If we wish to invert this quadratic equation to solve for R_{bulk} using an experimental ρ value, there will be two solutions. Defining

$$S = \sqrt{\frac{3}{5}\left(\frac{1}{\rho} - \frac{4}{3}\right)}$$

it can be shown [80] that

$$R_{\text{bulk}} = \frac{S+1}{S-2} \tag{10.38a}$$

or

$$R_{\text{bulk}} = \frac{S-1}{S+2}, \tag{10.38b}$$

where the two solutions result from the sign ambiguity due to the square appearing in Eq. 10.37. Values of $\alpha^{(2)}_{aac}$ and $\alpha^{(2)}_{ccc}$ are obviously critically important to any study that seeks to comment on the surface orientation distribution. One experimental approach is therefore to measure the Raman depolarization ratio ρ for an isotropic bulk distribution, and then assume that $R_{\text{surface}} \approx R_{\text{bulk}}$. An interesting point here is that this technically doesn't assume that $\langle 0|\hat{\alpha}_{aa}|1\rangle$ and $\langle 0|\hat{\alpha}_{cc}|1\rangle$ have the same values at the surface and the bulk, only that their ratios are the same. In other words, they could both be say 10% larger at the surface, and this wouldn't matter as far as R is concerned.

We now arrive one step shy of our goal, to be able to comment on the orientation distribution at the surface from our experimental knowledge of $A^{(2)}_{xxz}$, $A^{(2)}_{xzx}$ and $A^{(2)}_{zzz}$. Recall that, when polarized IR spectra were collected, we illustrated how the dichroic ratio could be used to solve for $\langle \cos^2\theta \rangle$, without the need for any assumptions about the orientation distribution. Similarly, for a uniaxial orientation distribution, three unique Raman tensor elements could be obtained, from which normalization resulted in two unique ratios that could be used to solve for $\langle \cos^2\theta \rangle$ and $\langle \cos^4\theta \rangle$. Now we find ourselves in the midst of a very similar discussion about SFG data, and it might appear that two ratios of the three equations (Eqs. 10.34) would yield $\langle \cos\theta \rangle$ and $\langle \cos^3\theta \rangle$. However, as I have already mentioned, the three elements of $\boldsymbol{\chi}^{(2)}$ are not independent. We are therefore able to write $A^{(2)}_{zzz}$ as a linear combination of the other two elements in the form of

$$A^{(2)}_{zzz} = \frac{1}{R_{\text{surface}}} A^{(2)}_{xxz} - \left(2 + \frac{1}{R_{\text{surface}}}\right) A^{(2)}_{xzx}. \tag{10.39}$$

This means that, if we want to work with only the ratios of SFG amplitudes obtained in different polarization schemes, $\langle \cos\theta \rangle$ and $\langle \cos^3\theta \rangle$ are out of reach. However this leads to an interesting consequence: If $A^{(2)}_{xxz}$, $A^{(2)}_{xzx}$ and $A^{(2)}_{zzz}$ can be extracted from fitting spectra in SSP, SPS, and PPP polarization schemes, we can solve for

$$R_{\text{surface}} = \frac{A_{xxz} - A^{(2)}_{xzx}}{A^{(2)}_{zzz} + 2A^{(2)}_{xzx}} \tag{10.40}$$

independent of any aspect of the orientation distribution [47, 79]. When combined with bulk Raman measurements, this now provides a route to directly comparing R_{surface} with R_{bulk}. One must keep in mind that, while $A^{(2)}_{xxz}$ and $A^{(2)}_{xzx}$ can be easily accessed from SSP and SPS spectra, respectively, $A^{(2)}_{zzz}$ is buried inside the PPP response. This is not a significant barrier, as there are some approaches to fitting SFG spectra [81, 82], though that topic is a detour from our current discussion.

Since we can't use all three tensor elements towards our goal of describing the orientation distribution at the surface, we now pick any two. For simplicity, we choose $A^{(2)}_{xxz}$ as would be obtained by fitting the SSP spectrum, and $A^{(2)}_{xzx}$ from the SPS spectrum. One can show that

$$\begin{aligned}\frac{\langle\cos^3\theta\rangle}{\langle\cos\theta\rangle} &= \frac{(1-R_{\text{surface}})A^{(2)}_{xxz}-(1+R_{\text{surface}})A^{(2)}_{xzx}}{(1-R_{\text{surface}})(A^{(2)}_{xxz}-A^{(2)}_{xzx})} \\ &= \frac{2}{5}\frac{\langle P_3\rangle}{\langle P_1\rangle}+\frac{3}{5}\end{aligned} \tag{10.41}$$

and, if $A^{(2)}_{zzz}$ was additionally available, the value of R_{surface} can be known from Eq. 10.40.[2]

Recall that the value of $\langle\cos^2\theta\rangle$ obtained from IR spectra can be expressed in terms of the order parameter $\langle P_2\rangle$. Likewise, $\langle\cos^2\theta\rangle$ and $\langle\cos^4\theta\rangle$ from Raman spectra can be used to determine $\langle P_2\rangle$ and $\langle P_4\rangle$ for a uniaxial distribution. However, for the SFG case, a consequence of Eq. 10.39 is that $\langle P_1\rangle$ and $\langle P_3\rangle$ are not separable when ratios of amplitudes are used. $\langle P_1\rangle$ and $\langle P_3\rangle$ can be isolated from calibrated measurements, but the magnitude and phase of $\boldsymbol{\chi}^{(2)}$ and $\boldsymbol{\alpha}^{(2)}$ are required in absolute units, in addition to knowledge of the surface number density N_A, so that is a significant hurdle that incorporates additional uncertainties. However, if the $\boldsymbol{\chi}^{(2)}$ amplitude ratios are used, the ratio of order parameters

$$\frac{\langle P_3\rangle}{\langle P_1\rangle} = \frac{5}{2}\left[\frac{\dfrac{A^{(2)}_{xxz}}{A^{(2)}_{xzx}}-\dfrac{1+R_{\text{surface}}}{1-R_{\text{surface}}}}{\dfrac{A^{(2)}_{xxz}}{A^{(2)}_{xzx}}-1}\right]-\frac{3}{2} \tag{10.42}$$

can be determined. We can therefore plot $A^{(2)}_{xxz}/A^{(2)}_{xzx}$ as a function of $\langle P_3\rangle/\langle P_1\rangle$ or $\langle\cos^3\theta\rangle/\langle\cos\theta\rangle$ for various values of R_{surface} as shown in Fig. 10.10. Taking a terminal methyl group at the end of a long aliphatic chain as an example, where $R_{\text{surface}} \approx 3.4$, we can see that if the fitting of the experimental spectra produced a CH_3 symmetric stretch amplitude ratio of $A^{(2)}_{xxz}/A^{(2)}_{xzx} = -7.5$, this corresponds to $\langle P_3\rangle/\langle P_1\rangle \approx 0.175$ or equivalently $\langle\cos^3\theta\rangle/\langle\cos\theta\rangle \approx 0.67$, as indicated with the dashed crosshairs.

[2] Note that some authors [81] analyse the SFG response in terms of $\langle\cos\theta\rangle/\langle\cos^3\theta\rangle$, the reciprocal of the ratio expressed in Eq. 10.41.

Code for Fig. 10.10

Note that it doesn't matter whether we calculate $\langle\cos^3\theta\rangle/\langle\cos\theta\rangle$ or $\langle P_3\rangle/\langle P_1\rangle$ since the two ratios are easily converted. In this code, we start with the definition of the amplitudes A_{xxz} and A_{xzx} from Eqs. 10.34 and then plot both ratios.

```
import numpy as np
from matplotlib import pyplot

def exptratio(R):
    xxz = (1. + R) - (1. - R)*Dinv
    xzx = (1. - R) -  (1. - R)*Dinv
    ratio = xxz/xzx
    line = ax.plot(P3P1[P3P1<1], ratio[P3P1<1], label=r'$R_{\rm surface} = %3.1f$' % R)
    ax.plot(P3P1[P3P1>1], ratio[P3P1>1], color=line[0].get_color())

# setup
pyplot.rc('font', size=8)
fig = pyplot.figure(figsize=(3.25, 3))
ax = fig.add_subplot()
Dinv = np.linspace(-2, 3, 10000)
P3P1 = 2.5*Dinv - 1.5
exptratio(.1)
exptratio(.5)
exptratio(3.4)

# pretty
ax.set_ylabel(r'$A_{xxz}/A_{xzx}$')
ax.set_xlim(-4, 4)
ax.set_ylim(-10, 15)

def order2cos(x):
    return 2*x/5. + 3./5.

def cos2order(x):
    return 2.5*x - 1.5

ax2 = ax.secondary_xaxis('top', functions=(order2cos, cos2order))
# ax.axvline(cos2order(np.cos(np.radians(39.2))**2), color='k', ls='-', lw=1)
ax.axvline(0, color='k', ls='-', lw=1)
ax.axhline(-7.5, color='k', ls='--', lw=1)
ax.axvline(cos2order(1/1.5), color='k', ls='--', lw=1)
ax.set_xlabel(r'$\dfrac{\langle P_3 \rangle}{\langle P_1 \rangle}$', labelpad=1)
ax2.set_xlabel(r'$\dfrac{\langle \cos^3 \theta \rangle}{\langle \cos \theta \rangle}$',
    labelpad=12)
ax2.set_xticks([-1, 0, 1, 2])

ax.legend(loc='upper left', frameon=False, fontsize=7)
fig.set_tight_layout(True)
pyplot.savefig('xxzxzxR.png', dpi=600)
```

This is a nice opportunity to show off the `secondary_xaxis` axis property. Creating the top axis in this way requires us to specify two functions for converting between $\langle\cos^3\theta\rangle/\langle\cos\theta\rangle$ and $\langle P_3\rangle/\langle P_1\rangle$ in both directions.

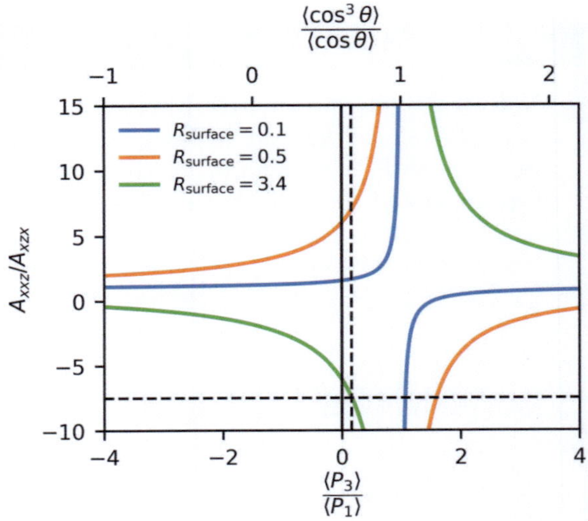

Fig. 10.10 Plots of $A^{(2)}_{xxz}/A^{(2)}_{xzx}$ as a function of the ratio $\langle P_3\rangle/\langle P_1\rangle$, shown for various values of the parameter R_{surface}. The dashed crosshairs correspond to our example of $A^{(2)}_{xxz}/A^{(2)}_{xzx} = -7.5$ yielding $\langle P_3\rangle/\langle P_1\rangle \approx 0.175$. The solid line at $\langle P_3\rangle/\langle P_1\rangle = 0$ marks the SFG magic angle

In order to make physical sense of the experimental result, we resort to some assumptions about the shape of the orientation distribution. The most common is that the molecules are distributed according to a Gaussian function. Figure 10.11a illustrates the values of $\langle P_1\rangle$ that correspond to a range of Gaussians with a mean tilt angle anywhere in the first quadrant, i.e. $0° < \theta_0 < 90°$, and with widths up to $\sigma = 50°$. Here it can be seen that the narrowest and most upright distributions (smallest values of σ and θ_0) have the largest value of $\langle P_1\rangle$, with $\langle\cos\theta\rangle = 1$ for a Dirac delta function with all molecules aligned along the surface normal, $\theta_0 = 0°$. We also see that $\langle P_1\rangle = 0$ for all Gaussians centred at $\theta_0 = 90°$. Even though this takes on the same value as for an isotropic solution with $\langle P_1\rangle = 0$, in this case, there is anisotropy, but no breaking of the centrosymmetry. No matter how wide the distribution, if the mean tilt angle is in the plane of the surface, half of the molecules are pointed up, and the other half are pointed down. A similar analysis is done for $\langle P_3\rangle$ in Fig. 10.11b, and the limiting cases have the same behaviour. The difference in the shapes of the $\langle P_1\rangle$ and $\langle P_3\rangle$ contours is most noticeable in Fig. 10.11c, where their ratio is plotted. This ratio can be pictured for the left-hand side of Eq. 10.42. In our example of the methyl group orientation, we can identify the $\langle P_3\rangle/\langle P_1\rangle = 0$ contour. This corresponds to a situation where all of the methyl groups at the surface are tilted precisely at $\theta_0 = 35.2°$ (situated along the horizontal line at $\sigma = 0°$), or with a Gaussian distribution with any combination of θ_0 and σ along the zero-value contour. For example ($\theta_0 = 30°, \sigma = 15°$) or ($\theta_0 = 10°, \sigma = 32°$). Although there are many possibilities along this contour, the experimental measurement of $\langle P_3\rangle/\langle P_1\rangle = 0.175$ tells us that θ_0 must be less than 32° and $\sigma < 35°$, provided the distribution is Gaussian and $R_{\text{surface}} = 3.4$.

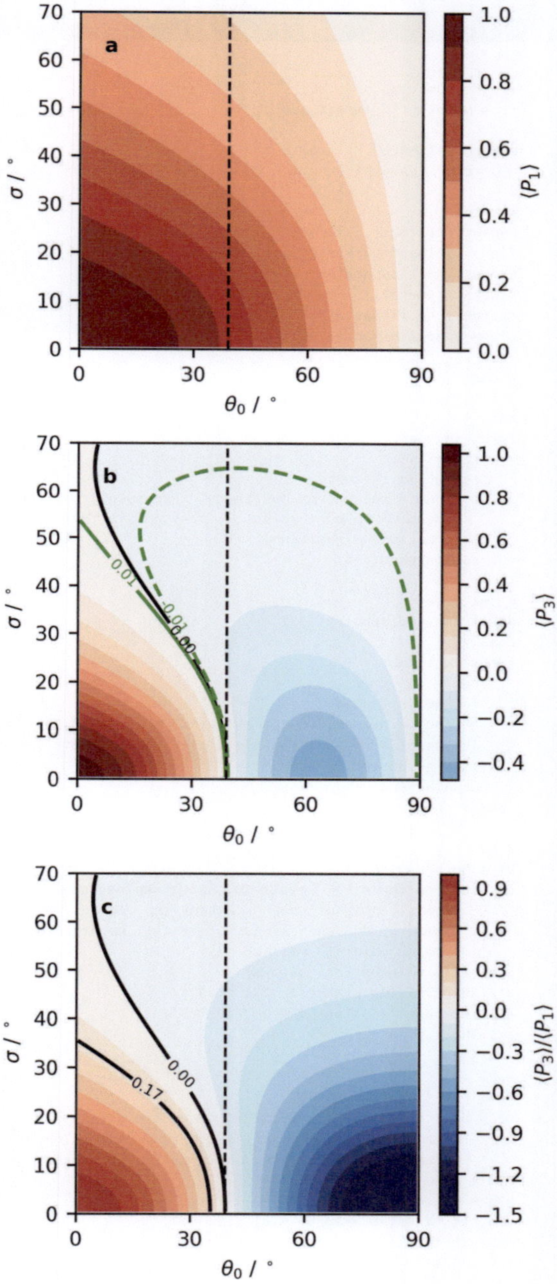

Fig. 10.11 Plots of the values of **a** $\langle P_1 \rangle$ and **b** $\langle P_3 \rangle$ that result from a range of Gaussian distributions centred between $\theta_0 = 0$–$90°$ and with widths up to $\sigma = 70°$. The solid black contour indicates values of θ_0 and σ for which $\langle P_3 \rangle = 0$. The green contours indicate $\langle P_3 \rangle \pm 0.01$. **c** The ratio $\langle P_3 \rangle / \langle P_1 \rangle$ resulting from these distributions. The vertical dashed line indicates the SFG magic angle at $\theta_0 = 39.2°$

Code for Fig. 10.11

```
import numpy as np
from matplotlib import pyplot

def gauss(theta0, sigma):
    return np.exp(-(theta - theta0)**2 / (2*sigma**2))

# setup
pyplot.rc('font', size=8)
fig = pyplot.figure(figsize=(3.25, 7.5))
theta = np.linspace(0, np.pi, 1000)
dtheta = theta[1] - theta[0]
theta0 = np.linspace(0, 0.999*np.pi/2, 100)
sigma = np.linspace(0.001, np.radians(70.), 100)

# cos
ax1 = fig.add_subplot(311)
cos1 = []
for s in sigma:
    row = []
    for q0 in theta0:
        num = np.sum( np.cos(theta) * gauss(q0, s) * np.sin(theta)) * dtheta
        denom = np.sum( gauss(q0, s) * np.sin(theta)) * dtheta
        row.append(num/denom)
    cos1.append(row)

# cos cubed
ax2 = fig.add_subplot(312)
cos3 = []
for s in sigma:
    row = []
    for q0 in theta0:
        num = np.sum( np.cos(theta)**3 * gauss(q0, s) * np.sin(theta)) * dtheta
        denom = np.sum( gauss(q0, s) * np.sin(theta)) * dtheta
        row.append(num/denom)
    cos3.append(row)
cos3 = np.array(cos3)

# P1
P1 = np.array(cos1)
c1 = ax1.contourf(P1, levels=10, aspect=np.degrees(np.max(theta0))/np.degrees(np.max(
      sigma)), origin='lower', extent=[np.degrees(np.min(theta0)), np.degrees(np.max(
      theta0)), np.degrees(np.min(sigma)), np.degrees(np.max(sigma))], cmap=pyplot.cm.
      RdBu_r, vmin=-1, vmax=1)
c1b = fig.colorbar(c1, ax=ax1)
c1b.set_ticks(np.arange(0, 1.1, 0.2))
c1b.set_label(r'$\langle P_1 \rangle$')
ax1.label = 'a'

# P3
P3 = 0.5*(5*cos3 - 3*P1)
c3 = ax2.contourf(P3, levels=20, aspect=np.degrees(np.max(theta0))/np.degrees(np.max(
      sigma)), origin='lower', extent=[np.degrees(np.min(theta0)), np.degrees(np.max(
      theta0)), np.degrees(np.min(sigma)), np.degrees(np.max(sigma))], cmap=pyplot.cm.
      RdBu_r, vmin=-1, vmax=1)
c3b = fig.colorbar(c3, ax=ax2)
c3b.set_ticks(np.arange(-0.4, 1.1, 0.2))
c3b.set_label(r'$\langle P_3 \rangle$')
ax2.label = 'b'

# show P3 = 0
```

```
an1 = ax2.contour(P3, levels=[0], colors='k', aspect=np.degrees(np.max(theta0))/np.
      degrees(np.max(sigma)), origin='lower', extent=[np.degrees(np.min(theta0)), np.
      degrees(np.max(theta0)), np.degrees(np.min(sigma)), np.degrees(np.max(sigma))])
ax2.clabel(an1, inline=True, fmt='%4.2f', fontsize=6)

an2 = ax2.contour(P3, levels=[-0.01, 0.01], colors='g', aspect=np.degrees(np.max(theta0)
      )/np.degrees(np.max(sigma)), origin='lower', extent=[np.degrees(np.min(theta0)),
      np.degrees(np.max(theta0)), np.degrees(np.min(sigma)), np.degrees(np.max(sigma))])
ax2.clabel(an2, inline=True, fmt='%4.2f', fontsize=6)

# c1/c3
ax3 = fig.add_subplot(313)
ratio = cos3/cos1
c1c3 = ax3.contourf(P3/P1, levels=30, aspect=np.degrees(np.max(theta0))/np.degrees(np.
      max(sigma)), origin='lower', extent=[np.degrees(np.min(theta0)), np.degrees(np.max
      (theta0)), np.degrees(np.min(sigma)), np.degrees(np.max(sigma))], cmap=pyplot.cm.
      RdBu_r, vmin=-1.2, vmax=1.2)

# highlight example contour
an3 = ax3.contour(P3/P1, levels=[0, 0.1667], colors='k', aspect=np.degrees(np.max(theta0
      ))/np.degrees(np.max(sigma)), origin='lower', extent=[np.degrees(np.min(theta0)),
      np.degrees(np.max(theta0)), np.degrees(np.min(sigma)), np.degrees(np.max(sigma))])
ax3.clabel(an3, inline=True, fmt='%4.2f', fontsize=6)

cr = fig.colorbar(c1c3, ax=ax3)
cr.set_label(r'$\langle P_3 \rangle / \langle P_1 \rangle$')
ax3.label = 'c'

# pretty
for ax in [ax1, ax2, ax3]:
    ax.set_xticks(range(0, 91, 30))
    ax.set_yticks(range(0, 71, 10))
    ax.axvline(39.2, color='k', ls='--', lw=1)
    ax.set_xlabel(r'$\theta_0$ / $^\circ$')
    ax.set_ylabel(r'$\sigma$ / $^\circ$')
    ax.text(0.07, 0.88, ax.label, fontweight='bold', transform=ax.transAxes)
fig.set_tight_layout(True)
pyplot.savefig('GaussianP1P3.png', dpi=600)
```

From the plots of the Legendre polynomials in Fig. 8.19 we can see that, with the exception of P_0 (which is not a function of θ) and $P_1(\cos\theta)$, the rest have nodes within the range $0° < \theta < 90°$ (and the equivalent number of nodes within $90° < \theta < 180°$). This means that, although the order parameters $\langle P_n(n > 0)\rangle \to 0$ as the distributions become isotropic (for example, as σ increases in a Gaussian distribution), there are also other angles at which those functions $P_n(\cos\theta)$ and therefore the corresponding order parameters $\langle P_n\rangle$ become zero. As we discussed in Chap. 8, those angles are referred to as magic angles [83, 84], although their appearance is not always magical. For $P_3(\cos\theta)$ this node appears at the angle for which $5\langle\cos^3\theta\rangle = 3\langle\cos\theta\rangle$. In the case of a narrow distribution, this corresponds to $\theta_0 = \arccos\sqrt{3/5} \approx 39.2°$. We can get a practical sense of the situation by returning to Fig. 10.11. As the Gaussian distribution widens, the contours in Fig. 10.11a illustrate that $\langle P_1\rangle \to 0$, although this requires σ to be considerably wider than the 70° limit shown. The same limiting behaviour appears in Fig. 10.11 where $\langle P_3\rangle \to 0$ as $\sigma \to \infty$, although it is evident that $\langle P_3\rangle$ becomes small much faster than $\langle P_1\rangle$. However, the key here is that, although there is nothing wrong with those order parameters tending towards zero for wide distribution, the $\langle P_3\rangle = 0$ contour extends all the way down to $\sigma = 0$, where the distribution is essentially a delta function $f(\theta) = \delta(\theta - \theta_0)$ with $\theta_0 = 39.2°$ (indicated by the vertical dashed black line). In this representation, we can see that it is not possible to distinguish between isotropic distributions and progressively narrower distributions centred at 39.2° based on $\langle P_3 \rangle$ information alone. Even though the magic angle is technically restricted to this single contour, a sense of how wide the space is in the vicinity of $\langle P_3 \rangle = 0$ can be realized by the $\langle P_3 \rangle = 0.01$ (solid green) and $\langle P_3 \rangle = -0.01$ (dashed green) contours in Fig. 10.11b. The ratio $\langle P_3 \rangle / \langle P_1 \rangle$ inherits much of this wide space as seen in Fig. 10.11c.

10.8 SFG at Thin Film Interfaces

10.8.1 Background

As we mentioned very briefly in our initial discussion of the **L** factors, the treatment so far has assumed that the interface we are studying is situated between two semi-infinite bulk phases. Often, the sample of interest has molecules adsorbed at the surface. In other commonly encountered situations, the substrate is a thin polymer film that has been prepared by spin coating. This is often employed whether the sample of interest is one of the surfaces of the film itself, or whether adsorption of a different species onto the film is targeted for analysis. The adsorbed layer and thin film have a lot in common, even though adsorbed layers are often monolayers, with a thickness of tens of nanometers, whereas spin-coated polymers can have thicknesses up to hundreds of nanometers. The salient feature of these "films" is that there are potentially two SFG active interfaces. It is tempting but confusing to label the two interfaces as "top" and "bottom" because solid–liquid interfaces can

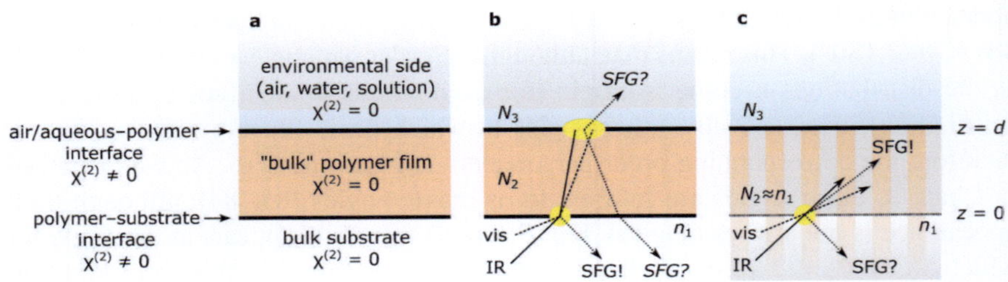

Fig. 10.12 **a** If the environmental phase (e.g. air or a solution), thin film, and substrate all have bulk inversion symmetry, then no SFG signal can be generated except for the interfaces. However, both the environment–film and film–substrate interfaces can generate SFG signals. **b** The film thickness d, angles of incidence, and beam diameters determine if there is sufficient overlap at the second interface to generate a transmitted and reflected SFG beam. **c** Even if we determine that the degree of spatial and temporal overlap is sufficient, the question of index-matching at either interface arises when the experiment is set up to detect reflected SFG signals

be studied in any orientation. Also consider that the plane of the surface can be either parallel or perpendicular to the optical table. Plus, for a given orientation, the beams can approach the surface from either side. We therefore refer to the $z = 0$ side of the film as the first one encountered by the incoming visible and IR beams in a (standard) co-propagating geometry. Since we will use d for the film thickness, the second interface is therefore located at $z = d$. This is illustrated in Fig. 10.12a. We are then faced with the situation where

$$\chi^{(2)}_{\text{eff},ijk} = \underbrace{\tilde{L}_{ii}\tilde{L}_{jj}\tilde{L}_{kk}\chi^{(2)}_{ijk}}_{z=0 \text{ interface}} + \underbrace{\tilde{L}_{ii}\tilde{L}_{jj}\tilde{L}_{kk}\chi^{(2)}_{ijk}}_{z=d \text{ interface}} . \tag{10.43}$$

Here we write $\tilde{\mathbf{L}}$ instead of $\mathbf{L}$ to indicate that something special needs to be considered. For a moment, just consider that $\tilde{\mathbf{L}}$ indicates a local field correction. To account for $\boldsymbol{\chi}^{(2)}_{\text{eff}}$ with contributions from both interfaces, we need to consider that interference between the two $\boldsymbol{\chi}^{(2)}$ terms depends on their relative magnitude and phase. In other words $|\boldsymbol{\chi}^{(2)}_{\text{eff,total}}|^2 = |\boldsymbol{\chi}^{(2)}_{\text{eff},z=0} + \boldsymbol{\chi}^{(2)}_{\text{eff},z=d}|^2 \neq |\boldsymbol{\chi}^{(2)}_{\text{eff},z=0}|^2 + |\boldsymbol{\chi}^{(2)}_{\text{eff},z=d}|^2$. As we will soon see, there needs to be some phase (and phase difference, i.e. Δk) considerations here, but we wrap all that up under the "effective" label in $\boldsymbol{\chi}^{(2)}_{\text{eff}}$. The full treatment of such situations is beyond the scope of the current discussion, but interested readers are referred to detailed accounts in the literature [85–90]. Our goal is to illustrate that the relative magnitude of $L_{ii}L_{jj}L_{kk}$ from each interface can be manipulated so that experimentally controlled conditions (film thickness and angle of incidence) can sometimes be selected in order to suppress the contribution from one of the surfaces [91, 92]. Throughout this discussion, we assume that $\boldsymbol{\chi}^{(2)} = 0$ in any of the bulk materials, as we have done so far in our discussion of surfaces. This treatment can be generalized to include multiple contributions to the measured SFG response [93–99].

Before presenting the formalism that enables selective probing of one of these two interfaces, it is instructive to first discuss three common misconceptions related to SFG spectra of thin films. The first, illustrated in Fig. 10.12b, is that the film is thick enough to prevent SFG from being generated at the undesired interface by enabling the beams to be overlapped only at the interface of interest. Although that might end up providing a means for probing a single interface in some unusual samples, the numbers need to be worked out to be sure. Practically, if the visible and infrared beams have diameters of 200 μm, and approach at an angle of approximately 65°, spatial mismatch would occur only for films with a thickness greater than 2 mm [91]. If such a thick sample can be prepared, then internal or external reflection SFG is indeed an excellent way to achieve interface specificity. However, one of the reasons that much thinner samples, i.e. films, are often used is that it is challenging to prepare thick surfaces that are also smooth enough to have a good specular reflection. Thin coating is therefore used that typically results in good quality films with a thickness of less than ≈1000 nm.

The second misconception has to do with near index-matching. Many polymers have the real part of their refractive index close to that of glass. The question then arises whether an appreciable amount of SFG can be reflected at glass–polymer interfaces. For example, consider an internal reflection geometry as shown in Fig. 10.12c where the beams are incident from a fused silica prism with refractive index n_1, encounter a polymer film with index n_2, and then a bulk material with index n_3. Here, even though $n_2 \approx n_1$, transmitted and reflected SFG beams can still be generated from the glass–polymer interface, so long as the molecules are ordered there such that the centrosymmetry is broken. We will later see that by modelling this situation, we can arrive at a quantitative prediction of the intensity of that beam, and this can be verified experimentally [91, 92].

The next consideration to discuss is whether the $z = d$ interface may be ignored as a result of the IR beam being sufficiently attenuated due to absorption as it travels through the film. Figure 10.13 compares the intensity of an IR beam at the top and bottom surfaces of a silicone rubber (polydimethylsiloxane) film, where the IR beam approaches at an angle of 52°. In these plots, the horizontal dashed line indicates the condition at which the IR intensity is the same at both interfaces. The red line is the prediction when ignoring thin film interference, and the blue line is calculated taking thin film interference into account. While absorption from travelling through a film always causes a decrease in intensity, considering multiple reflections of the IR beam within a thin film sometimes results in peculiar situations where their constructive or destructive interference at the interfaces results in a higher mean squared field at the second interface than the first. It is interesting to also study the situation of a film of the same material with thickness of 350 nm on glass with the IR beam approaching from the glass. The strongest IR-active band in this region is the methyl antisymmetric stretch at 2958 cm^{-1}. Here Beer's law prediction shows that the intensity at the film–air interface is only attenuated to about 85% of its initial value at the film–substrate interface. When multiple reflections inside the film are taken into account (blue line), we see that, by the time the beam reaches the second interface, its intensity is reduced, but is still about 65% of its value when it first entered the film. This usually provides more than enough IR intensity to create SFG from the second interface.

If the silicone film were 20 μm thick (Fig. 10.13b), there is no IR light reaching the bottom surface in the strongly absorbing 2958 cm^{-1} region. Note that the intensity difference actually favours the far side of the film by a factor of roughly 1.5 in the 2908 cm^{-1} region, so extreme caution needs to be taken when interpreting the results. Finally, if the thickness of the material was 75 μm (Fig. 10.13c), the intensity in the 2958 cm^{-1} region is zero over a wider wavelength range as expected, and the intensity in the 2908 cm^{-1} region is less than 10% of its initial value. However, a film of such a large thickness is far from the regime of spin-coated thin films.

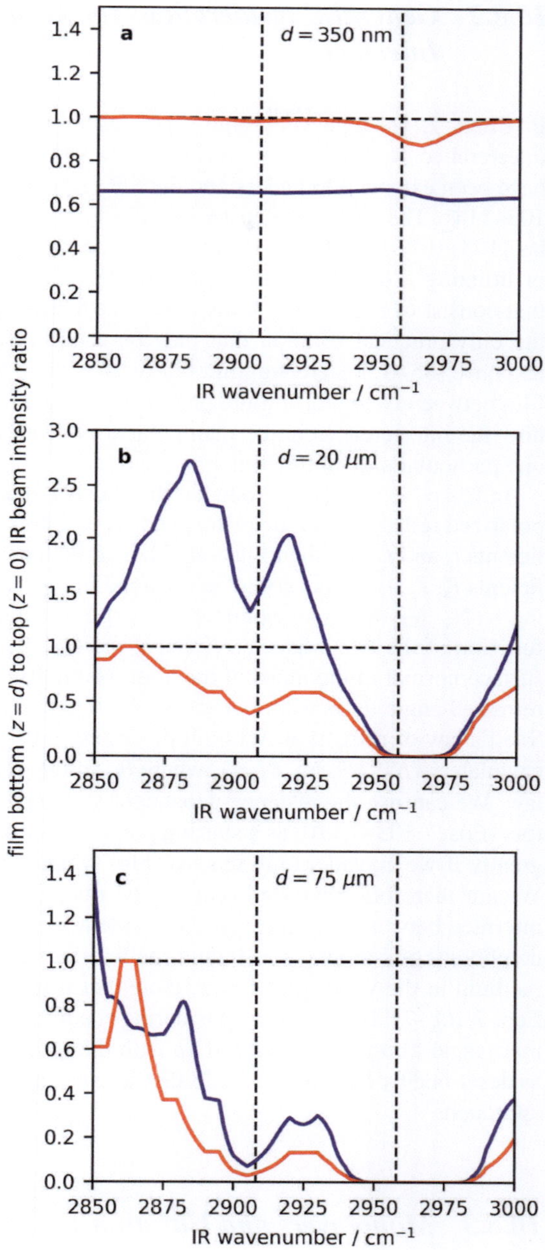

Fig. 10.13 The effect of absorption on the intensity of IR light incident at 52° that penetrates through a polymer film of thickness **a** 350 nm, **b** 20 μm, **c** 75 μm. The horizontal dashed line highlights the point where the intensity at both interfaces is the same. The two vertical dashed lines indicate the positions of the methyl symmetric and antisymmetric stretching resonances. The solution obtained ignoring thin film interference is plotted with the red line, and a more complete model of the field intensities is plotted with the blue line

10.8.2 General Considerations for Local Fields at Thin Film Interfaces

In Chap. 7, we reviewed some of the established methods for treating thin film interference, a classic topic in linear optics [100–102]. Many of those techniques have been extended to understand the SFG signal generated from thin films [85–87, 103–114]. The methods can be broadly grouped into those that use Airy formulas [103–107, 110, 112, 115–118], ones based on Abèles formalism [88–90, 119], or utilizing a transfer-matrix approach [85–87, 108, 114, 120–122]. For samples that consist of multi-layer films a transfer-matrix approach provides what is likely the only practical solution that may be used to model and interpret the net SFG response [85–87]. Here we limit our discussion to the simple situation a single thin film between two semi-infinite phases. Our goal is to identify conditions that reveal how the interference can be manipulated in order to use SFG to selectively probe one particular side of the film.

In Chap. 7, we also introduced the Fresnel equations and derived the s- and p-polarized reflection coefficients r_s and r_p, and the corresponding transmission coefficients t_s and t_p. We then illustrated how effective reflection and transmission coefficients $\tilde{r}_s$, $\tilde{r}_p$, $\tilde{t}_s$, and $\tilde{t}_p$ could be obtained in the case of thin film interference. Taking $r_s(N_1, N_2, \theta_1)$ as an example, our explicit notation here emphasizes that r_s is a function of the two adjacent refractive indices and the angle θ_1 with respect to the surface normal in the incident medium. Naturally, these expressions depend on the refracted angle θ_2 as well, but given N_1, N_2, and θ_1, we can always obtain θ_2 from Snell's law of refraction. With all this mind, note that the thickness d is needed to calculate $\tilde{r}_s(N_1, N_2, N_3, \theta_1, d)$, where θ_2 and θ_3 can both be obtained from Snell's law. We can use the distance-dependent solutions for the thin film local field factors (Eqs. 7.108–7.110) as a starting point for consideration. We can simplify things greatly if we consider only materials for which $\chi^{(2)} = 0$ in all of the bulk phases. We are therefore interested only in the two regions along z where $\chi^{(2)} \neq 0$: the interface between the film and the incident medium which we will refer to as the neighbourhood around $z = 0$, and the interface between the film and the refracted medium in the vicinity of $z = d$. Note that if we set $d = 0$ in addition to $z = 0$ in Eqs. 7.108–7.110 and then considering medium 2 is the same as medium 3 when there is no film present, we end up with the same solutions for the interfacial fields as described by Eqs. 10.26–10.28 that we arrived at using a more phenomenological approach.

10.8.3 Monolayers and Ultrathin Films

We now return to the full expressions of Eqs. 7.108–7.110 and, rather than setting $d = 0$ directly, we consider that $d \to 0$. This is the scenario pictured in Fig. 10.14c. Here we find that

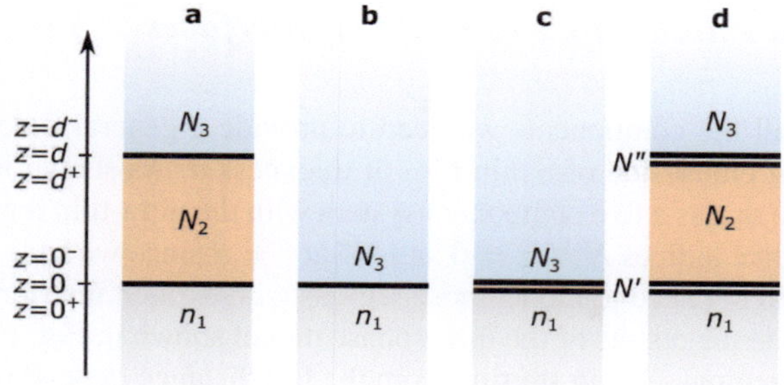

Fig. 10.14 Four different scenarios leading up to our ultimate description of the thin film system. **a** The standard three-phase model containing a film of thickness d, **b** the model collapsed to $d = 0$, **c** a three-phase model but for a middle phase with molecular dimensions, **d** a film with finite thickness d but two distinct interfacial regions in the vicinity of $z = 0$ and $z = d$. In all cases, the diagrams are aligned so that the beams approach the first interface from medium 1 at $z = 0$

$$L_{xx} = (1 - r_\text{p}) \cos\theta_1 \cos\gamma \tag{10.44}$$

and

$$L_{yy} = (1 + r_\text{s}) \sin\gamma \tag{10.45}$$

just as we found in the absence of the film, and also identical to what we arrived at in Eqs. 10.26 and 10.24. Note that these are still r rather than $\tilde{r}$, meaning that thin film interference need not be considered when the film has molecular dimensions. This important conclusion also holds true for the z-component, although we must now specify the location of interest. If we momentarily keep the molecular film's refractive index as N_2 we arrive at

$$L_{zz} = \begin{cases} (1 + r_\text{p}) \sin\theta_1 \cos\gamma & \text{in medium 1 at } z = 0^- \\ (1 + r_\text{p}) \left(\dfrac{n_1}{N_2}\right)^2 \sin\theta_1 \cos\gamma & \text{in medium 2 at } z = \frac{1}{2}d \\ (1 + r_\text{p}) \left(\dfrac{n_1}{N_3}\right)^2 \sin\theta_1 \cos\gamma & \text{in medium 3 at } z = d^+. \end{cases} \tag{10.46}$$

Even though we have used the classic thin film interference model to arrive at these expressions, we must now adapt the solution to that of a layer with molecular dimensions. Here the refractive index N_2 is better expressed with N' to account for the unique confined environment of a thin layer. If we further consider that the molecules generating the SFG response may in fact be the same as those in either of the adjacent bulk phases (Fig. 10.14c) this may be a better description than that in Fig. 10.14b. From this perspective, a good choice for the z-component of the local field correction for any single interface (no explicit film required) is that presented in Eq. 10.28.

10.8.4 Thin Films with Two Distinct Interfaces

We now have all the components we need to provide a general description of the electric fields at either side of a thin film of thickness d. As shown in Fig. 10.14d, we can describe this as a five component system with the ultra-thin regions given the effective refractive indices N' at $z = 0$ and N'' at $z = d$ since we have demonstrated above that we do not need to model those regions with explicit interference formula. We now combine the results of the three-phase model shown in Fig. 10.14a with the two interfacial regions. Again starting from the full distance-dependent expressions given in Eqs. 7.108–7.110, we can evaluate the fields at the first interface to arrive at

$$L_{xx}(z=0) = (1 - \tilde{r}_{\mathrm{p}}) \cos\theta_1 \cos\gamma \tag{10.47}$$

and

$$L_{yy}(z=0) = 1 + \tilde{r}_{\mathrm{s}} \sin\gamma. \tag{10.48}$$

Although these expressions appear similar to those described for the no-film and ultra-thin film cases, the values of $L_{xx}(z=0)$ and $L_{yy}(z=0)$ are unique as a result of the thin film interference effects contained in $\tilde{r}$. The two z-component solutions of interest located right before the film and immediately afterwards are given by

$$\frac{E_z(z=0^-)_1}{E_{\mathrm{i}}(z=-\infty)} = (1 + \tilde{r}_{\mathrm{p}}) \sin\theta_1 \cos\gamma \tag{10.49a}$$

$$\frac{E_z(z=0^+)_2}{E_{\mathrm{i}}(z=-\infty)} = (1 + \tilde{r}_{\mathrm{p}}) \left(\frac{n_1}{N_2}\right)^2 \sin\theta_1 \cos\gamma. \tag{10.49b}$$

Using the result from our previous model of the simple interface, we arrive at

$$L_{zz}(z=0) = \frac{E_z(z=0)}{E_{\mathrm{i}}(z=-\infty)} = (1 + \tilde{r}_{\mathrm{p}}) \left(\frac{n_1}{N'}\right)^2 \sin\theta_1 \cos\gamma. \tag{10.50}$$

At the second film interface located at $z = d$ we find that

$$\begin{aligned} \frac{E_x(z=d)_2}{E_{\mathrm{i}}(z=-\infty)} &= \cos\theta_1 (1 - \tilde{r}_{\mathrm{p}}) \cos\left(\frac{2\pi d N_2 \cos\theta_2}{\lambda}\right) \cos\gamma \\ &\quad + i \frac{n_1 \cos\theta_2}{N_2} (1 + \tilde{r}_{\mathrm{p}}) \sin\left(\frac{2\pi d N_2 \cos\theta_2}{\lambda}\right) \cos\gamma \end{aligned} \tag{10.51}$$

is equivalent to

$$L_{xx}(z=d) = \frac{E_x(z=d)_3}{E_{\mathrm{i}}(z=-\infty)} = \tilde{t}_{\mathrm{p}} \cos\theta_3 \cos\gamma \tag{10.52}$$

Table 10.2 The six local field correction factors needed to model SFG signal selectivity from a film with thickness d that would probe either the substrate ($z = 0$) or environmental ($z = d$) sides of the film

Local field correction	Substrate ($z = 0$) interface	Environmental ($z = d$) interface
$\tilde{L}_{xx}$	$(1 - \tilde{r}_\mathrm{p}) \cos\theta_1 \cos\gamma$	$\tilde{t}_\mathrm{p} \cos\theta_3 \cos\gamma$
$\tilde{L}_{yy}$	$1 + \tilde{r}_\mathrm{s} \sin\gamma$	$\tilde{t}_\mathrm{s} \sin\gamma$
$\tilde{L}_{zz}$	$(1 + \tilde{r}_\mathrm{p}) \left(\frac{n_1}{N'}\right)^2 \sin\theta_1 \cos\gamma$	$\frac{n_1 N_3}{(N'')^2} \tilde{t}_\mathrm{p} \sin\theta_1 \cos\gamma$

and we select Eq. 10.52 as the definition of $L_{xx}(z = d)$. The same situation occurs for the y-polarized component of the field where direct substitution into Eqs. 7.109 and 7.110 results in

$$\begin{aligned}\frac{E_y(z = d)_2}{E_\mathrm{i}(z = -\infty)} &= (1 + \tilde{r}_\mathrm{s}) \cos\left(\frac{2\pi d N_2 \cos\theta_2}{\lambda}\right) \sin\gamma \\ &\quad + i\frac{n_1 \cos\theta_1}{N_2 \cos\theta_2}(1 - \tilde{r}_\mathrm{s}) \sin\left(\frac{2\pi d N_2 \cos\theta_2}{\lambda}\right) \sin\gamma\end{aligned} \tag{10.53}$$

that is equivalent to

$$L_{yy}(z = d) = \frac{E_y(z = d)_3}{E_\mathrm{i}(z = -\infty)} = \tilde{t}_\mathrm{s} \sin\gamma. \tag{10.54}$$

As expected, the z-component is discontinuous across the interface, so we end up with

$$\begin{aligned}\frac{E_z(z = d^-)_2}{E_\mathrm{i}(z = -\infty)} &= \left(\frac{n_1}{N_2}\right)^2 (1 + \tilde{r}_\mathrm{p}) \cos\left(\frac{2\pi d N_2 \cos\theta_2}{\lambda}\right) \sin\theta_1 \cos\gamma \\ &\quad + i\frac{n_1 \cos\theta_1}{N_2 \cos\theta_2}(1 - \tilde{r}_\mathrm{p}) \sin\left(\frac{2\pi d N_2 \cos\theta_2}{\lambda}\right) \sin\theta_1 \cos\gamma\end{aligned} \tag{10.55a}$$

$$\frac{E_z(z = d^+)_3}{E_\mathrm{i}(z = -\infty)} = \frac{n_1}{N_3} \tilde{t}_\mathrm{p} \sin\theta_1 \cos\gamma. \tag{10.55b}$$

We note that

$$\frac{E_z(z = d^+)_3}{E_\mathrm{i}(z = -\infty)} = \left(\frac{N_2}{N_3}\right)^2 \frac{E_z(z = d^-)_2}{E_\mathrm{i}(z = -\infty)}$$

that enables us to write

$$L_{zz}(z=d) = \frac{n_1 N_3}{(N'')^2} \tilde{t}_p \sin\theta_1 \cos\gamma \tag{10.56}$$

where N'' could also be set to any value, including the compromise $N'' = \frac{1}{2}(N_2 + N_3)$. We now have all $\mathbf{L}$ factors that we require, namely, L_{xx}, L_{yy} and L_{zz} at the $z = 0$ interface, and another set of L_{xx}, L_{yy} and L_{zz} at the $z = d$ interface. Values of these local field correction factors are summarized in Table 10.2. Note that in the absence of a film (when $d = 0$), the expressions for L_{ii} in both columns (calculated with r and t in place of $\tilde{r}$ and $\tilde{t}$) are equivalent.

10.8.5 Model Predictions

The description to this point has been generally applicable to any thin film sandwiched between two bulk phases. However, as we transition from the topic of calculating local electric fields to achieving *selectivity*, i.e. finding combinations of film thickness and beam angles for which the fields at one surface dominate, we need to mention an important point about the experimental setup and choice of substrate. For dielectric phases, achieving such selectivity requires an internal reflection geometry [91]. This means that one cannot spin coat a film onto glass and have selectivity for the film–air interface using any combination of film thickness and beam angles in an external reflection geometry (beams approaching from air) [91]; that would require a more detailed treatment where the interference between two interfaces is considered if they are both SFG active [85–87]. In the discussion that follows, keep in mind that, in an internal reflection experiment, the $z = 0$ interface is between the thin film and the substrate, and the "environmental" interface is located at $z = d$.

As a demonstration, we evaluate the local electric fields for an experimental setup where the beams are incident from a silica prism, encounter a silicone rubber (polydimethylsiloxane, PDMS) film at the $z = 0$ PDMS–glass interface, and then reach the $z = d$ PDMS–air interface. The $\tilde{\mathbf{L}}$ model was evaluated using $\omega_{\text{vis}} = 18797$ cm^{-1} ($\lambda_{\text{vis}} = 532$ nm), $\omega_{\text{IR}} = 2910$ cm^{-1} ($\lambda_{\text{IR}} = 3.44$ μm), and $\omega_{\text{SFG}} = \omega_{\text{vis}} + \omega_{\text{IR}}$ providing $\lambda_{\text{SFG}} = 461$ nm. Refractive index values for IR-grade fused silica [123] and PDMS [124–126] were taken from the literature. We first consider the case where $\theta_{\text{vis}} = 44°$, $\theta_{\text{IR}} = 53°$, and θ_{SFG} is calculated from Eq. 10.19. The first column in Fig. 10.15 considers the silica–PDMS interface at $z = 0$, displaying $|\tilde{L}_{xx}|^2$ (blue), $|\tilde{L}_{yy}|^2$ (green), and $|\tilde{L}_{zz}|^2$ (red) as a function of the PDMS film thickness for the SFG (top row), visible (second row), and IR beams (third row). In the bottom row we have considered the SSP local field intensities defined by $|\tilde{L}_{yy}\tilde{L}_{yy}\tilde{L}_{zz}|^2$. In each case, points indicate the model prediction with $N' = N'' = N_2$, i.e. the refractive index of the interfacial region is the same as the bulk PDMS film. The results plotted with solid lines correspond to the output obtained with $N' = \frac{1}{2}(n_1 + N_2)$ and $N'' = \frac{1}{2}(N_2 + N_3)$. In the middle column, we have repeated this, but for the PDMS–air ($z = d$) interface. The SFG wavelength is slightly shorter than the visible beam wavelength and so there are more oscillations

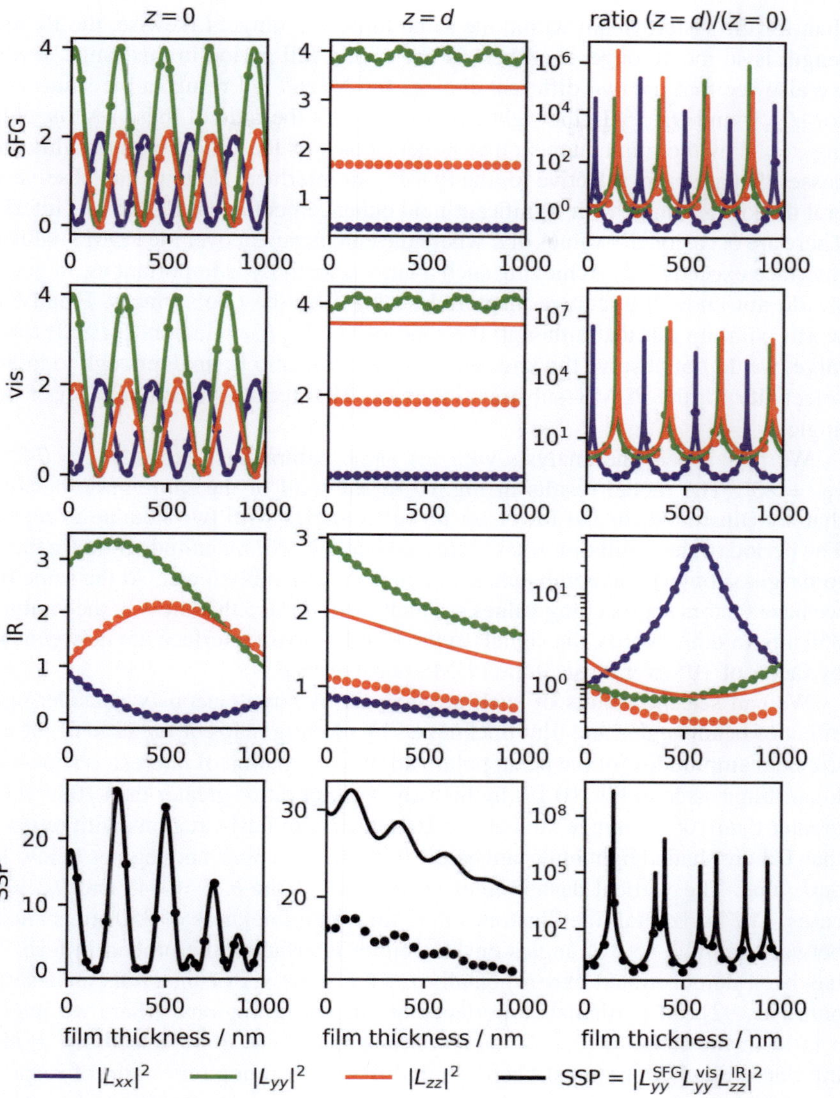

Fig. 10.15 Local fields at the film–substrate ($z = 0$) interface (left column), film–air ($z = d$) interface (middle column), and the air/substrate ratio (right column) for the SFG beam (top row), visible beam (second row, $\lambda_{\text{vis}} = 532$ nm, $\theta_{\text{vis}} = 44°$), IR beam (third row, $\omega_{\text{IR}} = 2910\ \text{cm}^{-1}$, $\theta_{\text{IR}} = \theta_{\text{vis}} + 9°$). In each of these nine subplots, $|\tilde{L}_{xx}|^2$ is plotted in blue, $|\tilde{L}_{yy}|^2$ is plotted in green, and $|\tilde{L}_{zz}|^2$ is plotted in red. In the bottom row, the combination of these factors that go into the SSP intensity is constructed. In each of the subplots, the data in points corresponds to $N' = N'' = N_2$; the solid lines of the same colour come from setting $N' = \frac{1}{2}(n_1 + N_2)$ and $N'' = \frac{1}{2}(N_2 + N_3)$

than for the visible beam within the same thickness range. Likewise, the IR wavelength is so much longer that there is not even a full period in this range. Overall, we also see that the two different choices for N' and N'' result in the same values for $|\tilde{L}_{xx}|^2$ and $|\tilde{L}_{yy}|^2$. In the right column, we plot the ratio of these results, choosing $z = d$ as the numerator so that larger values of the ratio highlight film thicknesses that are more selective for the PDMS–air interface. Here we can observe several thickness ranges with significant field enhancement at the PDMS–air interface. There are occasionally values of d where the enhancement over the PDMS–substrate interface exceeds 10^6. Achieving such a large selectivity is important as, in general, we do not know the relative magnitudes of $\boldsymbol{\chi}^{(2)}$ at the two surfaces. Finally, it is worth pointing out that although the ratio of $|\tilde{L}_{ii}\tilde{L}_{jj}\tilde{L}_{kk}|^2$ factors for SSP can get large, we do not observe thicknesses for which the ratio is small enough to indicate selectivity for the PDMS–substrate interface. That requires a different set of beam angles.

We now repeat the analysis with the same parameters, except $\theta_{\text{vis}} = 74°$ and $\theta_{\text{IR}} = 83°$. The model results in Fig. 10.16 show all of the same basic behaviour that we illustrated for the lower set of beam angles with two notable exceptions. The period of the oscillations has increased and the SSP magnitude ratio (in the bottom right subplot) maximum enhancement is considerably lower. At the same time, we have factors approaching values as small as 10^{-4} for a thickness in the vicinity of 350 nm. In other words, the signals at the PDMS–prism interface are now enhanced by factor of 10^4 over those at the PDMS–air interface.

We can see the results of such an analysis by simultaneously considering the effect of beam angles and film thicknesses by plotting maps of the ratio of the electric field amplitudes for the beam polarization combination of interest as shown on a logarithmic scale in Fig. 10.17. In this way, we plot ratios greater than 10 in yellow, greater than 10^2 in orange, and above 10^5 in red. Similarly, regions with ratios less than 0.1 are shaded light blue, and below 0.01 medium blue, and regions below 10^{-4} dark blue. The vertical dashed lines correspond to the $\theta_{\text{vis}} = 44°$ and $\theta_{\text{vis}} = 74°$ cases. The horizontal line illustrates that, for a film thickness of 350 nm, switching between the two sets of angles enables either interface to be probed in turn. This has been demonstrated experimentally for the same set of materials as described here [91, 92]. Of particular importance are the white regions, where we interpret a ratio in the range of 0.1–10 as a warning that there is no selectivity to either interface. Finally, one must keep in mind that even though the ratio of responses may be favourable for the interface we wish to target, we must also make sure that the actual signal is large enough to be measured that results from a combination of $|\tilde{L}_{ii}\tilde{L}_{jj}\tilde{L}_{kk}|^2$ and the magnitude of the nonlinear susceptibility tensor element(s) [92].

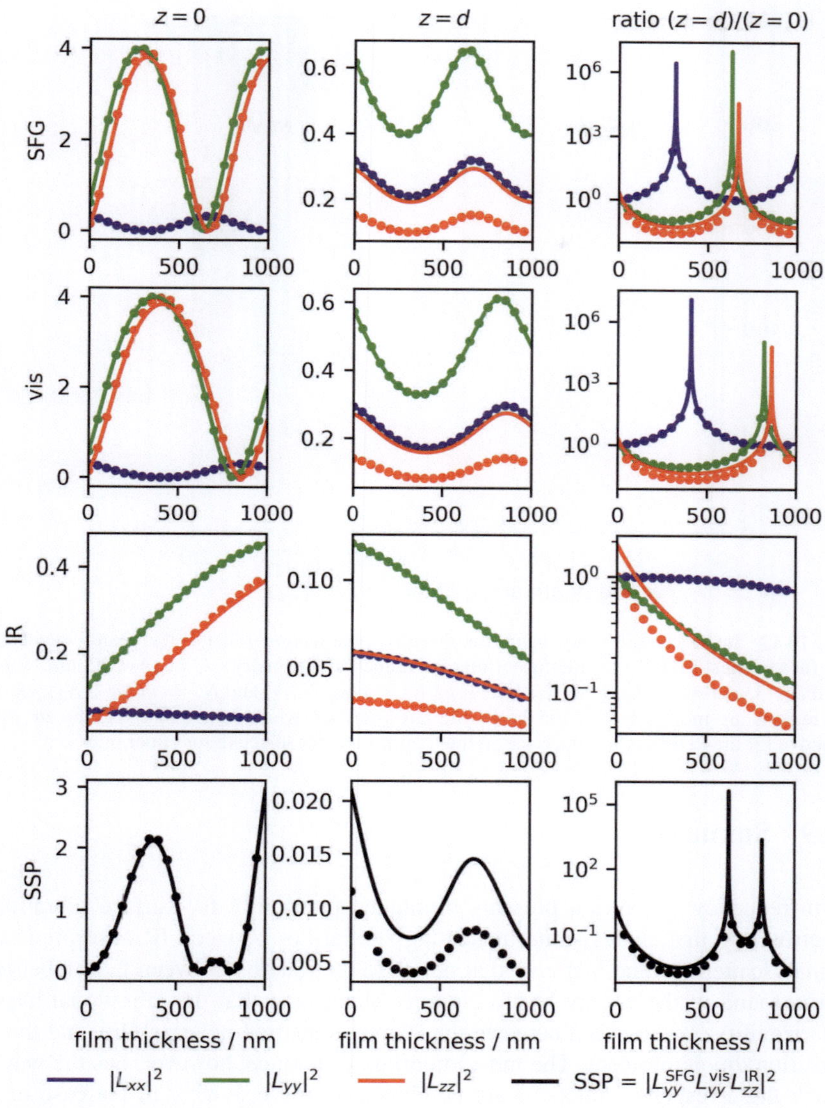

Fig. 10.16 Same parameters as in Fig. 10.15, except that θ_{vis} is now 74° and $\theta_{\mathrm{IR}} = 83°$

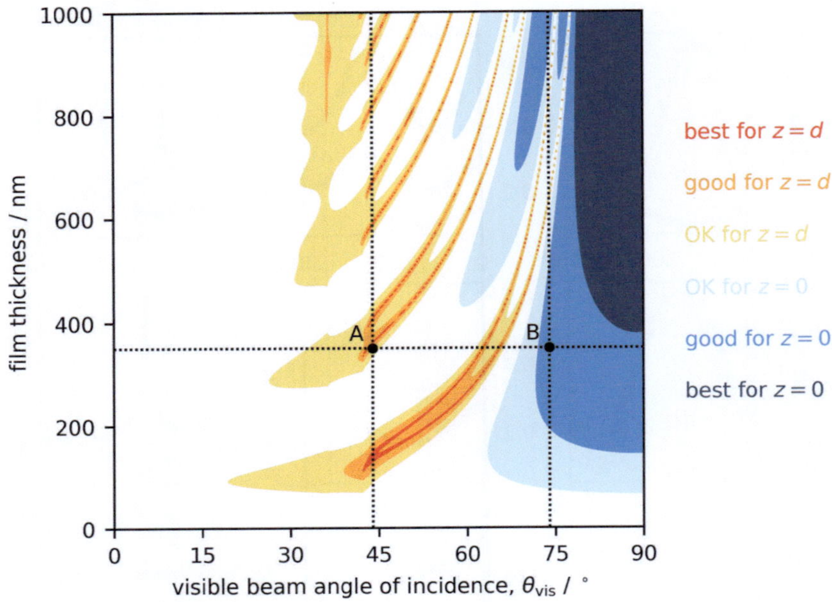

Fig. 10.17 Interface selectivity maps for the SSP polarization scheme for beams incident at the fused silica–PDMS–air interface (internal reflection geometry) at $\lambda_{\mathrm{vis}} = 532$ nm, $\omega_{\mathrm{IR}} = 2910\ \mathrm{cm}^{-1}$, and $\theta_{\mathrm{IR}} = \theta_{\mathrm{vis}} + 9°$. We have used $N' = \frac{1}{2}(n_1 + N_2)$ and $N'' = \frac{1}{2}(N_2 + N_3)$. Darker red regions are more selective for the PDMS–air interface, while darker blue regions are more selective for the PDMS–glass interface. White regions are not selective for either interface

10.9 Summary

Sum-frequency generation presents a unique opportunity for surface vibrational spectroscopy that shares some interesting similarities between IR absorption and Raman scattering. The first clue that points to the relation between these phenomena is found in the energy level diagrams, suggesting that the vibrational hyperpolarizability is a coupling between the Raman transition polarizability and the IR transition dipole moment. The most important difference, however, lies the way in which these response functions are combined to arrive at $\boldsymbol{\alpha}^{(2)}$. In the case of IR absorption, the polarizability is rank-2 tensor that arises from the direct product of two IR transition dipole moments. In Raman scattering, the scattering cross section is related to $\boldsymbol{\alpha}^{(3)}$, a rank-4 tensor that is assembled from the outer product of two transition polarizability tensors. In SFG, exactly one (rank-2) transition polarizability is coupled with a single (rank-1) transition dipole moment to create the rank-3 $\boldsymbol{\alpha}^{(2)}$ tensor. Although this discussion is somewhat overly focused on the shapes, the important consequence is that, unlike IR and Raman spectroscopy, our spectroscopic observable $\boldsymbol{\chi}^{(2)}$ is an even-rank tensor, and therefore has a zero isotropic average (for achiral tensor elements) under the electric dipole approximately. For this reason, a niche application of SFG spectroscopy is to study surfaces and buried interfaces, as it offers an extreme specificity to molecules in environments that break the centrosymmetry of the adjacent bulk phase(s). As with the other topics in previous chapters, our discussion of SFG has just scratched the surface of what is possible to explore. Each of the subsections in this chapter presents opportunities for current research into materials, surfaces, and fundamental nonlinear optical interactions.

10.10 Exercises

1. Einstein notation uses an implied summation over the listed indices. This is a time-saver as one doesn't need to write the summation explicitly. (I generally prefer not to use this as it is easy to forget about the contraction and inadvertently drop terms.) In tensor form, the second-order contribution to the induced dipole moment is

$$\boldsymbol{\mu}^{(2)} = \boldsymbol{\alpha}^{(2)}\mathbf{E}^2.$$

In Einstein notation, a particular element (i.e. one of the three Cartesian components) of $\boldsymbol{\mu}^{(2)}$ is given by

$$\mu_i^{(2)} = \alpha_{ijk}^{(2)} E_j E_k.$$

a. Using explicit summations, how would you obtain the element $\mu_y^{(2)}$?
b. How many terms are there that contribute to $\mu_y^{(2)}$?

---

2. Say you want to study CH_3 stretching on a surface based on the second-order response $\boldsymbol{\alpha}^{(2)}$ using a visible beam $\lambda_{\text{vis}} = 532$ nm ($\omega_{\text{vis}} = 18797$ cm^{-1}) and an infrared beam with $\omega_{\text{IR}} = 2840$ cm^{-1} (resonant with the methyl symmetric stretch). There are four signals generated in your experiment that depend on $\boldsymbol{\alpha}^{(2)}$. These are the

- second harmonic, $2\omega_{\text{vis}}$ (or $\omega_{\text{vis}} + \omega_{\text{vis}}$);
- second harmonic, $2\omega_{\text{IR}}$ (or $\omega_{\text{IR}} + \omega_{\text{IR}}$);
- sum frequency, $\omega_{\text{vis}} + \omega_{\text{IR}}$;
- difference frequency, $\omega_{\text{vis}} - \omega_{\text{IR}}$.

List possible pros and/or cons of each of these measurements from a practical standpoint. You'll want to also consider the wavelength of the measured response in each case.

---

3. An SFG experiment has a visible beam at $\lambda_{\text{vis}} = 532$ nm incident to the sample at $\theta_{\text{vis}} = 55°$, and an infrared beam at $\omega_{\text{IR}} = 2870$ cm^{-1} incident at $\theta_{\text{IR}} = 60°$.

a. If the beams approach a thick polystyrene sample from air, at what angle do you expect the reflected SFG beam? You may need to look up refractive index information for the bulk phases.
b. What if the beams approach this same polystyrene sample but incident from a CaF_2 hemispherical prism?
c. Now the CaF_2 hemisphere is replaced with a 1-mm-thick CaF_2 flat window. The incident angles specified above are for the air–CaF_2 interface. You'll need to use Snell's law to determine the refracted beam angles in polystyrene. At what angle (with respect to the surface normal in air) do you expect the SFG beam now?

---

4. You perform a reflection SFG experiment of sodium dodecyl sulphate adsorbed at the air–water interface.

a. Explain why the resulting SFG spectrum shows an intense CH_3 symmetric stretch but the CH_2 symmetric stretch is barely noticeable in the SSP polarization.
b. Do you expect the situation to be different with a different polarization combination?
c. In which polarization combinations do you expect to be able to observe either of these two modes?
d. Explain why this situation is different from the case of IR absorption or Raman scattering, where the CH_2 symmetric stretch is must stronger than the methyl symmetric stretch.

---

5. You have prepared a monolayer of octadecylsilane (OTS) on glass, and record an SFG spectrum at the air–glass (IR-grade fused silica) interface. You record how many counts of SFG is obtained in the SSP polarization scheme on resonance with the methyl symmetric stretch at $2875\,\text{cm}^{-1}$ for your set of beam angles $\theta_{\text{vis}} = 55°$ and $\theta_{\text{IR}} = 62°$.

a. Relative to the above number, how many counts are obtained if all experimental parameters (laser powers, etc.) are the same, but the incident beam angles are changed to $\theta_{\text{vis}} = 45°$ and $\theta_{\text{IR}} = 55°$? I.e. 1.2× the original value? 0.85× the original value? etc. (Is it possible to determine this without knowing anything about $\boldsymbol{\alpha}^{(2)}$ or $\boldsymbol{\chi}^{(2)}$? *Hint:* Yes, it has to do with the magnitude squared of the product of three **L** factors.)
b. Let's say you want to optimize the beam angles you will use in your experiment. To what extent can you do this in advance (i.e. without any prior knowledge of $\boldsymbol{\chi}^{(2)}$)?

c. What if the OTS monolayer were created on the flat surface of a hemicylindrical prism made of the same IR-grade fused silica? Can you predict the optimal set of beam angles now?
d. If you work in this field, it may be useful to create a plot of the relevant term you are calculating as a function of the beam angles for both external and internal reflection geometries. If your experiment offers a simple way to change the beam angles independently, you'll want to consider making a 3D plot where θ_{vis} appears along one axis, and θ_{IR} is along another, with colours or contours indicating the value of your calculated quantity. If the relative angles between the beams is fixed (as it is in my lab), then you can make a simpler plot. Either way, comment on the result. The large differences may be surprising to you. The point(s) at which the best signal is obtained may also be of interest.
e. Can the same analysis be performed for other polarization schemes? SPS? PSS? PPP? Explain why or why not. If the analysis is not possible in some of the cases, then can additional information be provided that permits the analysis? For cases where the analysis can proceed in the exact same way as for SSP, I encourage you to look the results.
f. Typically, people don't want to change their beam angles in the middle of a single experiment. If spectra are desired in multiple beam polarizations (e.g. SSP and SPS, or SSP, SPS and PPP) can your results be used to predict the best set of experimental conditions?
g. I've used OTS on silica as an example since the SFG response on resonance with OTS modes originates from a monolayer. What if the sample was thicker than a monolayer? Does this change or limit the applicability of any of your results?

---

6. Imagine that you want to study the structure of a different polymer at the surface of CaF_2. Unlikely, but let's assume that this polymer as the exact same refractive index as CaF_2 at all wavelengths.

a. Will there be a reflected SFG beam? Why or why not?
b. Calculate the **L** factors for the visible, infrared, and SFG beam.

---

7. In this chapter we have considered orientation distributions that are ordered with respect to θ only. Some discussion about more complex distributions where chromophores are oriented about multiple Euler angles appears in the next chapter. However, there is a simple case that can be described now, pertaining to crystals. If we assume that the structure of the crystal is perfect (no defects, infinitely narrow distribution, i.e. perfect ordering, no *distribution*). In this case, what we have been describing as the laboratory frame for $\boldsymbol{\chi}^{(2)}$ is the surface (crystallographic)

frame, and then the lab frame $\boldsymbol{\chi}^{(2)}$ elements are obtained by a simple rotation about the surface normal, z.

a. Write some SymPy code that provides all possible $\boldsymbol{\chi}^{(2)}$ elements in the laboratory frame when the surface frame $\boldsymbol{\chi}^{(2)}_{\text{surf}}$ contains an arbitrary number of unique or non-zero elements. Your expressions for the lab frame elements will contain the surface frame elements and the rotation angle (about z) that we can call γ.
b. A common approximation for quartz is that, far from any resonance, its surface frame second-order susceptibility tensor contains only $\chi^{(2)}_{xxx} \neq 0$ and all 26 other elements are zero. Work out the laboratory-frame $\boldsymbol{\chi}^{(2)}$ elements in this case.
c. If we limit ourselves to beams that are s- or p-polarized, and to the detection of the s or p-polarized SFG, then there are eight possible polarization schemes (SSS, SSP, SPP, etc.). Using your above result, provide expressions for $\boldsymbol{\chi}^{(2)}_{\text{eff}}$ (which is always defined in the laboratory frame) for these eight schemes.

---

8. Chiral chromophores have non-zero values of $\alpha^{(2)}_{\ell mn}$ for which $\ell \neq m \neq n$.

a. If you had an isotropic distribution of such moieties, which elements of $\boldsymbol{\chi}^{(2)}$ would you expect to be non-zero? What polarization schemes would be used to probe them?
b. Now picture a sample where the surface consists of an ordered arrangement of such chiral moieties. For simplicity, assume they are confined there, and the bulk does not contain these species. Assuming a simple θ-only distribution, answer the same questions as posed above.
c. Now picture the same molecules to be present at the surface and the bulk. Is there any way you could perform your experiment or analysis so as to isolate the contribution of the surface species? For example, is there a polarization scheme in which the bulk is silent?

---

9. Although we have not described such an approach here, a common way of performing orientation analysis using SFG data is to model the predicted ratio between two polarization schemes (for example, the SSP/SPS intensity ratio) and plot this as a function of one or parameters in an orientation distribution function. For example, if a Gaussian distribution of tilt angles is assumed (and isotropy about the ϕ and ψ orientations), the SSP/SPS ratio can be plotted versus θ_0 for some fixed values of the distribution width σ.

a. Write such a code. Make a note of what parameters you need (beam angles, refractive indices, $\boldsymbol{\alpha}^{(2)}$ values, etc.).
b. How sensitive is your analysis to the refractive index values you have used? This is an important question, as there is a paucity of refractive index data in

the mid-infrared. (When values of n and κ are reported in the infrared, they are often not direct measurements, i.e. not from ellipsometry. They are more typically obtained through modelling IR absorbance data using Kramers–Kronig or similar models.) Being able to understand how sensitive the results of your orientation analysis are to these uncertainties is extremely valuable.

c. Similarly, how sensitive are your results to changes in the N' value that appears in the L_{zz} local field correction?

d. What about the $\boldsymbol{\alpha}^{(2)}$ values? Even though only relative values of $\boldsymbol{\alpha}^{(2)}$ (for example, $\alpha^{(2)}_{aac}/\alpha^{(2)}_{ccc}$) may be required in your model, such ratios are often not known with a high degree of certainty.

e. If you can think of any other uncertainties in parameters that may critically affect your results, explore those. Would you consider that the amount of information required to arrive at molecular orientation information in SFG is more than that required in IR or Raman experiments, less, or about the same? It may be helpful to address this is a quantitative way.

---

10. a. What are the ranges of the possible values of the order parameters $\langle P_1\rangle$, $\langle P_2\rangle$, $\langle P_3\rangle$, $\langle P_4\rangle$?

    b. What is the range of the ratio $\langle P_3\rangle/\langle P_1\rangle$?

References

1. R.W. Boyd, *Nonlinear Optics*, 4th edn. (Academic Press, San Diego, 2020)
2. Y.R. Shen, *The Principles of Nonlinear Optics* (Wiley, New York, 1984)
3. Y.R. Shen, in *Proceedings of the International School of Physics, Enrico Fermi*, ed. by T.W. Hansch, M. Inguscio (North Holland, Amsterdam, 1994), pp. 139–165
4. P.N. Butcher, D. Cotter, *The elements of nonlinear optics* (Cambridge University Press, New York, 1990)
5. Y.R. Shen, *Fundamentals of Sum-Frequency Spectroscopy* (Cambridge University Press, Cambridge, 2016)
6. A. Morita, *Theory of Sum Frequency Generation Spectroscopy* (Springer, Singapore, 2018)
7. Y.R. Shen, V. Ostroverkhov, Chem. Rev. **106**, 1140 (2006)
8. Q. Du, E. Freysz, Y.R. Shen, Phys. Rev. Lett. **72**, 238 (1994)
9. R. Superfine, J.Y. Huang, Y.R. Shen, Opt. Lett. **15**, 1276 (1990)
10. R. Superfine, J.Y. Huang, Y.R. Shen, Chem. Phys. Lett. **172**, 303 (1990)
11. N. Ji, V. Ostroverkhov, C. Chen, Y.R. Shen, J. Am. Chem. Soc. **129**, 10056 (2007)
12. L. Zhang, C. Tian, G. Waychunas, Y.R. Shen, J. Am. Chem. Soc. **130**, 7686 (2008)
13. C. Tian, N. Ji, G.A. Waychunas, Y.R. Shen, J. Am. Chem. Soc. **130**, 13033 (2008)
14. C. Tian, Y.R. Shen, J. Am. Chem. Soc. **131**, 2790 (2009)
15. N. Ji, V. Ostroverkhov, C.S. Tian, Y.R. Shen, Phys. Rev. Lett. **100**, 096102 (2008)
16. K.C. Jena, P.A. Covert, D.K. Hore, J. Chem. Phys. **134**, 044712 (2011)
17. P.A. Covert, W.R. FitzGerald, D.K. Hore, J. Chem. Phys. **137**, 014201 (2012)
18. C.C. Rich, K.A. Lindberg, A.T. Krummel, J. Phys. Chem. Lett. **8**, 1331 (2017)
19. S. Nihonyanagi, S. Yamaguchi, T. Tahara, J. Chem. Phys. **130**, 204704 (2009)

20. J.E. Laaser, W. Xiong, M.T. Zanni, J. Phys. Chem. B **115**, 2536 (2011)
21. J.E. Laaser, W. Xiong, M.T. Zanni, J. Phys. Chem. B **115**, 9920 (2011)
22. R. Lu, Y. Rao, W.K. Zhang, H.F. Wang, Proc. SPIE Conf. Nonlinear. Spectrosc. **4812–4815**, 115 (2002)
23. S. Nihonyanagi, R. Kusaka, K.i. Inoue, A. Adhikari, S. Yamaguchi, T. Tahara, J. Chem. Phys. **143**, 124707 (2015)
24. H. Vanselous, P.B. Petersen, J. Phys. Chem. C **120**, 8175 (2016)
25. H. Wang, T. Gao, W. Xiong, ACS Photonics **4**, 1839 (2017)
26. I.V. Stiopkin, H.D. Jayathilake, A.N. Bordenyuk, A.V. Benderskii, J. Am. Chem. Soc. **130**, 2271 (2008)
27. N. Takeshita, M. Okuno, T.A. Ishibashi, J. Phys. Chem. C **121**, 25206 (2017)
28. E.C.Y. Yan, L. Fu, Z. Wang, W. Liu, Chem. Rev. **114**, 8471 (2014)
29. L. Fu, J. Liu, E.C. Yan, J. Am. Chem. Soc. **133**, 8094 (2011)
30. L. Fu, Z. Wang, E.C. Yan, Int. J. Mol. Sci. **12**, 9404 (2011)
31. L.M. Haupert, G.J. Simpson, Ann. Rev. Phys. Chem. **60**, 345 (2009)
32. M.G. Kuzyk, *Nonlinear Optics: A Student's Perspective: With Python Problems and Examples* (CreateSpace Independent Publishing Platform, North Charleston, SC, 2017)
33. I. Chae, S. Ahmed, H. Ben Atitallah, J. Luo, Q. Wang, Z. Ounaies, S.H. Kim, Macromolecules **50**, 2838 (2017)
34. V. Rodriguez, F. Lagugné Labarthet, C. Sourisseau, Appl. Spectrosc. **59**, 322 (2005)
35. M. Moskovits, J. Chem. Phys. **77**, 4408 (1982)
36. N.J. Harrick, *Internal Reflection Spectroscopy* (Interscience Publisher, Wiley, New York, NY, 1967)
37. N.J. Harrick, J. Opt. Soc. Am. **55**, 851 (1965)
38. J.A. Ekhoff, K.L. Rowlen, Anal. Chem. **74**, 5954 (2002)
39. P. Ye, Y.R. Shen, Phys. Rev. B **28**, 4288 (1983)
40. R.W. Munn, M.M. Shabat, J. Chem. Phys. **99**, 10052 (1993)
41. M. in het Panhuis, R.W. Munn, J. Chem. Phys. **113**, 10685 (2000)
42. R.W. Munn, J. Chem. Phys. **103**, 850 (1995)
43. C.C. Yu, T. Seki, Y. Wang, M. Bonn, Y. Nagata, Phys. Rev. Lett. **128**, 226001 (2022)
44. C.C. Yu, T. Seki, K.Y. Chiang, F. Tang, S. Sun, M. Bonn, Y. Nagata, J. Phys. Chem. B **126**, 6113 (2022)
45. K.Y. Chiang, T. Seki, C.C. Yu, T. Ohto, J. Hunger, M. Bonn, Y. Nagata, Proc. Nat. Acad. Sci. USA **119**, e2204156119 (2022)
46. X. Yu, K.Y. Chiang, C.C. Yu, M. Bonn, Y. Nagata, J. Chem. Phys. **158**, 044701 (2023)
47. P. Yang, A. Kumarasiri, D. Hore, J. Chem. Phys. **161**, 054703 (2024)
48. K.A. Briggman, J.C. Stephenson, W.E. Wallace, L.J. Richter, J. Phys. Chem. B **105**, 2785 (2001)
49. S.A. Hall, A.D. Hickey, D.K. Hore, J. Phys. Chem. C **114**, 9748 (2010)
50. K.S. Gautam, A.D. Schwab, A. Dhinojwala, D. Zhang, S.M. Dougal, M.S. Yeganeh, Phys. Rev. Lett. **85**, 3854 (2000)
51. G.P. Harp, H. Rangwalla, M.S. Yeganeh, A. Dhinojwala, J. Am. Chem. Soc. **125**, 11283 (2003)
52. C.S.C. Yang, P.T. Wilson, L.J. Richter, Macromolecules **37**, 7742 (2004)
53. Y. Liu, M. Messmer, J. Am. Chem. Soc. **124**, 9714 (2002)
54. D. Zhang, S.M. Dougal, M.S. Yeganeh, Langmuir **16**, 4528 (2000)
55. A.D. Curtis, A.R. Calchera, M.C. Asplund, J.E. Patterson, Vibr. Spectrosc. **68**, 71 (2013)
56. X. Lu, J.N. Myers, Z. Chen, Langmuir **30**, 9418 (2014)
57. A.D. Curtis, S.B. Reynolds, A.R. Calchera, J.E. Patterson, J. Phys. Chem. Lett. **1**, 2435 (2010)
58. K. Wandelt, S. Thurgate, *Solid-liquid interfaces: Macroscopic phenomena–microscopic understanding* (Springer-Verlag, Berlin, 2003)
59. C. van Oss, R. Giese, J. Disper. Sci. Tech. **25**, 631 (2004)
60. C. O'Connor, P. Wiggins, Biocatalysis **1**, 249 (1988)

61. P. Fenter, N.C. Sturchio, Prog. Surf. Sci. **77**, 171 (2004)
62. V. Gunko, V.V. Turov, V.M. Bogatyrev, V.I. Zarko, R. Leboda, E.V. Goncharuk, A.A. Novza, A.V. Turov, A.A. Chuiko, Adv. Coll. Interf. Sci. **118**, 125 (118)
63. B. Jayaram, T. Jain, Annu. Rev. Biophys. Biomol. Struct. **33**, 343 (2004)
64. U. Heinzmann, S. Holloway, A.W. Kleyn, R.E. Palmer, K.J. Snowdon, J. Phys.: Condens. Matter **8**, 3245 (1996)
65. Y. Levy, J.N. Onuchic, Proc. Nat. Acad. Sci. USA **101**, 3325 (2004)
66. M. Ahmad, W. Gu, T. Geyer, V. Helms, Nat. Commun. **2**, 261 (2011)
67. K. Shiraga, Y. Ogawa, N. Kondo, Biophys. J . **111**, 2629 (2016)
68. K.C. Jena, D.K. Hore, Phys. Chem. Chem. Phys. **12**, 14383 (2010)
69. K.D. Krause, S. Roy, D.K. Hore, Biointerphases **12**, 02D407 (2017)
70. T.G. Trudeau, D.K. Hore, Langmuir **26**, 11095 (2010)
71. S.C. Hong, M. Oh-e, X. Zhuang, Y.R. Shen, J.J. Ge, F.W. Harris, S.Z.D. Cheng, Phys. Rev. E **63**, 051706 (2001)
72. D. Kim, M. Oh-e, Y.R. Shen, Macromolecules **34**, 9125 (2001)
73. T.T. Nguyen, H.T.T. Khuat, S. Asakura, G. Mizutani, Y. Murakami, T. Okada, J. Chem. Phys. **155**, 084702 (2021)
74. M. Oh-e, S.C. Hong, Y.R. Shen, J. Phys. Chem. B **104**, 7455 (2000)
75. W. Chen, M.B. Feller, Y.R. Shen, Phys. Rev. Lett. **63**, 2665 (1989)
76. M.B. Feller, W. Chen, Y.R. Shen, Phys. Rev. A **43**, 6778 (1991)
77. Y. Ouchi, M.B. Feller, T. Moses, Y.R. Shen, Phys. Rev. Lett. **68**, 3040 (1992)
78. X. Wei, S. Hong, X. Zhuang, T. Goto, Y. Shen, Phys. Rev. E **62**, 5160 (2000)
79. A. Kumarasiri, P. Yang, D.K. Hore, J. Phys. Chem. Lett. **14**, 4449 (2023)
80. G. Placzek, *Handbuch der Radiologie* (Akademische Verlagsgesellschaft, Leipzig, 1934), p.205
81. H.F. Wang, W. Gan, R. Lu, Y. Rao, B.H. Wu, Int. Rev. Phys. Chem. **24**, 191 (2005)
82. A. Yousefi, D.K. Hore, J. Phys. Chem. C **128**, 7252 (2024)
83. G.J. Simpson, K.L. Rowlen, J. Am. Chem. Soc. **121**, 2635 (1999)
84. Z. Li, J. Wang, Y. Li, W. Xiong, J. Phys. Chem. C **120**, 20230 (2016)
85. D. O'Brien, A. Massari, J. Chem. Phys. **138**, 154708 (2013)
86. D.B. O'Brien, A.M. Massari, J. Opt. Soc. Am. B: Opt. Phys. **30**, 1503 (2013)
87. D.B. O'Brien, A.M. Massari, J. Chem. Phys. **142**, 024703 (2015)
88. B. Busson, J. Chem. Phys. **159**, 034705 (2023)
89. B. Busson, J. Chem. Phys. **159**, 034706 (2023)
90. B. Busson, J. Chem. Phys. **159**, 034707 (2023)
91. M.S. Azam, C. Cai, D.K. Hore, J. Phys. Chem. C **123**, 23535 (2019)
92. C. Cai, M.S. Azam, D.K. Hore, J. Phys. Chem. C **125**, 12382 (2021)
93. S. Yamaguchi, K. Shiratori, A. Morita, T. Tahara, J. Chem. Phys. **134**, 184705 (2011)
94. S. Sun, C. Tian, Y.R. Shen, Proc. Nat. Acad. Sci. USA **112**, 5883 (2015)
95. H. Held, A.I. Lvovsky, X. Wei, Y.R. Shen, Phys. Rev. B **66**, 205110 (2002)
96. D. Maikhuri, S.P. Rurohit, K.C. Mathur, AIP Adv. **5**, 047115 (2015)
97. X. Wei, S.C. Hong, A.I. Lvovsky, H. Held, Y.R. Shen, J. Phys. Chem. B **104**, 3349 (2000)
98. K. Matsuzaki, S. Nihonyanagi, S. Yamaguchi, T. Nagata, T. Tahara, J. Phys. Chem. Lett. **4**, 1654 (2013)
99. R. Zheng, W.M. Weo, Q. Shi, Phys. Chem. Chem. Phys. **17**, 9068 (2015)
100. M. Born, E. Wolf, *Principles of Optics*, 4th edn. (Pergamon Press Ltd., Oxford, 1970)
101. E. Hecht, A. Zajac, *Optics* (Addision-Wesley, Reading, MA, 1979)
102. R.M.A. Azzam, N.M. Bashara, *Ellipsometry and Polarized Light* (North-Holland Publishing Company, New York, 1977)
103. A.G. Lambert, D.J. Neivandt, A.M. Briggs, E.W. Usadi, P.B. Davies, J. Phys. Chem. B **106**, 5461 (2002)
104. S.J. McGall, P.B. Davies, D.J. Neivandt, J. Phys. Chem. B **108**, 16030 (2004)
105. Y. Tong, Y. Zhao, N. Li, M. Osawa, P.B. Davies, S. Ye, J. Chem. Phys. **133**, 034704 (2010)

106. Y. Tong, Y. Zhao, N. Li, Y. Ma, M. Osawa, P.B. Davies, S. Ye, J. Chem. Phys. **133**, 034705 (2010)
107. G. Li, A. Dhinojwala, M.S. Yeganeh, J. Phys. Chem. C **115**, 7554 (2011)
108. A.P. Moon, R. Pandey, J.A. Bender, D.E. Cotton, B.A. Renard, S.T. Roberts, J. Phys. Chem. C **121**, 18653 (2017)
109. X. Li, B. Li, X. Zhang, C. Li, Z. Guo, D. Zhou, X. Lu, Macromolecules **49**, 3116 (2016)
110. X. Lu, M.L. Clarke, D. Li, X. Wang, G. Xue, Z. Chen, J. Phys. Chem. C **115**, 13759 (2011)
111. P.T. Wilson, K.A. Briggman, W.E. Wallace, J.C. Stephenson, L.J. Richter, Appl. Phys. Lett. **80**, 3084 (2002)
112. C. Hirose, H. Ishida, K. Iwatsu, N. Watanabe, J. Kubota, A. Wada, K. Domen, J. Chem. Phys. **108**, 5948 (1998)
113. P. Yeh, *Optical Waves in Layered Media*, vol. 95 (Wiley, Hoboken, NJ, 1988)
114. J. Sipe, J. Opt. Soc. Am. B: Opt. Phys. **4**, 481 (1987)
115. W.R. FitzGerald, K.C. Jena, D.K. Hore, J. Mol. Struct. **1084**, 368 (2014)
116. E.H.G. Backus, N. Garcia-Araez, M. Bonn, H.J. Bakker, J. Chem. Phys. C **116**, 23351 (2012)
117. D. Wilk, D. Johannsmann, C. Stanners, Y. Shen, Phy. Rev. B **51**, 10057 (1995)
118. J. Wang, Z. Paszti, M.A. Even, Z. Chen, J. Phys. Chem. B **108**, 3625 (2004)
119. M. Ramsay, C. Beutier, G.B. McGarvey, D.K. Hore, J. Chem. Phys. **150**, 014702 (2019)
120. Y. Otsuki, T.S.T. Ishiyama, A. Morita, K. Watanabe, Y. Matsumoto, Phys. Rev. B **96**, 115405 (2017)
121. P. Yeh et al., *Optical waves in layered media*, vol. 95 (Wiley, New York, New York City, NY, 1988)
122. D.E. Cotton, S.T. Roberts, J. Chem. Phys. **154**, 114704 (2021)
123. I.H. Malitson, J. Opt. Soc. Am. **55**, 1205 (1965)
124. F. Schneider, J. Draheim, R. Kamberger, U. Wallrabe, Sens. Actuators, A **151**, 95 (2009)
125. M.R. Querry, Optical constants of minerals and other materials from the millimeter to the ultraviolet. Technical report 88009, Chemical Research, Development, and Engineering Center (1987)
126. Dow Corning Corporation, *Sylgard 184 silicone elastomer product information* (Technical report, Dow Corning Corporatio, Midland, MI, 1999)

Chapter 11
More Complex and Realistic Situations

11.1 Outline

The previous three chapters provided an idea of the strengths and limitations of different spectroscopic techniques applied to the analysis of a single vibrational mode (whose amplitude is determined by fitting) observed under different beam polarizations. In each case, the goal was to reveal the underlying properties of the molecular response in the most direct way possible. As a result, we have often made drastic assumptions about the symmetry of the vibrational mode, often simplifying it down to one or two elements of the underlying response tensor $\boldsymbol{\alpha}^{(n)}$. In a realistic situation, the degree to which such simplifications can be made depends on the nature of the mode you are studying. The stretching mode of a nitrile group is obviously simpler than any of the modes associated with a phenyl ring. Another aspect that differentiates real materials from those that we discussed is that the constituent molecules may be arranged in more complicated ways, most importantly in ways that depend on Euler angles other than just the tilt orientation specified by θ. You may think that we quickly run into a situation where the problem is too complex to analyse in a quantitative way, since we don't have enough information available in the experimental data. In general that is correct, and practically speaking assumptions are often made to reduce the complexity of the problem. However, a major additional source of information that we have not considered is that multiple vibrational modes can be analysed in tandem. Finally, the results of different experiments can be combined. This is especially useful when the techniques provide complementary order parameters. In the last section of this chapter, we'll consider how the interpretation of experimental vibrational spectra can be assisted with data from molecular dynamics (MD) simulations. In many ways, the tools we have explored for decomposing the vibrational response readily lend themselves to modelling the vibrational response from atomic/molecular coordinates as afforded by an MD trajectory.

D. K. Hore, *Molecular Orientation Analysis Using Polarized Vibrational Spectroscopy*,
https://doi.org/10.1007/978-3-032-04874-5_11

11.2 More Complex Vibrational Mode Responses

In our discussion of IR, Raman, and SFG spectra, a primary goal was to introduce the simplest molecular $\boldsymbol{\alpha}^{(n)}$ response that demonstrated the salient features of the respective $\chi^{(n)}$ response. In the case of IR absorption, this was assuming that all 9 $\alpha_{\ell m}$ were zero expect for α_{cc}. In the case of Raman scattering, we assumed that only 1 out of 81 $\boldsymbol{\alpha}^{(3)}$ elements was non-zero, that being $\alpha^{(3)}_{cccc}$ which in turn can be expressed as a single element of the Raman tensor $\boldsymbol{\alpha}^{\mathrm{R}}_{cc}$. In our SFG discussion, we considered only $\alpha^{(2)}_{aac} = \alpha^{(2)}_{bbc}$ and $\alpha^{(2)}_{ccc}$. In doing so, we were able to reveal some important characteristics: that IR absorption provides information on the order parameter $\langle P_2 \rangle$; that Raman spectra can be analysed to obtain $\langle P_2 \rangle$ and $\langle P_4 \rangle$, and that the Raman depolarization ratio $\rho = I_{\perp}/I_{\parallel}$ can used to relate to molecular-frame properties in solution from the isotropic average. In the case of SFG spectra, we saw that the response is proportional to $\langle P_1 \rangle$ and $\langle P_3 \rangle$, but that these two order parameters cannot be separated in the simplest uniaxial case when using amplitude ratios.

Although such drastic simplification of the molecular response is reasonable for the IR transition dipole moment and hence the polarizability, a more accurate description of molecular vibrations, even in the high symmetry case, must consider at least one additional element in the Raman response, typically α^{R}_{aa} in addition to α^{R}_{cc}. We dipped into this regime near the end of our discussion of SFG spectroscopy as we considered $\alpha^{(2)}_{aac}$ and $\alpha^{(2)}_{ccc}$ for a vibration mode with C_{3v} or $C_{\infty v}$ symmetry. However, some modes naturally have even lower symmetry, and additional elements of all $\boldsymbol{\alpha}^{(n)}$ tensors come into play. The consequence is that, even when the problem reduces itself to a θ-only situation (isotropic ϕ and ψ distribution) it is not easy or necessarily possible to obtain the $\langle P_n \rangle$ order parameters without knowledge of the $\alpha^{(n)}$ elements. When only two non-zero unique $\boldsymbol{\alpha}^{(n)}$ elements are present, this can be expressed as a ratio, for example, $R = \alpha^{\mathrm{R}}_{aa}/\alpha^{\mathrm{R}}_{cc}$ and sometimes this still enables $\langle P_2 \rangle$ and $\langle P_4 \rangle$ to be obtained from Raman data. But, in general, it is not always possible to avoid knowledge of the $\boldsymbol{\alpha}^{(n)}$ elements. Those then need to be obtained through experimental methods, for example, by collecting spectra with known orientation distributions, or through quantum chemical calculations, either for the vibrational mode directly or in conjunction with a bond-additivity approach. As one might expect, when additional $\boldsymbol{\alpha}^{(n)}$ elements are present, the expressions for the lab frame $\boldsymbol{\chi}^{(n)}$ elements become more complicated, with additional terms, and additional trigonometric functions introduced from the direction cosine elements that are used in the projections. At this stage, it is highly recommended that some Python code be constructed for this task, rather than tedious and error-prone manual derivations that involve many sums, especially for high-rank tensors. One should consider that, for many problems, numerical solutions will suffice, and so the NumPy module introduced in Chap. 3 offers everything that is required. However, when analytical expressions are desired, it really pays off to use a tool like SymPy (Chap. 5), especially inside a Jupyter notebook environment [1] where equations can be displayed nicely. A good starting point is to set up some code that reproduces some of the expressions that have been provided in the previous chapters. If analytical expressions are used, being able to convert between the SymPy expressions and numerical NumPy equivalents is very powerful.

11.3 More Complex Orientation Distributions

I don't have too much guilt associated with knocking out 80 $\alpha^{(3)}_{\ell mno}$ elements to leave only $\alpha^{(3)}_{cccc}$, regardless of how irresponsible and unrealistic that might be. As confessed, it got the point across, and enabled us to easily compare and contrast the different vibrational phenomena. However, what I do feel sheepish about is constantly integrating ϕ and ψ to leave only θ behind. The truth is, when it comes to θ-only distributions, there is no shortage of interesting and important problems one can study that fall under this category. However, it is most realistic only in the situation pictured in Fig. 11.1a, where we consider a vibrational mode with cylindrical symmetry, and therefore there can be no twist angle ψ. In this case, applying treatments such as those discussed in the preceding three chapters requires only that the assumption (or experimental verification) of an orientation distribution that is uniform in the xy-plane (i.e. no ϕ-dependence) is good.

However, just because those situations exist doesn't imply that many materials have molecules aligned in such a way. In fact, θ-only situations may well be important and interesting, but they represent a small subset of the systems that can be studied for all of the techniques we have been discussing. The important closely analogous situation is depicted in Fig. 11.1b, where the chemical group of interest is planar, and is therefore highly capable of being oriented about its c-axis in a manner described by the angle ψ. To use a uniform ψ model, one must either be reasonably confident that no ψ preference is exhibited, or be willing to make this drastic assumption and accept the consequences of interpreting the parameters of the sought distribution that may not describe the true system. In any case, θ-only distributions are limited to what is shown in Fig. 11.1a and b.

We will now take a quick look at the situations that arise when the molecular orientation must be described in terms of two or all three Euler angles. We will see that when we deal with a system that is ordered in a more complex way, with preference for the azimuthal angle ϕ or twist angle ψ, things become more complicated rather quickly.

11.3.1 Orientation Distributions Involving Two Euler Angles

Moving up in complexity, we first consider systems that are oriented about both θ and ϕ, that is, the molecules or functional groups of interest are tilted with respect to the z-direction (described by θ) and are oriented in the xy-plane (according to

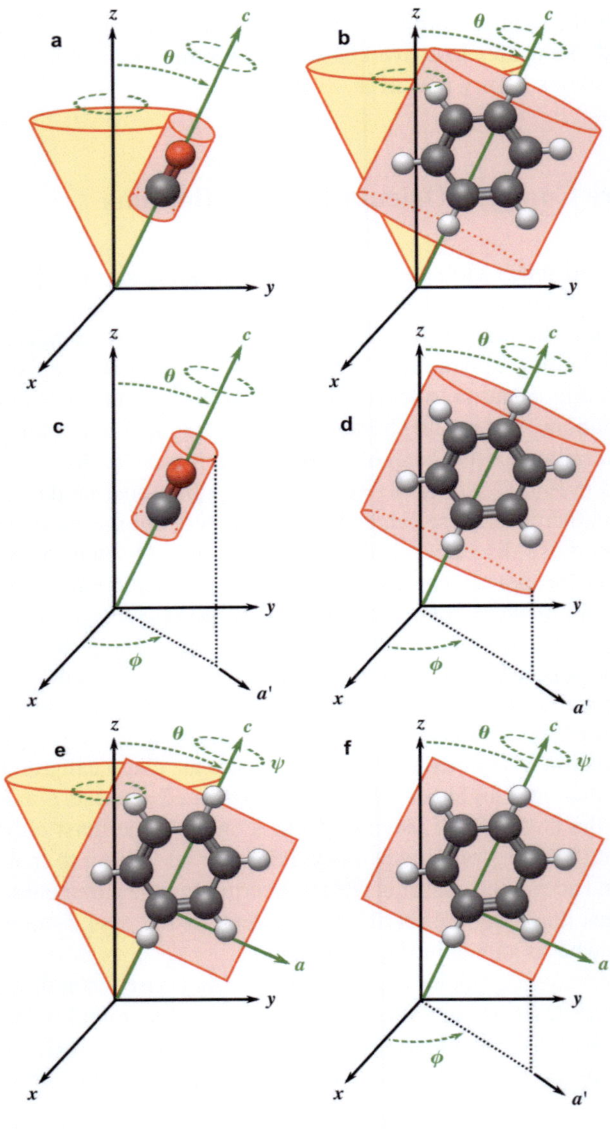

Fig. 11.1 Examples of functional group oriented in their tilt angle θ when **a** ϕ is uniformly distributed and there is no twist angle ψ due to the symmetry of the vibrational mode; **b** a uniform distribution of ϕ and ψ, regardless of the mode symmetry; **c** a cylindrical mode oriented with a ϕ preference; **d** an arbitrary mode assumed to have a ϕ orientation but no ψ preference; **e** twist orientation, but no alignment in ϕ; **f** an orientation involving all three Euler angles. Definitions of the Euler angles are indicated using dashed green arrows in each case, but the distribution about each angle is isotropic unless it is labelled. Cylinders indicate isotropic twist (ψ) distributions; cones indicate isotropic azimuthal (ϕ) distributions. a' is the projection of the molecular a-axis onto the lab xy-plane

their ϕ distribution) [2–4]. Again, it is worth mentioning that this assumption of a uniform twist distribution is either strictly necessary (for vibrational modes that exhibit cylindrical symmetry as shown in Fig. 11.1c) or known from some additional information, or assumed (Fig. 11.1d). One option for describing this is a product of two delta functions

$$f(\theta, \phi) = \delta(\theta - \theta_0) \cdot \delta(\psi - \phi_0) \tag{11.1}$$

where we assume all molecules are perfectly oriented at (θ_0, ϕ_0). While this may be far from realistic, it has the succinct advantage that this double-angle ODF has only two parameters. Another possibility is to consider a Gaussian distribution of tilt and azimuthal angles

$$f(\theta, \phi) = c_N \exp\left[-\frac{(\theta - \theta_0)^2}{2\sigma_\theta^2} - \frac{(\phi - \phi_0)^2}{2\sigma_\phi^2}\right] \tag{11.2}$$

where θ_0 is the mean tilt angle, ϕ_0 is the mean azimuthal angle, σ_θ and σ_ϕ are the corresponding half widths of the distributions, and c_N is a normalization constant defined as

$$\frac{1}{c_N} = \int_0^{2\pi} \int_0^{\pi} f(\theta, \phi) \sin\theta \, d\theta \, d\phi. \tag{11.3}$$

However, there is a physical problem with Eq. 11.2 that is not a major issue provided that σ_ϕ is relatively narrow, and/or that ϕ_0 is sufficiently far from the bounds that we have defined as 0–2π. Otherwise, care must be taken to ensure that $f(\theta, \phi) = f(\theta, \phi \pm 2\pi)$. One way of achieving this continuity is to use a wrapped distribution function such as

$$f(\theta, \phi) = c_N \sum_{k=-\infty}^{\infty} \exp\left[-\frac{(\theta - \theta_0)^2}{2\sigma_\theta^2} - \frac{(\phi - \phi_0 + 2\pi k)^2}{2\sigma_\phi^2}\right]. \tag{11.4}$$

We can keep going, and create a large family of distribution functions, just like we did in the case for θ-only problems, creating functions that are applicable to the material we are studying, perhaps incorporating some previous knowledge of how the molecules are oriented. If there is no insight into the shape of the distribution, then there is little reason to use a specific form of an ODF other than a Gaussian-like function (considering a wrapped Gaussian in the case of ϕ). However, another possibility is to use a generalized ODF based on order parameters. The formalism we have illustrated for the tilt-only case can be extended to include tilt and twist distributions using [5, 6]

$$f(\theta, \phi) = \frac{1}{4\pi^2} \sum_{\ell=0}^{\infty} \sum_{m=-\ell}^{\ell} \left[\frac{2\ell + 1}{2} \left(\frac{(\ell - m)!}{(\ell + m)!}\right) \langle Y_{\ell m} \rangle Y_\ell^m(\theta, \phi)\right]. \tag{11.5}$$

Instead of using the Legendre polynomials to expand $f(\theta)$, we use spherical harmonics $Y_\ell^m(\theta, \phi)$ as a complete set of orthonormal functions that can be used to express $f(\theta, \phi)$. These functions can be obtained from

$$Y_\ell^m(\theta, \phi) = N' P_\ell^m(\cos\theta) e^{im\phi} \tag{11.6}$$

where N' is a normalization constant and $P_\ell^m(\cos\theta)$ are known as associated Legendre polynomials. These functions of θ are related to the Legendre polynomials $P_\ell(\theta)$ (that we have previously referred to as $P_n(\cos\theta)$) by

$$P_\ell^m(\cos\theta) = (-1)^m (1 - \cos^2\theta)^{m/2} \frac{\mathrm{d}^m}{\mathrm{d}\cos^m\theta} P_\ell(\cos\theta). \tag{11.7}$$

Note that the associated Legendre polynomials are available as numerical functions called `special.lpmn` in NumPy and also as `sympy.functions.special.polynomials.assoc_legendre`. Forms of the spherical harmonics are also available as `special.sph_harm` in SciPy and analytical versions as `functions.special.spherical_harmonics.Ynm` in SymPy. In Eq. 11.5, the associated Legendre polynomials $P_\ell^m(\cos\theta)$ form the polar part of the distribution, and the azimuthal component is described by the complex exponential $e^{im\phi}$. Although Eq. 11.5 is valid, as written it includes the possibility of complex-valued order parameters $\langle Y_{\ell m} \rangle$. Although there is nothing technically wrong with that, it does mean that additional experimental data is required to make their extraction tractable. This can alternatively be formulated in terms of real spherical harmonics (thereby producing real order parameters) by forming linear combinations of the spherical harmonics [7].

There is something interesting and unique about systems that lack full rotational symmetry about z. At first glance, even the situation with uniform twist ($m = 0$) seems hopeless from an experimental perspective, since we already mentioned that we are limited in the number of independent $\chi^{(n)}$ observables. However, when samples are oriented in such a way that the distribution of molecules is not uniform in their ϕ distribution, it makes sense to collect spectra, in all possible polarization combinations, while rotating the sample about z. We now have three coordinate systems:

1. The molecule-fixed (microscopic) coordinate system, for which we have been using ℓ, m, and n as placeholders for the a, b, c Cartesian coordinates. This is the reference frame in which the $\boldsymbol{\alpha}^{(n)}$ elements are defined, and any symmetries in the $\boldsymbol{\alpha}^{(n)}$ tensor come from the nature of the vibrational mode.
2. The surface-fixed coordinate system that we refer to as (x', y', z'). This is a macroscopic coordinate system that moves when we rotate the sample. Picture a polymer whose surface is rubbed to create grooves along $\hat{x}'$. We can define the direction of the groves with respect to the lab frame, i.e. the angle between $\hat{x}$ and $\hat{x}'$, as γ. The central question is then how the molecules are oriented with respect to these grooves. Any $f(\theta, \phi, \psi)$ we construct therefore has the purpose of describing this orientation.

3. The usual laboratory-frame coordinate system, as we have been discussing over the course of this book. This is where the i, j, k are used as placeholders for any of the Cartesian x-, y-, and z-directions. This is also the coordinate system in which the s- and p-polarizations are defined.

If samples have no preference for their orientation about z, then the surface-fixed and laboratory frames are the same. Many authors use (x, y, z) for the surface frame, and (X, Y, Z) for the lab frame. Although this is certainly a logical choice of symbols, we have used lowercase letters throughout, as all of the discussion up until this point considered $C_\infty(z)$ symmetry in the laboratory frame.

The other common scenario for a two-angle distribution is one where the molecules are oriented in their tilt angles θ, and have a preference for their twist ψ about the molecular c-axis as shown in Fig. 11.1e. This would be the most general situation for a surfactant adsorbed at the air–water interface. Since such molecules typically have a charged head group in the aqueous phase, and a long aliphatic tail that points towards the air, the θ distribution can be rather narrow. Some functional groups (such as carbonyl) may fall into the cylindrical symmetry category (Fig. 11.1a) but the general situation is a (θ, ψ)-distribution, since there is no orientation in the plane of the surface, so the distribution is uniform with respect to ϕ. Mathematically, the functions that can be used to describe this case are the same as for (θ, ϕ) illustrated above, where we literally substitute ψ for ϕ and one typically uses the index n rather than m. As we require any ψ distribution to be continuous at the 2π boundary, the same wrapping considerations also apply for functions such as Gaussians. Finally, although spherical harmonics are typically thought of as pertaining to θ and ϕ, we can define an analogous set of functions $Y_\ell^n(\theta, \psi)$ that are identical in appearance to $Y_\ell^m(\theta, \phi)$.

One remaining situation that is not depicted in Fig. 11.1 is one where molecules are oriented about ϕ and ψ, but in a manner such that the orientation distribution is isotropic about θ. Although such a distribution is not encountered at surfaces (where we define θ as the tilt angle with respect to the surface normal), it could occur in materials that are ordered in their bulk.

11.3.2 Orientation Distributions that Display Anisotropy About All Three Euler Angles

The final situation we describe is one where the orientation distribution is not isotropic about any of the angles, so we need a way to specify how molecules or parts of molecules are ordered about θ, ϕ, and ψ. We can readily adapt the above approaches to cover this scenario, where a narrow distribution centred at $(\theta_0, \phi_0, \psi_0)$ becomes

$$f(\theta, \phi, \psi) = \delta(\theta - \theta_0) \cdot \delta(\phi - \phi_0) \cdot \delta(\psi - \psi_0) \tag{11.8}$$

and broader distributions that can be described by a single peak for each of the angles can use a Gaussian for θ and two wrapped Gaussians, one for ϕ and other for ψ in

$$f(\theta, \phi, \psi) = c_N \sum_{k=-\infty}^{\infty} \exp\left[-\frac{(\theta-\theta_0)^2}{2\sigma_\theta^2} - \frac{(\phi-\phi_0+2\pi k)^2}{2\sigma_\phi^2} - \frac{(\psi-\psi_0+2\pi k)^2}{2\sigma_\psi^2}\right]. \tag{11.9}$$

Likewise, a generalized form that makes use of orthogonal basis functions and their associated order parameters is given by [8–13]

$$\begin{aligned} f(\theta, \phi, \psi) = \frac{1}{8\pi^2} \sum_{\ell=0}^{\infty} \sum_{m=-\ell}^{\ell} \sum_{n=-\ell}^{\ell} \Bigg[& \frac{2\ell+1}{2} \left(\frac{(\ell-m)!}{(\ell+m)!}\right)\left(\frac{(\ell-n)!}{(\ell+n)!}\right) \\ & \times \langle P_{\ell mn}\rangle P_\ell^m(\cos\theta) e^{-im\phi} e^{-in\psi}\Bigg]. \end{aligned} \tag{11.10}$$

One notices that Eq. 11.10 is not expressed in terms of spherical harmonics (we have one too many angles to use them exclusively), but instead uses the associated Legendre polynomials directly, multiplied by a complex exponential for each of the ϕ and ψ angles. We also now return to using the letter "P" for the order parameter, but with three indices in $\langle P_{\ell mn}\rangle$. In studies of complex orientation distributions (anything more than a single angle), it is customary to always keep the three indices. Therefore a distribution that is uniform in ϕ only (such as that depicted in Fig. 11.1e) could have order parameters written as $\langle P_{\ell 0n}\rangle$ rather than $\langle Y_{\ell n}\rangle$. Substituting a value of 0 for m in Eq. 11.10 automatically reduces the expressions to the conventional spherical harmonics (albeit in terms of θ and ψ rather than θ and ϕ). Similarly, a distribution uniform in ψ only (Fig. 11.1d, and technically also Fig. 11.1c that has no twist angle) could have order parameters written as $\langle P_{\ell m0}\rangle$. While less common, tilt-only distributions (uniform in ϕ and ψ as in Fig. 11.1a and b) sometimes use order parameters written as $\langle P_{\ell 00}\rangle$ rather than $\langle P_\ell\rangle$.[1]

11.3.3 Experimental Considerations

Now that we have finished introducing the idea of orientation distributions that are more complex than the θ-only cases we worked on in our discussion of IR, Raman, and SFG spectroscopy, we need to consider why any of these distributions are not more commonly employed, especially in the case of θ and ψ if we have a sample that has an assured isotropy about ϕ.

If any of this applies to samples that you are interested in, the first thing I recommend is to work out which elements of $\chi^{(n)}$ are non-zero and, among those,

[1] In the previous chapters, we have written $\langle P_n\rangle$ as θ-only folks tend to use the letter n instead of ℓ when only a single index is involved.

which ones are unique for the particular type of spectroscopy that you are planning to perform, or have data from. Even though we have not explicitly developed those equations, the formalism introduced in Chaps. 8–10 can be easily extended to the case of distributions involving two or three Euler angles using any one of the above expressions. In doing so, you will also need to incorporate the nature of the vibrational mode in the form of its $\boldsymbol{\alpha}^{(n)}$ elements. Some approximations may be made here to simplify the problem. For example, if some $\boldsymbol{\alpha}^{(n)}$ elements are very small compared to others, those can be taken as zero. That may substantially simplify the resulting $\boldsymbol{\chi}^{(n)}$ expressions. In the end, it comes down to how many unique spectra (utilizing different polarization schemes) can be measured in comparison to how many parameters there are in your multi-angle orientation distribution function. If there are too many unknown $\boldsymbol{\alpha}^{(2)}$ this is usually not the biggest problem, as all of those elements can be estimated from electronic structure calculations or from bond-additivity approaches that take the symmetry of the vibrational mode into account.

The bigger issue is that the number of ODF parameters to be determined is often greater than the number of unique $\boldsymbol{\chi}^{(n)}$ elements than can be accessed. This is frequently the case in IR spectroscopy where a single dichroic ratio can be used to determine a value for $\langle P_2 \rangle$ for a distribution that is appropriately described by $\langle P_{200} \rangle$ but does not provide any insight into other $\langle P_{2mn} \rangle$ if ϕ and/or ψ are not uniformly distributed. A similar issue is encountered in SFG spectroscopy. In Chap. 10 we have described that for θ-only distributions, it is only the ratio $\langle P_{300} \rangle / \langle P_{100} \rangle$ that can be determined if ratios of band intensities are used in the analysis. However, a somewhat counter-intuitive result is that, when the orientation distribution gets more complex in the form of non-uniform ϕ, some additional possibilities for analysis open up.

Among the techniques discussed here, the most interesting is spontaneous Raman scattering, as a result of two reasons. The first is that the Raman response is proportional to $\langle P_{2mn} \rangle$ and $\langle P_{4mn} \rangle$ and 4 is a rather high value of ℓ. This means that Raman is extremely capable in determining the shape of the orientation distribution. In Chap. 9, we have seen that the most probable distribution constructed from experimentally determined values of $\langle P_{200} \rangle$ and $\langle P_{400} \rangle$ can simultaneously account for the width of the tilt angle distribution and the mean tilt angle. From IR data alone, it would be impossible to comment on the two properties of the θ distribution. Now consider the power of Raman when applied to distributions with molecular entities ordered about two or three of the Euler angles and the advantages of a $\boldsymbol{\chi}^{(3)}$-based method become apparent. The second point is that Raman spectroscopy may be applied to systems such as stretched polymer films. Even though SFG ($\boldsymbol{\chi}^{(2)}$-based) probes a higher order response than IR ($\boldsymbol{\chi}^{(1)}$), no SFG spectra can be generated from (bulk) orientations whose in-plane anisotropy is symmetric about $\pm x$ in the lab frame, so such an OD is centrosymmetric. In these cases (1) Raman is able to probe these samples; (2) high-order parameters with $\ell = 4$ can be extracted; (3) such symmetric distributions have all $\ell_{\text{odd}} = 0$. The last point interesting, as it means that for samples such as stretched polymers $\langle P_{1mn} \rangle = 0$, $\langle P_{3mn} \rangle = 0$, and $\langle P_{5mn} \rangle = 0$. So we have a means of knowing *all* order parameters up to $\ell = 5$, with no omissions. This provides an option that wasn't on the table before. We can construct a truncated ODF using order parameters that ends at $\ell = 4$ and contains only even values of ℓ. From what we know about ODFs, this doesn't guarantee that we have a well-behaved non-negative ODF, but it does often provide a reasonable approximation of the true OD.

11.4 Simultaneous Analysis of Multiple Vibrational Modes

11.4.1 General Considerations

The tone of the preceding section makes it clear that one is often data-starved in the sense that the number of parameters that can be obtained in an experiment exceed the number of parameters in a desired or reasonable orientation distribution function. I mentioned that one approach to circumvent this is to simplify the molecular response $\boldsymbol{\alpha}^{(n)}$, and another is to make some assumptions about the shape of the distribution function that are hopefully warranted. However, we must keep in mind that the treatment presented until now has focused exclusively on the use of information pertaining to a single vibrational mode. In other words, we have considered amplitudes of *one* mode obtained in many different polarization schemes, and we have been talking about the $\boldsymbol{\alpha}^{(n)}$ elements of this single mode.

For a large and flexible molecule, if information from another vibrational mode is obtained that is localized to a chemical functional group at some distance from the first group, and there are likely several torsional angles in between so that the information from the two parts of the molecule are sufficiently decoupled. Whatever orientation information can be gleaned from each mode then contributes to an understanding of the *conformation* of the molecule, as we have determined the relative orientation between two different parts [14].

However, when the two vibrations correspond to bonds that are in close proximity, their orientation information is linked. Imagine the scenario of two unit vectors $\hat{c}_1$ and $\hat{c}_2$ corresponding to the C_3 symmetry axes of two different methyl groups on the same molecule. Let's say that we know that the ϕ and ψ angles are uniformly distributed for each group, so there is a (assumed to be narrow) cone of angles represented by θ_1 and another cone described by θ_2. If these two methyl groups are in close proximity, perhaps separated by a single bond, we might know the angle between them, β. This information is extremely helpful as it restricts the range of possibilities for θ_2 given θ_1 and vice versa. We bring this point up here because it offers a route to get additional information about the orientation distribution of each group (and therefore a larger part of the whole molecule) as a result of this β constraint. One can imagine that, in the case of IR spectral data, this might enable the width of the (combined) orientation distribution to be determined in addition to its mean value, unlike the case where only two individual $\langle P_{200} \rangle$ values are obtained. In the case of Raman or SFG data, this might enable additional order parameters to be determined. The method by which such a scheme may be implemented would depend on the shape of the molecule and the nature of the vibrational modes.

11.4.2 Nested Orientation Distributions

Another common scenario occurs when the axis of interest in the molecule is at angle to that probed in the optical response. This can occur in any spectroscopic technique, but we consider the transition dipole moment as a vector indicated by the blue arrow in Fig. 11.2. An experiment would therefore probe the average orientation (or associated order parameters) of this arrow with respect to the z-axis. However, let's assume that the transition dipole moment makes an angle β with respect to the axis of interest, c. This could be the long chain of an alkyl group, for example. Although any type of averaging is possible about each of the axes in Fig. 11.2, a common situation is one where the azimuthal angle of the chain ϕ is uniformly distributed about z (as in the case of all the "θ-only problems we have described") and that the optical response vector (the transition dipole moment in blue) is uniformly distributed about c, i.e. the angle indicated as α in Fig. 11.2 has no preferential value.

We can use the spherical law of cosines to write

$$\cos\gamma = \cos\theta\cos\beta + \sin\theta\sin\beta\cos\alpha \tag{11.11}$$

from which it is clear that the projection of the optical response onto the laboratory z-axis depends on the value of α. However, since our distribution of interest is uniform in α, we can simplify this expression by integrating about α

$$\begin{aligned}\cos\gamma\int_0^{2\pi} \mathrm{d}\alpha &= \cos\theta\cos\beta\int_0^{2\pi} \mathrm{d}\alpha + \sin\theta\sin\beta\int_0^{2\pi} \cos\alpha\,\mathrm{d}\alpha\\ \cos\gamma &= \cos\theta\cos\beta.\end{aligned} \tag{11.12}$$

We can then make use of the Legendre addition theorem [15–17] under the special case of azimuthal symmetry about z that provides

$$\langle P_\ell(\cos\gamma)\rangle = \langle P_\ell(\cos\theta)\rangle\langle P_n(\cos\beta)\rangle. \tag{11.13}$$

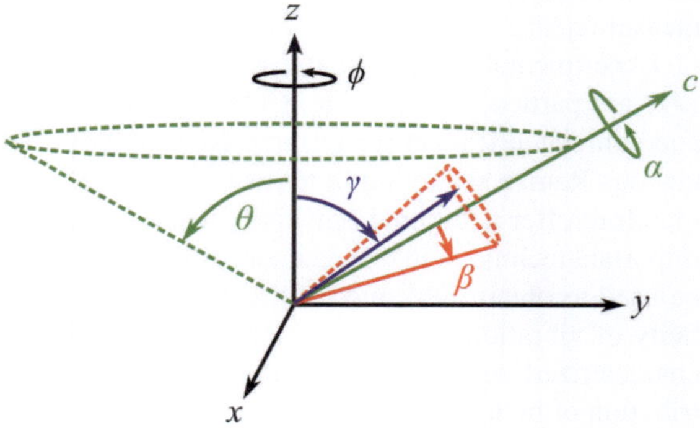

Fig. 11.2 A situation where the molecular response vector (blue) is tilted at an angle β with respect to a reference direction c, but is uniformly distributed about α. The molecular c-axis in turn is uniformly distributed about z. An experimental measurement provides order parameters related to the angle γ

Since the distribution about β is typically very narrow, as it is determined based on molecular bond angles, $\langle P_n(\cos\beta)\rangle \approx P_n(\cos\beta_0)$. The order parameter of interest related to the chain axis c is then obtained from

$$\langle P_n(\cos\theta)\rangle = \frac{\langle P_n(\cos\gamma)\rangle}{P_n(\cos\beta_0)}. \tag{11.14}$$

This has been particularly well explored in the case of $\langle P_2\rangle$ determination from infrared dichroic ratios [18–20].

This is very closely linked to the case where multiple vibrational modes correspond to the same set of molecular coordinates for a single chemical moiety. A classic example is the aromatic C–H stretching modes of a phenyl group. All the vibrations are in a single plane, such as the ac-plane as illustrated in Fig. 11.1e and f. Since the symmetry of the corresponding $\boldsymbol{\alpha}^{(n)}$ tensors differ for the different C–H modes, this information can be combined to provide additional parameters on the orientation distribution of the ac-plane of the phenyl group. The Legendre addition theorem is again applicable in such situations. Once again, any codes developed to tackle such situations would be molecule and mode specific. However, the general building blocks of those algorithms would largely follow the same steps as illustrated in the previous three chapters.

11.5 Combining Orientation Information from Different Experiments

A powerful way of addressing some of the limitations that arise as the complexity of the orientation distribution increases is to combine results from multiple experiments. One possibility is to use the same technique, but in multiple geometries. For example, IR absorption spectra of thin films on transparent substrates can be measured in transmission and reflection, where the combined data can be used to solve for the optical constants and comment on the molecular orientation [21–24]. A similar type of analysis can be performed when varying the angle of incidence [25, 26], a common technique in ellipsometry but underutilized other types of spectroscopic measurements.

Experimental methods that can provide high-order parameters, such as $\langle P_4\rangle$ and $\langle P_6\rangle$ are particularly valuable. Coherent anti-Stokes Raman spectroscopy (CARS) and stimulated Raman scattering (SRS) overcome some of the limitations of spontaneous Raman scattering in terms of which elements of $\chi^{(3)}$ can be accessed, and therefore offer additional sensitivity to molecular orientation [14, 27–31]. Another important technique in this category is fluorescence, as polarized measurements can be used to obtain $\langle P_2\rangle$ and $\langle P_4\rangle$ [32–38], albeit not with the chemical bond specificity of vibrational methods. Nevertheless, if some axis in a chromophore can be characterized wit fluorescence, this can provide additional information on the distribution of bond angles.

As we have seen, orientation distributions that are not symmetric about $\theta = 90°$ pose a special challenge, as $\langle P_2 \rangle$ and/or $\langle P_4 \rangle$ provide extremely valuable information about the orientation distribution, but even-order parameters are not able to capture the asymmetry in the tilt angle distribution. Probes such as vibrational SFG can provide this, even if $\langle P_1 \rangle$ and $\langle P_3 \rangle$ are not easily isolated from the SFG data. Figure 11.3 presents a hypothetical situation where we assume that a collection of molecules is ordered about θ only, and that all of the requisite information (number of available polarization schemes, choice of reference sample, additional calibration steps, full knowledge of the molecular $\boldsymbol{\alpha}^{(n)}$ response, etc.) is available in order to extract $\langle P_2 \rangle$ from the IR data, that and $\langle P_4 \rangle$ from Raman data, and $\langle P_1 \rangle$ and $\langle P_3 \rangle$ from SFG data. In order to evaluate the success of our efforts, we present the true orientation distribution generated with a model with the thick grey line, but then assume that this information is not available to the experimentalist, as is the case in the real situation where we wish to characterize the unknown orientation distribution. The same results are shown in a Cartesian and polar representation in the top and bottom panels of Fig. 11.3 as some aspects of the shape of the functions are easier to visualize in one of the representations. First we can see that if only IR data was available, the most probable distribution ($f_{MP}(\theta)$, shown in red) is centred at $\theta = 90°$. When only $\langle P_2 \rangle$ information is available, $f_{MP}(\theta)$ must be centred at either $\theta = 90°$ or $\theta = 0$ *and* $\theta = 180°$. Comparing this red curve to the unknown but true grey curve, this approximation is not terrible, as it more or less captures the average width of the distribution, even though the true distribution (which doesn't necessarily have a functional form) actually has a sharp minimum in the population of molecules with $\theta = 90°$. The results using data extracted from a Raman experiment are also shown, with $f_{MP}(\theta)$ plotted in dark blue. Here the match to the true distribution is impressive both in terms of the location of the two maxima and in the width of each half of the distribution. However any distribution constructed using information from only even-order parameters cannot possess asymmetry about $\theta = 90°$. We next consider SFG data, first without help from the other experiments. Even if $\langle P_1 \rangle$ and $\langle P_3 \rangle$ could be isolated from the polarized SFG spectra, the resulting distribution, plotted in green, is a rather poor approximation to the true distribution (grey), largely because it incorrectly predicts a sizeable broad population near $\theta = 180°$. However, as expected, this is the first distribution that is asymmetric. If we now combine these two odd-order parameters with $\langle P_2 \rangle$ obtained from the IR spectra, and use the three order parameters to construct $f_{MP}(\theta)$, the result (plotted in magenta) somewhat resembles the IR-only result (and is therefore a big improvement over the SFG-only result), and is correctly shifted to lower tilt angles. It is arguable whether this is better that what was obtained with Raman only but, to be fair, this comparison is with respect to the true distribution which is unknown. Furthermore, the relative success of each of these techniques on their own, and in combination, is not generalizable to any shape of orientation distribution. Nevertheless, we take a look at the final combination where the most information available comes from constructing $f_{MP}(\theta)$ using all four order parameters. This amounts to combining the SFG and Raman data, with the result shown in cyan being a rather faithful representation of the true

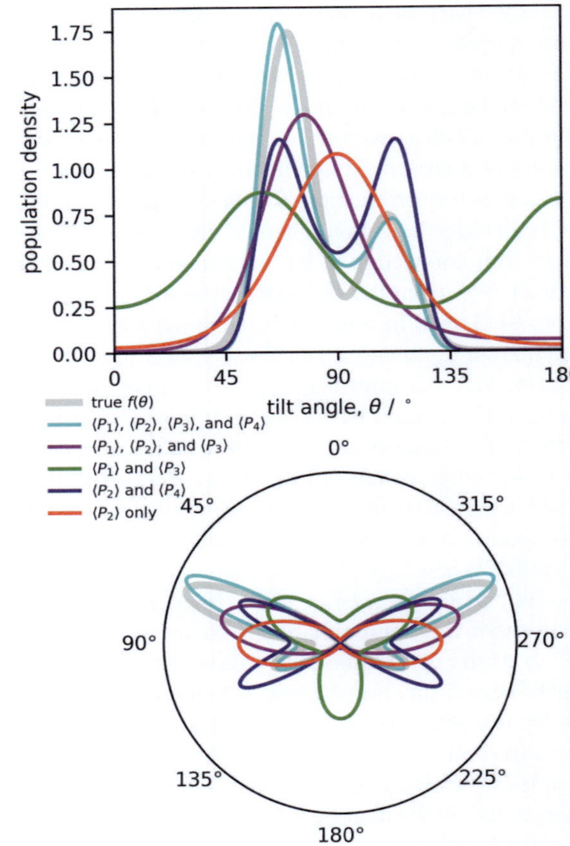

Fig. 11.3 A hypothetical orientation distribution plotted in grey using both a Cartesian (top panel) and polar representation (bottom panel), followed by attempts to capture this distribution using only $\langle P_2 \rangle$ from IR data (red), $\langle P_2 \rangle$ and $\langle P_4 \rangle$ from Raman data (dark blue), only $\langle P_1 \rangle$ and $\langle P_3 \rangle$ from SFG data (green), SFG combined with IR (magenta), and SFG combined with Raman (cyan)

distribution. Any further improvement would require additional input from higher order parameters ($\langle P_5 \rangle$, $\langle P_6 \rangle$, etc.):

Code to generate Fig. 11.3

```
import numpy as np
from scipy.optimize import brute
from matplotlib import pyplot

def Pn(n):
    P = [1.]
    P.append(np.cos(theta))
    for i in range(1, n):
        P.append(((2.*i + 1.)*np.cos(theta)*P[i] - i*P[i-1])/(i + 1.))
    return P[n]
```

```
def f_g():
    frac = 0.3
    num = (1.0 - frac)*np.exp(-(theta - theta0)**2/(2*sigma*sigma)) + frac*
         np.exp(-(theta - (np.pi - theta0))**2/(2*sigma*sigma))
    norm = sum(num*np.sin(theta)*np.pi/200.)
    return num/norm

def f_mp(l):
    if np.size(l) == 1:
        l = [l]
    first = np.zeros(len(theta))
    for i in range(len(l)):
        first += l[i]*Pn(ops[i])
    num = np.exp(first)
    norm = sum(num*np.sin(theta)*np.pi/200.)
    return num/norm

def op_g(n):
    return sum(Pn(n)*f_g()*np.sin(theta)*np.pi/200.)

def op_mp(n, l):
    return sum(Pn(n)*f_mp(l)*np.sin(theta)*np.pi/200.)

def fit(l):
    e = np.zeros(len(theta))
    for i in ops:
        e += abs(op_mp(i, l) - op_g(i))**2
    return sum(e)

def mirror(v):
    return np.hstack((v, v[::-1]))

# define the Gaussian target function
theta0 = np.radians(70.)
sigma = np.radians(10.)

# setup
pyplot.rc('font', size=8)
fig = pyplot.figure(figsize=(3.25, 5))
ax1 = fig.add_subplot(211)
ax2 = fig.add_subplot(212, projection='polar')
thetadeg = np.linspace(0., 180., 200)
theta = np.radians(thetadeg)
thetaPolar = np.linspace(0., 2*np.pi, 400)
colors = 'r b g m c'.split()
names = [r'$\langle P_2 \rangle$ only', r'$\langle P_2 \rangle$ and $\
     langle P_4 \rangle$', r'$\langle P_1 \rangle$ and $\langle P_3 \
     rangle$', r'$\langle P_1 \rangle$, $\langle P_2 \rangle$, and $\langle
      P_3 \rangle$', r'$\langle P_1 \rangle$, $\langle P_2 \rangle$, $\
     langle P_3 \rangle$, and $\langle P_4 \rangle$']
orders = [[2], [2,4], [1,3], [1,2,3], [1,2,3,4]]
ax2.set_theta_zero_location('N')

# look at the actual distribution
ax1.plot(thetadeg, f_g(), color='0.8', lw=3, label=r'true $f(\theta)$')
ax2.plot(thetaPolar, mirror(f_g()), color='0.8', lw=3)
```

```
# iterate over possibilities
for experiment in range(len(names)-1, -1, -1):
    ops = orders[experiment]

    # do the fitting
    lbounds = []
    for i in range(len(ops)):
        lbounds.append((-50.0, 50.0))
    opt = brute(fit, lbounds, Ns=5)
    ax1.plot(thetadeg, f_mp(opt), color=colors[experiment],
           label=names[experiment])
    ax2.plot(thetaPolar, mirror(f_mp(opt)), color=colors[experiment])

# pretty
ax1.set_xlabel(r'tilt angle, $\theta$ / $^\circ$')
ax1.set_ylabel('population density')
ax1.set_xlim(0, 180)
ax1.set_ylim(bottom=0)
ax1.set_xticks(range(0,181,45))
ax2.yaxis.grid(False)
ax2.set_rticks([])
ax2.grid(False)
ax2.tick_params(axis='both', which='major', pad=-1)
fig.legend(loc=(0.05, 0.39), fontsize=6, frameon=False)
fig.set_tight_layout(True)
pyplot.savefig('P1P2P3P4.png', dpi=600)
```

Such schemes have been experimentally demonstrated, combining $\langle P_2 \rangle$ from polarized IR and/or UV–visible data with $\langle P_4 \rangle$ from Raman experiments [39–41], polarized fluorescence [42], and X-ray data [9, 10, 43–45], including the use of NMR data [46, 47]. Asymmetric distributions have been characterized incorporating $\langle P_1 \rangle$ and $\langle P_3 \rangle$ from second harmonic data with results from polarized UV–vis [41] and Raman data [41, 48].

11.6 Using Data from Molecular Dynamics Simulations

The tools we have developed in the previous chapters can be used in two ways. First, they can be used to analyse experimental data in order to arrive at parameters that describe the molecular orientation in a sample. In the case of IR and Raman spectroscopy, these might be generalized order parameters that describe the tilt angle distribution [49]. In the case of SFG spectroscopy, we can make assumptions about the shape of the orientation distribution and then use the experimental data to determine the parameters in those functions. However it is also possible to model the spectral response by considering different types of distributions and studying what types of experimental results (ratios of amplitudes in different polarizations) would be obtained. In a sense, these are just two sides of the same coin, and any data fitting tools will make use of the same functions that model the response.

If the goal is to describe the molecular orientation distributions, then the results of molecular dynamics simulations are a very interesting addition to this modelling effort. The basic idea is to model molecular motions as a function of time. The simulations can be atomistic in the sense that the trajectory of each atom of a molecule is obtained, and therefore conformational as well as orientational changes can be studied. It is also possible to perform atomistic simulations but fix the bond lengths and/or bond angles, in which case the conformation of the molecules are frozen, but their orientation evolves with time. Another possibility is to perform more coarse-grained simulations where some chemical moieties are described as a single entity. This could be a relatively small unit such as methyl group, where the approach is often referred to as a united atom treatment. Or, at an even coarser level, proteins can be modelled with each amino acid represented as a single entity for which the dynamics are studied. The choice of approach or the degree of coarse-graining depends on the problem that is being studied. When molecular interactions at surfaces are of interest, all-atom simulations are frequently necessary as chemical interactions occur at the level of individual atoms. Finally, for atomistic simulations, the interaction energy that determines the next time point of the trajectory can be evaluated using classical mechanics (where the force fields are parameterized based on a combination of experimental data and quantum mechanical modelling) or using ab initio approaches in the so-called AIMD or ab initio molecular dynamics techniques.

Regardless of what approach is taken, what is of interest to us here is that these simulations return the Cartesian coordinates of all atoms (or groups of atoms when the detail is not atomistic) during each time step that is recorded. In many cases, the actual time evolution of the system is not of direct interest, but we are after the equilibrium structure of the system that can be understood from the average over many time snapshots. There is always a risk that such data may not represent the real system as a result of the approximations and necessary parameterizations. Nevertheless, for those interested in gaining insight into molecular orientation in complex molecular systems, such information is golden. A huge advantage of having information about every molecule in the ensemble is that we don't need to approximate the orientation distribution with a function (ODF) such as $f(\theta)$. Instead, we can write some code to loop over all the molecules of interest in each time step and over each frame in the trajectory, and access each θ value. One can then plot a histogram of these tilt angles, thereby visualizing the actual orientation distribution (OD) as shown in Fig. 8.17. We have seen expressions such as

$$
\begin{aligned}
\chi_{ij}(\theta) = \frac{N_\mathrm{V}}{\varepsilon_0}\langle\alpha_{ij}(\theta)\rangle &= \frac{1}{\varepsilon_0 V}\sum_{\text{molecules}}\alpha_{ij}(\theta) \\
&= \frac{N_\mathrm{V}}{\varepsilon_0}\int_0^\pi f(\theta)\alpha_{ij}(\theta)\,\mathrm{d}\theta
\end{aligned}
\tag{11.15}
$$

and analogous versions in our discussion of $\chi^{(2)}$ and $\chi^{(3)}$ tensor elements. Each time we have said that the sum over molecules is what we dream about, and the parameterization of the weighting function, the ODF $f(\theta)$, is what we settle for.

When data from molecular dynamics simulations is available, the averages of all needed quantities (in order to simulate the spectral response) can be determined from a sum over molecules, and the orientation distribution can be assessed.

11.6.1 Obtaining the Direction Cosine Matrix Elements Without Explicit Euler Angles

In Chaps. 8, 9, and 10 all coordinate transformations were performed using the Euler angles θ, ϕ, and ψ. This was a logical choice because we were often parameterizing orientation distribution functions, and the Euler angles are easily visualized quantities. The procedure involved constructing the direction cosine matrix **D** from the Euler angles as was illustrated in Chap. 8. However, since this section deals with handling data from molecular dynamics simulations, we will see how the DCM can be obtained without having to first obtain the Euler angles. Even though the values of the Euler angles are of interest in their own right, there is a significant benefit to not constructing the DCM from θ, ϕ, and ψ that stems from issues with the definitions of the Euler angles themselves, namely, the degeneracy of ϕ and ψ when $\theta = 0°$ or $180°$.

Recall that we use i, j, and k to represent any of the lab frame Cartesian coordinates x, y, or z. Similarly ℓ, m, and n serve as placeholders for any of the molecular-frame coordinates a, b, or c. The lab frame unit vectors are defined as

$$\hat{\mathbf{x}} = \begin{bmatrix} x_1 \\ x_2 \\ x_3 \end{bmatrix} = \begin{bmatrix} 1 \\ 0 \\ 0 \end{bmatrix} \tag{11.16a}$$

$$\hat{\mathbf{y}} = \begin{bmatrix} y_1 \\ y_2 \\ y_3 \end{bmatrix} = \begin{bmatrix} 0 \\ 1 \\ 0 \end{bmatrix} \tag{11.16b}$$

$$\hat{\mathbf{z}} = \begin{bmatrix} z_1 \\ z_2 \\ z_3 \end{bmatrix} = \begin{bmatrix} 0 \\ 0 \\ 1 \end{bmatrix}. \tag{11.16c}$$

If the molecule of interest is anything close to linear, or is at least much longer than it is wide, the definition of the tilt angle is the most important, as it will be confusing to choose something that is too different from the molecules' long axis when referring to a tilt. Such an orientation would then be more appropriately described by one of the other two angles. We define this long axis as the molecular c-axis. The tilt angle, θ, is the angle that brings the unit vector **z** onto **c**.

If the molecule or chemical group has cylindrical symmetry, with all properties along **a** the same as those along **b** (such as the carbonyl group pictured in Fig. 11.1a), then it doesn't matter how one defines the **a** and **b** molecular axes, so long as **a**, **b**, and **c** are all perpendicular to each other and form a right-handed coordinate system with

$$\hat{\mathbf{a}} \times \hat{\mathbf{b}} = \hat{\mathbf{c}}. \tag{11.17}$$

If the molecule or moiety has a flat part that is of interest (such as benzene pictured in Fig. 11.1e), the vector **a** is typically chosen so that the plane of interest is the *ac*-plane. In addition to defining logical unit vectors, one must ensure that **c** and **a** are perpendicular.

It's all fine to verbally provide such definitions of **a** and **c**, but let's take a look at how one goes about identifying these vectors in a computer code that is looping through molecular coordinate data from an MD trajectory. Obviously the easiest case is when any two vectors are can be obtained directly from the coordinates of three or four unique atoms. For example

```
atom1x, atom1y, atom1z = coords(atom1)
atom2x, atom2y, atom2z = coords(atom2)
atom3x, atom3y, atom3z = coords(atom3)
atom4x, atom4y, atom4z = coords(atom4)

ax = atom1x - atom2x
ay = atom1y - atom2y
az = atom1z - atom2z

bx = atom3x - atom4x
by = atom3y - atom4y
bz = atom3z - atom4z
```

where a user-defined `coords` function does the parsing. In case the trajectory files are in plain text format, any of the tools introduced in Chap. 2 such as `readlines` or Chap. 3 such as `np.loadtxt` would work. But, MD trajectory files are typically huge, so are more conveniently stored in a compressed binary format. In such cases, the programs that create the trajectories (in case they are not homemade) often are bundled with functions that parse them to retrieve the coordinates.

This approach to defining two of the molecular-frame vectors assumes that the rigid molecule contains two sets of atoms (with potentially one atom in common) that can be connected to form two perpendicular vectors, i.e. the angle between **a** and **c** should be exceeding close to 90°. Note that this convenient situation occurs in the case of benzene, as two carbon atoms or two hydrogen atoms can be used to define **c**, and a perpendicular set of two C or H atoms can be used to define **a**—so long as the molecule remains rigid over the course of the simulation, i.e. we were not modelling explicit vibrations that might distort the geometry. The only thing that remains is to take these molecular vectors and turn them into unit vectors by normalizing them

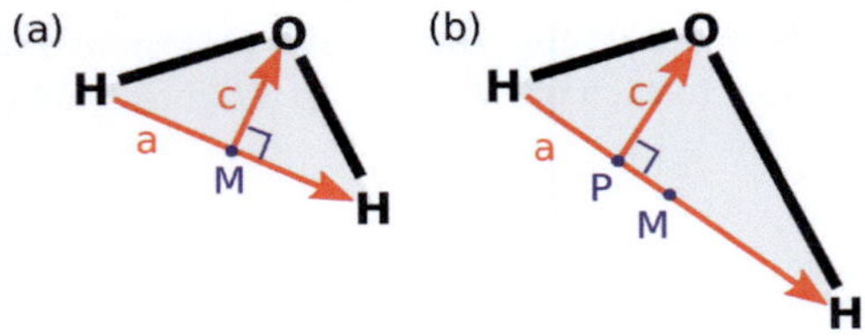

Fig. 11.4 Illustration of vectors that may be used to define the molecular-frame unit vectors for a **a** rigid water molecule with equal bond lengths and **b** a flexible water molecule with unequal bond lengths

$$\hat{\mathbf{a}} = \frac{\mathbf{a}}{\|\mathbf{a}\|} \tag{11.18a}$$

$$\hat{\mathbf{c}} = \frac{\mathbf{c}}{\|\mathbf{c}\|} \tag{11.18b}$$

and then using the cross product to arrive at the last vector

$$\hat{\mathbf{b}} = \hat{\mathbf{c}} \times \hat{\mathbf{a}}. \tag{11.18c}$$

Note that since $\hat{\mathbf{a}}$ and $\hat{\mathbf{c}}$ already have unit length, there is no need to normalize $\mathbf{b}$ after this operation. Also note that the order of the cross product is important, as we maintain a right-handed coordinate system. Alternatively, $\hat{\mathbf{a}} \times \hat{\mathbf{c}} = -\hat{\mathbf{b}}$, with the direction of $\hat{\mathbf{b}}$ then needing reversal.

However, more realistically, the molecule of interest will not have such perpendicular vectors that are immediately defined by existing atomic coordinates. This is especially the case when we need to define the $\hat{\mathbf{c}}$- and $\hat{\mathbf{a}}$-directions corresponding to the chemical functional group of interest. Or, the equilibrium geometry does present such convenient locations with reported Cartesian coordinates, but a flexible molecule is vibrating during the simulation, so the bond lengths and bond angles are not fixed. A simple example is discussed for the two water molecules shown in Fig. 11.4. Here we consider that $\mathbf{a}$ is the vector that runs between the two H atoms in any direction.

```
H1x, H1y, H1z = coords(H1)
H2x, H2y, H2z = coords(H2)
Ox, Oy, Oz = coords(O)

ax = H2x - H1x
ay = H2y - H1y
az = H2z - H1z
```

If both O–H bond lengths are known to be fixed and equal (for example, when using a rigid water molecule with this geometry specifically), one can take a convenient shortcut and simply define **c** as the vector that runs from the midpoint M of the two H atoms to the O atom

```
Mx = 0.5*(H1x + H2x)
My = 0.5*(H1y + H2y)
Mz = 0.5*(H1z + H2z)
```

and then use this point along with the oxygen coordinates to define **c**

```
cx = Ox - Mx
cy = Oy - My
cz = Oz - Mz
```

All that remains is to normalize the vectors and find $\hat{\mathbf{b}}$ from the cross product as before. This was a simple example of using an additional point, in this case the midpoint, that is derived from atomic coordinates but is not explicitly one of the recorded coordinates.

The case where the bond lengths may be unequal represents the most general scenario, and is straightforward in a simple example of H_2O shown in Fig. 11.4b. There are many equivalent approaches here, including one where **c** might be chosen as the H–O–H bisector. In this case, however, we keep our definition of **a** as running between the two H atoms. We are now looking for a **c** vector that originates from a point P along **a**, such that the vector defined between P and the O atom is perpendicular to **a**. The procedure for obtaining the point P can be implemented in a variety of ways, but once those coordinates are obtained, **c** come from

```
cx = Ox - Px
cy = Oy - Py
cz = Oz - Pz
```

and the rest follows from normalization and the cross product. In the most complex case, neither of the initial two vectors is readily obtained from any of the atomic coordinates, so "construction lines" need to be created to place the necessary points to enable the subsequent operations.

Whatever needed to happen to arrive at this stage, we are now in a position where we have found

$$\hat{\mathbf{a}} = \begin{bmatrix} a_1 \\ a_2 \\ a_3 \end{bmatrix} \tag{11.19a}$$

$$\hat{\mathbf{b}} = \begin{bmatrix} b_1 \\ b_2 \\ b_3 \end{bmatrix} \tag{11.19b}$$

$$\hat{\mathbf{c}} = \begin{bmatrix} c_1 \\ c_2 \\ c_3 \end{bmatrix} \tag{11.19c}$$

and, together with Eq. 11.16, have a complete set of 6 unit vectors, with all coordinates specified in the lab frame.

As we described in Chap. 8, the direction cosine matrix (DCM), **D**, is an operator that can be used to transform Eq. 11.16 into Eq. 11.19, or any property from the molecular frame to the lab frame. Recall that this 3×3 matrix gets its name as each of its nine elements is the cosine between a vector in the lab frame and a vector in the molecular frame (Eq. 8.33)

$$\mathbf{D} = \begin{bmatrix} \cos\gamma_{xa} & \cos\gamma_{xb} & \cos\gamma_{xc} \\ \cos\gamma_{ya} & \cos\gamma_{yb} & \cos\gamma_{yc} \\ \cos\gamma_{za} & \cos\gamma_{zb} & \cos\gamma_{zc} \end{bmatrix}$$

where $\gamma_{i\ell}$ is the angle between the $\hat{\mathbf{i}}$ vector in the lab frame and the $\hat{\boldsymbol{\ell}}$ vector in the molecular frame. In this above equation provides a fast route for determining all elements of the DCM, now that we have all six unit vectors that define both frames. Taking the top left element $D_{1,1}$ as an example

$$\begin{aligned} \cos\gamma_{xa} &= \frac{\hat{\mathbf{x}} \cdot \hat{\mathbf{a}}}{\|\hat{\mathbf{x}}\|\,\|\hat{\mathbf{a}}\|} \\ &= \frac{x_1 a_1 + x_2 a_2 + x_3 a_3}{1 \cdot 1} \\ &= a_1 \end{aligned} \tag{11.20}$$

and, as a result, the entire DCM may be written in a very compact form:

$$\mathbf{D} = \begin{bmatrix} a_1 & b_1 & c_1 \\ a_2 & b_2 & c_2 \\ a_3 & b_3 & c_3 \end{bmatrix}. \tag{11.21}$$

This may be formed in Python with

```
D = array([a, b, c]).T
```

assuming `a`, `b`, and `c` are already defined as arrays, and `.T` is the transpose method of an array object.

As everything from here onwards relies on this DCM, it is instructive to verify if we have the proper matrix before continuing. (When assembling the DCM via Eq. 11.21, it is especially easy to mistakenly form the transpose of this matrix.) By definition, the DCM provides the transformation from the molecular frame to the lab frame, and so we can verify that

$$\mathbf{D} \cdot \begin{bmatrix} 1 \\ 0 \\ 0 \end{bmatrix} = \begin{bmatrix} a_1 \\ a_2 \\ a_3 \end{bmatrix}$$

where the molecular unit vector (in the molecular coordinate system) is now written using lab coordinates. In order to do this, it is necessary to note $\hat{\mathbf{a}}$, $\hat{\mathbf{b}}$, and $\hat{\mathbf{c}}$ for one or more molecules of interest, and see if those same vectors may be reproduced with this check.

The inverse transformation, one that takes properties in the molecular frame and projects them into the lab frame, is accomplished with the inverse of the DCM. Since the DCM is an orthogonal matrix, its inverse is also its transpose

$$\mathbf{D}^{-1} = \begin{bmatrix} a_1 \; a_2 \; a_3 \\ b_1 \; b_2 \; b_3 \\ c_1 \; c_2 \; c_3 \end{bmatrix} = \mathbf{D}^T. \tag{11.22}$$

This could also be obtained using `inverseD = array([a, b, c])` but we do not recommend this, as it is likely to result in confusion at some point. When the inverse transformation is necessary, it is probably better practice to use `D.T`, and not define that array as a separate variable. Then the only transformation operator that is defined is the DCM itself. Also note that $\mathbf{D}^{-1}$ provides a slightly more convenient way of performing the verification, as it is not necessary to compare recorded values of $\hat{\mathbf{a}}$, $\hat{\mathbf{b}}$, and $\hat{\mathbf{c}}$. Knowing the molecular-frame unit vectors (in lab frame coordinates), one can see if they can be transformed into the more easily identifiable molecular frame unit vectors

$$\mathbf{D}^{-1} \cdot \begin{bmatrix} a_1 \\ a_2 \\ a_3 \end{bmatrix} = \begin{bmatrix} 1 \\ 0 \\ 0 \end{bmatrix}$$

etc. In summary, an analytical or symbolic form of the DCM incorporates the Euler angles in trigonometric functions. However, when a numerical version of the DCM is desired in terms of the nine numbers that make up the D_{ij} elements, and atomic coordinates are available, the Euler angles themselves are not necessary.

11.6.2 Obtaining Euler Angles from Molecular Coordinates

Although we have gone through lengths to avoid determining the Euler angles, we have also mentioned that when the orientation distribution is of interest, one of the primary reasons to perform MD simulations is to view the actual orientation distribution (OD), and not some guessed or approximate function (ODF). In order to do this, one first constructs the $\hat{a}$, $\hat{b}$, and $\hat{c}$ unit vectors as described in the previous section. That is the critical step, as changing the (arbitrary) definition of those vectors will change the definition of the Euler angles! Another way to state this is that if two researchers choose a different set of molecular-frame unit vectors, then they will present different-looking orientation distribution plots for the three Euler angles. But we should remember that this is okay, as they represent the same OD. The molecules aren't rearranged just because we choose a different coordinate system. All this to say, one should decide what angles one wishes to study based on the molecular structure, and then choose the $\hat{a}$, $\hat{b}$, and $\hat{c}$ unit vectors accordingly. Afterwards, we can compare our definitions of the DCM

$$\begin{aligned}\mathbf{D} &= \begin{bmatrix} \cos\phi\cos\theta\cos\psi - \sin\phi\sin\psi & -\cos\phi\cos\theta\sin\psi - \sin\phi\cos\psi & \sin\theta\cos\phi \\ \sin\phi\cos\theta\cos\psi + \cos\phi\sin\psi & -\sin\phi\cos\theta\sin\psi + \cos\phi\cos\psi & \sin\theta\sin\phi \\ -\cos\psi\sin\theta & \sin\psi\sin\theta & \cos\theta \end{bmatrix} \\ &= \begin{bmatrix} \cos\gamma_{xa} & \cos\gamma_{xb} & \cos\gamma_{xc} \\ \cos\gamma_{ya} & \cos\gamma_{yb} & \cos\gamma_{yc} \\ \cos\gamma_{za} & \cos\gamma_{zb} & \cos\gamma_{zc} \end{bmatrix} \\ &= \begin{bmatrix} a_1 & b_1 & c_1 \\ a_2 & b_2 & c_2 \\ a_3 & b_3 & c_3 \end{bmatrix}\end{aligned} \tag{11.23}$$

to see that the Euler angles, defined fully by the unit vector Cartesian coordinates in the last line of Eq. 11.23, may be obtained from

$$\theta = \cos^{-1} D_{3,3} \tag{11.24a}$$

$$\phi = \tan^{-1} \frac{D_{2,3}}{D_{1,3}} \tag{11.24b}$$

$$\psi = \tan^{-1} \frac{D_{3,2}}{-D_{3,1}}. \tag{11.24c}$$

Note that quadrant preservation is particularly important in the arctangent evaluation as both ϕ and ψ are defined in the interval $(-\pi, \pi)$. Hence, the explicit association of the negative sign with the denominator of Eq. 11.24b. The corresponding Python syntax is

```
theta = np.arccos(D[2,2])
  phi = np.arctan2(D[1,2], D[0,2])
  psi = np.arctan2(D[2,1], -D[2,0])
```

where `arctan2` is the quadrant-preserving arctangent, and the array indices have been adjusted from Eqs. 11.24 since Python indices start at zero.

11.7 Summary

We have taken a few steps towards expanding the scope of the model problems presented in Chaps. 8–10 to consider how they may be applied to more realistic situations. First on the list is that the chemical functional group whose orientation we wish to characterize may not be uniformly distributed about the ϕ and ψ Euler angles. That assumption was convenient when we introduced orientation analysis in the context of IR, Raman, and SFG spectroscopy, but in practice tilt-only distributions are rather rare. In this discussion, we see that opening up the possibility of orientation about two or more Euler angles leads to two immediate complications: (1) there are often more parameters in the distribution functions than unique pieces of experimental data available, and (2) elements of the corresponding molecular response tensor $\boldsymbol{\alpha}^{(n)}$ don't necessarily cancel out, and so molecular properties are needed, even though ratios of experimental quantities are used. One of the ways out of this situation is to make simultaneous use of multiple vibrational modes that may be linked to the same moiety (for example, the symmetric and antisymmetric stretch of a methyl group) and therefore share the same local coordinate system. Another is to solve for the orientation of two vectors corresponding to two directions in different function groups whose relative angle is constrained due to the molecular geometry. We have not provided explicit examples of such cases, as the required codes quickly become rather molecule-specific and are tailored to the data acquired. The hope is that enough information has been presented here to facilitate embarking on such projects. Finally, we have discussed how the results of experiments and molecular dynamics simulations can go hand-in-hand, either to confirm the origin of spectroscopic observables, or to get a sense of the shape of an orientation distribution in helping to come up with a reasonable functional approximation.

11.8 Exercises

1. As we have discussed, we have made a drastic simplification of the $\boldsymbol{\alpha}$ tensor for IR absorption spectroscopy (only $\alpha_{cc} \neq 0$), $\boldsymbol{\alpha}^{(2)}$ for SFG (only $\alpha_{cc} \neq 0$), and the Raman tensor (only $\alpha^{\mathrm{R}}_{cc} \neq 0$ stemming from only $\alpha^{(3)}_{cccc} \neq 0$ far from electronic resonance).
   a. Keeping the orientation distribution simple (anisotropic about only θ), rework all relevant expressions for IR, Raman, and SFG orientation analysis when additional $\boldsymbol{\alpha}^{(n)}$ tensor elements are non-zero. Start with only two (for example, α_{aa} and α_{cc}, or $\alpha^{(2)}_{aac}$ and $\alpha^{(2)}_{ccc}$, etc.).
   b. The primary advantage of having only a single molecular-frame tensor element is that it cancels when experimental ratios are used. However, when there are two non-zero unique $\boldsymbol{\alpha}^{(n)}$ tensor elements, your expressions can be cast in terms of their ratio. Referring to this ratio as R (for example, $R = \alpha^{\mathrm{R}}_{aa}/\alpha^{\mathrm{R}}_{cc}$, and then later $\alpha^{(2)}_{aac}/\alpha^{(2)}_{ccc}$), simplify your expressions. Do you find that ratios of $\boldsymbol{\chi}^{(n)}$ elements can be expressed in terms of your R values alone? I.e. Do the $\boldsymbol{\chi}^{(n)}$ ratios depend on the specific values of the $\boldsymbol{\alpha}^{(n)}$ elements? This question is relevant as there are some instances when molecular property ratios are easier to obtain accurately than their individual values.
   c. If for some reason you have preferred to work these out by hand, now is the time to write some SymPy code for this. Now you can make the $\boldsymbol{\alpha}^{(n)}$ tensor as dense or sparse as you like. This is a perfect opportunity to play around with different vibrational mode symmetries.

d. As a specific example of the above point, say you wanted to determine the orientation of a methyl group. The methyl symmetric stretch is typically found to have C_{3v} symmetry, with only two non-zero unique tensor elements in $\boldsymbol{\alpha}^{(n)}$ as you have modelled. But in some situations, rotation about the C_3-axis is hindered, and the symmetry of the symmetric stretch is reduced to C_s. If you have written your codes in the above parts, you can explore the consequences on the IR, Raman, and SFG responses. In particular, is there any experimental indication of this symmetry lowering?

---

2. It's now fun to explore more complex orientation distributions, but it's best to return to our simple $\boldsymbol{\alpha}^{(n)}$ tensor first, with only one non-zero component.

   a. In this case, we have already worked out the expressions for θ-only distributions. What do the $\boldsymbol{\chi}^{(n)}$ elements look like for distributions that are oriented about only ϕ or ψ? How about 2 of the three Euler angles? What about all three Euler angles? To avoid having to handle complex distributions, you may want to consider narrow (δ-type) distributions where the molecules are oriented at θ_0, ϕ_0, and/or ψ_0.
   b. What do you notice about any of your orientations that are ordered about ψ? This is a consequence of our definition of the $\boldsymbol{\alpha}^{(n)}$ tensor.
   c. Now introduce a second unique element to $\boldsymbol{\alpha}^{(n)}$ and inspect how your results change. Do you still notice the same behaviour about ψ? Why or why not?

---

3. Using the above one or two (if you separated them) SymPy codes as starting points, write some code that predicts the $\boldsymbol{\chi}^{(n)}$ elements for an arbitrary $\boldsymbol{\alpha}^{(n)}$ tensor (any number of non-zero elements, with any symmetry among those elements).

   a. In the first round, restrict yourself to δ-type orientations. When you display the results with `sp.pprint(result)` or even `sp.pprint(sp.simply(result))` it is very likely that none of the expressions will be small enough to fit within your field of view, even on a widescreen monitor! Although symbolic expressions are powerful since they allow you to inspect the form of the functions, these will be quite messy. However, a useful next step is to convert the expressions into NumPy functions using `sp.lambdify`. Then you can set values of the parameters and make some plots of the results. Give this a go.
   b. We now want to introduce the possibility of including an arbitrary orientation distribution function. This will enable us to use things like Gaussians for θ or wrapped Gaussians (von Mises) for ϕ and ψ, or any functions of any shape. When coding this, an important consideration is that it is often not practical to arrive at the final symbolic expressions before converting to numerical versions. In the case of the sharp distributions, the results were simple to obtain, merely long. But now you will run into the situation where some of the integrals are difficult to evaluate. Consider an approach for solving this.

One suggestion is to convert some symbolic expressions into numerical versions earlier, prior to obtaining the final result. You can also make use of the SymPy `result.evalf` feature. If you have written codes to perform some of the above exercises, you already have nearly all of the building blocks to put this together.

c. Once things are in a numerical form, you may wish to experiment with some just-in-time (JIT) compilers for Python such as Numba. There is not much to be gained from one-time execution of the functions (since there is some code translation and compilation overhead), but there are huge benefits (close to two orders of magnitude speed-up) when the functions need to be called repeatedly. This is of interest since your code is now one small step away from being useful in fitting experimental data.

---

4. Imagine that your molecule has a methyl group with C_{3v} symmetry, and a carbonyl group ($C_{\infty v}$ symmetry) with an angle β between the C_3 and C_∞ vectors.

   a. Does this simplify any of your orientation analysis? If yes, how so?
   b. Does this restriction of β enable you to solve the problem with less information? For example, can you have only partial $\boldsymbol{\alpha}^{(n)}$ information between the two vibrational modes, compared to what you would need if knowledge of β was not available. Assume that the actual value of the angle β is known. You can either tackle this problem entirely symbolically, or pick a value for β and use numerical versions of your functions to address this.
   c. If θ_1 is the angle between the methyl C_3-axis and the surface normal z, and θ_2 is the angle between the carbonyl axis and z, what does knowledge of the fixed β tell you about the two orientation distributions. For example, if θ_1 can be described by a δ function, can θ_2 also be a δ function?

References

1. T. Kluyver, B. Ragan-Kelley, F. Pérez, B. Granger, M. Bussonnier, J. Frederic, K. Kelley, J. Hamrick, J. Grout, S.C. adn Paul Ivanov, D. Avila, S. Abdalla, C. Willing, *Positioning and Power in Academic Publishing: Players, Agents and Agendas* (IOS Press, Amsterdam, 2016), chap. Jupyter Notebooks—A Publishing Format for Reproducible Computational Workflows, pp. 87–90
2. N.J. Everall, A. Bibby, Appl. Spectrosc. **51**, 1083 (1997)
3. K.C. Cole, H.B. Daly, B. Sanschagrin, K.T. Nguyen, A. Ajji, Polymer **40**, 3505 (1999)
4. H. Ben Daly, K.C. Cole, K.T. Nguyen, B. Sanschagrin, J. Appl. Polymer Sci. **104**, 1319 (2007)
5. S. Nomura, H. Kawai, I. Kimura, M. Kagiyama, J. Poly. Sci. A **8**, 383 (1970)
6. C. Sourisseau, Chem. Rev. **104**, 3851 (2004)
7. G.B. Arfken, H.J. Weber, F.E. Harris, *Mathematical Methods for Physicists: A Comprehensive Guide*, 7th edn. (Academic Press, Waltham, MA, 2013)
8. R.J. Roe, W.R. Kirigbaum, J. Appl. Phys. **35**, 2215 (1964)
9. R.J. Roe, W.R. Krigbaum, J. Chem. Phys. **40**, 2608 (1964)

10. R.J. Roe, J. Appl. Phys. **36**, 2024 (1965)
11. D.I. Bower, J. Poly. Sci. **10**, 2135 (1972)
12. D.A. Jarvis, I.J. Hutchinson, D.I. Bower, I.M. Ward, Polymer **21**, 41 (1980)
13. D.I. Bower, J. Polymer Sci. **19**, 93 (1981)
14. Y.J. Lee, Opt. Express **23**, 29279 (2015)
15. D.E. Winch, P.H. Roberts, J. Austral. Math. Soc. Ser. B **37**, 212 (1995)
16. K. Maleček, Z. Nádeník, Studia Geoph. Geod. **45**, 1 (2001)
17. N.N. Beloozerov, Mathematical notes of the Academy of Sciences of the USSR **7**, 83 (1970)
18. L. Bokobza, T. Buffeteau, B. Desbat, Appl. Spectrosc. **54**, 360 (2000)
19. L. Bokobza, Polymers **14**, 1257 (2022)
20. C.P. Lafrance, A. Nabet, R.E. Prud'homme, M. Pézolet, Can. J. Chem. **73**, 1497 (1995)
21. N. Kumaran, C.L. Donley, S.B. Mendes, N.R. Armstrong, J. Phys. Chem. C **112**, 4971 (2008)
22. P.A. Chollet, J. Messier, C. Rosilio, J. Chem. Phys. **64**, 1042 (1976)
23. J. Umemura, T. Kamata, T. Kawai, T. Takenaka, J. Phys. Chem. **94**, 62 (1990)
24. E. Okamura, J. Umemura, T. Takenaka, Can. J. Chem. **69**, 1691 (1991)
25. H. Nakahara, K. Fukuda, J. Colloid Interface Sci. **69**, 24 (1979)
26. I. Pelletier, I. Laurin, T. Buffeteau, M. Pézolet, J. Phys. Chem. B **108**, 7162 (2004)
27. R.P. Davis, A.J. Moad, G.S. Goeken, R.D. Wampler, G.J. Simpson, J. Phys. Chem. B **112**, 5834 (2008)
28. J. Duboisset, P. Berto, P. Gasecka, F.Z. Bioud, P. Ferrand, H. Rigneault, S. Brasselet, J. Phys. Chem. B **119**, 3242 (2015)
29. S.H. Parekh, K.F. Domke, Chem. Eur. J. **19**, 11822 (2013)
30. C. Zhang, J. Wang, J. Jasensky, Z. Chen, J. Phys. Chem. Lett. **6**, 1369 (2015)
31. Y. Wang, Z.f. Cui, H.f. Wang, Chin. J. Chem. Phys. **20**, 449 (2007)
32. M.G. Bell, R.E. Dale, U.A. van der Heide, Y.E. Goldman, Biophys. J . **83**, 1050 (2002)
33. E. Frehland, Z. Naturforsch. **30**, 1241 (1975)
34. F. Jähnig, Proc. Nat. Acad. Sci. USA **76**, 6361 (1979)
35. R.P.H. Kooyman, Y.K. Levine, B.W. van der Meer, Chem. Phys. **60**, 317 (1981)
36. H. Pottel, W. Herreman, B.W. van der Meer, M. Ameloot, Chem. Phys. **102**, 37 (1986)
37. B. Schartel, V. Wachtendorf, M. Grell, D.D.C. Bradley, M. Hennecke, Phys. Rev. B **60**, 277 (1999)
38. G. van Rossum, J. de Boer, CWI Quarterly **4**, 283 (1991)
39. E. Anglaret, M. Brunet, B. Desbat, P. Keller, T. Buffeteau, Macromolecules **38**, 4799 (2005)
40. T. Buffeteau, F. Lagugné Labarthet, C. Sourisseau, S. Kostromine, T. Bieringer, Macromolecules **37**, 2880 (2004)
41. V. Rodriguez, F. Lagugné Labarthet, C. Sourisseau, Appl. Spectrosc. **59**, 322 (2005)
42. A.F. Runge, S.S. Saavedra, S.B. Mendes, J. Phys. Chem. B **110**, 6721 (2006)
43. B. Jérôme, Y.R. Shen, Phys. Rev. E **48**, 4556 (1993)
44. M. Sakata, M. Takata, High Pressure Res. **14**, 327 (1996)
45. M. Takata, B. Umeda, E. Nishibori, M. Sakata, Y. Saito, M. Ohno, H. Shinohara, Nature **377**, 46 (1995)
46. I.M. Ward, Adv. Poly. Sci. **66**, 81 (1985)
47. R. Berardi, F. Spinozzi, C. Zannoni, J. Chem. Phys. **109**, 3742 (1998)
48. F. Lagugné-Labarthet, C. Sourisseau, R.D. Schaller, R.J. Saykally, P. Rochon, J. Phys. Chem. B **108**, 17059 (2004)
49. C. Bolterauer, H. Heller, Eur. Biophys. J. **24**, 322 (1996)

MIX
Papier aus verantwortungsvollen Quellen
Paper from responsible sources
FSC® C105338

If you have any concerns about our products,
you can contact us on
ProductSafety@springernature.com

In case Publisher is established outside the EU,
the EU authorized representative is:
Springer Nature Customer Service Center GmbH
Europaplatz 3, 69115 Heidelberg, Germany

Printed by Libri Plureos GmbH
in Hamburg, Germany